Nanocellulose

Synthesis, Structure, Properties and Applications

Nanocellulose

Synthesis, Structure, Properties and Applications

editors

Guang Yang • Muhammad Wajid Ullah • Zhijun Shi

Huazhong University of Science and Technology, China

World Scientific

NEW JERSEY • LONDON • SINGAPORE • BEIJING • SHANGHAI • HONG KONG • TAIPEI • CHENNAI • TOKYO

Published by

World Scientific Publishing Europe Ltd.

57 Shelton Street, Covent Garden, London WC2H 9HE

Head office: 5 Toh Tuck Link, Singapore 596224

USA office: 27 Warren Street, Suite 401-402, Hackensack, NJ 07601

Library of Congress Cataloging-in-Publication Data

Names: Yang, Guang, 1968 April 18– editor. | Ullah, Muhammad Wajid, editor. | Shi, Zhijun, editor.

Title: Nanocellulose : synthesis, structure, properties and applications / Guang Yang, Muhammad Wajid Ullah, Zhijun Shi, Huazhong University of Science and Technology, China.

Description: New Jersey : World Scientific, [2021] | Includes bibliographical references and index.

Identifiers: LCCN 2020038674 (print) | LCCN 2020038675 (ebook) | ISBN 9781786349460 (hardcover) | ISBN 9781786349477 (ebook) | ISBN 9781786349484 (ebook other)

Subjects: LCSH: Cellulose--Biotechnology. | Nanobiotechnology.

Classification: LCC TP248.65.C45 N36 2021 (print) | LCC TP248.65.C45 (ebook) | DDC 660.6--dc23

LC record available at https://lccn.loc.gov/2020038674

LC ebook record available at https://lccn.loc.gov/2020038675

British Library Cataloguing-in-Publication Data

A catalogue record for this book is available from the British Library.

For any available supplementary material, please visit
https://www.worldscientific.com/worldscibooks/10.1142/Q0278#t=suppl

Desk Editors: Aanand Jayaraman/Michael Beale/Shi Ying Koe

Typeset by Stallion Press
Email: enquiries@stallionpress.com

https://doi.org/10.1142/9781786349477_fmatter

Dedication

The lead editor Professor Guang Yang would like to dedicate this book to Professor Lina Zhang (Academician of Chinese Academy of Sciences, China) on the occasion of her 80th Birthday and in the memory of her recent demise to value her contribution in Nanocellulose.

Dr. Muhammad Wajid Ullah would like to dedicate this book to his mother, late father, siblings, beloved wife Dr. Sehrish Manan, and cute daughter Mysha Wajid and son Zaviyar Wajid, for their love, care, and unconditional support.

Dr. Zhijun Shi would like to dedicate this book to his dear wife Lily and lovely sons Danny and Connie.

https://doi.org/10.1142/9781786349477_fmatter

Foreword

Although well explored, proposing the development of advanced synthesis strategies, surface modifications, and novel applications in new areas, *Nanocellulose* can serve as an impressive material to address several issues of immense human concerns, including in fields of healthcare, environmental hazards, energy crisis, food storage, and advanced materials.

This book presents the multidisciplinary aspects of the concept of nanoscience and nanotechnology and represents a typical example of their applications as nanocellulose and nanocellulose-based high-performance functional materials. The different aspects of nanocellulose research, such as the synthesis of different morphological forms by various methods, various surface modification methods (chemical, physical, solvent-exchange, and biochemical), composites synthesis with polymers, organic and inorganic nanoparticles, clay materials etc. by various methods, development of its hydrogels and aerogels by various methods, and its applications in biomedical materials, food industry, environment, paper industry, optoelectronics, renewable energy, and other fields that have greatly revolutionized human life, are discussed. The examples of different areas of applications of nanocellulose-based high-performance materials showcase the trends of their synthesis, characterization, structure–function relationship, and potential novel applications. This will take the reader into the realm of this impressive nanomaterial.

I believe this book will be a valuable reference source for researchers in the academia and industry working in the subject area and draw the attention of different classes of scientific communities by stimulating new thoughts and ideas. It is devised both for the beginners in the field, who will appreciate its comprehensive approach, and for the active researchers, who will find it a valuable source as it presents the recent advancements on topics of current research interest, including cellulose nanostructures, synthesis, structure–properties relationship, surface modification, and biomedical scaffolding. Particularly, several chapters will not only appeal to the biochemical engineers but also the biomedical, environmental, electronics, and agriculture engineers and the food biotechnologists. This book should be a great inspiration for the researchers to develop advanced materials with improved functionalities and applications in new areas as well as beginners to initiate their carrier in this interesting area of research.

Professor Lina Zhang
Academician of Chinese Academy of Sciences Professor,
College of Chemistry & Molecule Sciences,
Wuhan University, Wuhan, China
Phone: +86-27-87219274, 87216311
E-mail: zhangln@whu.edu.cn
October 2020

https://doi.org/10.1142/9781786349477_fmatter

Preface

Nanocellulose refers to the cellulose with at least one of its dimensions less than or equal to 100 nm. It is extracted from native plant cellulose as well as produced by different microorganisms and synthesized enzymatically by cell-free enzyme system. It exists in different morphological forms, including cellulose nanocrystals (CNCs), cellulose nanofibers (CNFs), and bacterial nanocellulose (BNC), depending upon the source and the synthesis methods. CNCs and CNFs are obtained from plant sources while BNC is produced through microbial fermentation. CNCs represent the pure crystalline structure, while CNFs and BNC contain both the crystalline and amorphous regions. Among these, BNC represents the purest form of cellulose. Over the last couple of decades, nanocellulose has received immense consideration owing to its abundance, renewability, remarkable physical properties, special surface chemistry, and excellent biological properties. The presence of free hydroxyl group on its surface blesses nanocellulose with hydrophilic behavior and tailored functionalities by making composites with other materials, including metals, metal oxide, polymers, and other materials. Its network-like fibrous and porous structure, stabilized by hydrogen bonding, enables it to serve as a carrier for immobilization and impregnation of different materials as well as for developing high-performance materials. The presence of large amount of crystalline regions blesses it with high mechanical strength. Due to its non-toxic nature, it is widely used as a dressing material, drug carrier, and in the development of artificial organs and biomedical devices. Taking advantage of such unique features, the current materials science aims at developing new materials with advanced and novel functionalities in an effort to meet the human requirements in various fields, including healthcare, environment, energy, food and agriculture, and paper and other industries.

As underlined by few recent reviews on "Nanocellulose", a comprehensive and timely overview of the synthesis, structure–properties relationship, and multidisciplinary applications is still lacking due to the rapid developments in the subject area. Therefore, in this book "*Nanocellulose: Synthesis, Structure, Properties, and Applications*", we have tried to provide a comprehensive and timely overview of different synthesis methods including physico-chemical, mechanical, and enzymatic synthesis of CNCs, CNFs, and BNC, and describe their surface modification and applications in different areas. This book consists of 16 chapters and covers three main areas of nanocellulose research: (1) synthesis of different morphological forms of nanocellulose by using various methods, (2) surface modification of nanocellulose by various methods, and (3) broad-spectrum applications of nanocellulose in biomedical field, food industry, environment, paper industry, optoelectronics, renewable energy, and others. The first chapter provides a comprehensive overview of different synthesis methods of nanocellulose such as mechanical, chemical, enzymatic, and through microbial fermentation, as well as different morphological forms of nanocellulose including CNCs, CNFs, and BNC, in terms of their features such as chemical structure, structural morphology, physico-mechanical, thermal and biological properties, rheology, optical behavior, and their interrelationships and applications. The following four chapters are devoted to the synthesis, surface modification, and development of nanocellulose-based composites. From here on, specific applications of nanocellulose in different areas are discussed, including in skin tissue repair, drug delivery, food industry, environment, paper industry, optoelectronics, and renewable energy. The last chapter summarizes various applications of nanocellulose in different areas.

Guang Yang
Muhammad Wajid Ullah
Zhijun Shi
October 2020

https://doi.org/10.1142/9781786349477_fmatter

About the Editors

Guang Yang is Full Professor at Huazhong University of Science and Technology, China. She received PhD degree in Chemistry from Wuhan University, China. She remained the Distinguished Young Chutian Scholar and Outstanding Talents in Hubei province, as well as Alexander von Humboldt and JSPS fellow. She was also a visiting scholar in Asahi Chemical Industry Co., Ltd., Japan and University of Akron, USA. Currently, she is serving as the Deputy Director of the Cellulose division of the Chinese Chemical Society and the council member of several committees, including the Nanocellulose and Materials Committee of China Paper Association, Polymer Characterization Committee of the Chinese Chemical Society, and the Biomedical Polymer Materials Branch of China Biomaterials Society. She has authored more than 150 publications in high impact international peer-reviewed journals, edited a book, authored several chapters, and registered more than 20 authorized patents. Her current research focuses on the development of nanocellulose-based functional materials, design and fabrication of novel nano-drug transporters, 3D printing, nano-assembly of ordered materials, and tissue engineering.

Muhammad Wajid Ullah is a post-doctoral researcher at Huazhong University of Science and Technology, China. He received his BS and MS degrees in Biotechnology from University of Peshawar and COMSATS University Islamabad, Pakistan, in 2009 and 2011, respectively, and PhD degree in Chemical Engineering from Kyungpook National University, South Korea, in 2016. He has received several honors and awards including KNU Honor Scholarship (KHS), Bachelor and Master Fellowship Awards, 'Best Researcher' and 'Most Contributed Researcher' awards. He has published more than 100 articles in peer-reviewed international journals (h-index: >30), authored more than 25 monographs, two authorized patents, and presented his work in more than 65 international conferences around the world. He is a member of several scientific societies, including Chinese Chemical Society (CCS), Korean Institute of Chemical Engineers (KIChE), Korean Society of Biotechnology and Bioengineering (KSBB), Asian Federation of Biotechnology (AFOB), and others. His research interests include the fabrication of bacterial cellulose and other natural polymers-based functional materials *via* 3D printing, advanced cell-free approach, and physico-chemical approaches for engineering of bone, skin, and other tissues as well as applications in drug delivery and biosensing.

Zhijun Shi is a Lecturer at Department of Biomedical Engineering, College of Life Science and Technology, Huazhong University of Science and Technology, Wuhan, China. He received his BS and PhD degrees in Biotechnology and Microbiology in 2008 and 2014, respectively, and worked as post-doctorate researcher (2014–2019) at Huazhong University of Science and Technology, China. He has authored more than 30 publications in peer-reviewed international journals as well as several chapters and authorized patents. His research interests include biofabrication based on microbes and fabrication of functional hydrogels based on natural polymers.

https://doi.org/10.1142/9781786349477_fmatter

About the Contributors

Abdul Khalil H.P.S. obtained his PhD degree from the University of Wales, Bangor, United Kingdom, in 1999. He is currently a Senior Professor in Biomass and Biocomposites at the School of Industrial Technology, Universiti Sains Malaysia, Penang, Malaysia. His fields of specialization cover biocomposite technology, fibre-reinforced polymer composite technology, hybrid composites, nano-based composites, fibre science (natural and man-made fibre), and biodegradable polymer as new materials for a new economy. He is a fellow of Institute of Materials, Minerals and Mining (FIMMM), United Kingdom. He is a member of the American Chemical Society and a member of Wood Science and Technology. He had been appointed to the editorial board of more than 5 Scopus/ISI listed journals. To date, he has generated over 250 scientific articles in Scopus with h-Index of 56 and over 15,000 citations. He has also excellently published 3 academic books, 41 chapters in books, as well as 4 non-academic books (community/coffee table books). As an expert researcher in his field, he has reviewed more than 200 articles in reputed journals. He was awarded as one of the Top Research Scientists in Malaysia by the Academy of Sciences of Malaysia in 2014 and received Malaysia's Rising Star Award (Thomson & Reuters) in 2015.

Ahmed Saadi Abrehem is an Associate Professor at the Department of Chemical Engineering, Dhofar University, Oman. He completed his PhD studies in Chemical Engineering from the University of Malaya, Malaysia. With speciality in chemical process design and control, Dr. Abrehem has actively contributed in renewable materials and waste recycling. He has published more than 100 journal and conference papers. Besides, he has contributed several book chapters and filed patents.

Amarachi Rosemary Osi is a PhD student at the University of Chinese Academy of Sciences, Beijing, and Ningbo Institute of Materials Technology and Engineering, Ningbo, China. She was awarded CSC MOFCOM scholarship in 2015 to obtain her Master's degree in Biomedical Engineering from Huazhong University of Science and Technology, Wuhan, China, as well as the CAS-TWAS scholarship for her PhD studies. Her research interests include production of bacterial cellulose, design of natural-based polymers as injectable hydrogels, and 3D bioprinting for tissue engineering applications. She is currently focused on injectable hydrogels for 3D printing and its applications for osteochondral regeneration. She is a co-author of several chapters and 2 publications in peer-reviewed international journals.

Aref Ahmad Wazwaz obtained his PhD degree in Solar Energy, from the Paul Sabatier University, Toulouse, France, in 2001 (Scholarship from European Community). He is currently an Associate Professor in Chemical Engineering, at the Dhofar University, Salalah, Oman. He secured and finished two TRC projects, one as PI and the other as Co-PI, and obtained the "Research and Innovation Award for Water Science — 4th Round" in 2019–2020 from TRC, Oman. Dr. Aref is currently serving as the editorial board member in several international journals and conferences. He has published a number of papers indexed in Web of Science and Scopus journals.

Bianza Moise Bakadia studied at the Institut Superieur des Techniques Medicales de Lubumbashi (Higher Institute of Medical Techniques of Lubumbashi), Lubumbashi, Democratic Republic of the Congo, where he received a Bachelor's degree in Medical Biology (2010) and started working as a Teaching Assistant from 2010 to 2016 and Research Fellow from 2016 till now. He was awarded a CSC scholarship to obtain his Master's degree (2017) in Biochemistry and Molecular Biology from the Huazhong University of Science and Technology, Wuhan, China, as well as for his PhD studies (2019). His research interests include the immune response to microorganisms, biomedical analysis, and the development of biomaterials for tissue engineering and regenerative medicine. He has published 9 articles in peer-reviewed international journals.

Biaou O. Ode Boni is currently working as a post-doctoral fellow at Huazhong University of Science and Technology, China, in the laboratory of Prof. Guang Yang, and under the supervision of Prof. Bifeng Liu. He carried out his PhD studies (2016–2016) at the College of Life Sciences and Technology of the same university in Biochemistry and Molecular Biology. He received his Master's (2012) and Bachelor's (2009) degrees from University of Abomey-Calavi, Benin, in Biochemistry and Molecular Biology. His research interests include the immune response to biomaterials and the development of biomaterials for vascular graft, nerve conduit, and wound healing applications.

Chao Jia received his PhD degree in Materials Science and Engineering from Beijing Institute of Technology in Beijing, China, in 2018. From 2015 to 2017, he worked with Prof. Liangbing Hu at University of Maryland, College Park as a visiting scholar. He is currently a post-doctoral researcher at the Department of Materials Science and Engineering, Tsinghua University, China. His research interests include cellulose nanomaterials, nanofibers, and wood nanotechnologies.

Chao Wang is an Associate Researcher Fellow in the Research Center of Functional Polymer Materials at Technical Institute of Physics and Chemistry, Chinese Academy of Sciences (TIPC, CAS), China. In 2006, she graduated from the same institute with a doctorate degree in physical chemistry. Her research interests are mainly focused on cellulose and nanocellulose materials for applications in developing conducting materials for sensing and other applications. To date, she has published more than 38 scientific papers and 10 patents.

Cheng Zhong is a Professor of State Key Laboratory of Food Nutrition and Safety, Tianjin University of Science and Technology, China. He has published more than 60 SCI-indexed articles (including 1 highly cited paper) and registered more than 60 patents (including 20 authorized patents). He received his PhD degree from Tianjin University, China, in 2007, and has been a visiting scholar at Universitat Leipzig in Germany (2007) and Michigan State University in the United States (2008 and 2012). His research interests include synthetic biology, cellulose biosynthesis and degradation mechanism, as well as cellulose-based materials and their modification.

Daniel Pasquini is an Associate Professor and Researcher at the Chemistry Institute of the Federal University of Uberlândia, Uberlândia-MG, Brazil. He has edited 7 books, and published more than 100 publications in indexed journals, 12 book chapters, and more than 85 works in conferences. His research interests include polymers from renewable sources, vegetal macromolecules, cellulosic fibers, composites, nanocellulose, and nanocomposites.

Deepu A. Gopakumar is a post-doctoral fellow and CNRS grant recipient at Universite de Bretagne-SUD, IRDL, France. He obtained his doctoral degree in Material Science and Engineering in 2017 from the University of South Britany, France. His research focusses on the the fabrication of polymer membranes for water purification, polymer nanocomposites for EMI shielding application, bionanocomposites, fiber-filled polymer composites, and polymers scaffolds for tissue engineering and wound healing. He has authored 26 research articles, 17 book chapters, and edited 2 books. He was formerly a post-doctoral fellow at the Universiti Sains Malaysia, Penang, Malaysia.

Fazli Wahid is working as an Associate Professor at the Department of Biomedical Sciences, Pak-Austria Fachhochschule: Institute of Applied Sciences and Technology, Haripur, Pakistan. He obtained his PhD degree from the School of Life Sciences and Biotechnology, Kyungpook National University, South Korea. He has published over 45 research articles in well-reputed international journals with an impact factor of over 100 and, to date, his work has been cited over 1900 times. He has contributed 6 chapters in international books and filed 2 patents. His research interests include biopolymers, tissue engineering, regenerative and nanomedicines, cancer and natural products. He is working on the *in vitro* and *in vivo* effects of nanomaterials, polymers, and natural products on the modulation of different signaling pathways in chronic diseases as well as the development of therapeutic biomaterials.

Fazli Wahid is a post-doctoral scholar in the State Key Laboratory of Food Nutrition and Safety, Tianjin University of Science and Technology (TUST), China. He received his PhD degree in 2017 in Light Chemical and Materials Science from TUST. His field of research is the fabrication of biopolymer-based functional materials in the form of hydrogels, films, and membranes for different biomedical and environmental applications.

Hak Lae Lee is a Professor at the Department of Forest Sciences, Seoul National University, Korea, and an Adjunct Professor of State Key Laboratory of Biobased Material and Green Papermaking, Qilu University of Technology, China. He received his BS and MS from Seoul National University, and PhD degree from the State University of New York (SUNY), College of Environmental Science and Forestry, USA. He worked as an Assistant Professor at Seoul National University in 1993, and served as Department Chair, Dean of the College of Agriculture and Life Sciences. He has served as the President of Korea TAPPI and actively participated in many international conferences as a speaker, chair, and program committee member. His research interests are wet end chemistry, paper coating, and nanocellulose. He has published more than 230 papers in refereed journals.

Hanif Ullah is a researcher in the field of drug delivery and an expert in pharmaceutical regulations. He received his PharmD degree from University of Malakand (2011) and MS degree from COMSATS University Islamabad, Pakistan (2013), where he also completed his PhD studies, with a focus on bacterial cellulose for drug delivery in biomedical and pharmaceutical fields. He has also been a visiting researcher at the University of Helsinki, Finland. He has published many research articles in his field of research. Moreover, he is also (co-author of several review articles and book chapters in the fields of his interest.

Hanuma Reddy Tiyyagura received his PhD degree in Metallurgical and Materials Engineering in 2018 at the National Institute of Technology, Warangal, India. He is currently working as a post-doctoral researcher at the Faculty of Mechanical Engineering, University of Maribor, Slovenia. His research interests focus mainly on gold-based inks for (inkjet printing/3D printing) paper-based flexible electronics, biopolymer coatings on metals/alloys, etc. He is the author/co-author of several scientific articles and book chapters.

Hao Ling received his Bachelor's degree in Polymer Materials and Engineering from South China University of Technology, China. Presently, he is a doctoral student with Prof. Xiaohui Wang at South China University of Technology. He mainly focuses on biomass, especially cellulose-based functional materials for the development of biomaterials, flexible electronics, and oil/water separation materials.

Hidayah Ariffin is a Professor at the Faculty of Biotechnology and Biomolecular Sciences, and a Deputy Director at the Institute of Tropical Forestry and Forest Products (INTROP), Universiti Putra Malaysia (UPM). Her research interests include biopolymers, biomass valorization, and utilization of lignocellulosic materials. Presently, her research focuses on the production and applications of nanocellulosic materials. To date, she has published more than 90 high-impact journal articles with more than half as senior author, edited one book, authored 8 book chapters, and participated in more than 30 conferences. Prof. Hidayah Ariffin has also invented multiple research products, 8 of them with intellectual property (IP) protection. At present, she is the editorial board member of *Journal of Renewable Materials, Academy of Science Malaysia (ASM) Science Journal*, and *Asian Federation of Biotechnology (AFOB) Newsletters*. She is also a member of the Research Committee, TAPPI Nano Division.

Hideya Kawasaki is a Professor at the Department of Chemistry and Materials Engineering, Kansai University, Japan. He received his PhD degree in Chemistry (1996) in Science from Kyushu University, Japan. He joined the Department of Chemistry, Kyushu University, as an Assistant Professor (1997–2006) and the Department of Chemistry and Materials Engineering, Kansai University, as an Associate Professor (2006–2012).

He received Awards for Younger Researcher from the Division of Colloid and Surface Chemistry from the Chemical Society of Japan (2003) and the Mass Spectrometry Society of Japan (2011). His current research interests include (1) copper-based inks for printed/flexible devices, (2) synthesis of photo-functionalized metal nanoclusters for their applications in catalysis, sensing, and biomedical fields, and (3) nanomaterials for mass spectrometry.

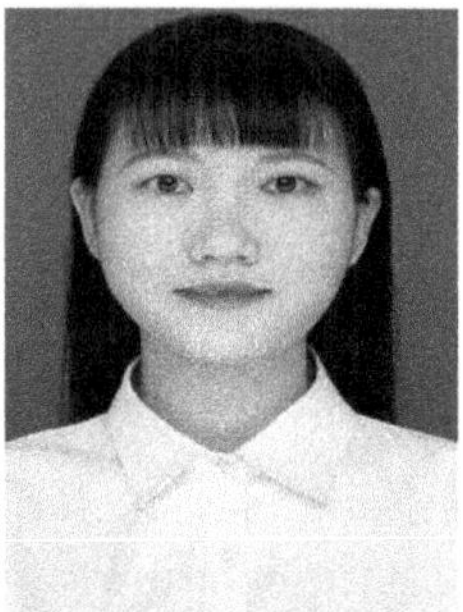

Huan Chen received her Bachelor's degree in 2014 from Wuhan University of Technology, China, where she continues learning as a Master's student under the guidance of Prof. Quanling Yang. Her research interests include biomass materials, dielectric materials, and their applications in dielectric energy storage and sensors.

Hye Jung Youn is a Professor at the Department of Forest Sciences, Seoul National University, Korea, and an Adjunct Professor of State Key Laboratory of Biobased Material and Green Papermaking, Qilu University of Technology, China. She received her BS, MS, and PhD degrees from Seoul National University, Korea. She worked for Hansol Paper Co. during 2000–2002 and worked as an Assistant Professor at Seoul National University in August 2002. She has published more than 130 papers in refereed journals on paper science and engineering, and recently on the preparation and application of cellulose nanomaterials.

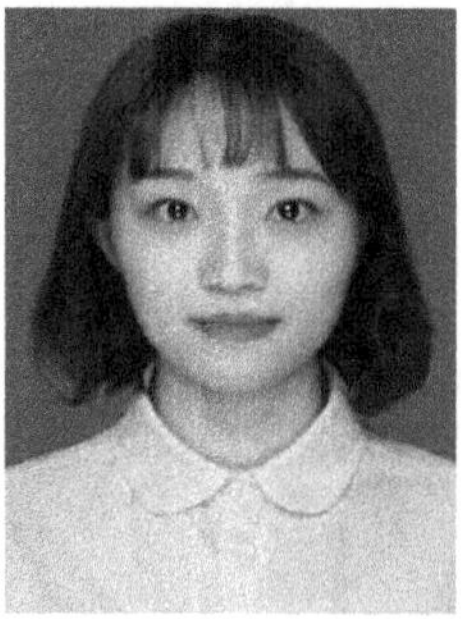

Jinyu Wang received her Bachelor's degree in Polymer Materials (2017) from Wuhan University of Technology, China, and Master's degree in Materials Science and Technology (2020) under the supervision of Prof. Quanling Yang. She has been working as a researcher for Wuhuan Engineering Corporation, China since August 2020. Her research interests include biomass materials, electronic functional materials, degradable plastic, etc.

Lallepak Lamboni currently works in the Molecular Biology and Virology Laboratory at National Institute of Hygiene (Institut National d'Hygiène), Togo. Previously, she carried out a three-year post-doctoral fellowship at the Department of Biomedical Engineering, Huazhong University of Science and Technology, China, under the supervision of Prof. Guang Yang. Under Chinese scholarship, she obtained her PhD in Biomedical Engineering from the same university in 2016. She received her BSc degree (2011) in Biology and Health from the Faculty of Science and Technology of Fez, Morocco. Her research work focuses on the development of biomimetic materials for tissue engineering and regenerative medicine, and the study of cell–biomaterial interactions and related biological processes. Other interests include the functional exploration of biomacromolecules and their biomedical applications, among others.

Liangbing Hu received his BS degree in Physics from the University of Science and Technology of China in 2002. He completed his PhD at UCLA, USA (2002–2007). In 2006, he joined Unidym, Inc. as a co-founding scientist. After that, he conducted post-doctoral research at Stanford University, USA, from 2009–2011. Currently, he is a Herbert Rabin Distinguished Professor at the University of Maryland, College Park, USA. His research group focuses on materials innovations, device integrations, and manufacturing in general, with ongoing research activities on wood nanotechnologies, 3000 K extreme materials, and beyond Li-ion batteries. Hu has published over 350 research papers (including in *Science* and *Nature*) and given more than 150 invited talks. He has received many awards, including Highly Cited Researchers by Clarivate Analytics (2016–2019), Young Innovator Award (2019, *Small* — Wiley Journal); Blavatnik Awards for Young Scientists (2020 Finalist); TAPPI Nano Middle Career Award (2019), and several others.

Mazhar Ul-Islam is an Associate Professor at the Department of Chemical Engineering, Dhofar University, Salala, Oman. He obtained his PhD degree in Chemical Engineering from Kyungpook National University, South Korea, in 2013, and worked at the same institution as a post-doctorate fellow and Contract Professor for one year each. He has published more than 100 journal papers, authored several chapters, and participated in numerous international conferences. He is the recipient of several TRC and other research grants. His research areas include biopolymers, nanomaterials, biofuels, and nano-biopolymer composites.

Min Wu is the Professor and Head of Applied Biomass Resources Laboratory, Research Center of Functional Polymer Materials at Technical Institute of Physics and Chemistry, Chinese Academy of Sciences (TIPC, CAS), China. She received her PhD degree in Materials Engineering in 2002 from Nagoya Institute of Technology, Japan. After working as a post-doctoral researcher in National Institute of Advanced Industrial Science and Technology (AIST Shikoku), Japan, during 2002–2004, she accepted an appointment as JSPS Fellow at the University of Tokyo from 2004 to 2006. In 2007, Prof. Wu returned to China and served as an Associate Professor at Institute of Chemistry, CAS, China. In 2009, she was offered the role of Associate Professor, in TIPC, CAS, China. In July 2012, she accepted her current position as a Professor and Lab Leader at TIPC, CAS, China.

Mingwei Zhu received his PhD degree from Zhejiang University, Hangzhou, China. From 2015 to 2016, he was a visiting professor at the University of Maryland, USA, where he worked on transparent wood, ultra-strong wood, wood-based solar steam, and transparent paper. He received post-doctoral training at Nanjing University, Nanjing, China. Currently, he is a Professor at the College of Engineering and Applied Sciences at Nanjing University, China. His research interests include disruptive materials, technologies, and applications based on natural materials.

Mohd Ali Hassan is a Full Professor at the Department of Bioprocess Technology, Universiti Putra Malaysia. He graduated from University of Leeds, UK, in Chemical Engineering in 1980. He received his MS in Food Engineering from the same university in 1982, Masters of Philosophy from University of Birmingham, UK, in 1987, and PhD from Okayama University, Japan, in 1997. His current research interest includes the sustainable utilization of oil palm biomass for value-added products. Together with his counterpart from Japan, he has successfully developed a showcase oil palm biorefinery together with a palm oil industry under SATREPS project, which is a joint project between Malaysia and Japan. To date, he has published more than 200 journal papers, 5 book chapters, and more than 100 conference papers. He sits in the Board of Directors, Universiti Putra Malaysia. He is also a Vice President of Asian Federation of Biotechnology (AFOB), and a fellow of the Academy of Sciences Malaysia.

Mohd Nor Faiz Norrahim received his PhD degree in Nanotechnology. He is currently a postdoctoral fellow at Research Center for Chemical Defence, Universiti Pertahanan Nasional Malaysia. His research areas include nanotechnology, biomaterials, composites, and biotechnology.

Munair Badshah is currently working as an Associate Professor at Islam College of Pharmacy, Islam Group, Sialkot, Pakistan. He has vast teaching experience in various courses in the discipline of Pharmacy since 2008 at the undergraduate (Doctor of Pharmacy) level. In addition, Dr. Badshah has been part of the pharmaceutical manufacturing industry with responsibilities for management and supervision of the production of various dosage forms. In the year 2019, he completed his PhD degree from COMSATS University Islamabad, Abbottabad Campus, Pakistan, under the supervision of Prof. Dr. Taous Khan. His research work theme was the application of bacterial cellulose in the development of novel drug delivery system. He has contributed as an author to many research papers and as a reviewer to international journals.

Pei Huang received his PhD in Polymer Chemistry and Physics from Institute of Chemistry, Chinese Academy of Sciences (ICCAS), in 2013. He conducted his post-doctoral research on the development of aerogel for thermal insulation in College of Civil Engineering, Brunel University (UK), from 2013 to 2015. In 2015, Dr. Huang joined the group of Prof. Shao-Yun Fu in College of Aerospace Engineering, Chongqing University. He mainly works on nanomaterial synthesis and multi-functional composites, and has published 35 scientific papers and 4 patents so far.

Preetha Balakrishnan is the Principal Scientist, QA, QC, ADSO Naturals, India, and CureSupport, Netherlands. She did her graduation in Chemistry from Calicut University, Kerala, India, and post-graduation from Mahatma Gandhi University, Kerala, India, with Gold medal and first rank. She is a recipient of the prestigious INSPIRE Fellowship from Government of India. She completed her PhD in Chemistry from Mahatma Gandhi University, India, under the guidance of Prof. Sabu Thomas, the vice-chancellor and a renowned scientist in this area. She visited many foreign universities as a part of her research activities and published 20+ book chapters and around 15 research articles in peer-reviewed international journals. She edited 10 books with leading publishers like Elsevier, Springer, Wiley, and RSC across the globe. Dr. Preetha has received a number of national and international presentation awards. She worked as a post-doctoral researcher in the research group of Prof. Thomas and did enormous works in biomaterials for tissue engineering. She has also worked as a Guest Lecturer in Chemistry at Department of Chemistry, Morning Star Home Science College, Angamaly, Kerala, India.

Qi Li is a Master's student at Huazhong Agricultural University, China, and a PhD candidate at Kyushu University, Japan. Her research is focused on the investigation of food-grade functional Pickering emulsions stabilized by cellulose nanofibrils, especially bacterial cellulose nanofibrils. She has published 7 academic papers.

Quanling Yang is a Professor at the School of Materials Science and Engineering, Wuhan University of Technology, Wuhan, China. He received his Bachelor's degree in Applied Chemistry and a Master degree in Polymer Chemistry and Physics from Wuhan University, China, under the supervision of Prof. Lina Zhang in 2008 and 2010, respectively. After three years of doctoral course in Biomaterials Science at The University of Tokyo, Japan, under the supervision of Prof. Akira Isogai, he received his PhD degree in 2013. From 2013 to 2015, he worked as JSPS post-doctoral research fellow at The University of Tokyo. He has been honored Chutian Scholar of Hubei Province, Hubei Chenguang Talented Youth Development Foundation, Fellow of IAAM, IAAM Medal Award, and JSPS Research Fellowship for Young Scientists. He has published more than 100 scientific papers, 25 patents, and delivered over 30 presentations in international academic conferences. His main research focuses on environment-friendly biomass materials, functional polymer materials, 1D/2D nanomaterials, and energy storage and conversion materials.

Revin Victor Vasilyevich is the Dean of Faculty of Biotechnology and Biology and Professor at the Department of Biotechnology, Bioengineering, and Biochemistry, Ogarev Mordovia State University, Russia. He graduated from Ogarev Mordovia State University, Russia, in 1972. During his career, he worked at different positions like Senior Engineer (1973), Head of Laboratory (1974–1979), Lecturer (1979), Senior Lecturer (1981), Associate Professor

(1983–1990), and Full Professor (1990–Present). He has received several honors and awards, including 'Professor of the Year' (2018), gold medals (2010, 2012, 2013) for his innovation works, State Award of the Republic of Mordovia (2005), and several others. He is the member of several scientific communities such as European Biotechnological Association, Russian Academy of Natural Sciences, and several others. His research interests include the study of microbial polysaccharides, bacterial cellulose, bioethanol production, wood biodegradation, preparation of new materials by using different microorganisms, and study of the role of lipids and lipid-based enzymes in functioning of nervous cells.

Runcang Sun is a Professor at the Centre for Lignocellulose Science and Technology and Liaoning Key Laboratory of Pulp and Papermaking Engineering at Dalian Polytechnic University, China. He graduated from Northwestern University in 1982, received his PhD degree from the University of Wales in 1996, and was specially appointed to the Yangtze River Scholars Award Program in 2000. He is a fellow of the Royal Society of Chemistry (FRSC, 2003), and was awarded the Anselme Payen Award in 2020. Dr. Sun has been engaged in the efficient fractionation and high-value utilization of lignocellulosic materials for production of sustainable biofuels and biomaterials. Several of his novel techniques have been successfully applied at industrial scale. Dr. Sun currently holds 119 Chinese National patents on the conversion of lignocelluloses into biomaterials and biofuels, published 835 peer-reviewed articles, and 35 books/book chapters, with more than 36,000 citations. He is a co-Editor-in-Chief of *Industrial Crops and Products*, Associate Editor of *Carbohydrate Polymers* and *Journal of Biobased Materials* and *Bioenergy*, and a member of the editorial board at *ChemSusChem*, *Journal of Agricultural and Food Chemistry*, *Cellulose Chemistry and Technology*, *BioResources*, *Bioresources and Bioprocessing*, etc.

Sabu Thomas is the vice-chancellor of Mahatma Gandhi University, Kottayam, Kerala, India. He has been a Full Professor of Polymer Science and Engineering from 1997 at the School of Chemical Sciences and the Founding Director of the International and Inter University Centre for Nanoscience and Nanotechnology, Mahatma Gandhi University, as well as the fellow of the Royal Society of Chemistry, London. He received his PhD degree in 1987 in Polymer Engineering from the Indian Institute of Technology (IIT), Kharagpur, India. He has more than 800 publications, 5 patents, 50,000+ citations with an h-index of >100. He has edited around 100 books and supervised around 100 PhD theses. He was cited in the list of Most Productive Researchers in India, 5th Position in 2008, and in the list of the most cited researchers in Materials Science and Engineering by Elsevier Scopus data in 2016. His areas of research include Nanoscience, Polymer Science and Engineering, Polymer Nanocomposites, Elastomers, Polymer Blends, Interpenetrating Polymer Networks, Polymer Membranes, Green Composites and Nanocomposites, Nanomedicine, and Green Nanotechnology.

Sehrish Manan is a post-doctoral researcher at Huazhong University of Science and Technology, China. She received her PhD degree in Crop Genetics and Breeding from Huazhong Agricultural University, China, in 2017. She worked as a Lecturer in the Department of Public Health, University of Haripur, Pakistan, during 2012–2013. She has published more than 25 articles in peer-reviewed international journals and authored several book chapters. Her research interests include the development of mycelium and cellulose-based functional materials for wound healing and food packaging applications in addition to plant transformation and oil synthesis through genetic engineering approaches.

Shaukat Khan obtained his Master's degree in Organic Chemistry from the Institute of Chemical Sciences, University of Peshawar, Pakistan, in 2008. He worked as a Lecturer in Chemistry at the same institute and joined the School of Materials Science, Gwangju Institute of Science and Technology, South Korea, as the research assistant in 2011. He later joined the School of Chemical Engineering, Kyungpook National University, South Korea, as a PhD scholar and graduated in February 2017. Since then, he has been serving as the post-doctoral research fellow at the School of Chemistry, Sun Yat-Sen University, Guangzhou, China. His research interests include biopolymeric scaffolds for tissue engineering applications and nanocarbon materials for energy storage applications.

Shilin Liu is a Professor at the College of Food Science and Technology at Huazhong Agricultural University, Wuhan, China. He received his PhD degree with major in Polymer Chemistry and Physics from Wuhan University, China, in 2009. Following a 24-month post-doctoral training at Tokyo University, Japan, as a JSPS fellow, he joined Huazhong Agricultural University in 2013 to establish a research program in food biophysics and nanotechnology. Dr. Liu's philosophy is to apply interdisciplinary approaches to solve the complex problems in the food industry. He has published more than 100 academic papers, which have been cited over 3000 times, and he has also participated in the compilation of 4 academic books. His specific research areas include the study of physical properties of food biopolymers (proteins and polysaccharides) and food products, delivery systems as novel food ingredients, and interfacial phenomena and engineering.

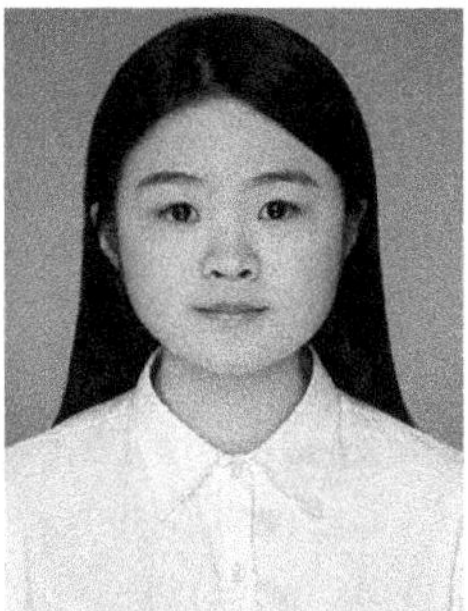

Siling Chen received her Bachelor's degree (2014) and continues learning as a Master's student under the supervision of Prof. Quanling Yang at the Wuhan University of Technology, China. Her research interests include biomass materials and piezoelectric materials and their applications in nanogenerators and sensors.

Sreerag Gopi is Chief Scientific Officer and Vice President of Research in ADSO Naturals, India, and CureSupport, Netherlands, respectively. He received his graduation in Chemistry from Calicut University, Kerala, India, and post-graduation from Madras Christian College, Chennai, India. He is a recipient of the prestigious ERASMUS MUNDUS Fellowship from the European Union for his PhD studies. He is a material chemist and nanomaterial scientist, and has expertise in nanomaterial synthesis, characterization, biocomposites for natural products, biomedical, and water purification experiments. He has published over 20 peer-reviewed international papers and several book chapters. He is also working on book projects with several publishers like RSC, Springer, Wiley, etc. He was selected into the Royal Society of Chemistry in 2018 and awarded the AMRSC. He is also a chartered member of the Royal Australian Chemical Institute.

Tahseen Kamal is an Assistant Professor at King Abdulaziz University, Jeddah, Saudi Arabia (professional period: 2015–2020). His research interests include the modification of polysaccharides and synthetic polymers with metals for utilization in dealing with aqueous-based pollutants. Before availing the current position, he worked on the structures of polymeric materials, primarily focusing on structure–property relationships in the solid and solution states using X-ray and neutron scatterings at Kyungpook National University (Daegu, South Korea), Pohang Accelerator Laboratory at POSTECH (Pohang, South Korea), HANARO at Korea Atomic Energy Research Institute (Daejeon, South Korea), and NSRRC (Hsinchu Science Park, Taiwan) (PhD period: 2009–2014).

Tao Wu received his Bachelor's degree from Hubei University in 2016. In 2019, he received his Master's degree in Materials Science and Engineering from Wuhan University of Technology, China, under the supervision of Prof. Quanling Yang. His research interests include natural polymer materials and piezoelectric functional materials and devices.

Taous Khan is a Professor and Chairperson at the Department of Pharmacy, COMSATS University Islamabad, Abbottabad Campus, Pakistan. He obtained his Bachelor's degree in Pharmacy and M. Phil degree from University of Peshawar, Pakistan, in 1994 and 2000, respectively. He obtained his first PhD degree in Pharmacy from Karachi University, Pakistan, and his second PhD degree in 2008 in Chemical Engineering from the Department of Chemical Engineering, Kyungpook National University (KNU), South Korea. He joined KNU as a post-doctoral fellow from 2011-2012. He has supervised more than 20 graduate students in their Master's and PhD studies. Prof. Khan has filed 3 patents and has published more than 100 research papers and several chapters in leading journals and books. He is also an editorial board member of various journals. His research interests include biomedical and pharmaceutical applications of biopolymers and isolation of bioactive natural products.

Tengku Arisyah Tengku Yasim-Anuar completed her PhD and undergraduate studies at Universiti Putra Malaysia (UPM). She was a post-doctoral researcher at Faculty of Biotechnology and Biomolecular Sciences, UPM, and currently is a researcher at NextGreen Pulp & Paper Sdn Bhd. Her research interests lie in the areas of nanocellulose, biomaterials, and bionanocomposites.

Vanja Kokol received a PhD degree in the area of Textile Chemistry in 2001 from the University of Maribor, Faculty of Mechanical Engineering (UM-FS). She has been employed at UM-FS from 1994, currently as a research counsellor with the habilitation of Associate Professor. Her research work in the last decade has been oriented toward modification and functionalization of fibers and biopolymers, and their processing in highly engineered materials for different applications (from technical to biomedical). Special attention is attributed to the development of biopolymeric 2D and 3D materials with targeted and biocompatible antimicrobial activity. She is the author of around 100 papers, 1 book, 4 book chapters, 3 patents, and has supervised several doctoral (7) and post-doctoral (5) students and been active in research programmes on Textile Chemistry (from 1999) and Center of Excellence (from 2010) for advanced materials and technologies and in the area of soft biomaterials. She has also been involved (leading or collaborating) in many national, bilateral, international funded research projects.

Waleed Ahmad Khattak completed his PhD in Chemical Engineering from the Department of Chemical Engineering, Kyungpook National University, South Korea, in 2014. His areas of specialization include bioethanol production from waste materials, biopolymers, and biopolymer composites. Dr. Waleed has published several articles in reputed journals and has filed patents. Besides academic achievements, he has been associated with Certas

Energy, Hyperdrive Lubricants as Technical Supervisor. His responsibilities include the development of new products, modifying the existing products, and improving the overall quality of the products produced at Hyperdrive.

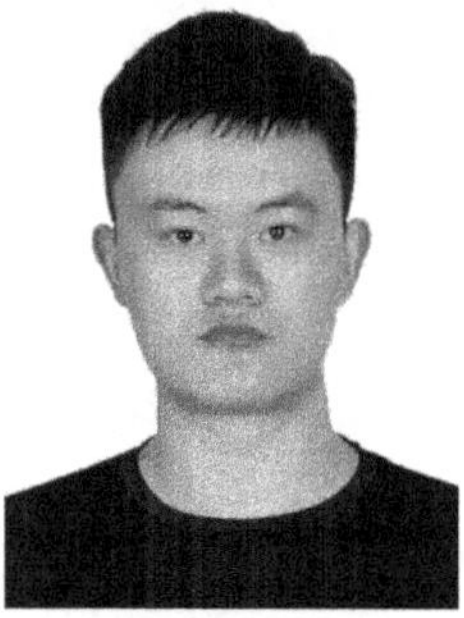

Wenchao Yu received his Bachelor's degree (2018) from Nanchang Hangkong University, China, and his Master's degree in Materials Engineering (2020) from Wuhan University of Technology, China, under the supervision of Profs. Guohua Hu and Quanling Yang. His research interests include biomass materials and supercapacitors. He is currently a Hardware Technical Engineer at Huawei Technologies Co., Ltd., China.

Xiaohui Wang is a Professor and Vice Director of the State Key Laboratory of Pulp and Paper Engineering at South China University of Technology, China. She received her PhD from the School of Resources and Environmental Science, Wuhan University, China, in 2006. She studied at University of Washington and the Public Health Laboratory of Washington State at Seattle from 2006–2009. She joined South China University of Technology in 2009 and was promoted to Full Professor in 2014. Her research interests include using biomacromolecular self-assembly to create ordered arrangements of biomass-based materials for nanomedicine, sensing and bioimaging, and developing biomass-based functional carbon materials and functional paper-based materials. She has authored more than 90 publications, 2 monographs, and is a co-inventor for over 20 patents and invention disclosures.

Xinshuo Liang is a postgraduate student at the Department of Mechanical Engineering, School of Biomaterial and Tissue Engineering, University College London (UCL), London, United Kingdom. He received his Bachelor's degree in Material Science and Engineering from Sichuan University in 2019 and Master's degree in Biomaterial and Tissue Engineering from UCL in 2020. His research interests include biomedical applications of cellulose-based hydrogels and bio-smart wearables for people.

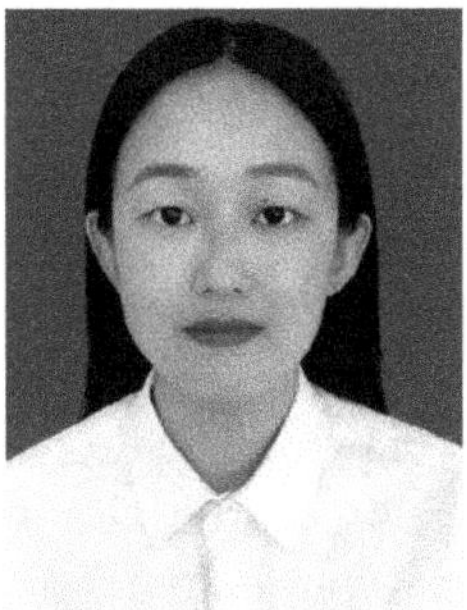

Yanan Yin received her Bachelor's degree (2014) and continues to learn as a Master's student under the supervision of Prof. Quanling Yang at the Wuhan University of Technology, China. Her research interests include biomass materials and dielectric materials and their applications in dielectric capacitors.

Yasir Beeran Pottathara is currently working as an assistant with a doctorate at the Faculty of Mechanical Engineering, University of Maribor, Slovenia. He obtained his Master's degree in Physics from University of Calicut, India, and an Associate Master's degree in Nanoscience and Nanotechnology from the University of Kerala, India. He received his PhD degree in Engineering Sciences in 2017 from the University of South Brittany, Lorient, France. He also worked as a postdoctoral fellow at the Mahatma Gandhi University, Kottayam, India, and Universiti Teknologi Mara, Malaysia. He is the author/co-author of several scientific articles and book chapters, and edited 3 books in the field of nanomaterials. His research interests focus mainly on nanocellulose multifunctional composites for energy storage, flexible electronics, and sensors.

Yong Huang is Professor at Technical Institute of Physics and Chemistry, Chinese Academy of Sciences, China. He received the PhD degree in Material Science from South China University of Technology, China, in 1996. In 2006, he worked as a visiting professor in Dortmund University, Germany. Prior to his current position as a Professor in Polymer Science at Technical Institute of Physics and Chemistry (TIPC, CAS), Prof. Huang was appointed as a Professor at Institute of Chemistry, CAS from 2000 to 2007. His research interests are cellulose chemistry and materials and materials from natural and renewable resources. He has published more than 150 papers and 112 patents of which 65 have been authorized. Simultaneously, he has devoted himself to the commercialization of cellulose materials and significant progress has been made so far.

Zakiah Ahmad is the Dean of Faculty of Civil Engineering, Universiti Teknologi Mara, Malaysia. Dr. Zakiah is specializing in timber engineered composites and nano-polymer composites. She has secured and leads more than 50 research grants worth more than Euro 5 million and published more than 300 papers in journals and proceedings and 8 books. She has received several gold and silver awards on her research projects at national and international research competitions. She is the recipient of Top Research Scientist Malaysia, Standard Malaysia Award, Best Researcher Award, Gold Consultant Award, and the Best Academic Award. She has also delivered keynote lectures at numerous seminars and conferences at national and international levels. Dr. Zakiah has served as a consultant, committee member, and advisor to numerous public agencies and institutions and has been appointed as visiting scientist at BRE Research Institute, University of Bath, United Kingdom, and Centre for Nanoscience and Nanotechnology at the Mahatma Gandhi University, India.

Zhiqiang Fang received his PhD degree from South China University of Technology (SCUT), Guangzhou, China. As a visiting scholar at the University of Maryland from 2012 to 2013, he worked on transparent paper for flexible electronics. He did his post-doctoral research in Materials Science and Engineering at SCUT. Currently, he is an Associate Professor at the School of Light Industry and Engineering at SCUT. His research interests include multiscale cellulose materials for flexible electronics.

Zhuqun Shi received her Master's degree in Chemical Engineering from Wuhan University, China, under the supervision of Prof. Jie Cai in 2011. After three years of doctoral course in Biomaterials Science at The University of Tokyo, Japan, under the supervision of Prof. Yuji Matsumoto, she received her PhD degree in September 2014. From October 2014, she worked as a post-doctoral researcher with Prof. Akira Isogai at The University of Tokyo. In 2016, she joined the School of Chemistry, Chemical Engineering and Life Sciences, Wuhan University of Technology, China, as an Associate Professor. Her main research interests focus on biomaterials, e-skin, nanomaterials, etc.

https://doi.org/10.1142/9781786349477_fmatter

Acknowledgments

The editors Guang Yang, Muhammad Wajid Ullah, and Zhijun Shi are thankful to the first and corresponding authors of all chapters, i.e., Hidaya Ariffin, Pei Huang, Min Wu, Preetha Balakrishnan, Sabu Thomas, Lallepak Lamboni, Munair Badshah, Taous Khan, Fazli Wahid, Qi Li, Shilin Liu, Mazhar Ul-Islam, Hye Jung Youn, Hao Ling, Xiaohui Wang, Mingwei Zhu, Liangbing Hu, Jinyu Wang, Zhuqun Shi, Quanling Yang, Yasir Beeran Pottathara, Deepu A. Gopakumar, and Cheng Zhong (according to the order of the chapters) as well as the co-authors of all chapters, for their valuable contribution in different areas related to Nanocellulose. The editors are especially thankful to Professor Victor V. Revin (Mordovian State University, Saransk, Russia) and Professor Sabu Thomas (Mahatma Gandhi University, Kottayam, India) for their valuable suggestions in organizing this book. We are also thankful to the copyeditors and production team of World Scientific Publishing, UK.

Finally, we would like to acknowledge the BRICS STI Framework Programme 3rd call 2019 (2018YFE0123700), National Natural Science Foundation of China (21774039, 51973076), China Postdoctoral Science Foundation (2016M602291), and the Fundamental Research Funds for Central Universities, Open Research Fund of State Key Laboratory of Polymer Physics and Chemistry, and Changchun Institute of Applied Chemistry, Chinese Academy of Sciences.

https://doi.org/10.1142/9781786349477_fmatter

Contents

https://doi.org/10.1142/9781786349477_0001

Chapter 1

Introduction to Nanocellulose

Muhammad Wajid Ullah[*], Sehrish Manan[*], Mazhar Ul-Islam[†], Revin Victor Vasilyevich[‡], Sabu Thomas[§], and Guang Yang[*,¶]

[*]*Department of Biomedical Engineering, College of Life Science and Technology, Huazhong University of Science and Technology, Wuhan, PR China*

[†]*Department of Chemical Engineering, College of Engineering, Dhofar University, Salalah, Sultanate of Oman*

[‡]*Federal State-Financed Academic Institution of Higher Education, National Research Ogarev Mordovia State University, Saransk, Russia*

[§]*International and Inter University Centre for Nanoscience and Nanotechnology, Mahatma Gandhi University, Kottayam, Kerala, India*

[¶]*yang_sunny@yahoo.com*

Abstract

Cellulose with at least one of its dimensions less than or equal to 100 nm is termed as nanocellulose. It is a unique and promising natural material extracted from native cellulose and produced by certain microbial cells and cell-free systems. Nanocellulose has received immense consideration in last couple of decades owing to its abundance, renewability, remarkable physical properties, special surface chemistry, and excellent biological features (biocompatibility, biodegradability, and non-toxicity). Taking advantage of the structure and properties of nanocellulose, the current science of biomaterials aims at developing new and formerly non-existing materials with novel and multifunctional properties, in an attempt to meet current requirements in different fields such as biomedicine, the environment, energy, pharmaceutics, agriculture, food, etc. This chapter provides an overview of different synthesis methods of nanocellulose: mechanical approaches by applying high-pressure, grinding, crushing, sonication, and milling; chemical synthesis involving alkaline, acidic, oxidation, and enzymatic treatment; as well as by using bacteria and cell-free systems. It further discusses different morphological forms of nanocellulose including cellulose nanocrystals (CNCs), cellulose nanofibers (CNFs), bacterial nanocellulose (BNC), and cellulose produced by cell-free systems, in terms of their features such as chemical structure, macrostructural morphology, physico-mechanical properties, thermal and biological properties, rheology, optical behavior, and their interrelationships and applications.

Keywords: Nanocellulose; synthesis; structure; properties; functional materials

1. Introduction

Cellulose is the most abundant and ubiquitous biopolymer on earth. It is an inexhaustible source of raw material produced by plants [1, 2]; microorganisms such as certain bacteria, algae (Valonia, marine algae), fungi, invertebrates, several marine animals (tunicates); and cell-free systems [3–7]. It is a long-chain homopolymer, comprised of repeated dimers or disaccharides of anyhydro-*D*-glucopyranose units, known as cellobioses, linked by β-1,4-glycosidic linkage [1, 3, 8, 9]. The individual units are rotated at 180 degree with respect to its neighboring units. It possesses high molecular weight, ranging from several hundred to 20,000 Daltons, depending on the number of repeated anyhydro-*D*-glucopyranose units, which in turn depends upon its source and the method of synthesis [1, 10]. The cellulose chain is asymmetric in nature, with one terminus possessing reducing functionality, containing the hydroxyl (−OH) group, while the other is non-reducing in character [1, 11]. Regardless of source, its chemical structure is the same, but with varying morphological, physiological, mechanical, and biological properties. Owing to its diverse array of properties, cellulose can meet the demand for a wide spectrum of materials and products for diverse applications in biomedicine [12, 13], energy [14, 15], pharmaceutics [16–18], infrastructure [19], additive manufacturing [20–22], paper [23], textile [24], environment [25], and food industries [26]. Nanocellulose-based materials have received considerable research attention in recent times, as indicated by the increasing number of publications in the last ten years (Fig. 1a). Further, its potential for diverse applications have been explored, as indicated by the growing number of publications in different research areas (Fig. 1b).

Cellulose with fiber diameter at the nanoscale lattice is known as nanocellulose, and this form exists in different morphologies such as cellulose nanocrystals (CNCs), cellulose nanofibrils (CNFs) or cellulose nanowhiskers (CNWs), and bacterial nanocellulose (BNC). Nanocellulose possesses unique morphological, structural, physical, chemical, mechanical, and biological features such as reticulate three-dimensional network structure, high crystallinity and mechanical strength, high water holding capacity, hydrophilicity, biocompatibility, biodegradability, non-toxicity, optical transparency, high surface area, moldability, and polyfunctionality [24, 27, 28]. These properties can be tuned by incorporating different materials (polymers, nanoparticles, and clay, etc.) in the cellulose matrix through *in situ* synthesis, *ex situ* blending or mixing, and through regeneration methods [28] for specialized applications in tissue engineering, regenerative medicines, wound dressing, cosmetics, optoelectronics, energy storage, drug delivery, bioprinting, biosensing, enzyme immobilization, environmental monitoring, and food storage. [2, 27, 29–34].

Wood biomass is mainly comprised of cellulose, hemicellulose, and lignin. The soft and hard wood from higher plants, possessing a hierarchical structure, have been largely exploited as the sources of plant nanocellulose, specifically CNCs (Fig. 2). However, higher plants are not the only sources of nanocellulose, since it is also obtained from other plants such as hemp, ramie [35], sisal [36], rice [37], palm, flax, jute [38], sugarcane jaggery (gurr) [39, 40], bamboo [41, 42], eucalyptus, wheat straw, starch [43], and cotton [35]. The cellulose microfibrils in higher plants are formed by a biogenetic bottom-up process that involves the assembly of cellulose chains and their subsequent stabilization through hydrogen bonding and van der Waals forces, which further built up into higher hierarchical structures through assemblies of fibrils [44]. The microfibrils are several micrometers in length and 10–50 nm in diameter and contain

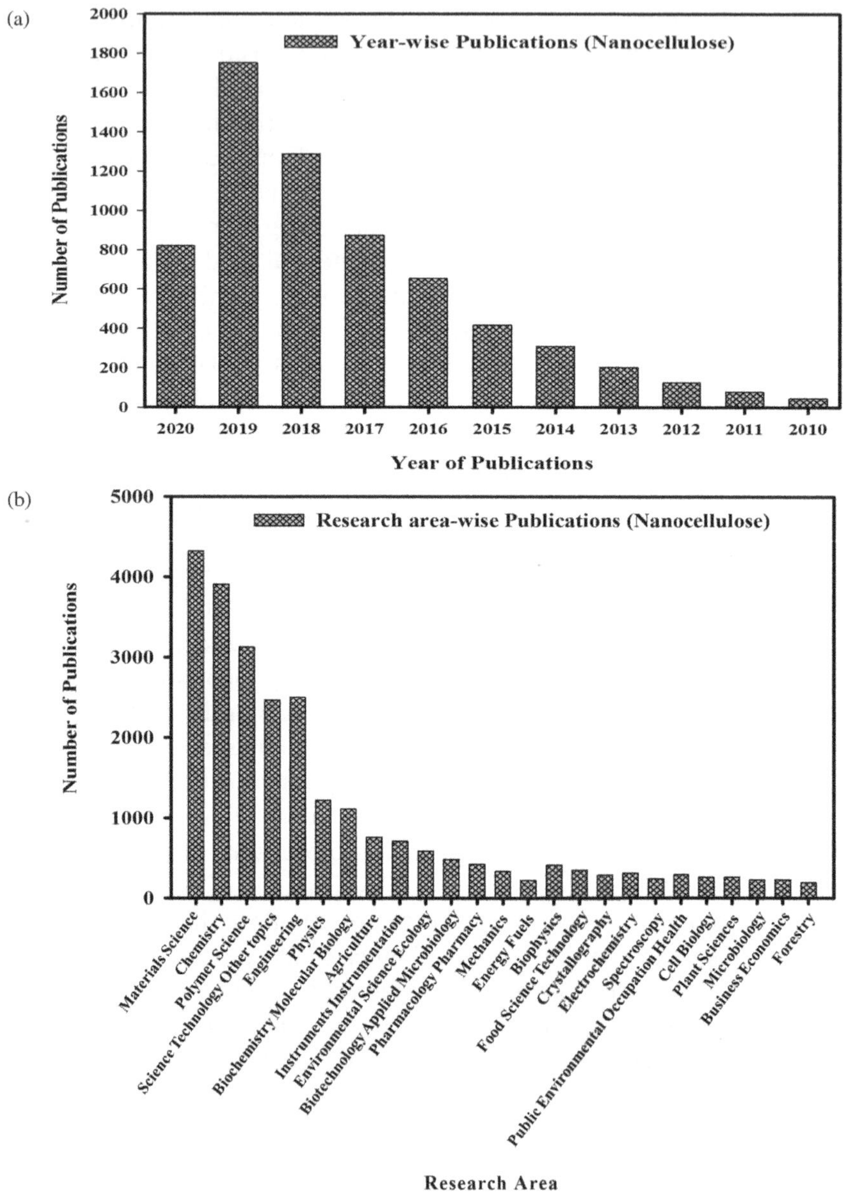

Fig. 1. Number of publications based on nanocellulose with respect to (a) year of publication and (b) area of research.

Source: Web of Science; search keyword (title): nanocellulose.

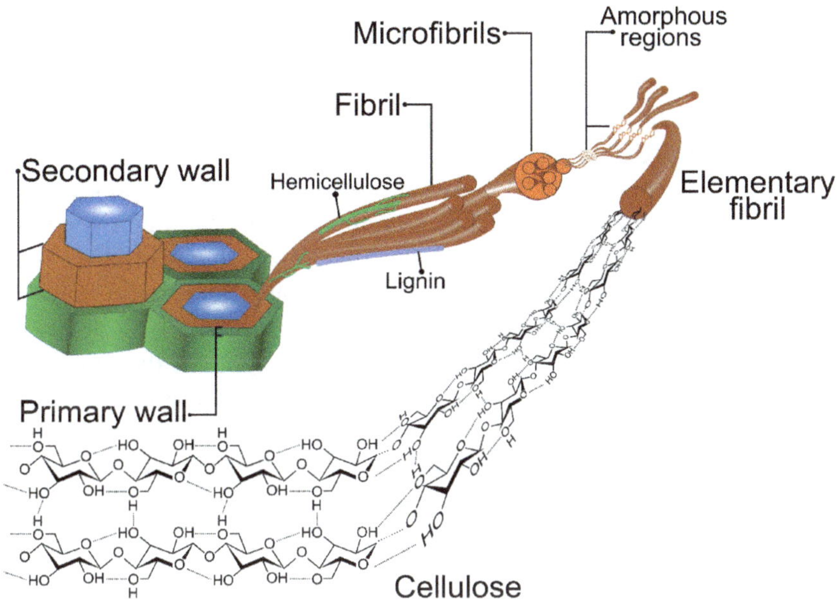

Fig. 2. Graphical illustration of the hierarchical structure of plant-derived nanocellulose.

Source: Reproduced with permission from Ref. [58] under open access Creative Common CC BY license.

highly ordered and random domains, representing the crystalline and amorphous regions, respectively. The crystalline domains can be separated from the mixed cellulose through a top-down approach by using various mechanical, chemical (or combined), chemical, and/or enzymatic treatments to form CNCs [11, 45]. Different mechanical procedures such as refining and high-pressure homogenization [46], grinding [37], cryocrushing [37, 47], sonication [48], and ball milling [49], etc. are used for the preparation of CNCs while chemical treatment involves acid hydrolysis [50–52]. In acid hydrolysis, cellulose is treated with different acids such as phosphoric acid [53], hydrobromic acid [54], and 1-butyl-3-methylimidazolium hydrogen sulfate ionic liquid [55]. The most preferred approach, among all methods, is the treatment with concentrated sulfuric acid and hydrochloric acid, which facilitates the cleaving of accessible glycosidic linkages within the cellulose microfibrils and results in the formation of characteristic rod-shaped CNCs. In addition to cleaving of glycosidic linkages, sulfuric acid hydrolysis also introduces anionic sulfate ester groups onto the surface of CNCs that produces electrostatic repulsion among the cellulose particles in aqueous dispersion and yields stable CNCs over a wide pH range. Nevertheless, a large portion of reactive −OH groups on CNCs are preserved, allowing for the grafting or attachment of other small molecules on them, thus enabling a wide array of chemical modifications [56]. Aside from plants, nanocellulose is also obtained from microbial sources — commonly known as microbial nanocellulose (MNC) or bacterial nanocellulose (BNC) — as well as synthesized by cell-free systems [6, 7], which represent the purest form of cellulose, thus not requiring any physical, chemical, or enzymatic pretreatment. Cellulose production by microbial cells is an aerobic process where it is produced at the air-medium interface in the form of a hydrogel at the cost of utilizing chemically defined media or different waste materials such as agricultural waste, industrial waste, and beer fermentation waste [57].

This chapter provides an overview of different types of nanocellulose with emphasis on its synthesis, properties, and applications. Various mechanical, chemical, enzymatic, bacterial, and cell-free approaches used for the synthesis of different forms of nanocellulose, such as CNF, CNC, CNWs, microfibrillated cellulose (MFC), and bacterial nanocellulose, are summarized. Further, different structural features of nanocellulose in term of its source and synthesis methods are discussed. Finally, the applications of nanocellulose in different areas are overviewed.

2. Synthesis Methods of Nanocellulose

The methods of preparation of nanocellulose vary based on its source and final application. For example, plant-based nanocellulose preparation usually involves destructuring through mechanical means such as high-pressure homogenization, grinding/crushing, and sonication, as well as chemical treatment through different acids, or a combination of both (Fig. 3). In addition, enzymatic digestion is also used for the preparation of wood-based nanocellulose. On the contrary, BNC is produced by different microorganisms and cell-free systems. The following sections provide an overview of the different methods used for the preparation of nanocellulose from multiple sources.

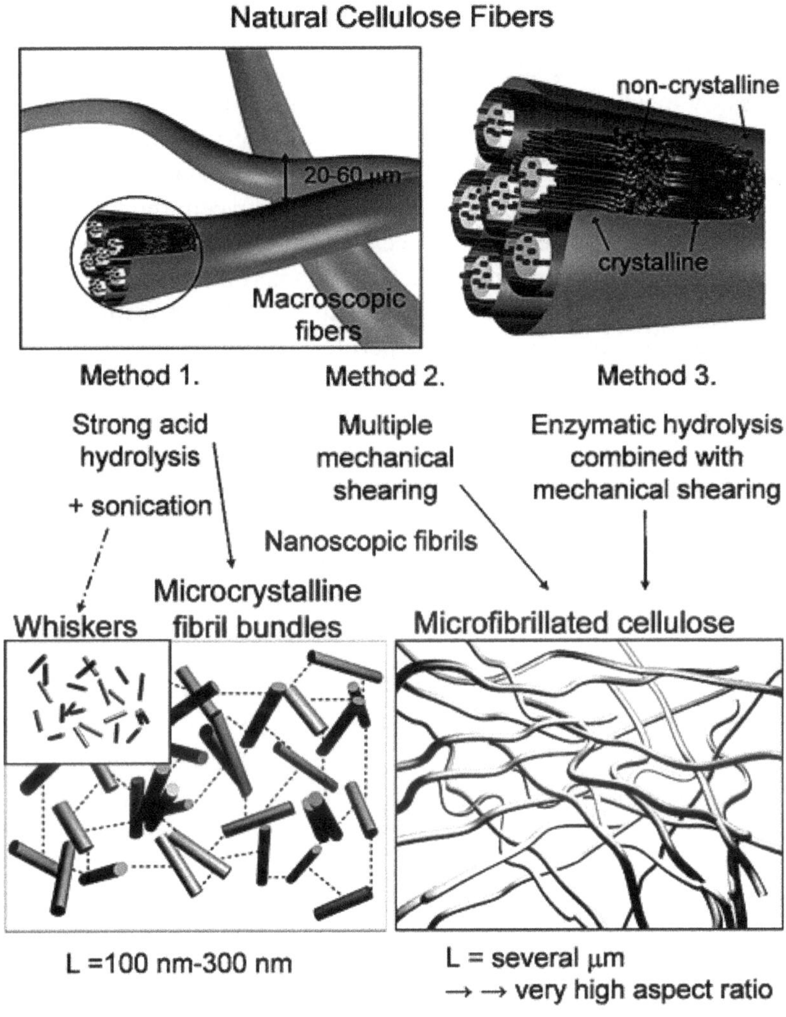

Fig. 3. Schematic illustration of the disintegration of macroscopic cellulose fibers into nanoscale fibrils by three different methods: Method 1. Successive acid hydrolysis and formation of colloidal suspension of aggregated, highly crystalline, and low aspect ratio fibril aggregates, and their subsequent hydrolysis and sonication to form cellulose fibrils and whiskers. Method 2. Mechanical shearing of MFCs into highly entangled and inherently connected fibrils and fibril aggregates. Method 3. Enzymatic hydrolysis of natural cellulose fibers (crystalline and non-crystalline regions) to form cellulose I fibrils (5 nm) and fibril aggregates (10–20 nm).

Source: Reproduced with permission from Ref. [59]. Copyright (2007) American Chemical Society.

2.1 *Mechanical*

The nanocellulose preparation (i.e., defibrillation) from lignocellulosic biomass by mechanical means requires chemical pretreatment with acids or enzymes and physical treatment such as decrystallization, grinding, or derivatization of plant material. The chemical treatment eases the defibrillation process by widening the space between the −OH groups, increase the inner surface, alters the crystallinity, and breaks the hydrogen bonds, which leads to enhanced surface area and improves the activity of cellulose fibrils [60]. The chemical treatment of lignocellulose is also carried out with ionic liquids that permeate through the microstructure of cellulose and subsequently attack the hydrogen bonds between the cellulose fibers. A high-speed blending or grinding also results in defibrillation [61]. Similarly, high-pressure homogenization coupled with ultrasonication also results in the relaxation of fibers by breaking the intra- and inter-molecular hydrogen bonds [62, 63]. A study has also reported the acid-steam coupled treatment for preparation of nanocellulose from pineapple leaf fibers where high steam explosion with alkali hydrolyzed the hemicellulose and depolymerized the lignin, which was subsequently oxidized by chlorine to accelerate its degradation. The chemical treatment resulted in the formation of various groups such as −OH, C=O, and −COOH on the lignin surface, which facilitated its solubilization in the alkaline medium [64]. The following sections cover the different mechanical methods for preparation of nanocellulose from lignocellulosic biomass.

2.1.1 *Refining and high-pressure homogenization*

High-pressure homogenization is a simple, safe, and highly efficient method for refining cellulose fibers from wood pulp. In this process, the wood pulp is passed through a narrow-sized nozzle by applying different types of pressures such as high-velocity pressure, impact, and shear forces, which results in shear rates in the stream and decreases the size of the fiber to the nanoscale level [65]. This method is also used for the preparation of nanocellulose from other raw materials such as bleached sugar beet [66], extract of prickly pear [67], and kenaf bast fiber [68]. The efficiency of this process is reduced by clogging of nanocellulose in the nozzle, which can be bypassed or somehow minimized through chopping of cellulose fibers before applying high pressure.

2.1.2 *Grinding*

Grinding involves the application of shear forces on the pulp by passing it through two stones, where one stone is fixed while the other is continuously rotating. Grinding results in breaking of hydrogen bonds and the cell wall structure that converts the cellulose into nano-sized fibers [37]. A study reported the preparation of CNFs from bleached eucalyptus pulp by using the commercial stone grinder and studied the relationship between energy consumption and fibrillation time as a function of crystallinity. During the grinding process, the friction force between the stones generates heat to increase the temperature of the system and helps in the evaporation of water content; this results in increased solid content.

2.1.3 *Cryocrushing*

In this technique, the wood cellulose is placed in water for some water to allow the absorption of water into its cavity followed by its immersion in liquid nitrogen, which immediately solidifies the water content. The solidified wood cellulose is immediately crushed by a mortar and pestle to convert the cellulose into nano-size fibers [47]. In fact, the applied high force on frozen cellulose fibers causes their rupturing due to the applied pressure through ice crystals that results in the conversion of cellulose microfibers to CNF [43]. This method has been applied in the preparation of nanocellulose with fiber diameter 50–100 nm from soybean stock [69].

2.1.4 *Ultrasonication*

In this method, the hydrodynamic forces of ultrasound with oscillation power are utilized for the isolation of cellulose fibrils from wood sources [48]. The cellulose molecules absorb the ultrasonic energy of the high-intensity waves and form, expand, and implode the microscopic gas bubbles. Several studies have reported the preparation of nanocellulose through sonication [65, 70] and have studied the effect of different parameters, such as temperature, concentration, power, size, time, and distance from probe tip, on the degree of fibrillation of cellulose fibrils [71]. The improved fibrillation of cellulose is achieved at high temperature, high power, and small fiber size. Small fiber size can be achieved through the chopping of cellulose fibers prior to ultrasonication. For further improved fibrillation, high-pressure homogenization and ultrasonication can also be used together [72]. Ultrasonication produces nanocellulose with high yield and with improved properties such as small diameter, high fibrillation, and surface functionality (e.g., carboxylation) [73–75].

2.1.5 *Ball milling*

Ball milling is an advanced mechanical process in which cellulose is broken down in a hollow cylindrical container filled with balls made up of ceramics, zirconia, or metal. The cellulose suspension is placed in the cylindrical container that rotates and breaks the cell wall by high energy collision between the balls. The preparation of nanocellulose through ball milling is greatly affected by the ball size, weight ratio of balls and material, grinding time, moisture content, and carboxylic charge [49]. Although this process is very efficient in preparing nano-sized cellulose fibrils, homogeneity is still a challenge.

2.2 *Chemical*

The CNFs isolated by mechanical means consist of alternating crystalline and amorphous regions within the single cellulose domain [76]. Therefore, in order to dissolve the amorphous domain and permit longitudinal cuttings of cellulose microfibrils, a strong but controlled acid hydrolysis is required for the cleavage and destruction of more readily accessible regions to liberate rod-like CNCs [77]. The isolation of crystalline cellulose regions through acid hydrolysis requires pretreatment with chemicals such as alkali, oxidative, enzymatic, and finally with acids, for removal of hemicellulose, lignin, fat, wax, and pectins surrounding the cellulose structure.

2.2.1 *Alkaline pretreatment*

The alkaline pretreatment removes a considerable amount of lignin, wax, and oils covering the external surface of the cell wall [69, 78]. The alkaline treatment disrupts the lignin structure and separates the structural linkages between lignin and cellulose. Treatment of cellulose with 17–18% NaOH usually solubilizes lignin, pectins, and hemicelluloses. The isolated intact fibers require that alkaline treatment be carried out with care to avoid undesirable cellulose degradation, and further it ensures that acid hydrolysis is carried out only at the surface [69, 79].

2.2.2 *Oxidative pretreatment*

Oxidative pretreatment, such as the TEMPO-oxidative process, introduced by Isogi *et al.* [80] occurs at the surface of microfibrils and creates a negative charge that causes repulsion of nanofibrils and causes fibrillation. Oxidative pretreatment solves the issue of aggregation caused due to the presence of −OH groups and introduces new functional groups such as carboxylic (−COOH) and aldehyde (−COH) onto the cellulose surface under aqueous and mild conditions [81].

2.2.3 *Enzymatic pretreatment*

The enzymatic pretreatment of cellulose is considered a "green approach." Cellulose is degraded not by a single enzyme but rather by a group of enzymes categorized as type-A and type-B cellulases, collectively called as cellobiohydrolases, and type-C and type-D cellulases, also known as endoglucanases. Cellobiohydrolases directly degrade the highly crystalline cellulose, while endoglucanases attack the disordered structure in cellulose and degrade it. Both these classes of cellulose-degrading enzymes show a synergistic effect during cellulose fibrillation. The produced MFC possesses a favorable structure with high average molar mass and large aspect ratio as compared to nanofibers produced through hydrolysis by strong acids [82]. Additionally, ionic-liquid (IL)-based catalysts are also used for the degradation of cellulose biomass, offering several advantages such as electrochemical stability, electrical conductivity, ionic mobility, selective dissolution to different organic and inorganic solvents, and excellent chemical and thermal stabilities [83]. A study also reported the nanocellulose preparation in an anaerobic microbial consortium through the controlled hydrolysis of MCC from cotton fibers. This approach offered several advantages such as simplicity, low energy consumption, green approach (cellulase), biocompatibility, non-toxicity, and scalability [84].

2.2.4 *Acid hydrolysis*

Acid hydrolysis refers to the degradation of polysaccharides such as cellulose, starch, or hemicellulose into simple sugars using a strong acid treatment [85]. This is the most common and routinely used method for the preparation of nanocellulose from lignocellulosic biomass [86–88].

Prior to acid hydrolysis, the lignocellulosic biomass obtained after the removal of hemicellulose is treated with different chemicals such as dimethyl sulfoxide to allow the swelling of the matrix and facilitate the diffusion of acid into the domain structure of lignocellulosic biomass and disintegrate the nanowhiskers [39]. The efficiency of acid hydrolysis in terms of yield of nanocellulose depends upon several factors such as the content of lignocellulosic biomass and reaction conditions, including acid concentration, treatment time, and temperature, which in turn determine the morphology of the obtained nanocellulose [86]. Studies have reported that the nano-dimension of nanocellulose decreases with prolonged acid treatment [89].

2.3 *Bacterial Synthesis*

Bacterial nanocellulose (BNC) is produced by certain bacterial genera belonging to *Acetobacter*, *Rhizobium*, *Agrobacterium*, *Aerobacter*, *Achromobacter*, *Azotobacter*, *Salmonella*, *Escherichia*, and *Sarcina* [90, 91], as well as few algae such as *Valonia*, *Chaetamorpha* spp., and fungi, by utilizing glucose and other carbon sources [92–94]. Bacterial nanocellulose is produced by different microbial strains, as described above, and represents the purest form of cellulose. It is produced in the form of β-1,4-glucan chains within the microbial cells which protrude out as protofibrils through terminal complexes (TCs), located in the cell wall of microbial cells, and crystallize to form the ribbon-shaped microfibrils and ultimately form pellicles comprised of bundles [95, 96] (Fig. 4a).

BNC possesses unique structural, physico-chemical, mechanical, and biological features as compared to plant cellulose, most importantly its ultra-purity, high crystallinity (60–90%), high degree of polymerization (DP) varying between 2000 to 8000, and high water holding capacity (>100 times its dry weight) [98, 99]. Currently, BNC research is mainly focused on its low-cost production [57], exploration and genetic engineering of novel microbial strains for improved productivity [6], and improving its existing structural features along with imparting additional features by developing its composites with different functional materials [28]. For low-cost production, researchers are exploring different waste materials [100], such as fruit juices [101, 102], sugarcane molasses [40], agricultural wastes [103, 104], industrial wastes [105], food wastes [26, 106], and beer wastes [107, 108]. To improve the production of BNC

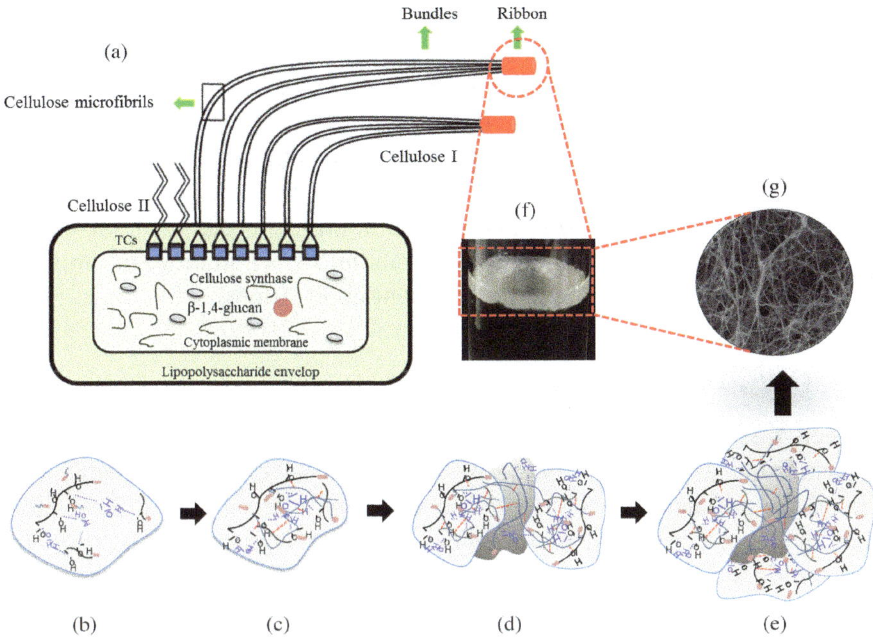

Fig. 4. Schematic illustration of (a) synthesis of β-1,4-glucan chains and their secretion from the bacterial cells across the cell wall through TCs, involving the (b) synthesis and aggregation of fibrils, (c) formation of pellicles, (d) movement of pellicle towards the air–medium interface due to the density gradient (cell-free system), and (e) and formation of BC sheet at the air–medium interface in the form of a (f) hydrogel, which is seen as a (g) reticulated fibrous structure forming a network of cellulose fibers.

Source: Reproduced with permission from [95, 97].

by microbial cells, different strategies such as static and submerged fermentation, genetic engineering strategies, strain improvement, co-culturing, as well as advanced reactors such as rotating disc reactors, rotary biofilm contactor, spin filter bioreactors, and silicon membrane reactors, have been designed [57]. Studies have also been carried out to investigate the effect of different additives, such as carboxymethyl cellulose (CMC) [109], organic acids [110], lignosulfonate [111], sugar-linked glucuronic acid-based oligosaccharides [112], agar [113], xanthan [114], sodium alginate [115, 116], alcohols [117], acetan [118], dextran [114], and yeast extracts [119], etc. on the production and structural features of BNC. These studies have demonstrated that by selecting appropriate substrates, cultivation conditions, suitable additives, and active bacterial strains, it is possible to control the molar mass, molar mass distribution, supramolecular structure, structural features, kinetics, yield, and production of other metabolic products alongside BNC production during the fermentation process [85]. Pristine BNC lacks antimicrobial, antioxidant, magnetic, and electroconductive properties. Therefore, different strategies, such as *in situ*, *ex situ*, and regeneration [28, 120, 121], have been developed to introduce different materials into BNC matrix and impart it with such additional features for different applications.

2.4 *Cell-Free Synthesis*

A cell-free system, comprised of a mixture of enzymes and cofactors constituting a metabolic pathway, represents an efficient, feasible, and reliable approach for the synthesis of bioproducts [122, 123]. Unlike cellulose production by microbial cells, the β-1,4-glucan chains (Fig. 4b) are produced inside the culture medium by the enzymes in a cell-free system and then self-aggregate and form micro/macrofibrils (Fig. 4c) and pellicles (Fig. 4d), which move to the top of the medium due to the density gradient and form a cellulose hydrogel (Fig. 4e). The thickness of the hydrogel increases when more pellicles are produced with the passage of time and get attach through self-aggregation (Fig. 4f) until all available glucose is consumed [95]. The hydrogel is comprised of reticulated fibrous arrangement as observed under a scanning electron microscope (Fig. 4g). Nanocellulose synthesis by a cell-free system follows the same chemical route of BNC synthesis by microbial cells where each biochemical reaction for nanocellulose synthesis is regulated by specific enzymes, substrate, and/or cofactors. The absence of any of these components retards or abolishes the synthesis pathway. During the first step, the sugar nucleotides provide monosaccharides by means of interconversion through epimerization, dehydrogenation, and decarboxylation reactions [7]. In the second step, the cellulose backbone is formed by the sequential addition of *D*-glucose-1-phosphate and UDP-glucose, each catalyzed by UDP-glucose pyrophosphorylase and cellulose synthase, respectively. In the third step, specific sugar transferases carry out the transfer of any acyl group and form acetyl Co-A. The second and third steps determine the structure of nanocellulose. In the final step, cellulose nanocellulose is excreted across the cell wall/membrane into the medium, and this is the step where the microbial and cell-free systems differ from each other. It is

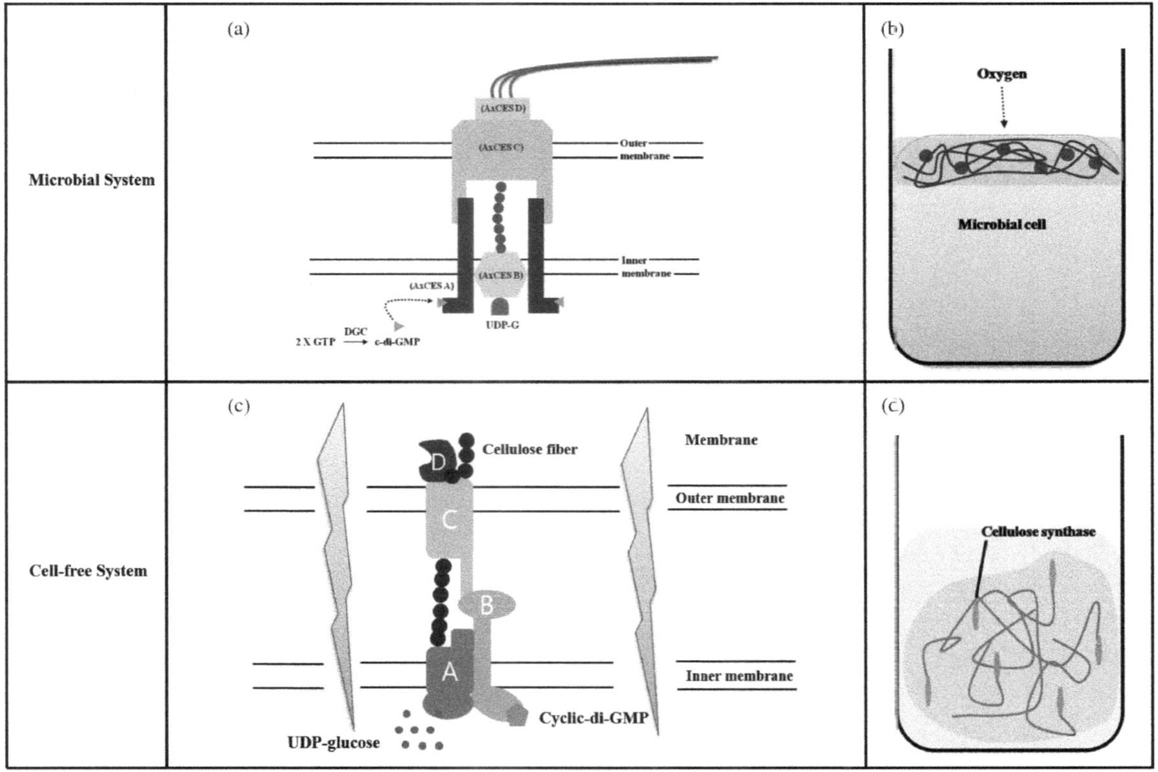

Fig. 5. Illustration of schematic models of (a) cellulose synthesizing machinery in a microbial system and (b) production of cellulose in a microbial system, (c) cellulose synthesizing machinery in a cell-free system, and (d) synthesis of bio-cellulose in a cell-free system.

Source: Reproduced from Ref. 95 with permission from Elsevier.

an energy-dependent step in microbial cells that utilizes ATP molecules in a well-organized system. In contrast, the cell-free system bypasses the complex process because it does not contain any barrier in the form of a cell wall or membrane [95]. Further, the cell-free synthesis of nanocellulose differs from the microbial system in several other aspects, including: (1) high yield (~58%), (2) anaerobic process, (3) simplicity of composition of production medium (carbon source only), (4) energy-efficient process, (5) capability of batch operation, (6) formation of newly synthesized nanofibrils at the bottom surface with the older ones deposited at the top surface (independent of oxygen supply), and (7) ability to potentially synthesize different polymorphic structures of cellulose with varying structural features by changing the experimental conditions [7, 95]. A comparative analysis of nanocellulose synthesis by microbial and cell-free systems is shown in Fig. 5. Ullah *et al.* showed that compared to BNC, nanocellulose synthesized by a cell-free system showed clear variation in thickness, density, diameter, and compactness of fibrils. Further, cell-free cellulose showed strong hydrogen bonding, indicating its high water holding capacity and slow water release rate as well as better mechanical features such as tensile strength, strain, and Young's modulus [96]. Such features are important for biomedical applications such as wound healing, tissue regeneration, and scaffold development.

3. Nanocellulose Types

The different types of nanocellulose vary based on their source raw material, synthesis methods, and structural features. These are named and abbreviated differently by different researchers (Table 1).

3.1 *Cellulose Nanocrystals*

CNCs, also known as CNWs, are the most commonly used and renewable nanomaterials obtained from amorphous sections of wood pulp through acid hydrolysis [124]. The acid hydrolysis process consists of two steps: the alkaline treatment of wood pulp to remove polysaccharide at the fibril surface followed by its heat treatment in an acidic environment for several hours to cleave and destruct the more readily accessible amorphous regions of cellulose fibers. At the appropriate level of glucose-chain depolymerization, the acidic mixture is diluted, and residual acids and

Table 1. The description of nanocellulose family in terms of types, synonyms and abbreviations, source materials, synthesis methods, and salient features.

Type	Synonyms and abbreviations	Source	Synthesis method	Features	Refs.
Nanocrystalline cellulose	Cellulose nanocrystals (CNCs), cellulose nanowhiskers (CNWs), Rod-like cellulose (RC), cellulose microcrystals (CMCs)	Wood, cotton, hemp, flax, wheat straw, mulberry bark, ramie, avicel, tunicin, cellulose from algae and bacteria	Acid hydrolysis, TEMPO-oxidation, APS-treatment, and bio-based enzymes	Rod-like structure, high dispersion in organic solvents (DMSO, ethylene glycol), varied length and diameter depending on the source material: Cotton (length: 200–300 nm; diameter: 10–30 nm); Ascidian (diameter: >1 μm) Plants (length 100–250 nm; diameter: 5–70 nm); Algae and bacteria: 100 nm to several micrometers	[59, 77, 81, 124, 126–128, 130, 131]
Nanofibrillated cellulose	Microfibrillated cellulose (MFC), Cellulose nanofirils (CNFs), cellulose microfirils (CMFs)	Wood, sugar beet, potato tuber, hemp, flax, tunicate, cotton fibers	Mechanical (grinding, homogenization, and ultrasonication), chemical, mechanical/ chemical, enzymatic	Long, flexible, and interconnected fibrils, contains amorphous and crystalline chains, high aspect ratio, specific surface area, high mechanical properties, high crystallinity, and pre-colation threshold 5–60 nm in diameter and several micrometers in length	[146–148]
Bacterial nanocellulose	Bacterial cellulose (BC), microbial cellulose (MC), cell-free cellulose (CFC), regenerated bacterial cellulose (RBC) bio-cellulose	Chemically defined media, fruit juices, sugarcane molasses, agricultural wastes, industrial wastes, and beer fermentation wastes	Microbial fermentation, cell-free synthesis	Ultra-purity, high crystallinity (60–90%), high degree of polymerization (DP: 2000– 8000), high water holding capacity (>100 times its dry weight), reticulate fibrous structure, high mechanical strength, optical transparency, biocompatibility, high thermal stability, polyfunctionality, moldability, high surface area, and non-toxicity; 20–100 nm in diameter	[7, 24, 27, 28, 95, 96, 98]

impurities are completely removed by repeated centrifugation and dialysis to obtain purified and highly crystalline CNCs [77]. Morphologically, the CNCs exist in rod or needle-like structures when observed under a scanning electron microscope. The structure, properties, phase-separation-behavior, and dimensions, including length and diameter, of these CNCs vary depending upon the type of raw material, acid, hydrolysis temperature, time, and intensity of ultrasonic irradiation [125]. For example, CNCs with a diameter of 200–300 nm and length 10–30 nm were obtained when cotton fibers were hydrolyzed by sulfuric acid at 65°C. In contrast, changing the source to ascidian produced CNCs with a diameter of more than 1 μm under the same experimental conditions. Further, the hydrolysis of cotton with HCl at 65°C produced CNCs with 100–200 nm length and 10 nm diameter [126, 127]. Besides acid hydrolysis, CNCs are also produced by treating wood pulp with other reagents such as tetramethylpiperidin-1-oxyl (TEMPO) [81], ammonium persulfate (APS) [128], and some bio-based enzymes [59]. The TEMPO-oxidized CNCs are produced by treating cellulose fibers with TEMPO, NaBr, and NaClO, where the reaction is carried out in an alkaline environment by constantly adding NaOH solution to the reaction mixture until a constant pH is obtained [81]. The enzymatic hydrolysis of microcrystalline cellulose (MCC) by using cellulolytic fungus yielded 20% CNCs; however, the high cost of this process restricted its commercial application [129].

CNCs show considerable dispersion in aqueous-based mixtures as well as organic solvents with dielectric constants, such as dimethyl sulfoxide (DMSO) and ethylene glycol; however, these form agglomerates in highly hydrophobic solutions. The agglomerated CNCs can be redispersed by sonication, for example in *N,N*-dimethylformamide (DMF) [130, 131]. NCCs act as fillers with different materials to form functional materials for diverse applications due to their renewability, biocompatibility, biodegradability, low density, high crystallinity, and high mechanical properties. As a reinforcement material, CNCs are blended with different polymeric materials, such as polyacrylamide, poly (*N,N*-dimethylacrylamide) [132], poly (acrylic acid) [133], poly (acrylic acid-coacrylamide) [134], poly (vinyl alcohol) [135], alginate [136], BC [137], and gelatin [138, 139], etc., to improve their physico-chemical, mechanical, and biological properties, such as stiffness, crystallinity, mechanical strength, swelling, biocompatibility, etc. The addition of different salts such as Na^+, K^+, Mg^+, and Ca^+, etc. into CNCs-based suspensions form nematic-ordered hydrogels by aggregating them [140]. Using

CNCs as reinforcement materials, researchers have developed different functional materials such as double-membrane hydrogels [141], self-healing hydrogels [142, 143], and macroporous hydrogels [144]. These CNC-reinforced composites hydrogels find applications as photonics, sensors, filters, and as templates in different fields [145].

3.2 *Cellulose Nanofibrils*

CNFs are prepared from different materials such as wood, cotton fibers, tunicate, etc., and consist of long, flexible, and interconnected fibrils. The length and diameter of CNFs vary depending on the source material; however, these generally range from 5–50 nm in diameter and a few micrometers in length [146]. These are obtained through homogenization and contain both crystalline and amorphous chains. CNFs possess high aspect ratio, specific surface area, high mechanical properties, high crystallinity, and pre-colation thresholds. These features allow the formation of a highly rigid network of CNFs in different matrices and improve their mechanical properties. The preparation of CNFs is very simple, involving the physical separation of cellulose fibers through grinding, homogenization, ultrasonication, etc., and does not essentially require any modification in the chemical structure of the cellulose chain [147, 148]. In addition to the physical process, CNFs can also be produced by chemical methods such as oxidation of raw material by TEMPO under magnetic stirring [81]. Additionally, both physical and chemical processes can be employed together for the preparation of individual CNFs; for example, carboxymethylation and high-pressure homogenization produce uniformly distributed CNFs [128]. As compared with the formation of CNCs-based hydrogels through self-aggregation, CNFs-based hydrogels are formed through the mechanical treatment of their aqueous suspensions with alkali followed by neutralization [149, 150]. Studies have shown improved compactness of the 3D network and mechanical properties of polyacrylamide [151] and gelatin [139] hydrogels.

3.3 *Bacterial Nanocellulose*

BNC synthesis by microbial cells is a four-step process: (1) Activation of monosaccharides through the formation of sugar nucleotides, (2) assembly of repeating units by sequential addition of sugar units through polymerization, (3) simultaneous addition of acyl groups (if present), and (4) excretion of cellulose fibrils through the wall or membrane complex at the cell surface (i.e., TCs), into the extracellular environment [152]. This entire process of BNC synthesis and formation of high-order structures is regulated by at least four genes, *axcess A*, *axcess B*, *axcess C*, and *axcess D*, that are arranged in an operon and encode respective proteins that perform different functions [153]. The thickness of the pellicle increases downward, with newly synthesized nanofibrils located at the top while the older ones are at the bottom surface, until all cells entrapped in the pellicle become inactive or die due to depletion of oxygen or the available nutrients are exhausted [28, 154] (Fig. 5). The structural properties such as nano-scale diameter (20–100 nm), arrangement of fibrils, physico-mechanical characteristics, and biological properties of BNC strictly depend on the microbial strain, the synthesis method, and culture conditions such as carbon and nitrogen source [155]. Pristine BNC lacks antibacterial, antifungal, antioxidant, and magnetic properties and possesses limited biocompatibility and biodegradability. Therefore, different composites of BNC are prepared with different biopolymers including collagen [156], silk-sericin [157], gelatin [158], alginate [159], chitosan [160], etc., and different nanoparticles such as silver [161], zinc [120, 162], titanium dioxide [121, 163, 164], gold [158], etc., as well as clay materials such as montmorillonite and modified montmorillonite [165] to impart it with additional properties. Both pristine BNC and BNC-based composites have found potential applications in bone and cartilage tissue engineering [166], artificial blood vessels [167], artificial skin tissues and burn skin substitute [168, 169], scaffolds [170], specialty membranes [171], drug release [16, 172], enzyme immobilization [173], 3D printing [20], and wound dressing [27], where these are used in the form of a hydrogel, membrane, porous structure, fiber, or solution [20, 27, 158, 174]. Natural cellulose is insoluble in common organic or aqueous solvents; therefore, different solvent systems including LiCl/dimethylacetamide (DMAc) [175], *N*-methylmorpholine-*N*-oxide (NMMO) [176], and alkali/urea (or thiourea) aqueous systems [177] have been developed to dissolve cellulose.

4. Properties of Nanocellulose

The structural, physico-chemical, mechanical, and biological properties of nanocellulose are largely dependent on its source, synthesis methods, and pre- and post-synthesis processing conditions, which in turn account for different applications of nanocellulose in various fields. The following sections provide an overview of different properties of nanocellulose and their relationship with each other.

4.1 *Structural Features and Porosity*

Nanocellulose is produced as a three-dimensional arrangement of web-shaped fibrils [178, 179], interconnected through inter- and intra-molecular hydrogen bonding. However, an amorphous agglomeration is observed when AC is used [167]. The nature and arrangement of these fibrils control the different properties of cellulose. The cross-sectional observation demonstrates the layered-arrangement of cellulose fibers. A similar web-like structure is also obtained in vegetal NFC, containing micro- and nano-pores [180]. In contrast, the NCC is presented as a suspension of whiskers in a chiral nematic ordered phase [3].

BC possesses certain levels of porosity that largely depend on different parameters, such as synthesis methods, chemical composition of the production medium, and post-synthesis treatment methods. The porous geometry of BC allows the absorption and penetration of liquid and solid molecules such as polymer solution, water, medium components, wound exudates, nanoparticles, and cells [28, 97, 157, 163, 181]. Regardless of the penetration of these materials into the BC matrix, its current porosity is still below par. The limited porosity is advantageous in that it prevents the invasion of microbial cells; however, it is disadvantageous from a medical perspective in that it bars the infiltration of mammalian cells [182] and large size particles [163] into its matrix, thus limiting its broad-spectrum biomedical applications. To address this issue of limited porosity, efforts have been made to improve the porosity of BNC by using porogens. A porogen, such as salt, paraffin, ice, gelatin, or sugar, is added to the BNC matrix as a space holder and subsequently removed without disturbing the native chemical and molecular structure, to improve the porosity and interconnectivity in the matrix [97, 183].

4.2 *Macrostructure Control and Orientation*

Specifically talking about BNC, it presents exciting features both in dry and wet forms. In the form of a hydrogel, BNC is mostly utilized in biomedical applications and for separation purposes. The surface chemistry, porosity, fiber orientation, as well as physical structure of BNC can be controlled at the macro-, micro-, and even nano-scales. This can be achieved by incorporating additional functional groups and varying the growth and culture conditions of cellulose, producing microbial cells that can form flat sheets, pellets, and granules under static, shaking, and agitation conditions, respectively [28]. Such macrostructure control and orientation of BNC bless it with unique features to find potential applications in various biomedical applications [27, 174, 183]. In nature, BNC shows an anisotropic pattern in the culturing plane that results in a web-shaped arrangement of fibrils attributed to the motion of microbial cells within the culture medium [184]. Despite the anisotropic behavior, the overall structure of BNC is quite compact and strong. BNC can achieve a comparable isotropic alignment by controlling the motion of microbial cells, such as by using microfluidic channels [185], by using honeycomb-patterned microgrooves [186], and by applying electric or magnetic fields [184], using templates such as chitin [187] and poly-di-methyl-siloxane (PDMS) [188], thus finding potential application in tissue engineering.

4.3 *Morphology*

The structural morphology of plant-based nanocellulose is highly dependent on the efficient removal of non-cellulosic parts (lignin and hemicellulose) and dissolution of the amorphous region of domain structure [64, 86]. Thus, the structural morphology, physical properties, and dimension of nanocellulose vary significantly based on the different plant sources and extraction methods (i.e., physical, chemical, and enzymatic). For example, nanocellulose obtained through acid hydrolysis of sisal, pineapple leaf, coir, banana rachis, and kapok possesses long and flexible visible fibers that were well-separated with rod-like structures and entangled morphology [189]. In contrast, the BNC nanofibrils are about 100 times smaller in diameter than plant cellulose and are extruded from the cytoplasmic membrane of microbial cells in the form of β-1,4-glucan chains that self-aggregate into ribbon-like fibrils and finally form bundles [178, 179]. A single ribbon is composed of about 1,000 individual β-1,4-glucan chains [27]. The high aspect ratio (L/W) increased the surface of BNC and determines its potential to be used as a reinforcement material for an easy stress transfer [190]. Studies have shown that the typical sizes of BNC nanofibrils fall in the range of NFC, being longer than both vegetal NFC and NCC, whose length depends on the source of nanocellulose [180, 191].

4.4 *Crystallinity*

Nanocellulose obtained from plant and microbial sources vary greatly in their crystallinities. The highly crystalline nature of BNC is one of the most promising properties that also blesses it with excellent mechanical properties as compared to plant cellulose [190]. It is anticipated that the crystallinity of cellulose is due to the arrangement of fibers; thus, improving the alignment of cellulose fibers could potentially improve its crystallinity. The crystalline structure of BNC is comprised of cellulose type $I\alpha$ and $I\beta$, which correspond to triclinic and monoclinic unit cells, respectively [190]. The proportion of each unit cell varies in accordance with culture conditions; thus, the crystallinity of BNC can be modulated by changing the culture conditions such as synthesis method, nutrient source and composition, type of reactor (static, shaking, and agitation), pH, temperature, and type of microbial strain [190]. The degree of crystallinity of any material is calculated from the relative integrated area of the crystalline and amorphous peaks using the following equation:

$$X_c = \frac{A_{cr}}{\left(A_{cr} + A_{am}\right)} \times 100$$

The X-ray diffraction (XRD) analysis of cellulose reveals two distinct peaks at 2θ 14.2° and 22.7° along with a weak middle shoulder peak at 16.6°, corresponding to (1−10), (200), and (110) crystalline planes of the cellulose $I\beta$ structure, respectively [192]. In general, the crystallinity of BNC varies from 60–90% where cellulose $I\alpha$ accounts for 70–80% of the total crystallinity [190, 193–195]. The calculated crystallinity of cellulose usually deviates to some extent from the actual crystallinity due to the presence of small peaks that raise the background and, in turn, contribute to increasing the amorphous and total areas [192]. In contrast, the crystallinity of cellulose in plants and tunicates is mostly accounted for by the cellulose type $I\beta$ structure [24].

4.5 *Degree of Polymerization*

The degree of polymerization (DP) refers to the number of glucose units in a chain [196], which in turn accounts for the length and crystallinity of nanocellulose. The DP of BNC is higher owing to its long and crystalline fibers, followed by plant-based NFC and NCC [43, 197].

4.6 *Thermal Stability*

The commercial applications of nanocellulose are largely dependent on its thermal stability. The thermal degradation of cellulose depends on its source and extraction procedures, which generally involves three processes — dehydration (i.e., removal of water molecules), depolymerization (i.e., breakage of glycosidic linkages among glucose units), and decomposition (i.e., degradation of glycosyl units) — that ultimately result in the formation of charred residue [198]. The nanocellulose extracted from natural cellulose resources generally possesses better thermal properties than the cellulose present in original lignocellulose biomass [199]. This improved thermal stability is attributed to higher crystallinity, flexible structure, and removal of low-thermally-stable lignin materials [189]. Compared to plant cellulose, BNC is thermally more stable owing to its high crystallinity and absence of any impurities. Ullah *et al.* reported that cellulose produced by a cell-free system was thermally more stable even than BNC owing to its more compact and well-arranged fibril structure [96]. El-Saied *et al.* compared the thermal stability of BNC with different plant celluloses including cotton linters, viscose wood pulp (VWP), and MCC and showed that their degradation took place in two stages: decomposition of cellulose (volatilization stage) and rapid volatilization and oxidation of char and formation of a carbonaceous residue (carbonaceous stage). The results showed that although BNC started degrading first, its complete degradation took longer than both VMP and MCC and required comparatively small activation energy, indicating its higher thermal stability [200].

4.7 *Mechanical Strength*

The mechanical properties of nanocellulose are directly associated with its crystallinity and degree of polymerization (DP), which are in turn related to its source and arrangement of fibrils [201]. This indicates that nanocellulose with highly ordered fibrils (i.e., aligned by applying electric and magnetic fields, microfluidic channels, and templates) would possess high crystallinity and thus high mechanical properties. In line with this, BNC with comparatively ordered and long fibrils that possess mainly crystalline fibers is expected to possess better mechanical properties. The low fibril lengths in plant NFC reduce the surface area and lead to lower crystallinity [180]. Ullah *et al.* reported that cellulose produced by a cell-free system demonstrated better thermal properties (i.e., tensile strength, elongation strain, and Young's modulus) as compared with BNC, owing to more compact and well-arranged fibrils as shown by its cross-sectional analysis *via* scanning electron microscopy [96]. The compact and well-arranged fibrils favor a uniform response to the applied force [202, 203]. The typical tensile strength of BNC ranges between 200–2000 MPa while its Young's modulus usually falls in the range of 15–138 GPa [181, 196, 204]. The mechanical properties of nanocellulose are better than their lignocellulose source biomass owing to uniform morphological structure. Dufresne reported that the average modulus of nanocellulose is 100 GPa, which is much higher than the base cellulosic materials [205] and comparable with carbon fiber of TORAYCA T1000G, a material with one of the world's best mechanical properties, especially tensile strength (TORAYCA T1000G DATA SHEET). However, the limited compatibility of NCC with typical matrixes and their difficulty in dispersion limit their commercial applications.

4.8 *Water Holding and Release Properties*

The water holding and release properties of nanocellulose are of utmost importance in the biomedical field, where it is used for wound dressing and as a healing material. A wound dressing material requires an appropriate moisture content to favor the penetration of active substances, enable easy and painless dressing, accelerate the wound-healing process, and prevent damage to newly formed skin [158, 176, 206]. In general, cellulose possesses high water holding capacity (WHC), with BNC having the potential to absorb water more than 100 times its dry weight [207, 208]. The WHC of cellulose is attributed to the arrangement of fibrils that facilitate the penetration of water molecules to the inner matrix [209]. The water molecules interact with cellulose fibrils in two ways: either these are entrapped on its surface or reside inside the matrix of reticulated fibrils [210]. The difference in the WHC of cellulose from different sources varies due to the differences in porosity, fibril arrangement, and surface area [209]. The well-arranged fibrils offer a highly expanded surface area and highly porous matrix for the absorption of water molecules [211, 212]. Gelin *et al.* studied the water interaction with cellulose fibrils in a BC hydrogel and found that only 10% of the 99% (w/w) water contained in the hydrogel was present as free bulk-water while the remaining 89% was bound to the nanofibrils [213]. In addition to its source (i.e., plant and microbial cellulose), the WHC of cellulose varies greatly by the cultivation method in the case of BNC. Ullah *et al.* reported that compared to BNC, the cellulose produced by a cell-free system possessed high WHC [96]. In another study, Krystynowicz *et al.* reported that BNC produced in an RDB possessed five times more water than the one produced in SC [214]. In contrast, the vegetal NFC possesses low WHC due to lower surface area of fibrils to hold water molecules [193]. In addition to WHC, the water release rate (WRR) is also of immense importance in biomedical applications. The wound requires a moist environment during the healing process to allow the penetration and adsorption of drugs. Therefore, a wound dressing material with slow WRR is preferred for wound healing applications. As described earlier, since only about 10% of water molecules are present as free bulk-water while the remainder is tightly bound to nanofibrils [213], BNC is found to possesses a slow WRR. The WRR from cellulose from different sources indicates a similar trend initially due to evaporation of unbound water from the surface; however, the trend changes with time and a slow WRR is observed for cellulose with a more compact and well-arranged fibrous network, which tends to prevent the escape of water molecules from within the matrix [96, 209].

4.9 *Rheological Properties*

The rheological properties of nanocellulose are greatly affected by its surface properties, which affect the shear thinning rheology of suspension [215]. Therefore, an increase in shear rate breaks the nanocellulose network into individual entities, resulting in a reduction in the viscosity of hydrogels [74, 216]. Several studies have investigated the rheological properties of nanocellulose with the aim to discuss the microstructural changes and the associated outcomes during processing, handling, and applications. CNCs possess unique rheological features owing to their rod-like structure and surface charge of particles and are found as liquid crystals above a critical concentration [217, 218]. Shafiei-Sabet *et al.*

reported that the CNC suspension prepared by acid hydrolysis was isotropic and, at low concentrations (3 wt%), phase-separated into liquid crystalline form, but formed isotropic biphasic suspension at high concentration. The viscosity profile presented a typical three-region behavior of rheological gels, which showed a shear thinning behavior at even higher concentrations [217]. A comparative analysis of the rheological behavior of three CNF-suspensions obtained, including the one obtained through enzymatic treatment, CMC, and CMC-grafted cellulose, showed completely different rheological properties. In a similar study, Moberg *et al.* compared the rheological properties of aqueous suspensions of three different nanocelluloses where one system was developed *via* acid hydrolysis (i.e., CNC) while the other two were developed by mechanical shearing (CNF and intermediate between CNC and CNC); however, these were from different origins and were subjected to different parameters. The three systems showed different dimensions, indicating that not only the surface character but also the dimension can affect and control the rheological properties of nanocellulose within a broad range [219]. The rheological properties of CNCs are significantly affected by various factors. Temperature affects the rheological properties by changing the relative fractions of isotropic and liquid crystalline regions [220, 221]. Similarly, other factors such as CNC concentration, ionic strength, and pH of suspension also affect the flow and viscoelastic properties of amorphous cellulose suspensions [222].

4.10 *Optical Transparency*

Optical transparency (i.e., light transmittance) is an interesting property of nanocellulose. Nanocellulose allows the transference of light because the wavelength of visible light is higher than the size of nanofibers. BNC is semi-transparent in nature [223] whereas its natural optical transparency is due to its high refractive index (n) determined using the following equation:

$$n = \frac{Sin\theta\, i}{Sin\theta\, r}$$

The optical transparency of cellulose is improved by different chemical methods such as acetylation. Acetylation is a two-step process: first, access of acetylating agent into the proximity of −OH moieties of BC;

second, the substitution of acetyl group (CH_3—CO) with the H-atom of the −OH moiety, generally progressing from top surface towards the inner core [224]. As the acetyl functional group occupies more space due to its large size, it pushes the nearby cellulose fibrils and increases the volume [225] as well as weakens the hydrogen bonding, which leads to relaxation of fibrils. As a result, the refractive index of BC is enhanced, which allows greater transmittance of light [225]; thus, acetylated BC looks more transparent than pristine BC. The optical transparency of cellulose can also be improved by adding other molecules. In a study, Di *et al.* developed a BC whiskers-based hydrogel with poly (2-hydroxyethyl methacrylate) coated with silver nanoparticles as a transparent wound dressing material and showed that the addition of BCWs improved the equilibrium water content and light transmittance by about 20−40% and 10%, respectively [137]. In contrast, the optical transparency of vegetal NFC can be improved by forming larger fiber fragments and aggregates [43]. In addition, the optical transparency of NFC can also be improved by improving the yield of homogenization as the less homogenized materials may contain other compounds that absorb light and result in overall low transmittance (Spence *et al.*, 2010). In general, NCC possesses better transmittance than NFC due to its high crystallinity and short length [226, 227]. The transparent cellulose-based composite materials can also find potential applications in display devices, coatings, and lenses [190].

4.11 *Biocompatibility*

Biocompatibility of a material refers to its ability to remain in contact with living tissues without causing any toxic, immunogenic, or allergic effects [97, 228]. According to Williams's definition of biocompatibility, it is "the ability of a material to perform with an appropriate host response in a specific application" [229]. BNC is regarded as a moderate-to-strong biocompatible material as it supports the adhesion, growth, proliferation, and infiltration of different types of cells into its matrix owing to its high porosity, fibrous structure, and high WHC [230, 231]. BNC has shown considerable biocompatibility both *in vitro* and *in vivo,* both alone and in the form of composites with other materials. An *in vivo* study in rats demonstrated the biocompatibility of BNC for up to 12 weeks when subcutaneously implanted and showed no immunogenic response, inflammation, swelling, or formation of exudates around the implantation sites [232].

Although several studies have demonstrated the biocompatibility of BNC, both *in vitro* and *in vivo*, it is not up to the desired levels in some cases; thus, different strategies have been developed to improve its biocompatibility. Khan *et al.* fabricated porogen (gelatin)-induced 3D microporous regenerated BC scaffolds by using oil-in-water emulsion methods for skin tissue regeneration application. The *in vitro* studies showed adhesion and proliferation of human keratinocytes (HaCaT) as well as infiltration into the scaffolds up to 300 μm. Further, *in vivo* studies showed complete wound healing and skin regeneration after two weeks in an experimental mouse model [158].

4.12 *Biodegradability*

BNC is rarely biodegradable owing to its high crystallinity and compact structure. Further, humans lack the cellulase enzyme required for degradation of cellulose. Several efforts have been made to overcome these discrepancies and meet the requirements of BNC for biomedical applications. Li *et al.* improved the *in vitro* degradation of BC through periodate oxidation, which enabled its degradation in water, PBS, and SBF, while still retaining its network structure [233]. The periodate oxidation causes specific cleavage of the C2–C3 bond of the glucopyranoside to produce two aldehydes per glucose [234, 235] unit and yields a biodegradable and biocompatible 2,3-dialdehyde cellulose (DAC). DAC is degradable at physiological pH both *in vitro* and *in vivo* and produces glycolic acid and 2,4-dihydroxybutyric acid [235]; glycolic acid is excreted through urine or enters the tricarboxylic acid cycle while 2,4-dihydroxybutyric acid participates in the metabolism of *L*-homoserine in the liver [236].

5. Conclusions and Future Perspective

Nanocellulose is an interesting material obtained from plant as well as microbial sources. It has received immense consideration in diverse applications owing to its abundance, structural features, surface chemistry, renewability, and non-toxicity. However, the broad-spectrum applications of nanocellulose in different fields are limited by the requirement of pretreatment processing to remove impurities such as hemicellulose and lignin, which also leads to an increase in the overall production cost. Bacterial nanocellulose, representing the purest form of cellulose, is widely used for biomedical applications; however, its high production cost and lack of certain important features such as antimicrobial and antioxidant properties restrict its direct applications. Efforts have been devoted towards the cost-effective production of nanocellulose from different low-cost raw materials so as to impart additional features to it by forming composites with different functional materials. However, the structural morphology and mechanical features of nanocellulose produced from different materials need to be further improved. Further, various physical approaches have been developed to prepare nanocellulose with the aim to overcome the energy issue; however, these approaches compromise its crystalline and other structural features. It is generally perceived that a hydrophilic polymer requires a hydrophilic nanocellulose while a hydrophobic polymer requires a hydrophobic modified nanocellulose to improve their structural features and biocompatibility. Thus, the addition of hydrophilic/hydrophobic nanocellulose to the hydrophilic/hydrophobic polymers is expected to affect the structural features accordingly, either positively or negatively. Further, cell-free strategies have been developed at a low-cost and low energy; however, these require extended incubation time and are, thus, beyond the scope of industrialization.

In addition to plant and microbial sources, the nanocellulose synthesis by cell-free systems can be effectively improved to run in parallel with them through upregulation of the cellulose synthesis pathway, blocking endogenous transcription–translation of competitive pathways, varying the substrate, improving the availability of cofactors, developing a hybrid cell-free system, purifying the enzymes from different hosts, and using biomimetic cofactors. The interaction of nanocellulose with different polymer matrices affects its crystalline structure and other physical properties. As it is known that nanocellulose form a network linked by hydrogen bonds, this feature can be utilized to control the flexibility and performance of nanocellulose-based functional materials; thus, further efforts are required to explore the nature of the interaction between the nanocellulose surface and functional groups of different materials. It has also been anticipated that the type of linkages, such as hydrogen bonding and electrostatic interaction, between nanocellulose and the interacting material could affect the performance of the composite material. However, it is yet to be explored how the surface modification strategy affects the mechanical properties.

Acknowledgments

This work was supported by the National Natural Science Foundation of China (21774039, 51973076), BRICS STI Framework Programme 3rd call 2019 (2018YFE0123700), China Postdoctoral Science Foundation (2016M602291), and the Fundamental Research Funds for Central Universities, Open Research Fund of State Key Laboratory of Polymer Physics and Chemistry, and Changchun Institute of Applied Chemistry, Chinese Academy of Sciences.

References

[1] Y. Habibi, L.A., Lucia, and O.J. Rojas, Cellulose nanocrystals: Chemistry, self-assembly, and applications. *Chem. Rev.*, 110 (2010) 3479–3500. doi:10.1021/cr900339w.

[2] D. Klemm, B. Heublein, H.P. Fink, and A. Bohn, Cellulose: Fascinating biopolymer and sustainable raw material. *Angew. Chemie — Int. Ed.*, 44 (2005) 3358–3393. doi:10.1002/anie.200460587.

[3] B.L. Peng, N. Dhar, H. L. Liu, and K.C. Tam, Chemistry and applications of nanocrystalline cellulose and its derivatives: A nanotechnology perspective. *Can. J. Chem. Eng.*, (2011). doi:10.1002/cjce.20554.

[4] R.M.A. Domingues, M.E. Gomes, and R.L. Reis, The potential of cellulose nanocrystals in tissue engineering strategies. *Biomacromolecules*, (2014). doi:10.1021/bm500524s.

[5] M. Kaushik and A. Moores, Review: Nanocelluloses as versatile supports for metal nanoparticles and their applications in catalysis. *Green Chem.*, (2016). doi:10.1039/c5gc02500a.

[6] M.W. Ullah, M. Ul Islam, S. Khan, N. Shah, and J.K. Park, Recent advancements in bioreactions of cellular and cell-free systems: A study of bacterial cellulose as a model. *Korean J. Chem. Eng.*, 34 (2017) 1591–1599. doi:10.1007/s11814-017-0121-2.

[7] M.W. Ullah, M. Ul-Islam, S. Khan, Y. Kim, and J.K. Park, Innovative production of bio-cellulose using a cell-free system derived from a single cell line. *Carbohydr. Polym.*, 132 (2015) 286–294. doi:10.1016/j.carbpol.2015.06.037.

[8] N. Lin, J. Huang, and A. Dufresne, Preparation, properties and applications of polysaccharide nanocrystals in advanced functional nanomaterials: A review. *Nanoscale*, (2012). doi:10.1039/c2nr30260h.

[9] Y. Mao, K. Liu, C. Zhan, L. Geng, B. Chu, and B.S. Hsiao, Characterization of nanocellulose using small-angle neutron, X-ray, and dynamic light scattering techniques. *J. Phys. Chem. B.*, (2017). doi:10.1021/acs.jpcb.6b11425.

[10] M.A.S. Azizi Samir, F. Alloin, and A. Dufresne, Review of recent research into cellulosic whiskers, their properties and their application in nanocomposite field. *Biomacromolecules*, (2005). doi:10.1021/bm0493685.

[11] M.S. Islam, L. Chen, J. Sisler, and K.C. Tam, Cellulose nanocrystal (CNC)-inorganic hybrid systems: Synthesis, properties and applications. *J. Mater. Chem. B.*, (2018). doi:10.1039/c7tb03016a.

[12] L. Fu, J. Zhang, and G. Yang, Present status and applications of bacterial cellulose-based materials for skin tissue repair. *Carbohydr. Polym.*, 92 (2013) 1432–1442. doi:10.1016/j.carbpol.2012.10.071.

[13] M. Ul-Islam, F. Subhan, S.U. Islam, S. Khan, N. Shah, S. Manan, M.W. Ullah, and G. Yang, Development of three-dimensional bacterial cellulose/chitosan scaffolds: Analysis of cell-scaffold interaction for potential application in the diagnosis of ovarian cancer. *Int. J. Biol. Macromol.*, 137 (2019) 1050–1059. doi:10.1016/j.ijbiomac.2019.07.050.

[14] S. Li, D. Huang, B. Zhang, X. Xu, M. Wang, G. Yang, and Y. Shen, Flexible supercapacitors based on bacterial cellulose paper electrodes. *Adv. Ener. Mater.*, 4 (2014) 1301655. doi:10.1002/aenm.201301655.

[15] S. Li, D. Huang, J. Yang, B. Zhang, X. Zhang, G. Yang, M. Wang, and Y. Shen, Freestanding bacterial cellulose-polypyrrole nanofibres paper electrodes for advanced energy storage devices. *Nano Energy.*, 9 (2014) 309–317. doi:10.1016/j.nanoen.2014.08.004.

[16] S. Li, A. Jasim, W. Zhao, L. Fu, M.W. Ullah, Z. Shi and G. Yang, Fabrication of pH-electroactive bacterial cellulose/polyaniline hydrogel for the development of a controlled drug release system. *ES Mater. Manuf.*, (2018) 41–49. doi:10.30919/esmm5f120.

[17] M. Badshah, H. Ullah, A.R. Khan, S. Khan, J.K. Park, and T. Khan, Surface modification and evaluation of bacterial cellulose for drug delivery. *Int. J. Biol. Macromol.*, 113 (2018) 526–533. doi:10.1016/j.ijbiomac.2018.02.135.

[18] H. Ullah, F. Wahid, H.A. Santos, and T. Khan, Advances in biomedical and pharmaceutical applications of functional bacterial cellulose-based nanocomposites. *Carbohydr. Polym.*, (2016). doi:10.1016/j.carbpol.2016.05.029.

[19] M. Nasir, R. Hashim, O. Sulaiman, and M. Asim, Nanocellulose, in: Cellul. *Nanofibre Compos. Elsev.*, (2017) 261–276. doi:10.1016/B978-0-08-100957-4.00011-5.

[20] W. Aljohani, M.W. Ullah, X. Zhang, and G. Yang, Bioprinting and its applications in tissue engineering and regenerative medicine. *Int. J. Biol. Macromol.*, 107 (2018) 261–275. doi:10.1016/j.ijbiomac.2017.08.171.

[21] R.R. McCarthy, M.W. Ullah, E. Pei, and G. Yang, Antimicrobial inks: The anti-infective applications of bioprinted bacterial polysaccharides. *Tren. Biotechnol.*, 37 (2019) 1153–1155. doi:https://doi.org/10.1016/j.tibtech.2019.05.004.

[22] R.R. McCarthy, M.W. Ullah, P. Booth, E. Pei, and G. Yang, The use of bacterial polysaccharides in bioprinting. *Biotechnol. Adv.*, 37 (2019) 107448. doi:10.1016/j.biotechadv.2019.107448.

[23] J. George, S.N. Sabapathi, Cellulose nanocrystals: Synthesis, functional properties, and applications. *Nanotechnol. Sci. Appl.*, (2015) 45. doi:10.2147/NSA.S64386.

[24] R.J. Moon, A. Martini, J. Nairn, J. Simonsen, and J. Youngblood, Cellulose nanomaterials review: Structure, properties and nanocomposites. *Chem. Soc. Rev.*, 40 (2011) 3941–3994. doi:10.1039/c0cs00108b.

[25] A. Shoukat, F. Wahid, T. Khan, M. Siddique, S. Nasreen, G. Yang, M.W. Ullah, and R. Khan, Titanium oxide-bacterial cellulose bioadsorbent for the removal of lead ions from aqueous solution. *Int. J. Biol. Macro.*, 129 (2019) 965–971. doi:10.1016/j.ijbiomac.2019.02.032.

[26] Z. Shi, Y. Zhang, G.O. Phillips, and G. Yang, Utilization of bacterial cellulose in food. *Food Hydro.*, 35 (2014) 539–545. doi:10.1016/j.foodhyd.2013.07.012.

[27] W. Czaja, A. Krystynowicz, S. Bielecki, and R.M. Brown, Microbial cellulose — The natural power to heal wounds. *Biomaterials*, 27 (2006) 145–151. doi:10.1016/j.biomaterials.2005.07.035.

[28] N. Shah, M. Ul-Islam, W.A. Khattak, and J.K. Park, Overview of bacterial cellulose composites: A multipurpose advanced material. *Carbohydr. Polym.*, 98 (2013) 1585–1598. doi:10.1016/j.carbpol.2013.08.018.

[29] N. Lin and A. Dufresne, Nanocellulose in biomedicine: Current status and future prospect. *Eur. Polym.*, J. 59 (2014) 302–325. doi:10.1016/j.eurpolymj.2014.07.025.

[30] Z. Shi, X. Gao, M.W. Ullah, S. Li, Q. Wang, and G. Yang, Electroconductive natural polymer-based hydrogels. *Biomaterials*, 111 (2016). doi:10.1016/j.biomaterials.2016.09.020.

[31] C. Yan, J. Wang, W. Kang, M. Cui, X. Wang, C.Y. Foo, K.J. Chee, and P.S. Lee, Highly stretchable piezoresistive graphene-nanocellulose nanopaper for strain sensors. *Adv. Mater.*, 26 (2014) 2022–2027. doi:10.1002/adma.201304742.

[32] A. Jasim, M.W. Ullah, Z. Shi, X. Lin, and G. Yang, Fabrication of bacterial cellulose/polyaniline/single-walled carbon nanotubes membrane for potential application as biosensor. *Carbohydr. Polym.*, 163 (2017) 62–69. doi:10.1016/j.carbpol.2017.01.056.

[33] M. Ul-Islam, M.W. Ullah, S. Khan, T. Kamal, S. Ul-Islam, N. Shah, and J.K. Park, Recent advancement in cellulose based nanocomposite for addressing environmental challenges. *Recent Pat. Nanotechnol.*, 10 (2016). doi:10.2174/1872210510666160429144916.

[34] U. Farooq, M.W. Ullah, Q. Yang, A. Aziz, J. Xu, L. Zhou, and S. Wang, High-density phage particles immobilization in surface-modified bacterial cellulose for ultra-sensitive and selective electrochemical

detection of Staphylococcus aureus. *Biosens. Bioelect.*, 157 (2020) 112163. doi:10.1016/j.bios.2020.112163.

[35] M. Ioelovich and E. Morag, Study of enzymatic hydrolysis of pretreated biomass at increased solids loading. *BioResources*, (2012). doi:10.15376/biores.7.4.4672-4682.

[36] D. Klemm, B. Philipp, T. Heinze, U. Heinze, and W. Wagenknecht, Comprehensive cellulose chemistry: Volume I: Fundamentals and analytical methods. *Compr. Cellul. Chem.*, (1998). doi:10.1002/3527601929.

[37] I. Siró and D. Plackett, Microfibrillated cellulose and new nanocomposite materials: A review. *Cellulose*, (2010). doi:10.1007/s10570-010-9405-y.

[38] K. Okubo, T. Fujii, and N. Yamashita, Improvement of interfacial adhesion in bamboo polymer composite enhanced with micro-fibrillated cellulose. *Jsme Int. J.*, (2005). doi:10.1299/jsmea.48.199.

[39] A. Mandal and D. Chakrabarty, Isolation of nanocellulose from waste sugarcane bagasse (SCB) and its characterization. *Carbohydr. Polym.*, (2011). doi:10.1016/j.carbpol.2011.06.030.

[40] W.A. Khattak, T. Khan, M. Ul-Islam, M.W. Ullah, S. Khan, F. Wahid, and J.K. Park, Production, characterization and biological features of bacterial cellulose from scum obtained during preparation of sugarcane jaggery (gur). *J. Food Sci. Technol.*, 52 (2015) 8343–8349. doi:10.1007/s13197-015-1936-7.

[41] A.P. Deshpande, M. Bhaskar Rao, and C. Lakshmana Rao, Extraction of bamboo fibers and their use as reinforcement in polymeric composites. *J. Appl. Polym. Sci.*, (2000). doi:10.1002/(SICI)1097-4628(20000404)76:1<83::AID-APP11>3.0.CO;2-L.

[42] K.P. Rajan, N.R. Veena, H.J. Maria, R. Rajan, M. Skrifvars, K. Joseph, Extraction of bamboo microfibrils and development of biocomposites based on polyhydroxybutyrate and bamboo microfibrils, *J. Compos. Mater.*, (2011). doi:10.1177/0021998310381543.

[43] I. Siró and D. Plackett, Microfibrillated cellulose and new nanocomposite materials: A review. *Cellulose*, (2010). doi:10.1007/s10570-010-9405-y.

[44] M. Ul-Islam, W.A. Khattak, M. Kang, S.M. Kim, T. Khan, and J.K. Park, Effect of post-synthetic processing conditions on structural variations and applications of bacterial cellulose. *Cellulose*, 20 (2013) 253–263. doi:10.1007/s10570-012-9799-9.

[45] S. Xie, X. Zhang, M.P. Walcott, and H. Lin, Applications of cellulose nanocrystals: A review. *Eng. Sci.*, (2018). doi:10.1007/s11244-010-9598-1.

[46] F.W. Herrick, R.L. Casebier, J.K. Hamilton, and K.R. Sandberg, Microfibrillated cellulose: Morphology and accessibility. *J. Appl. Polym. Sci. Appl. Polym. Symp.*, (1983). doi:10.1109/ICDMW.2007.60.

[47] A.N. Frone, D.M. Panaitescu, and D. Donescu, Some aspects concerning the isolation of cellulose micro- and nano-fibers. *UPB Sci. Bull. Ser. B Chem. Mater. Sci.*, (2011).

[48] Q. Cheng, S. Wang, and T.G. Rials, Poly(vinyl alcohol) nanocomposites reinforced with cellulose fibrils isolated by high intensity ultrasonication. *Compos. Part A Appl. Sci. Manuf.*, (2009). doi:10.1016/j.compositesa.2008.11.009.

[49] L. Zhang, T. Tsuzuki, and X. Wang, Preparation of cellulose nanofiber from softwood pulp by ball milling. *Cellulose*, (2015). doi:10.1007/s10570-015-0582-6.

[50] J. Araki, M. Wada, S. Kuga, and T. Okano, Flow properties of microcrystalline cellulose suspension prepared by acid treatment of native cellulose. *Collo. Surfa. A Physicochem. Eng. Asp.*, (1998). doi:10.1016/S0927-7757(98)00404-X.

[51] I. Filpponen and D.S. Argyropoulos, Regular linking of cellulose nanocrystals *via* click chemistry: Synthesis and formation of cellulose nanoplatelet gels. *Biomacromolecules*, (2010). doi:10.1021/bm1000247.

[52] D. Liu, T. Zhong, P.R. Chang, K. Li, and Q. Wu, Starch composites reinforced by bamboo cellulosic crystals. *Bioresour. Technol.*, (2010). doi:10.1016/j.biortech.2009.11.058.

[53] S. Camarero Espinosa, T. Kuhnt, E.J. Foster, and C. Weder, Isolation of thermally stable cellulose nanocrystals by phosphoric acid hydrolysis. *Biomacromolecules*, (2013). doi:10.1021/bm400219u.

[54] H. Sadeghifar, I. Filpponen, S.P. Clarke, D.F. Brougham, and D.S. Argyropoulos, Production of cellulose nanocrystals using hydrobromic acid and click reactions on their surface. *J. Mater. Sci.*, (2011). doi:10.1007/s10853-011-5696-0.

[55] Z. Man, N. Muhammad, A. Sarwono, M.A. Bustam, M.V. Kumar, and S. Rafiq, Preparation of cellulose nanocrystals using an ionic liquid. *J. Polym. Environ.*, (2011). doi:10.1007/s10924-011-0323-3.

[56] S. Eyley and W. Thielemans, Surface modification of cellulose nanocrystals. *Nanoscale*, (2014). doi:10.1039/c4nr01756k.

[57] M.U. Islam, M.W. Ullah, S. Khan, N. Shah, and J.K. Park, Strategies for cost-effective and enhanced production of bacterial cellulose. *Int. J. Biol. Macromol.*, 102 (2017) 1166–1173. doi:10.1016/j.ijbiomac.2017.04.110.

[58] M.A. Cader Mhd Haniffa, Y.C. Ching, L.C. Abdullah, S.C. Poh, and C.H. Chuah, Review of bionanocomposite coating films and their applications. *Poly. (Basel).*, (2016). doi:10.3390/polym8070246.

[59] M. Pääkko, M. Ankerfors, H. Kosonen, A. Nykänen, S. Ahola, M. Österberg, J. Ruokolainen, J. Laine, P.T. Larsson, O. Ikkala, and T. Lindström, Enzymatic hydrolysis combined with mechanical shearing and high-pressure homogenization for nanoscale cellulose fibrils and strong gels. *Biomacromolecules*, (2007). doi:10.1021/bm061215p.

[60] S. Chauhan, Vipul, K. Chakrabarti, Swapan, Use of nanotechnology for high performance cellulosic and papermaking products. *Cellul. Chem. Technol.*, (2012).

[61] F. Jiang and Y. Lo Hsieh, Chemically and mechanically isolated nanocellulose and their self-assembled structures. *Carbohydr. Polym.*, (2013). doi:10.1016/j.carbpol.2013.02.022.

[62] H. Wang, X. Zhang, Z. Jiang, W. Li, and Y. Yu, A comparison study on the preparation of nanocellulose fibrils from fibers and parenchymal cells in bamboo (*Phyllostachys pubescens*). *Ind. Crop. Prod.*, (2015). doi:10.1016/j.indcrop.2015.03.086.

[63] H. Wei, K. Rodriguez, S. Renneckar, W. Leng, and P.J. Vikesland, Preparation and evaluation of nanocellulose-gold nanoparticle nanocomposites for SERS applications. *Analyst*, (2015). doi:10.1039/c5an00606f.

[64] B.M. Cherian, A.L. Leão, S.F. de Souza, S. Thomas, L.A. Pothan, and M. Kottaisamy, Isolation of nanocellulose from pineapple leaf fibres by steam explosion. *Carbohydr. Polym.*, (2010). doi:10.1016/j.carbpol.2010.03.046.

[65] A.N. Frone, D.M. Panaitescu, D.D. Spataru, C. Radovici, R. Trusca, and R. Somoghi, Preparation and characterization of PVA composites with cellulose nanofibers obtained by ultrasonication. *BioResources*, (2011). doi:10.15376/biores.6.1.487-512.

[66] J. Leitner, B. Hinterstoisser, M. Wastyn, J. Keckes, and W. Gindl, Sugar beet cellulose nanofibril-reinforced composites. *Cellulose*, (2007). doi:10.1007/s10570-007-9131-2.

[67] Y. Habibi, M. Mahrouz, and M.R. Vignon, Microfibrillated cellulose from the peel of prickly pear fruits. *Food Chem.*, (2009). doi:10.1016/j.foodchem.2008.12.034.

[68] M. Jonoobi, J. Harun, A.P. Mathew, M.Z.B. Hussein, and K. Oksman, Preparation of cellulose nanofibers with hydrophobic surface characteristics. *Cellulose*, (2010). doi:10.1007/s10570-009-9387-9.

[69] B. Wang and M. Sain, Dispersion of soybean stock-based nanofiber in a plastic matrix. *Polym. Int.*, (2007). doi:10.1002/pi.2167.

[70] R.K. Johnson, A. Zink-Sharp, S.H. Renneckar, and W.G. Glasser, A new bio-based nanocomposite: Fibrillated TEMPO-oxidized celluloses in hydroxypropylcellulose matrix. *Cellulose*, (2009). doi:10.1007/s10570-008-9269-6.

[71] S. Wang and Q. Cheng, A novel process to isolate fibrils from cellulose fibers by high-intensity ultrasonication, part 1: Process optimization. *J. Appl. Polym. Sci.*, (2009). doi:10.1002/app.30072.

[72] M.A. Mohammed Nasir, R. Hashim, and O. Sulaiman, Nanocellulose: Preparation methods and applications. *Cellul. Nanofi. Compos.*, (2017) 261–276.

[73] A. Müller, Z. Ni, N. Hessler, F. Wesarg, F.A. Müller, D. Kralisch, and D. Fischer, The biopolymer bacterial nanocellulose as drug delivery system: Investigation of drug loading and release using the model protein albumin. *J. Pharm. Sci.*, (2013). doi:10.1002/jps.23385.

[74] E. Loranger, K. Jradi, and C. Daneault, Nanocellulose production by ultrasound-assisted TEMPO oxidation of Kraft pulp on laboratory and pilot scales. in: *IEEE Int. Ultrason. Symp. IUS*, 2012. doi:10.1109/ULTSYM.2012.0238.

[75] S.P. Mishra, A.S. Manent, B. Chabot, and C. Daneault, Production of nanocellulose from native cellulose — Various options utilizing ultrasound. *BioResources*, (2012).

[76] A. Dufresne, Nanocellulose: A new ageless bionanomaterial. *Mater. Today.*, 16 (2013) 220–227. doi:10.1016/j.mattod.2013.06.004.

[77] D. Klemm, F. Kramer, S. Moritz, T. Lindström, M. Ankerfors, D. Gray, and A. Dorris, Nanocelluloses: A new family of nature-based materials. *Angew. Chemie — Int. Ed.* (2011). doi:10.1002/anie.201001273.

[78] B. Wang and M. Sain, The effect of chemically coated nanofiber reinforcement on biopolymer based nanocomposites. *BioResources*, (2007).

[79] A. Bhatnagar and M. Sain, Processing of cellulose nanofiber-reinforced composites. *J. Reinf. Plast. Compos.*, (2005). doi:10.1177/0731684405049864.

[80] A. Isogai, T. Saito, and H. Fukuzumi, TEMPO-oxidized cellulose nanofibers. *Nanoscale*, (2011). doi:10.1039/c0nr00583e.

[81] T. Saito, S. Kimura, Y. Nishiyama, and A. Isogai, Cellulose nanofibers prepared by TEMPO-mediated oxidation of native cellulose. *Biomacromolecules*, (2007). doi:10.1021/bm0703970.

[82] M. Henriksson, G. Henriksson, L.A. Berglund, and T. Lindström, An environmentally friendly method for enzyme-assisted preparation of microfibrillated cellulose (MFC) nanofibers. *Eur. Polym. J.*, (2007). doi:10.1016/j.eurpolymj.2007.05.038.

[83] S.B. Abd Hamid, M. Al Amin, and M.E. Ali, Zeolite supported ionic liquid catalyst for the synthesis of nanocellulose from palm tree biomass. *Micro/Nano Sci. Eng.*, (2014). doi:10.4028/www.scientific.net/AMR.925.52.

[84] P. Satyamurthy and N. Vigneshwaran, A novel process for synthesis of spherical nanocellulose by controlled hydrolysis of microcrystalline cellulose using anaerobic microbial consortium. *Enzy. Microb. Technol.*, (2013). doi:10.1016/j.enzmictec.2012.09.002.

[85] M.T. Islam, M.M. Alam, A. Patrucco, A. Montarsolo, and M. Zoccola, Preparation of nanocellulose: A review. AATCC *J. Res.*, (2014). doi:10.14504/ajr.1.5.3.

[86] Z. Lu, L. Fan, H. Zheng, Q. Lu, Y. Liao, and B. Huang, Preparation, characterization and optimization of nanocellulose whiskers by simultaneously ultrasonic wave and microwave assisted. *Bioresour. Technol.*, (2013). doi:10.1016/j.biortech.2013.07.047.

[87] J.P.S. Morais, M.D.F. Rosa, M.D.S.M. De Souza Filho, L.D. Nascimento, D.M. Do Nascimento, and A.R. Cassales, Extraction and characterization

of nanocellulose structures from raw cotton linter. *Carbohydr. Polym.*, (2013). doi:10.1016/j.carbpol.2012.08.010.
[88] N. Yongvanich and P. Visuttpitukul, Isolation of nanocellulose from pomelo fruit fibers by chemical treatments. *Adv. Mater. Res.*, (2013). doi:10.4028/www.scientific.net/AMR.747.363.
[89] I. Shahabi-Ghahfarrokhi, F. Khodaiyan, M. Mousavi, and H. Yousefi, Preparation and characterization of nanocellulose from beer industrial residues using acid hydrolysis/ultrasound. *Fibers Polym.*, (2015). doi:10.1007/s12221-015-0529-4.
[90] C. Seo, H.W. Lee, A. Suresh, J.W. Yang, J.K. Jung, and Y.C. Kim, Improvement of fermentative production of exopolysaccharides from aureobasidium pullulans under various conditions. *Kore. J. Chem. Eng.*, 31 (2014) 1433–1437. doi:10.1007/s11814-014-0064-9.
[91] M. Shoda and Y. Sugano, Recent advances in bacterial cellulose production. *Biotechnol. Bioproc. Eng.*, 10 (2005) 1–8. doi:10.1007/bf02931175.
[92] H. Ahrem, D. Pretzel, M. Endres, D. Conrad, J. Courseau, H. Müller, R. Jaeger, C. Kaps, D.O. Klemm, and R.W. Kinne, Laser-structured bacterial nanocellulose hydrogels support ingrowth and differentiation of chondrocytes and show potential as cartilage implants. *Acta Biomater.*, 10 (2014) 1341–1353. doi:10.1016/j.actbio.2013.12.004.
[93] L. Fu, P. Zhou, S. Zhang, and G. Yang, Evaluation of bacterial nanocellulose-based uniform wound dressing for large area skin transplantation. *Mater. Sci. Eng. C.*, 33 (2013) 2995–3000. doi:10.1016/j.msec.2013.03.026.
[94] K. Kubiak, M. Kurzawa, M. Jedrzejczak-Krzepkowska, K. Ludwicka, M. Krawczyk, A. Migdalski, M.M. Kacprzak, D. Loska, A. Krystynowicz, and S. Bielecki, Complete genome sequence of Gluconacetobacter xylinus E25 strain-Valuable and effective producer of bacterial nanocellulose. *J. Biotechnol.*, (2014). doi:10.1016/j.jbiotec.2014.02.006.
[95] Y. Kim, M.W. Ullah, M. Ul-Islam, S. Khan, J.H. Jang, and J.K. Park, Self-assembly of bio-cellulose nanofibrils through intermediate phase in a cell-free enzyme system. *Biochem. Eng. J.*, 142 (2019) 135–144. doi:10.1016/j.bej.2018.11.017.
[96] M.W. Ullah, M. Ul-Islam, S. Khan, Y. Kim, and J.K. Park, Structural and physico-mechanical characterization of bio-cellulose produced by a cell-free system. *Carbohydr. Polym.*, 136 (2016) 908–916. doi:10.1016/j.carbpol.2015.10.010.
[97] M. Ul-Islam, S. Khan, M.W. Ullah, and J.K. Park, Bacterial cellulose composites: Synthetic strategies and multiple applications in bio-medical and electro-conductive fields. *Biotechnol. J.*, 10 (2015) 1847–1861. doi:10.1002/biot.201500106.

[98] S. Taokaew, S. Seetabhawang, P. Siripong, and M. Phisalaphong, Biosynthesis and characterization of nanocellulose-gelatin films. *Mater. (Basel).*, (2013). doi:10.3390/ma6030782.

[99] M. Ul-Islam, S. Khan, M.W. Ullah, and J.K. Park, Comparative study of plant and bacterial cellulose pellicles regenerated from dissolved states. *Int. J. Biol. Macromol.*, 137 (2019) 247–252. doi:10.1016/j.ijbiomac.2019.06.232.

[100] M. Ul-Islam, M.W. Ullah, S. Khan, and J.K. Park, Production of bacterial cellulose from alternative cheap and waste resources: A step for cost reduction with positive environmental aspects. *Korean J. Chem. Eng.*, 37 (2020) 925–937. doi:10.1007/s11814-020-0524-3.

[101] A. Kurosumi, C. Sasaki, Y. Yamashita, and Y. Nakamura, Utilization of various fruit juices as carbon source for production of bacterial cellulose by *Acetobacter xylinum* NBRC 13693. *Carbohydr. Polym.*, 76 (2009) 333–335. doi:10.1016/j.carbpol.2008.11.009.

[102] I. Algar, S.C.M. Fernandes, G. Mondragon, C. Castro, C. Garcia-Astrain, N. Gabilondo, A. Retegi, and A. Eceiza, Pineapple agroindustrial residues for the production of high value bacterial cellulose with different morphologies. *J. Appl. Polym. Sci.*, 132 (2015). doi:10.1002/app.41237.

[103] Z. Cheng, R. Yang, X. Liu, X. Liu, and H. Chen, Green synthesis of bacterial cellulose *via* acetic acid pre-hydrolysis liquor of agricultural corn stalk used as carbon source. *Bioresour. Technol.*, 234 (2017) 8–14. doi:10.1016/j.biortech.2017.02.131.

[104] P. Carreira, J.A.S. Mendes, E. Trovatti, L.S. Serafim, C.S.R. Freire, A.J.D. Silvestre, and C.P. Neto, Utilization of residues from agro-forest industries in the production of high value bacterial cellulose. *Bioresour. Technol.*, 102 (2011) 7354–7360. doi:10.1016/j.biortech.2011.04.081.

[105] W.A. Khattak, T. Khan, M. Ul-Islam, F. Wahid, and J.K. Park, Production, characterization and physico-mechanical properties of bacterial cellulose from industrial wastes. *J. Polym. Environ.*, 23 (2015) 45–53. doi:10.1007/s10924-014-0663-x.

[106] V. Revin, E. Liyaskina, M. Nazarkina, A. Bogatyreva, and M. Shchankin, Cost-effective production of bacterial cellulose using acidic food industry by-products. *Brazilian J. Microbiol.*, 49 (2018) 151–159. doi:10.1016/j.bjm.2017.12.012.

[107] O. Shezad, S. Khan, T. Khan, and J.K. Park, Production of bacterial cellulose in static conditions by a simple fed-batch cultivation strategy. *Korean J. Chem. Eng.*, (2009). doi:10.1007/s11814-009-0232-5.

[108] X. Fan, Y. Gao, W. He, H. Hu, M. Tian, K. Wang, and S. Pan, Production of nano bacterial cellulose from beverage industrial waste of citrus peel and pomace using *Komagataeibacter xylinus*. *Carbohydr. Polym.*, (2016). doi:10.1016/j.carbpol.2016.06.062.

[109] K.C. Cheng and J.M. Catchmark, A. Demirci, Effects of CMC addition on bacterial cellulose production in a biofilm reactor and its paper sheets analysis. *Biomacromolecules*, 12 (2011) 730–736. doi:10.1021/bm101363t.

[110] H. Lu, Q. Jia, L. Chen, and L. Zhang, Effect of organic acids on bacterial cellulose produced by *Acetobacter xylinum*. *J. Microbiol. Biotechnol.*, 5 (2015) 30–37. https://www.rroij.com/open-access/effect-of-organic-acids-on-bacterial-cellulose-produced-byacetobacter-xylinum-.php?aid=72595.

[111] S. Keshk and K. Sameshima, Influence of lignosulfonate on crystal structure and productivity of bacterial cellulose in a static culture. *Enzyme Microb. Technol.*, 40 (2006) 4–8. doi:10.1016/j.enzmictec.2006.07.037.

[112] J.H. Ha, N. Shah, M. Ul-Islam, T. Khan, and J.K. Park, Bacterial cellulose production from a single sugar alpha-linked glucuronic acid-based oligosaccharide. *Process Biochem.*, 46 (2011) 1717–1723. doi:10.1016/j.procbio.2011.05.024.

[113] S. Bae, Y. Sugano, and M. Shoda, Improvement of bacterial cellulose production by addition of agar in a jar fermentor. *J. Biosci. Bioeng.*, 97 (2004) 33–38. doi:10.1016/S1389-1723(04)70162-0.

[114] Y. Chao, M. Mitarai, Y. Sugano, and M. Shoda, Effect of addition of water-soluble polysaccharides on bacterial cellulose production in a 50-L airlift reactor. *Biotechnol. Prog.*, 17 (2001) 781–785. doi:10.1021/bp010046b.

[115] L.L. Zhou, D.P. Sun, L.Y. Hu, Y.W. Li and J.Z. Yang, Effect of addition of sodium alginate on bacterial cellulose production by *Acetobacter xylinum*. *J. Ind. Microbiol. Biotechnol.*, 34 (2007) 483–489. doi:10.1007/s10295-007-0218-4.

[116] K.C. Cheng, J.M. Catchmark, and A. Demirci, Effect of different additives on bacterial cellulose production by *Acetobacter xylinum* and analysis of material property. *Cellulose*, 16 (2009) 1033–1045. doi:10.1007/s10570-009-9346-5.

[117] Z. Lu, Y. Zhang, Y. Chi, N. Xu, W. Yao, and B. Sun, Effects of alcohols on bacterial cellulose production by *Acetobacter xylinum* 186. *World J. Microbiol. Biotechnol.*, 27 (2011) 2281–2285. doi:10.1007/s11274-011-0692-8.

[118] T. Ishida, Y. Sugano, T. Nakai, and M. Shoda, Effects of acetan on production of bacterial cellulose by *Acetobacter xylinum*. *Biosci. Biotechnol. Biochem.*, (2002). doi:10.1271/bbb.66.1677.

[119] M. Matsuoka, T. Tsuchida, K. Matsushita, O. Adachi, and F. Yoshinaga, A synthetic medium for bacterial cellulose production by *Acetobacter xylinum* subsp. *sucrofermentans*. *Biosci. Biotechnol. Biochem.*, (1996). doi:10.1271/bbb.60.575.

[120] M. Ul-Islam, W.A. Khattak, M.W. Ullah, S. Khan, and J.K. Park, Synthesis of regenerated bacterial cellulose-zinc oxide nanocomposite films for

biomedical applications. *Cellulose*, 21 (2014). doi:10.1007/s10570-013-0109-y.
[121] S. Khan, M. Ul-Islam, W.A. Khattak, M.W. Ullah, and J.K. Park, Bacterial cellulose-titanium dioxide nanocomposites: nanostructural characteristics, antibacterial mechanism, and biocompatibility. *Cellulose*, 22 (2015) 565–579. doi:10.1007/s10570-014-0528-4.
[122] M.W. Ullah, W.A. Khattak, M. Ul-Islam, S. Khan, and J.K. Park, Metabolic engineering of synthetic cell-free systems: Strategies and applications. *Biochem. Eng. J.*, 105 (2016). doi:10.1016/j.bej.2015.10.023.
[123] W.A. Khattak, M.W. Ullah, M. Ul-Islam, S. Khan, M. Kim, Y. Kim, and J.K. Park, Developmental strategies and regulation of cell-free enzyme system for ethanol production: A molecular prospective. *Appl. Microbiol. Biotechnol.*, 98 (2014) 9561–9578. doi:10.1007/s00253-014-6154-0.
[124] D. Trache, M.H. Hussin, M.K.M. Haafiz, and V.K. Thakur, Recent progress in cellulose nanocrystals: Sources and production. *Nanoscale*, (2017). doi:10.1039/c6nr09494e.
[125] Y. Chen, C. Liu, P.R. Chang, X. Cao, and D.P. Anderson, Bionanocomposites based on pea starch and cellulose nanowhiskers hydrolyzed from pea hull fibre: Effect of hydrolysis time. *Carbohydr. Polym.*, (2009). doi:10.1016/j.carbpol.2008.11.030.
[126] L. Heux, G. Chauve, and C. Bonini, Nonflocculating and chiral-nematic self-ordering of cellulose microcrystals suspensions in nonpolar solvents. *Langmuir*, (2000). doi:10.1021/la9913957.
[127] A. Pakzad, J. Simonsen, and R.S. Yassar, Gradient of nanomechanical properties in the interphase of cellulose nanocrystal composites. *Compos. Sci. Technol.*, (2012). doi:10.1016/j.compscitech.2011.11.020.
[128] L. Wågberg, G. Decher, M. Norgren, T. Lindström, M. Ankerfors, and K. Axnäs, The build-up of polyelectrolyte multilayers of microfibrillated cellulose and cationic polyelectrolytes. *Langmuir*, (2008). doi:10.1021/la702481v.
[129] P. Satyamurthy, P. Jain, R.H. Balasubramanya, and N. Vigneshwaran, Preparation and characterization of cellulose nanowhiskers from cotton fibres by controlled microbial hydrolysis. *Carbohydr. Polym.*, (2011). doi:10.1016/j.carbpol.2010.07.029.
[130] M.A.S.A. Samir, F. Alloin, J.Y. Sanchez, and A. Dufresne, Cross-linked nanocomposite polymer electrolytes reinforced with cellulose whiskers. *Macromolecules*, 37 (2004) 4839–4844. doi:10.1021/ma049504y.
[131] M.A.S. Azizi Samir, F. Alloin, J.Y. Sanchez, and A. Dufresne, Cellulose nanocrystals reinforced poly(oxyethylene). Polymer (Guildf). (2004). doi:10.1016/j.polymer.2004.03.094.

[132] J. Yang, C.R. Han, F. Xu, and R.C. Sun, Simple approach to reinforce hydrogels with cellulose nanocrystals. *Nanoscale*, (2014). doi:10.1039/c4nr01214c.

[133] T. Zhang, Q. Cheng, D. Ye, and C. Chang, Tunicate cellulose nanocrystals reinforced nanocomposite hydrogels comprised by hybrid cross-linked networks. *Carbohydr. Polym.*, (2017). doi:10.1016/j.carbpol.2017.04.007.

[134] T. Zhang, T. Zuo, D. Hu, and C. Chang, Dual physically cross-linked nanocomposite hydrogels reinforced by tunicate cellulose nanocrystals with high toughness and good self-recoverability. *ACS Appl. Mater. Interfaces.*, (2017). doi:10.1021/acsami.7b06219.

[135] T. Abitbol, T. Johnstone, T.M. Quinn, and D.G. Gray, Reinforcement with cellulose nanocrystals of poly (vinyl alcohol) hydrogels prepared by cyclic freezing and thawing. *Soft Matter.*, (2011). doi:10.1039/c0sm01172j.

[136] K. Wang, K.C. Nune, and R.D.K. Misra, The functional response of alginate-gelatin-nanocrystalline cellulose injectable hydrogels toward delivery of cells and bioactive molecules. *Acta Biomater.*, (2016). doi:10.1016/j.actbio.2016.03.016.

[137] Z. Di, Z. Shi, M.W. Ullah, S. Li, and G. Yang, A transparent wound dressing based on bacterial cellulose whisker and poly(2-hydroxyethyl methacrylate). *Int. J. Biol. Macromol.*, 105 (2017) 638–644. doi:10.1016/j.ijbiomac.2017.07.075.

[138] S.Y. Ooi, I. Ahmad, and M.C.I.M. Amin, Cellulose nanocrystals extracted from rice husks as a reinforcing material in gelatin hydrogels for use in controlled drug delivery systems. *Ind. Crops Prod.*, (2016). doi:10.1016/j.indcrop.2015.11.082.

[139] W. Wang, X. Zhang, A. Teng, and A. Liu, Mechanical reinforcement of gelatin hydrogel with nanofiber cellulose as a function of percolation concentration. *Int. J. Biol. Macromol.*, (2017). doi:10.1016/j.ijbiomac.2017.05.027.

[140] P. Bertsch, S. Isabettini, and P. Fischer, Ion-Induced hydrogel formation and nematic ordering of nanocrystalline cellulose suspensions. *Biomacromolecules*, (2017). doi:10.1021/acs.biomac.7b01119.

[141] N. Lin, A. Gèze, D. Wouessidjewe, J. Huang, and A. Dufresne, Biocompatible double-membrane hydrogels from cationic cellulose nanocrystals and anionic alginate as complexing drugs codelivery. *ACS Appl. Mater. Interfaces.*, (2016). doi:10.1021/acsami.6b00555.

[142] X. Yang, G. Liu, L. Peng, J. Guo, L. Tao, J. Yuan, C. Chang, Y. Wei, and L. Zhang, Highly efficient self-healable and dual responsive cellulose-based hydrogels for controlled release and 3D cell culture. *Adv. Funct. Mater.*, (2017). doi:10.1002/adfm.201703174.

[143] C. Shao, M. Wang, H. Chang, F. Xu, and J. Yang, A self-healing cellulose nanocrystal-poly(ethylene glycol) nanocomposite hydrogel *via* diels-

alder click reaction. *ACS Sustain. Chem. Eng.*, (2017). doi:10.1021/acssuschemeng.7b01060.

[144] S. Liu, M. Jin, Y. Chen, H. Gao, X. Shi, W. Cheng, L. Ren, and Y. Wang, High internal phase emulsions stabilised by supramolecular cellulose nanocrystals and their application as cell-adhesive macroporous hydrogel monoliths. *J. Mater. Chem. B.*, (2017). doi:10.1039/c7tb00145b.

[145] J.A. Kelly, A.M. Shukaliak, C.C.Y. Cheung, K.E. Shopsowitz, W.Y. Hamad, and M.J. MacLachlan, Responsive photonic hydrogels based on nanocrystalline cellulose. *Angew. Chemie — Int. Ed.*, (2013). doi:10.1002/anie.201302687.

[146] W. Chen, H. Yu, S.Y. Lee, T. Wei, J. Li, and Z. Fan, Nanocellulose: A promising nanomaterial for advanced electrochemical energy storage. *Chem. Soc. Rev.*, (2018). doi:10.1039/c7cs00790f.

[147] W. Chen, Q. Li, Y. Wang, X. Yi, J. Zeng, H. Yu, Y. Liu, and J. Li, Comparative study of aerogels obtained from differently prepared nanocellulose fibers. *Chem SusChem*, (2014). doi:10.1002/cssc.201300950.

[148] K. Uetani and H. Yano, Nanofibrillation of wood pulp using a high-speed blender. *Biomacromolecules*, (2011). doi:10.1021/bm101103p.

[149] K. Abe and H. Yano, Formation of hydrogels from cellulose nanofibers. *Carbohydr. Polym.*, (2011). doi:10.1016/j.carbpol.2011.03.028.

[150] K. Abe and H. Yano, Cellulose nanofiber-based hydrogels with high mechanical strength. *Cellulose*, (2012). doi:10.1007/s10570-012-9784-3.

[151] J. Niu, J. Wang, X. Dai, Z. Shao, and X. Huang, Dual physically cross-linked healable polyacrylamide/cellulose nanofibers nanocomposite hydrogels with excellent mechanical properties. *Carbohydr. Polym.*, (2018). doi:10.1016/j.carbpol.2018.03.086.

[152] I.W. Sutherland, Microbial polysaccharides from Gram-negative bacteria. *in: Int. Dairy J.*, 2001: pp. 663–674. doi:10.1016/S0958-6946(01)00112-1.

[153] A. Endler, C. Sánchez-Rodríguez, and S. Persson, Cellulose squeezes through. *Nat. Chem. Biol.*, 6 (2010) 883–884. doi:10.1038/nchembio.480.

[154] W. Borzani and S.J. de Souza, Mechanism of the film thickness increasing during the bacterial production of cellulose on non-agitaded liquid media. *Biotechnol. Lett.*, 17 (1995) 1271–1272. doi:10.1007/BF00128400.

[155] C. Castro, R. Zuluaga, C. Álvarez, J.L. Putaux, G. Caro, O.J. Rojas, I. Mondragon, and P. Gañán, Bacterial cellulose produced by a new acid-resistant strain of Gluconacetobacter genus. *Carbohydr. Polym.*, 89 (2012) 1033–1037. doi:10.1016/j.carbpol.2012.03.045.

[156] N. Takeda, K. Tamura, R. Mineguchi, Y. Ishikawa, Y. Haraguchi, T. Shimizu, and Y. Hara, *In situ* cross-linked electrospun fiber scaffold of collagen for fabricating cell-dense muscle tissue. *J. Artif. Organs.*, (2016). doi:10.1007/s10047-015-0871-8.

[157] L. Lamboni, Y. Li, J. Liu, and G. Yang, Silk sericin-functionalized bacterial cellulose as a potential wound-healing biomaterial. *Biomacromolecules*, 17 (2016) 3076–3084. doi:10.1021/acs.biomac.6b00995.

[158] S. Khan, M. Ul-Islam, M. Ikram, S.U. Islam, M.W. Ullah, M. Israr, J.H. Jang, S. Yoon, and J.K. Park, Preparation and structural characterization of surface modified microporous bacterial cellulose scaffolds: A potential material for skin regeneration applications *in vitro* and *in vivo*. *Int. J. Biol. Macromol.*, 117 (2018) 1200–1210. doi:10.1016/j.ijbiomac.2018.06.044.

[159] S. Kirdponpattara, A. Khamkeaw, N. Sanchavanakit, P. Pavasant, and M. Phisalaphong, Structural modification and characterization of bacterial cellulose–alginate composite scaffolds for tissue engineering. *Carbohydr. Polym.*, (2015). doi:10.1016/j.carbpol.2015.06.059.

[160] M. Ul-Islam, N. Shah, J.H. Ha, and J.K. Park, Effect of chitosan penetration on physico-chemical and mechanical properties of bacterial cellulose. *Korean J. Chem. Eng.*, 28 (2011) 1736–1743. doi:10.1007/s11814-011-0042-4.

[161] T. Maneerung, S. Tokura, and R. Rujiravanit, Impregnation of silver nanoparticles into bacterial cellulose for antimicrobial wound dressing. *Carbohydr. Polym.*, (2008). doi:10.1016/j.carbpol.2007.07.025.

[162] A. Khalid, R. Khan, M. Ul-Islam, T. Khan, and F. Wahid, Bacterial cellulose-zinc oxide nanocomposites as a novel dressing system for burn wounds. *Carbohydr. Polym.*, (2017). doi:10.1016/j.carbpol.2017.01.061.

[163] M.W. Ullah, M. Ul-Islam, S. Khan, Y. Kim, J.H. Jang, and J.K. Park, *In situ* synthesis of a bio-cellulose/titanium dioxide nanocomposite by using a cell-free system. *RSC Adv.*, 6 (2016) 22424–22435. doi:10.1039/C5RA26704H.

[164] A. Khalid, H. Ullah, M. Ul-Islam, R. Khan, S. Khan, F. Ahmad, T. Khan, and F. Wahid, Bacterial cellulose–TiO_2 nanocomposites promote healing and tissue regeneration in burn mice model. *RSC Adv.*, (2017). doi:10.1039/C7RA06699F.

[165] M. Ul-Islam, T. Khan, W.A. Khattak, and J.K. Park, Bacterial cellulose-MMTs nanoreinforced composite films: Novel wound dressing material with antibacterial properties. *Cellulose*, (2013). doi:10.1007/s10570-012-9849-3.

[166] A. Svensson, E. Nicklasson, T. Harrah, B. Panilaitis, D.L. Kaplan, M. Brittberg, and P. Gatenholm, Bacterial cellulose as a potential scaffold for tissue engineering of cartilage. *Biomaterials*, (2005). doi:10.1016/j.biomaterials.2004.02.049.

[167] D. Klemm, D. Schumann, U. Udhardt, and S. Marsch, Bacterial synthesized cellulose — Artificial blood vessels for microsurgery. *Prog. Polym. Sci.*, 26 (2001) 1561–1603. doi:10.1016/S0079-6700(01)00021-1.

[168] J.D. Fontana, A.M. De Souza, C.K. Fontana, I.L. Torriani, J.C. Moreschi, B.J. Gallotti, S.J. De Souza, G.P. Narcisco, J.A. Bichara, and L.F.X. Farah, Acetobacter cellulose pellicle as a temporary skin substitute. *Appl. Biochem. Biotechnol.*, (1990). doi:10.1007/BF02920250.
[169] W. Sajjad, T. Khan, M. Ul-islam, R. Khan, Z. Hussain, and A. Khalid, Development of modified montmorillonite-bacterial cellulose nanocomposites as a novel substitute for burn skin and tissue regeneration. *Carbohydr. Polym.*, 206 (2019) 548–556. doi:10.1016/j.carbpol.2018.11.023.
[170] S. Khan, M. Ul-Islam, M.W. Ullah, M. Ikram, F. Subhan, Y. Kim, J.H. Jang, S. Yoon, and J.K. Park, Engineered regenerated bacterial cellulose scaffolds for application in *in vitro* tissue regeneration. *RSC Adv.*, 5 (2015) 84565–84573. doi:10.1039/C5RA16985B.
[171] J.M. Rajwade, K.M. Paknikar, and J. V. Kumbhar, Applications of bacterial cellulose and its composites in biomedicine. *Appl. Microbiol. Biotechnol.*, (2015). doi:10.1007/s00253-015-6426-3.
[172] Y. Numata, L. Mazzarino, and R. Borsali, A slow-release system of bacterial cellulose gel and nanoparticles for hydrophobic active ingredients. *Int. J. Pharm.*, (2015). doi:10.1016/j.ijpharm.2015.03.068.
[173] T. Zhang, W. Wang, D. Zhang, X. Zhang, Y. Ma, Y. Zhou, and L. Qi, Biotemplated synthesis of gold nanoparticle-bacteria cellulose nanofiber nanocomposites and their application in biosensing. *Adv. Funct. Mater.*, 20 (2010) 1152–1160. doi:10.1002/adfm.200902104.
[174] W.K. Czaja, D.J. Young, M. Kawecki, and R.M. Brown, The future prospects of microbial cellulose in biomedical applications. *Biomacromolecules*, 8 (2007) 1–12. doi:10.1021/bm060620d.
[175] A. Potthast, T. Rosenau, R. Buchner, T. Röder, G. Ebner, H. Bruglachner, H. Sixta, and P. Kosma, The cellulose solvent system *N,N*-dimethylacetamide/lithium chloride revisited: The effect of water on physicochemical properties and chemical stability. *Cellulose*, (2002). doi:10.1023/A:1015811712657.
[176] M. Ul-Islam, W.A. Khattak, M.W. Ullah, S. Khan, and J.K. Park, Synthesis of regenerated bacterial cellulose-zinc oxide nanocomposite films for biomedical applications. *Cellulose*, 21 (2014) 433–447. doi:10.1007/s10570-013-0109-y.
[177] J. Cai, L. Zhang, S. Liu, Y. Liu, X. Xu, X. Chen, B. Chu, X. Guo, J. Xu, H. Cheng, C.C. Han, and S. Kuga, Dynamic self-assembly induced rapid dissolution of cellulose at low temperatures. *Macromolecules*, (2008). doi:10.1021/ma801110g.
[178] D.R. Ruka, G.P. Simon, and K.M. Dean, *In situ* modifications to bacterial cellulose with the water insoluble polymer poly-3-hydroxybutyrate. *Carbohydr. Polym.*, (2013). doi:10.1016/j.carbpol.2012.11.007.

[179] Y. Huang, C. Zhu, and J. Yang, Recent advances in bacterial cellulose, (2014) 1–30. doi:10.1007/s10570-013-0088-z.

[180] N. Lavoine, I. Desloges, A. Dufresne, and J. Bras, Microfibrillated cellulose — Its barrier properties and applications in cellulosic materials: A review. *Carbohydr. Polym.*, (2012). doi:10.1016/j.carbpol.2012.05.026.

[181] S. Vitta and V. Thiruvengadam, Multifunctional bacterial cellulose and nanoparticle-embedded composites. *Curr. Sci.*, (2012).

[182] S.A. Hutchens, R.S. Benson, B.R. Evans, H.M. O'Neill, and C.J. Rawn, Biomimetic synthesis of calcium-deficient hydroxyapatite in a natural hydrogel. *Biomaterials*, 27 (2006) 4661–4670. doi:10.1016/j.biomaterials.2006.04.032.

[183] H. Bäckdahl, M. Esguerra, D. Delbro, B. Risberg, and P. Gatenholm, Engineering microporosity in bacterial cellulose scaffolds. *J. Tissue Eng. Regen. Med.*, 2 (2008) 320–330. doi:10.1002/term.97.

[184] Z. Shi, X. Shi, M.W. Ullah, S. Li, V. V. Revin, and G. Yang, Fabrication of nanocomposites and hybrid materials using microbial biotemplates. *Adv. Compos. Hybr. Mater.*, (2017). doi:10.1007/s42114-017-0018-x.

[185] G. Wang, X.F. Chen, X.D. Shi, L.J. Yu, B.F. Liu, and G. Yang, Bio-Fabrication of patterned cellulose nano-fibers. *Multi-Funct. Mater. Struct.*, 47 (2008) 1359–1362. doi:10.4028/www.scientific.net/AMR.47-50.1359.

[186] Y. Uraki, J. Nemoto, H. Otsuka, Y. Tamai, J. Sugiyama, T. Kishimoto, M. Ubukata, H. Yabu, M. Tanaka, and M. Shimomura, Honeycomb-like architecture produced by living bacteria, *Gluconacetobacter xylinus*. *Carbohydr. Polym.*, 69 (2007) 1–6. doi:10.1016/j.carbpol.2006.08.021.

[187] T. Kondo, W. Kasai, M. Nojiri, Y. Hishikawa, E. Togawa, D. Romanovicz, and R.M. Brown, Regulated patterns of bacterial movements based on their secreted cellulose nanofibers interacting interfacially with ordered chitin templates. *J. Biosci. Bioeng.*, (2012). doi:10.1016/j.jbiosc.2012.02.020.

[188] A. Putra, A. Kakugo, H. Furukawa, J.P. Gong, Y. Osada, T. Uemura, and M. Yamamoto, Production of bacterial cellulose with well oriented fibril on PDMS substrate. *Polym. J.*, (2008). doi:10.1016/0029-554X(81)91167-8.

[189] M. Faria, S. Thomas, and L.A. Pothan, Utilization of various lignocellulosic biomass for the production of nanocellulose: A comparative study. *Cellulose*, (2015). doi:10.1007/s10570-015-0554-x.

[190] Y. Huang, C. Zhu, J. Yang, Y. Nie, C. Chen, and D. Sun, Recent advances in bacterial cellulose. *Cellulose*, (2014). doi:10.1007/s10570-013-0088-z.

[191] M. Jonoobi, R. Oladi, Y. Davoudpour, K. Oksman, A. Dufresne, Y. Hamzeh, and R. Davoodi, Different preparation methods and properties of nanostructured cellulose from various natural resources and residues: A review. *Cellulose*, (2015). doi:10.1007/s10570-015-0551-0.

[192] A.D. French, Idealized powder diffraction patterns for cellulose polymorphs. *Cellulose*, 21 (2014) 885–896. doi:10.1007/s10570-013-0030-4.
[193] B.V. Mohite and S.V. Patil, Physical, structural, mechanical and thermal characterization of bacterial cellulose by G. Hansenii NCIM 2529. *Carbohydr. Polym.*, (2014). doi:10.1016/j.carbpol.2014.02.012.
[194] Y.B. Park, C.M. Lee, K. Kafle, S. Park, D.J. Cosgrove, and S.H. Kim, Effects of plant cell wall matrix polysaccharides on bacterial cellulose structure studied with vibrational sum frequency generation spectroscopy and x-ray diffraction. *Biomacromolecules*, (2014). doi:10.1021/bm500567v.
[195] D. Sun, L. Zhou, Q. Wu, and S. Yang, Preliminary research on structure and properties of nano-cellulose. *J. Wuhan Univ. Technol. Mater. Sci. Ed.*, (2007). doi:10.1007/s11595-006-4677-7.
[196] D.J. Gardner, G.S. Oporto, R. Mills, and M.A.S.A. Samir, Adhesion and surface issues in cellulose and nanocellulose. *J. Adhes. Sci. Technol.*, (2008). doi:10.1163/156856108X295509.
[197] Y. Habibi, H. Chanzy, and M.R. Vignon, TEMPO-mediated surface oxidation of cellulose whiskers. *Cellulose*, (2006). doi:10.1007/s10570-006-9075-y.
[198] J. George, K.V. Ramana, and A.S. Bawa, Siddaramaiah, Bacterial cellulose nanocrystals exhibiting high thermal stability and their polymer nanocomposites. *Int. J. Biol. Macromol.*, 48 (2011) 50–57. doi:10.1016/j.ijbiomac.2010.09.013.
[199] E. Abraham, B. Deepa, L.A. Pothan, M. Jacob, S. Thomas, U. Cvelbar, and R. Anandjiwala, Extraction of nanocellulose fibrils from lignocellulosic fibres: A novel approach. *Carbohydr. Polym.*, (2011). doi:10.1016/j.carbpol.2011.06.034.
[200] H. El-Saied, A.I. El-Diwany, A.H. Basta, N.A. Atwa, and D.E. El-Ghwas, Production and characterization of economical bacterial cellulose. *BioResources*, (2008). doi:10.15376/biores.3.4.1196-1217.
[201] D. Le Bras, M. Strømme, and A. Mihranyan, Characterization of dielectric properties of nanocellulose from wood and algae for electrical insulator applications. *J. Phys. Chem. B.*, (2015). doi:10.1021/acs.jpcb.5b00715.
[202] O. Shezad, S. Khan, T. Khan, and J.K. Park, Physicochemical and mechanical characterization of bacterial cellulose produced with an excellent productivity in static conditions using a simple fed-batch cultivation strategy. *Carbohydr. Polym.*, 82 (2010) 173–180. doi:10.1016/j.carbpol.2010.04.052.
[203] M. Ul-Islam, T. Khan, and J.K. Park, Nanoreinforced bacterial cellulose-montmorillonite composites for biomedical applications. *Carbohydr. Polym.*, 89 (2012) 1189–1197. doi:10.1016/j.carbpol.2012.03.093.

[204] M. Iguchi, S. Yamanaka, and A. Budhiono, Bacterial cellulose — a masterpiece of nature's arts. *J. Mater. Sci.*, 35 (2000) 261–270. doi:10.1023/A:1004775229149.

[205] A. Dufresne, Nanocellulose: A new ageless bionanomaterial. *Mater. Today.*, 16 (2013) 220–227. doi:10.1016/j.mattod.2013.06.004.

[206] S. Khan, M. Ul-Islam, M.W. Ullah, M. Israr, J.H. Jang, and J.K. Park, Nano-gold assisted highly conducting and biocompatible bacterial cellulose-PEDOT:PSS films for biology-device interface applications. *Int. J. Biol. Macromol.*, 107 (2018) 865–873. doi:10.1016/j.ijbiomac.2017.09.064.

[207] S.T. Schrecker and P.A. Gostomski, Determining the water holding capacity of microbial cellulose. *Biotechnol. Lett.*, 27 (2005) 1435–1438. doi:10.1007/s10529-005-1465-y.

[208] S. Bin Lin, C.P. Hsu, L.C. Chen, and H.H. Chen, Adding enzymatically modified gelatin to enhance the rehydration abilities and mechanical properties of bacterial cellulose. *Food Hydrocoll.*, 23 (2009) 2195–2203. doi:10.1016/j.foodhyd.2009.05.011.

[209] M. Ul-Islam, T. Khan, and J.K. Park, Water holding and release properties of bacterial cellulose obtained by *in situ* and *ex situ* modification. *Carbohydr. Polym.*, 88 (2012) 596–603. doi:10.1016/j.carbpol.2012.01.006.

[210] K. Watanabe, M. Tabuchi, Y. Morinaga, and F. Yoshinaga, Structural features and properties of bacterial cellulose produced in agitated culture. *Cellulose*, 5 (1998) 187–200. doi:10.1023/A:1009272904582.

[211] A. Meftahi, R. Khajavi, A. Rashidi, M. Sattari, M.E. Yazdanshenas, and M. Torabi, The effects of cotton gauze coating with microbial cellulose. *Cellulose*, 17 (2010) 199–204. doi:10.1007/s10570-009-9377-y.

[212] Y. Dahman, Nanostructured biomaterials and biocomposites from bacterial cellulose nanofibers. *J. Nanosci. Nanotechnol.*, (2009). doi:10.1166/jnn.2009.1466.

[213] K. Gelin, A. Bodin, P. Gatenholm, A. Mihranyan, K. Edwards, and M. Strømme, Characterization of water in bacterial cellulose using dielectric spectroscopy and electron microscopy. *Polymer* (Guildf). 48 (2007) 7623–7631. doi:10.1016/j.polymer.2007.10.039.

[214] A. Krystynowicz, W. Czaja, A. Wiktorowska-Jezierska, M. Gonçalves-Miśkiewicz, M. Turkiewicz, and S. Bielecki, Factors affecting the yield and properties of bacterial cellulose. *J. Ind. Microbiol. Biotechnol.*, (2002). doi:10.1038/sj.jim.7000303.

[215] A. Rees, L.C. Powell, G. Chinga-Carrasco, D.T. Gethin, K. Syverud, K.E. Hill, and D.W. Thomas, 3D bioprinting of carboxymethylated-periodate oxidized nanocellulose constructs for wound dressing applications. *Biomed Res. Int.*, 2015 (2015). doi:10.1155/2015/925757.

[216] E. Loranger, A.O. Piche, and C. Daneault, Impact of transducers configuration in a pilot sonoreactor used for nanocellulose production by ultrasound-assisted TEMPO oxidation. *In: IEEE Int. Ultrason. Symp. IUS*, (2013). doi:10.1109/ULTSYM.2013.0481.
[217] S. Shafiei-Sabet, W.Y. Hamad, and S.G. Hatzikiriakos, Rheology of nanocrystalline cellulose aqueous suspensions. *Langmuir*, (2012). doi:10.1021/la303380v.
[218] E.E. Ureña-Benavides, G. Ao, V.A. Davis, and C.L. Kitchens, Rheology and phase behavior of lyotropic cellulose nanocrystal suspensions. *Macromolecules*, (2011). doi:10.1021/ma201649f.
[219] T. Moberg, K. Sahlin, K. Yao, S. Geng, G. Westman, Q. Zhou, K. Oksman, and M. Rigdahl, Rheological properties of nanocellulose suspensions: Effects of fibril/particle dimensions and surface characteristics. *Cellulose*, (2017). doi:10.1007/s10570-017-1283-0.
[220] D. Liu, X. Chen, Y. Yue, M. Chen, and Q. Wu, Structure and rheology of nanocrystalline cellulose. *Carbohydr. Polym.*, (2011). doi:10.1016/j.carbpol.2010.11.039.
[221] S. Shafeiei-Sabet, W.Y. Hamad, and S.G. Hatzikiriakos, Influence of degree of sulfation on the rheology of cellulose nanocrystal suspensions. *Rheol. Acta.*, (2013). doi:10.1007/s00397-013-0722-6.
[222] X. Jia, Y. Chen, C. Shi, Y. Ye, M. Abid, S. Jabbar, P. Wang, X. Zeng, and T. Wu, Rheological properties of an amorphous cellulose suspension. *Food Hydrocoll.*, (2014). doi:10.1016/j.foodhyd.2013.12.026.
[223] M. Nogi, K. Abe, K. Handa, F. Nakatsubo, S. Ifuku, and H. Yano, Property enhancement of optically transparent bionanofiber composites by acetylation. *Appl. Phys. Lett.*, 89 (2006) 233123. doi:10.1063/1.2403901.
[224] D.Y. Kim, Y. Nishiyama, and S. Kuga, Surface acetylation of bacterial cellulose. *Cellulose*, 9 (2002) 361–367. doi:10.1023/A:1021140726936.
[225] S. Ifuku, M. Nogi, K. Abe, K. Handa, F. Nakatsubo, and H. Yano, Surface modification of bacterial cellulose nanofibers for property enhancement of optically transparent composites: Dependence on acetyl-group DS. *Biomacromolecules*, 8 (2007) 1973–1978. doi:10.1021/bm070113b.
[226] J.F. Luna-Martínez, D.B. Hernández-Uresti, M.E. Reyes-Melo, C.A. Guerrero-Salazar, V.A. González-González, and S. Sepúlveda-Guzmán, Synthesis and optical characterization of ZnS-sodium carboxymethyl cellulose nanocomposite films. *Carbohydr. Polym.*, (2011). doi:10.1016/j.carbpol.2010.12.021.
[227] H. Yang, A. Tejado, N. Alam, M. Antal, and T.G.M. Van De Ven, Films prepared from electrosterically stabilized nanocrystalline cellulose. *Langmuir*, (2012). doi:10.1021/la2049663.

[228] J.M. Morais, F. Papadimitrakopoulos, and D.J. Burgess, Biomaterials/Tissue interactions: Possible solutions to overcome foreign body response. *AAPS J.*, 12 (2010) 188–196. doi:10.1208/s12248-010-9175-3.

[229] D.F. Williams, Moderator's papers. *In: Prog. Biomed. Eng. 4. Defin. Biomater.*, (1987).

[230] D.A. Schumann, J. Wippermann, D.O. Klemm, F. Kramer, D. Koth, H. Kosmehl, T. Wahlers, and S. Salehi-Gelani, Artificial vascular implants from bacterial cellulose: Preliminary results of small arterial substitutes. *Cellulose*, 16 (2009) 877–885. doi:10.1007/s10570-008-9264-y.

[231] D. Klemm, D. Schumann, U. Udhardt, and S. Marsch, Bacterial synthesized cellulose — Artificial blood vessels for microsurgery. *Prog. Polym. Sci.*, 26 (2001) 1561–1603. doi:10.1016/S0079-6700(01)00021-1.

[232] G. Helenius, H. Bäckdahl, A. Bodin, U. Nannmark, P. Gatenholm, and B. Risberg, *In vivo* biocompatibility of bacterial cellulose. *J. Biomed. Mater. Res. — Part A.*, 76 (2006) 431–438. doi:10.1002/jbm.a.30570.

[233] J. Li, Y. Wan, L. Li, H. Liang, and J. Wang, Preparation and characterization of 2,3-dialdehyde bacterial cellulose for potential biodegradable tissue engineering scaffolds. *Mater. Sci. Eng. C.*, 29 (2009) 1635–1642. doi:10.1016/j.msec.2009.01.006.

[234] Q.G. Fan, D.M. Lewis, and K.N. Tapley, Characterization of cellulose aldehyde using Fourier transform infrared spectroscopy. *J. Appl. Polym. Sci.*, (2001). doi:10.1002/app.1953.

[235] M. Singh, A.R. Ray, and P. Vasudevan, Biodegradation studies on periodate oxidized cellulose. *Biomaterials*, (1982). doi:10.1016/0142-9612(82)90055-2.

[236] P. RoyChowdhury and V. Kumar, Fabrication and evaluation of porous 2,3-dialdehydecellulose membrane as a potential biodegradable tissue-engineering scaffold. *J. Biomed. Mater. Res. — Part A.*, (2006). doi:10.1002/jbm.a.30503.

https://doi.org/10.1142/9781786349477_0002

Chapter 2

Synthesis of Cellulose Nanofiber from Oil Palm Biomass by High Pressure Homogenization and Wet Disk Milling

Hidayah Ariffin[*,†,§], **Tengku Arisyah Tengku Yasim-Anuar**[*,†], **Mohd Nor Faiz Norrrahim**[‡], **and Mohd Ali Hassan**[*]

[*]*Department of Bioprocess Technology, Faculty of Biotechnology and Biomolecular Sciences, Universiti Putra Malaysia, Serdang, Selangor, Malaysia*

[†]*Laboratory of Biopolymer and Derivatives, Institute of Tropical Forestry and Forest Products (INTROP), Universiti Putra Malaysia, Serdang, Selangor, Malaysia*

[‡]*Research Center for Chemistry Defence, Universiti Pertahanan Nasional Malaysia, Kem Perdana Sungai Besi, Kuala Lumpur, Malaysia*

[§]*hidayah@upm.edu.my*

Abstract

In this study, the characteristics of cellulose nanofiber (CNF) from oil palm mesocarp fiber (OPMF) produced by two different nanofibrillation methods — high-pressure homogenization (HPH) and wet disk milling (WDM) — were compared for their morphological, crystallinity, and thermal stability analyses. Results showed that CNF produced from

HPH and WDM had almost similar characteristics, with WDM being advantageous due to its ability to process cellulose at higher concentration without clogging. This may suggest the suitability of WDM for large-scale CNF production *via* mechanical processing route.

Keywords: Oil palm mesocarp fiber; cellulose nanofiber; high pressure homogenization; wet disk milling

1. Introduction

Malaysia is one of the top producers of palm oil in the world with total area of oil palm plantation of 50,000 km^2. The oil palm plantation covers almost 77% of agricultural land, or 15% of total land area in Malaysia (Alam *et al.*, 2016). During the palm oil extraction processes, approximately 44 million tons of solid oil palm biomass (OPB) are being produced, which comprises of 54% empty fruit bunch (OPEFB), 30% shell and 18% mesocarp fiber (OPMF) (Chiew and Shimada, 2013). This value is expected to rise exponentially by the year 2020 as the palm oil industry keeps expanding year by year. According to the report in Malaysia National Biomass Strategy 2020, carbon emission could be reduced by replacing non-renewable materials with renewable biomass as raw material. In addition, biomass-based industry will create more opportunities and social benefits for the domestic businesses.

OPMF is one of the OPB that can be collected from palm oil mills after oil extraction processing. Conventionally, OPMF is used as boiler fuel or simply being burnt to dispose it off (Warid, 2017). To promote efficient utilization of OPMF, it is proposed that the OPMF can be used as a starting material for the production of high-value bio-products, particularly cellulose nanofibers (CNFs). According to Chen *et al.* (2011a), CNF is a polymer with one dimension of 100 nanometer (nm) or less, with extremely high specific area and high porosity and with excellent pore interconnectivity. It has been demonstrated that CNF enhanced the mechanical strength of materials such as polypropylene (PP) as they have high crystallinity as well as high thermal resistance as compared to neat polymer (Iwamoto *et al.*, 2014). All these properties make CNF usable for several applications, ranging from consumer products to high-tech industrial applications. Therefore, with the uninterrupted supply of OPMF throughout the year, there is high possibility to commercially produce CNF from OPMF, thus indirectly ensuring maximum utilization of OPMF.

In general, CNF can be produced by various methods such as electrospinning (Yasim-Anuar *et al.*, 2017), ultrasonication (Chen *et al.*, 2011b; Wang *et al.*, 2015), TEMPO-oxidation (Soni *et al.*, 2015), steam explosion (Chirayil *et al.*, 2014) and high-pressure homogenization (Davoudpour *et al.*, 2015). All these methods have been proven to be able to produce CNF with diameter size less than 500 nm. Another method for producing CNF is by using a wet disk mill (WDM), which is a chemical-free, simple and fast nanofibrillation method. Based on the previous reports, the use of WDM successfully produced CNF from wood (Nogi *et al.*, 2009) and bamboo (Iwamoto *et al.*, 2008) with diameter size less than 100 nm. To date, the effectiveness of this method in producing CNF from OPMF has not been reported. In this study, the characteristics of CNF produced by WDM were compared with those produced by high-pressure homogenization in order to propose the best method for CNF production from OPMF cellulose.

2. Experimental

2.1 *Materials*

Samples of oil palm mesocarp fiber (OPMF) were obtained from Seri Ulu Langat Palm Oil Mill (Dengkil, Selangor, Malaysia). The OPMF samples were cleaned, dried and kept at room temperature prior to further use. Sodium chlorite ($NaClO_2$) and potassium hydroxide (KOH) used in cellulose extraction from OPMF were purchased from SIGMA-Aldrich.

2.2 *Methods*

2.2.1 *Cellulose extraction*

Cellulose from OPMF was extracted according to the procedure given in Iwamoto *et al.* (2008). Lignin in the OPMF was removed by soaking the OPMF in 5% (w/v) sodium chlorite ($NaClO_2$) at 70°C with pH ranging from pH 4–5 for 90 minutes. This was followed by hemicellulose extraction, in which the lignin-removed OPMF was treated with 6% (w/v) of potassium hydroxide (KOH) at room temperature for 24 hours. The resulting cellulose, referred to as OPMF-cellulose, was then ground with a hammer mill (Miki Pulley Co. Ltd., Japan) to reduce the size to approximately 0.5–1 mm.

2.2.2 *Preparation of cellulose nanofiber*

2.2.2.1 High-pressure homogenization

Nanofibrillation of OPMF-cellulose by a high-pressure homogenizer was conducted according to the method of Davoudpour *et al.* (2015). OPMF-cellulose was suspended in distilled water at 0.025 wt% for 48 hours. The cellulose suspension was homogenized using a high-pressure homogenizer (C50, Avestin Inc., Ottawa, Canada) at 50 MPa for 15, 30, 45 and 60 passes, respectively. The obtained CNF was then freeze dried prior to further analysis.

2.2.2.2 Wet disk milling

Nanofibrillation of OPMF-cellulose by using a wet disk mill (WDM) was conducted based on the method suggested by Iwamoto *et al.* (2008). The CNF was produced by using a high-shear ultrafine friction grinder, or WDM (Multi mill, Grow Engineering, Adachi-ku, Tokyo, Japan). The concentration of the fiber suspension was adjusted to 1 wt% by adding distilled water. The treated fibers were first soaked in distilled water for 72 hours before being fibrillated using WDM. The suspension was passed through the WDM for several cycles at 1,800 rpm. Similar to high-pressure homogenized CNF, the slurry was then freeze dried for further characterization.

2.2.3 *Characterization*

2.2.3.1 Scanning electron microscopy analysis (SEM)

The morphology of the CNF was observed using a field-emission scanning electron microscopy (FE-SEM) (Sirion, 200, FEI, Eindhoven, Netherlands) with accelerating voltage of 5 kV. The freeze-dried CNF were coated with platinum using a vacuum sputter coater prior to FE-SEM observation.

2.2.3.2 X-ray diffraction analysis

The crystallinity index of CNF was analyzed using an X-ray diffractometer (XRD). XRD measurement was performed on a Philips PW3040/00

X'Pert MPD system (Eindhoven, Netherlands). The X-ray pattern was recorded in the 2θ range of 5–50°, and the crystallinity index, *CrI*, was calculated using the following equation:

$$CrI = \frac{I_{002} - I_{am}}{I_{002}} \times 100\%$$

where I_{002} equals to $2\theta = 23°$, which refers to the peak of the crystalline portion of the biomass (i.e., cellulose), while I_{am} refers to the peak at about $2\theta = 18°$, which corresponds to the amorphous region (Nordin *et al.*, 2013).

2.2.3.3 Thermal stability analysis

Thermal stability and decomposition temperature of the CNF were determined using a thermogravimetric analyzer (TG 400, Perkin Elmer, Waltham, MA, USA). The CNF samples (approximately 10 mg) were placed in aluminum cups, and the analysis was carried out at a heating rate of 10°C/min under nitrogen flow of 20 mL/min from 50 to 550°C (Yasim-Anuar *et al.*, 2017).

3. Results and Discussion

3.1 *Characteristics of CNF Produced by High-Pressure Homogenization*

3.1.1 *Morphology*

The changes in the morphology of the CNF produced by high-pressure homogenizer were evaluated using FE-SEM. Figure 1 illustrates the CNF obtained after homogenization at 50 MPa for 15, 30, 45 and 60 passes using 0.025% (w/v) OPMF-cellulose suspension. The results as tabulated in Table 1 revealed that a greater number of passes further reduced the CNF diameter to the nanometer range markedly. The fibrillation occurred as early as 15 passes, as the diameter range of CNF reached approximately 130–600 nm at this stage. Homogenization with 30 passes produced CNF with diameter ranging from 40–75 nm. The fibrillation became increasingly obvious with subsequent homogenizations at 45 and 60 passes, with the diameter sizes recorded being approximately between 30–50 nm and 20–45 nm, respectively.

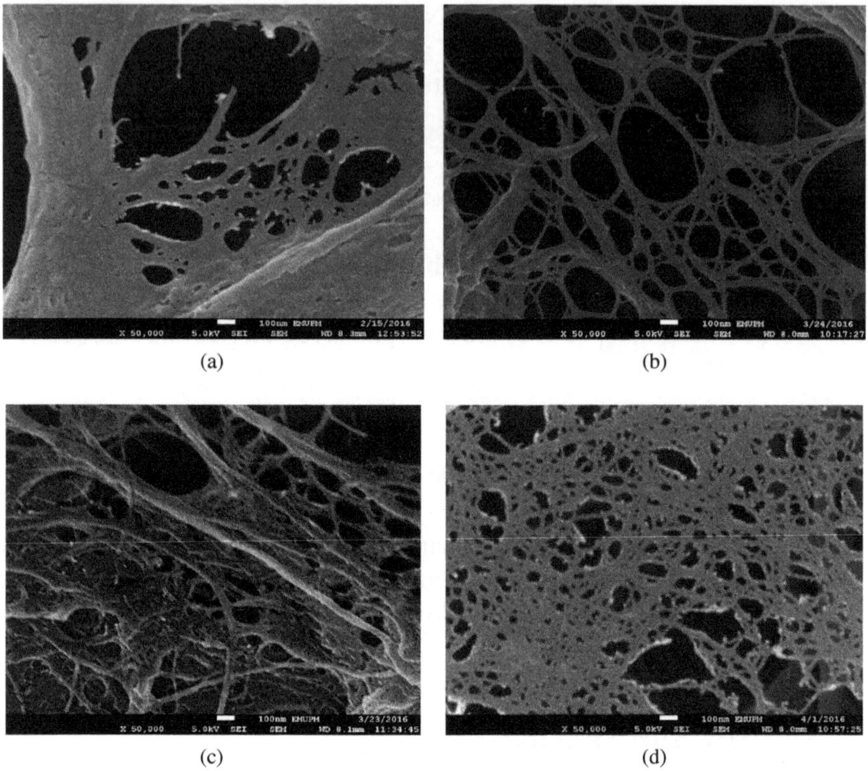

Fig. 1. The FE-SEM images of CNF prepared from OPMF-cellulose which were subjected to (a) 15, (b) 30, (c) 45, and (d) 60 passes; at 50,000× magnification.

Table 1. The average diameter of high-pressure homogenized CNF.

	Size distribution (%)				
Passes	**20–40 nm**	**40–60 nm**	**60–80 nm**	**80–100 nm**	**>100 nm**
0	0.00 ± 0.0	0.00 ± 0.0	0.00 ± 0.0	0.00 ± 0.0	0.00 ± 0.0
15	0.00 ± 0.0	0.00 ± 0.0	0.00 ± 0.0	0.00 ± 0.0	100 ± 2.8
30	0.00 ± 0.0	13.7 ± 4.2	82.9 ± 5.8	3.4 ± 1.1	0.00 ± 0.0
45	11.5 ± 8.6	86.4 ± 6.2	2.1 ± 1.2	0.00 ± 0.0	0.00 ± 0.0
60	95.6 ± 1.5	4.4 ± 3.1	0.00 ± 0.0	0.00 ± 0.0	0.00 ± 0.0

The size reduction was mainly due to the exposure to shear and impact forces during the entire homogenization process, which led to the disintegration of a large bundle of cellulose fibers into individual CNFs. This happens when the cellulose suspension passes into a vessel through a very narrow nozzle at a high pressure. High pressure and impact forces generated shear rate, reducing the size of cellulose fibers into the nanoscale (Nechyporchuk *et al.*, 2016). According to Abdul Khalil *et al.* (2014), hydrogen bonds are broken down during internal nanofibrillation mainly due to mechanical action, while abrasive actions cause external nanofibrillation on the surface of cellulose fiber. In this study, OPMF-cellulose was soaked in distilled water for 48 hours prior to high-pressure homogenization. This enhanced swelling of the OPMF-cellulose significantly eased the disintegration of microfibrils into individual nanofibrils during high-pressure homogenization.

3.1.2 *Crystallinity index*

CNF from OPMF exhibited an increase in crystallinity index compared to OPMF-cellulose after being subjected to 15 passes of homogenization, as shown in Table 2. Davoudpour *et al.* (2015) explains that homogenization which takes place using water as a medium might help in rearranging the crystal part of the fiber, which was disordered because of the increased movement of the fibers in water. Thus, homogenization leads to CNF reorganization in water and eventually increases the crystallinity index. Nevertheless, by increasing the number of homogenization passes, the crystallinity index starts to reduce. Marked reduction was recorded with increasing number of passes. This indicates that an increase in the number of homogenization passes may damage crystalline cellulose partially, as suggested by Du *et al.* (2016), and Saelee *et al.* (2016).

Table 2. Summary of crystallinity index of homogenized CNF.

Samples	**Crystallinity index (%)**
OPMF-cellulose	51
OPMF-cellulose 15 passes	54
OPMF-cellulose 30 passes	52
OPMF-cellulose 45 passes	48
OPMF-cellulose 60 passes	44

3.1.3 *Thermal stability of homogenized CNF*

Thermogravimetric analysis (TGA) was carried out to evaluate the thermal stability and degradation profiles of CNF produced by the high-pressure homogenization method as tabulated in Table 3. Slight weight loss at around 50–100°C, corresponding to the evaporation of water, was observed for all TG curves of the CNF produced by homogenization method (Fig. 2). An increase in the degradation temperature, $T_{d50\%}$, was recorded after OPMF-cellulose was subjected to 15 homogenization passes, and this increase in temperature was from 340°C (before homogenization) to 350°C. This shows that OPMF-CNF (15 passes) had higher thermal stability compared to OPMF-cellulose. Meanwhile, a higher number of homogenization passes led to reduction of the thermal degradation temperature.

Table 3. Summary of thermal degradation of homogenized CNF.

Samples	$T_{d10\%}$ (°C)	$T_{d50\%}$ (°C)
OPMF-cellulose	296	340
OPMF-cellulose 15 passes	268	350
OPMF-cellulose 30 passes	267	347
OPMF-cellulose 45 passes	266	345
OPMF-cellulose 60 passes	256	340

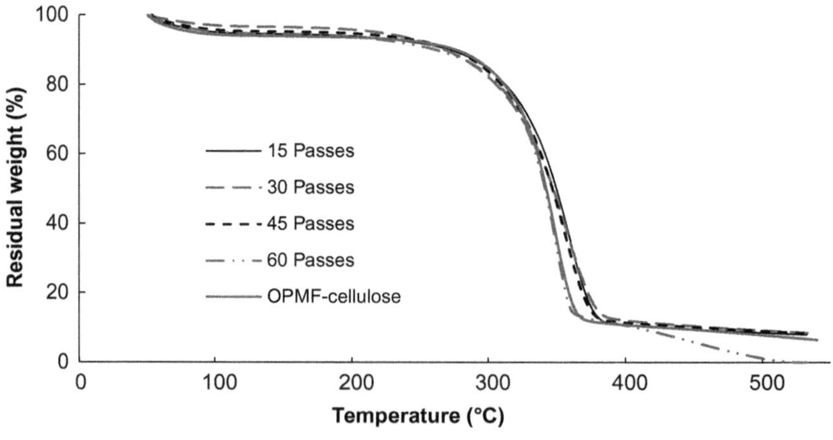

Fig. 2. Thermal stability of OPMF-cellulose and CNF prepared from high-pressure homogenization of OPMF-cellulose subjected to different number of passes.

The trend of CNF thermal stability could be related with its crystallinity index; a reduction in CNF crystallinity due to disruption of crystalline cellulose as a result of prolonged homogenization caused the CNF to have lower thermal stability. As for the OPMF-cellulose, lower thermal stability was recorded due to the presence of the amorphous region, which is more prone to thermal degradation as compared to the crystalline region.

3.2 *Characteristics of Wet Disk Milled CNF*

3.2.1 *Morphology of wet disk milled CNF*

Figure 3 and Table 4 show the morphology and size distribution of OPMF-CNF. It was observed that the OPMF-cellulose started to fibrillate at cycle 6. The degree of nanofibrillation was increased as the WDM cycles increased. At cycle 12, more than 85% of the CNF produced had a diameter less than 40 nm. This is comparable to CNF produced by homogenization, with WDM having an advantage in that it can process cellulose at higher concentration without facing the issue of clogging, unlike HPH.

3.2.2 *Crystallinity of wet disk milled CNF*

The crystallinity index of cellulose and WDM-CNF are shown in Table 5. Original OPMF cellulose had a crystallinity of 51%. Results showed that the crystallinity index of CNF samples decreased as the number of milling cycles increased, finally reaching 30% after 12 cycles of milling. This was due to the fibrillation of the cellulose, which led to the increase in the amorphous region (Iwamoto *et al.*, 2010), and this could be due to the disruption of crystalline region as a result of shear effect.

3.2.3 *Thermal stability of wet disk milled CNF*

Thermorgrams of WDM CNF are shown in Fig. 4. Similar to CNF prepared by HPH, a minor weight loss was found in the range of 50–100°C due to the evaporation of water of the materials or the low molecular weight compounds remaining during the isolation procedures. As can be seen from Table 6, as the WDM cycle increased, the thermal stability based on the $T_{d10\%}$ value of the OPMF CNF decreased. This was due to the increase in the degree of fibrillation and reduction of crystallinity. A more

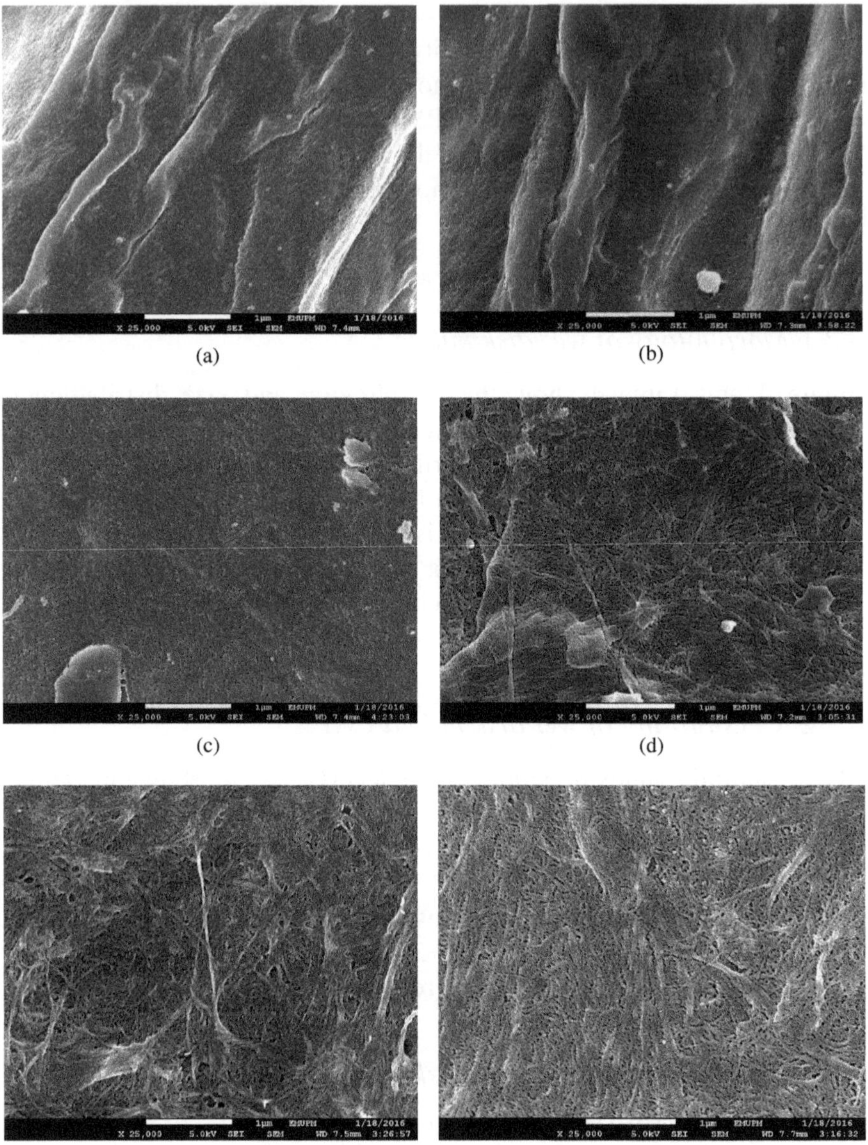

Fig. 3. FE-SEM micrographs (25,000× magnifications) of WDM OPMF-cellulose at (a) cycle 2, (b) cycle 4, (c) cycle 6, (d) cycle 8, (e) cycle 10 and (f) cycle 12.

Table 4. Size distribution of CNF for every WDM cycle.

	Size distribution (%)				
Samples	**20–40 nm**	**40–60 nm**	**60–80 nm**	**80–100 nm**	**>100 nm**
Cycle 4	0.00 ± 0.0	0.00 ± 0.0	0.00 ± 0.0	0.00 ± 0.0	100.00 ± 0.0
Cycle 6	0.00 ± 0.0	0.00 ± 0.0	0.95 ± 1.3	4.73 ± 1.4	94.32 ± 2.2
Cycle 8	0.00 ± 0.0	0.95 ± 1.3	15.88 ± 1.2	18.64 ± 3.1	52.33 ± 1.5
Cycle 10	60.00 ± 2.0	6.64 ± 1.2	2.83 ± 2.3	1.90 ± 1.3	28.60 ± 0.0
Cycle 12	85.28 ± 0.4	7.82 ± 1.2	3.92 ± 2.8	0.95 ± 1.3	2.00 ± 0.0

Table 5. Crystallinity index of the OPMF-cellulose and CNF obtained after WDM.

Samples	Crystallinity index (%)
OPMF-cellulose	51
OPMF-cellulose WDM cycle 2	40
OPMF-cellulose WDM cycle 4	36
OPMF-cellulose WDM cycle 6	34
OPMF-cellulose WDM cycle 8	32
OPMF-cellulose WDM cycle 10	30
OPMF-cellulose WDM cycle 12	30

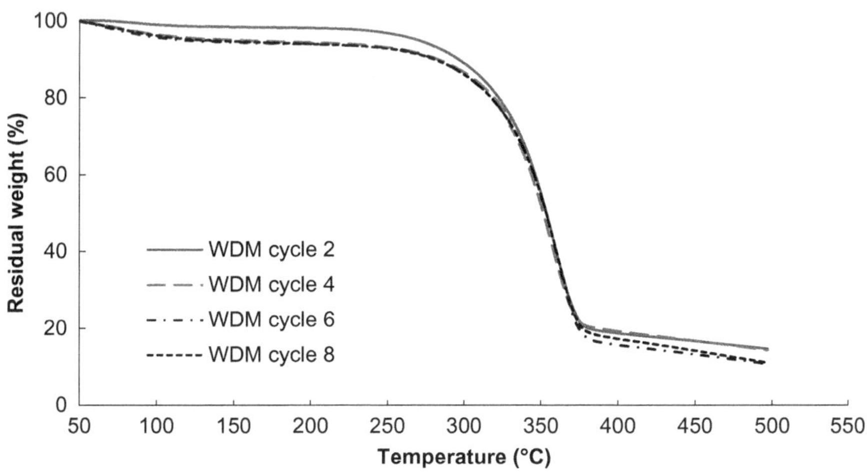

Fig. 4. Thermal stability of wet-disk milled OPMF-cellulose subjected to different number of milling cycles.

Table 6. Degradation temperature of the OPMF-cellulose and CNF obtained after WDM.

Samples	$T_{d10\%}$ (°C)	$T_{d50\%}$ (°C)
OPMF-cellulose	296	358
OPMF-cellulose WDM cycle 2	296	354
OPMF-cellulose WDM cycle 4	282	350
OPMF-cellulose WDM cycle 6	280	350
OPMF-cellulose WDM cycle 8	280	340

amorphous cellulose obtained after fibrillation of the fiber enhanced the decomposition of the cellulose fibrils.

4. Conclusions

CNFs were successfully produced using two different methods, HPH and WDM. The properties of the OPMF-CNF were highly attributed to their processing method. CNF produced by HPH and WDM had diameter size less than 100 nm. In terms of the thermal stability, a similar trend was observed for all prepared OPMF CNFs, with all samples showing a decrease in thermal stability with prolonged exposure to treatment methods, as indicated by number of passes or cycles. This could be related to the crystallinity of the OPMF CNF samples, in which the crystallinity decreased with a higher number of passes/cycles for both HPH and WDM. WDM is slightly more advantageous in that it can produce CNF at higher cellulose concentration without suffering from the issue of clogging. This may indicate the suitability of WDM for producing CNF *via* the mechanical approach, suitable for large-scale processing.

Acknowledgments

We would like to thank the Ministry of Higher Education Malaysia (MOHE) for funding this research through Science and Technology Research Partnership for Sustainable Development (SATREPS) project as well as to Seri Ulu Langat Palm Oil Mill for providing the oil palm mesocarp fiber.

References

Abdul Khalil, H.P.S., Davoudpour, Y., Islam, M.N., *et al.* (2014). Production and modification of nanofibrillated cellulose using various mechanical processes: A review. *Carbohydr. Polym.*, 99: 649–665. doi: 10.1016/j.carbpol.2013.08.069.

Alam, F.A.S.A., Er, A.C., Begum, B., *et al.* (2016). Smallholders the prominent contributor towards sustainable oil palm sector. *Int. J. Adv. Appl. Sci.*, 3(2): 20–24.

Chen, W., Yu, H., Liu, Y., *et al.* (2011a). Isolation and characterization of cellulose nanofibers from four plant cellulose fibers using a chemical-ultrasonic process. *Cellulose*, 18: 433–442. doi: 10.1007/s10570-011-9497-z.

Chen, W., Yu, H., Liu, Y., *et al.* (2011b). Individualization of cellulose nanofibers from wood using high-intensity ultrasonication combined with chemical pretreatments. *Carbohydr. Polym.*, 83: 1804–1811. doi: 10.1016/j.carbpol.2010.10.040.

Chiew, Y.L. and Shimada, S. (2013). Current state and environmental impact assessment for utilizing oil palm empty fruit bunches for fuel, fiber and fertilizer e A case study of Malaysia. *Biomed. and Bioengg.*, 51: 109–124. doi: 10.1016/j.biombioe.2013.01.012.

Chirayil, C.J., Joy, J., Mathew, L., *et al.* (2014). Isolation and characterization of cellulose nanofibrils from Helicteres isora plant. *Ind. Crops. Prod.*, 59: 27–34. doi: 10.1016/j.indcrop.2014.04.020.

Davoudpour, Y., Hossain, S., Khalil, H.P.S.A., *et al.* (2015). Optimization of high pressure homogenization parameters for the isolation of cellulosic nanofibers using response surface methodology. *Ind. Crops. Prod.*, 74: 381–387. doi: 10.1016/j.indcrop.2015.05.029.

Du, H., Liu, C., Zhang, Y., *et al.* (2016). Preparation and characterization of functional cellulose nanofibrils *via* formic acid hydrolysis pretreatment and the followed high-pressure homogenization. *Ind. Crops. Prod.*, 94: 736–745. doi: 10.1016/j.indcrop.2016.09.059.

Iwamoto, S., Abe, K., and Yano, H. (2008).The effect of hemicelluloses on wood pulp nanofibrillation and nanofiber network characteristics. *Biomacromolecules*, 9: 1022–1026. doi: 10.1021/bm701157n.

Iwamoto, S., Kai, W., Isogai, T., *et al.* (2010). Comparison study of TEMPO-analogous compounds on oxidation efficiency of wood cellulose for preparation of cellulose nanofibrils. *Polym. Degrad. Stab.*, 95: 1394–1398. doi: 10.1016/j.polymdegradstab.2010.01.017.

Iwamoto, S., Yamamoto, S., Lee, S.H., *et al.* (2014). Mechanical and thermal properties of polypropylene composites reinforced with lignocellulose nanofibers dried in melted ethylene-butene copolymer. *Materials (Basel)*, 7: 6919–6929. doi: 10.3390/ma7106919.

Nechyporchuk, O., Belgacem, M.N., and Bras, J. (2016). Production of cellulose nanofibrils: A review of recent advances. *Ind. Crops. Prod.*, 93: 2–25.

Nogi, M., Iwamoto, S., Nakagaito, A.N., and Yano, H. (2009). Optically transparent nanofiber paper. *Adv. Mater.*, 21: 1595–1598. doi: 10.1002/adma.200803174.

Nordin, N.I.A.A., Ariffin, H., Andou, Y., *et al.* (2013). Modification of oil palm mesocarp fiber characteristics using superheated steam treatment. *Molecules*, 18: 9132–9146. doi: 10.3390/molecules18089132.

Saelee, K., Yingkamhaeng, N., Nimchua, T., and Sukyai, P. (2016). An environmentally friendly xylanase-assisted pretreatment for cellulose nanofibrils isolation from sugarcane bagasse by high-pressure homogenization. *Ind. Crops. Prod.*, 82: 149–160. doi: 10.1016/j.indcrop.2015.11.064.

Soni, B., Hassan, E.B., and Mahmoud, B. (2015). Chemical isolation and characterization of different cellulose nanofibers from cotton stalks. *Carbohydr. Polym.*, 134: 581–589. doi: 10.1016/j.carbpol.2015.08.031.

Wang, H.Y., Chen, Y.Y., and Zhang, Y.Q. (2015). Processing and characterization of powdered silk micro- and nano-fibers by ultrasonication. *Mater. Sci. Eng. C*, 48: 444–452. doi: 10.1016/j.msec.2014.12.028.

Warid, M.N.M. (2017). Optimization of oil palm biomass superheated steam treatment for improvement of fiber characteristics and polypropylene biocomposite performance. Universiti Putra Malaysia.

Yasim-Anuar, T.A.T., Ariffin, H., Norrahim, M.N.F., and Hassan, M.A. (2017). Factors affecting spinnability of oil palm mesocarp fiber cellulose solution for the production of microfiber. *Bioresources*, 12: 715–734. doi: 10.15376/biores.12.1.715-734.

https://doi.org/10.1142/9781786349477_0003

Chapter 3

Surface Modification of Nanocellulose

Pei Huang[*,†], Chao Wang[*], Yong Huang[*], and Min Wu[*,‡]

[*]*Technical Institute of Physics and Chemistry, Chinese Academy of Sciences, Haidian District, Beijing, China*

[†]*College of Aerospace Engineering, Chongqing University, Chongqing, China*

[‡]*wumin@mail.ipc.ac.cn*

Abstract

Cellulose is the most abundant renewable natural biopolymer on earth and exists in a wide variety of living species including plants, animals and some bacteria [1]. The term "nanocellulose" generally refers to cellulosic materials having at least one dimension in the nanometer range. On the basis of their dimensions, functions, and preparation methods, nanocellulose can be categorized into different groups: cellulose nanocrystals (CNC), microfibrillated cellulose (MFC), cellulose nanofiber (CNF), bacterial cellulose (BC), etc. [2]. The qualities of large surface area, exceptional mechanical properties, surface versatility and biodegradability bring nanocellulose to the forefront as a functional or structural component with uses in a wide range of fields, e.g., composites, drug release, wearable devices. By carefully adjusting interactions with the surrounding medium, we may be able to engineer nanocellulose with unprecedented properties. As a result, the focus of present chapter is the surface modification of nanocellulose by surfactant coupling, co-dispersion, chemical modification and nano-deposition.

Keywords: Nanocellulose; surface modification; grafting; functionalization; applications

1. Surfactant Coupling

Surfactant modification, by its principle, depends on the interactions between different substances. The conventionally produced nanocellulose has abundant hydroxyls or other hydrophilic groups on the surface, resulting in the establishment of strong hydrogen bonds between nanocellulose, thus leading to rapid aggregation, especially in non-polar solvent or polymer matrix. Surfactant coupling, a simple and environmentally friendly approach, forms an "insulting" layer on the surface of nanocellulose through entropic effects, charge–charge interactions, hydrophobic interaction, solvency and hydrogen bonding, and therefore promotes the dispersion of nanocellulose. According to the surface property of nanocellulose, the surfactant used can be divided into non-ionic surfactant, cationic surfactant, anionic surfactant and amphiphilic polymer.

1.1 *Non-Ionic Surfactant*

The adsorption of nonionic surfactants at solid/liquid interfaces is of great practical importance in many fields. Gordeyeva *et al.* [3] used non-ionic surfactant (pluronic P123) to produce high-foamability nanocellulose/P123 foam, in which both the aliphatic and the hydrophilic segments of P123 are likely contributed to the adsorption affinity. The dissolution of $CaCO_3$ nanoparticles triggered by pluronic P123 delays the Ca-induced aggregation of nanocellulose, and therefore significantly improves the long-term stability of the wet composite foams. Torn *et al.* [4] suggested that the adsorption behavior of non-ionic surfactants onto cellulose shows three distinct regions (Fig. 1a). At low surfactant concentration, the headgroup and the tail both contribute to the adsorption. At higher concentration, lateral attraction becomes prominent and leads to the formation of aggregates on the surface. In case of Pluronic F68, Quennouz *et al.* [5] suggested that surfactant below a critical mass concentration produces an increase in the gel modulus with retention of optical clarity. Larger-than-critical concentration induces

(a) (b)

Fig. 1. (a) Adsorption kinetics of non-ionic surfactant (poly(ethylene oxide) alkylethers) near a cellulose surface [4]; (b) schematic illustration (not to scale) of the CNFs bridged by surfactant micelles [5].

significant fibril aggregation, leading to the loss of suspension stability and optical clarity, caused by micelle bridging of fibrils (Fig. 1b).

1.2 *Cationic Surfactant*

Carboxylated nanocellulose and other negatively charged nanocellulose are classical examples of nanoparticles carrying anionic groups on their surfaces. The negative charge facilitates the surface modification of nanocellulose by the adsorption of cationic surfactant *via* ionic bonding. Salajková *et al.* [6], reported that the adsorption of quaternary ammonium salts with long alkyl, phenyl, glycidyl and diallyl groups can modify the surface hydrophobicity of nanocellulose. The modified cellulose nanocrystal can be re-dispersed and individualized in organic solvents such as toluene and chloroform. Syverud *et al.* [7] used *n*-hexadecyl trimethylammonium bromide (CTAB) to enhance the hydrophobicity of carboxylated nanocellulose film, without affecting the tensile strength significantly. Xhanari *et al.* [8] used single- and double-chain cationic surfactants with different chain lengths (cetyltrimethyl-, didodecyl- and dihexadecyl-ammonium bromide) to alter the water wettability of NFC. Dhar *et al.* [9] observed a charge transformation of cellulose nanocrystal from negative to positive values with increasing tetradecyl trimethyl ammonium bromide (TTAB) concentration, confirming the formation of NCC–TTAB complexes in aqueous solutions (Fig. 2a and 2b). In terms of the mechanism, the interaction between cellulose nanocrystal and TTAB displays four distinct states (Fig. 2c). First, the electrostatic interaction drives the

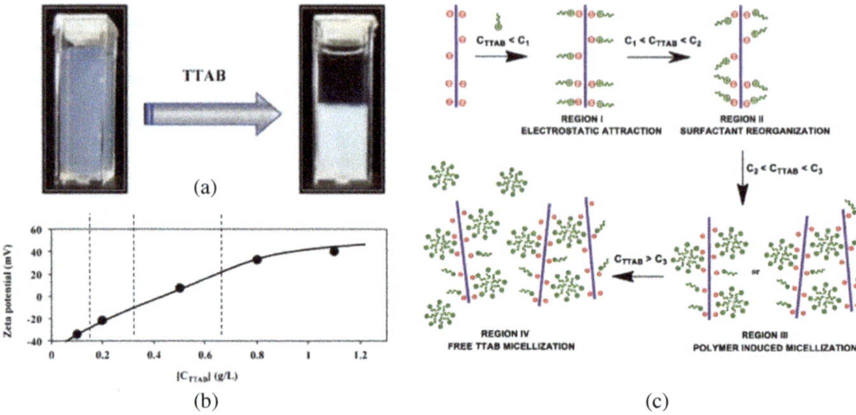

Fig. 2. (a) Transformation of NCC suspension upon the introduction of TTAB; (b) zeta potential of NCC–TTAB system; (c) interaction between NCC and TTAB (C_1 is the onset of reorganization of adsorbed TTAB molecules, C_2 is the onset of aggregation of nanorods, and C_3 is the onset of free TTAB micellization) [9].

TTAB adsorption onto the negatively charged NCC. Increasing the TTAB concentration leads to the reorganization of surfactant molecules on the NCC surface, induced by hydrophobic interactions of surfactant alkyl chains. Further increasing the concentration of surfactant promotes the formation of micelles on the surface of NCC, therefore bridging several NCCs. When the concentration of TTAB surfactant was further increased, free surfactant micelles begin to aggregate, resulting in the precipitation of NCCs [10].

1.3 *Anionic Surfactant*

The use of anionic surfactant to functionalize nanocellulose is attributed to the hydrophobic interactions between the hydrophobic groups of the surfactant and the nanocellulose. Bondeson *et al.* [11] used acid phosphate ester of ethoxylated nonylphenol, an anionic surfactant, to improve the dispersion of cellulose nanowhisker in PLA matrix. The results showed that the addition of surfactant enhanced the tensile strength and elongation at break but degraded the PLA matrix at the same time. Crawford *et al.* [12] used anionic surfactants, i.e., sodium dodecylsulfate (SDS), sodium lauryl ether sulfate (SLES), sodium lauroyl sarconsinate or sodium cocoyl

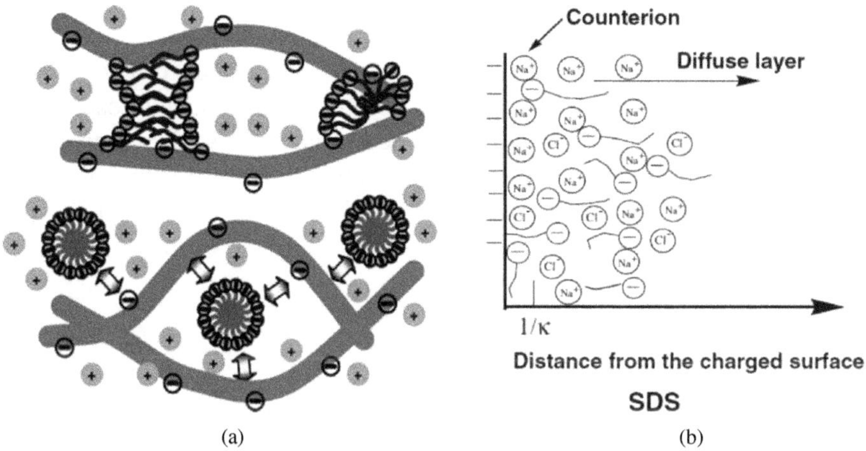

Fig. 3. (a) Illustrations of soft cross-links formed by micelles (top) and fibers forced into close proximity due to repulsion by micelles, known as depletion flocculation (bottom) [12]; (b) screening effect of electrolyte ions onto negatively charged cellulose [15].

sarconsinate within carboxylated cellulose nanofiber suspension to produce shear thinning gel, which could be used in personal care products with a thick consistency. The cross-link within the gel is formed by surfactant micelles interacting directly with the nanofiber, consistent with the strongly shear thinning behavior of these gels (Fig. 3a). Also, acid phosphate ester of ethoxylated nonylphenol (Beycostat A B09), phosphoric ester of poly(ethylene oxide) nonyl phenyl (PEPNP) and sorbitan monostearate (Span-60), etc., were employed to improve the dispersion of nanocellulose nanowhisker in apolar medium [13, 14]. Increasing ionic strength is expected to induce a screening effect in a colloidal system, which results in the reduction of the electrostatic repulsion between the negatively charged surfactant and nanocellulose. Divalent ions (e.g., Ca^{2+}, Al^{3+}, Mg^{2+}) are able to act like bridges Divalent ions can connect the surface of nano-cellulose and anionic surfactants, while they are absorbed to the hydrophobic sites of the nanocellulose in a tail-down configuration, with their head sticking out of the monolayer without the presence of electrolytes (Fig. 3b), increasing the hydrophobicity of nanocellulose [15]. As a result, the formation of interfibrillar cross-linkages *via* divalent ions increases the stiffness of the nanocellulose films and the resistance to humidity as well [16].

1.4 *Amphiphilic Polymer*

The interactions between nanocellulose can be weakened *via* polymer bridging of nanoparticles by adsorbing polymer onto nanocellulose surface *via* the interaction-multiplier effect, where relatively weak segment-level interaction is multiplied by the large number of identical segments in the polymer, making nanocellulose more stable and compatible to the surrounding medium (Fig. 4a) [17]. Wang *et al.* [18] synthesized an

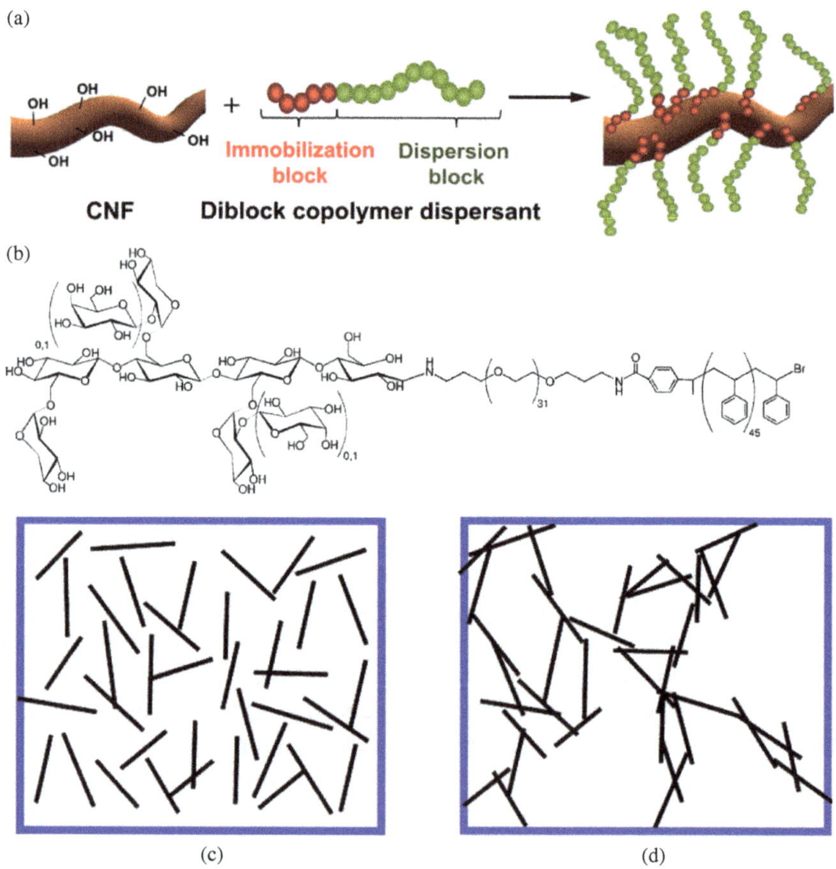

Fig. 4. (a) Schematic of adsorption of polymer dispersant onto (CNF) [17]; (b) structure of the xyloglucan oligosaccharide–poly(ethylene glycol)–polystyrene triblock copolymer [19]; (c–d) illustration of cellulose nanocrystal suspensions: (c) absence of non-adsorbing polymer; (d) percolated cellulose nanocrystal network induced by the depleted non-adsorbing polymer [22].

amphiphilic poly(1,2-butadiene)-block-poly(dimethylaminoethyl methacrylate) block copolymer (PB-b-PDMAEMA) as a dispenser to modify the surface property of CFC. The cationic part of the block copolymer serves as a binder to cellulose nanofiber, while the hydrophobic micellar cores facilitate energy dissipation and nanoscale lubrication between the CNF under deformation. Zhou *et al.* [19] employed xyloglucan oligosaccharide–poly(ethylene glycol)–polystyrene triblock copolymer to assist in the dispersion of cellulose nanocrystals in a nonpolar solvent (Fig. 4b). As a result, they found that cellulose nanocrystals can be easily re-dispersed in the desired amount of toluene following ultrasonic treatment for a few tens of seconds. In addition, diblock copolymer poly(lauryl methacrylate)-block-poly(2-hydroxyethyl methacrylate) (PLMA-b-PHEMA) [17], poly(amideamine) epichlorohydrin (PAE) [20], 3–6 ionene (Polybrene) with quaternary cationic nitrogen in polymer chain and polydimethyl diallylammonium chloride (PDMAC) [21] have been used to promote the dispersion of nanocellulose in an apolar resin without chemical modification.

Regarding to the non-adsorbing polymer, the exclusion of the polymer segments from the particulate volumes leads to an effective "entropic" attraction between nanocellulose, known as the depletion interaction, leading to the flocculation of nanocellulose [23]. Boluk *et al.* [22] indicated that non-adsorbing polymer hydroxyethyl cellulose (HEC) and carboxymethyl cellulose (CMC) caused depletion-induced interaction among NCC particles, leading to a weak gel structure. Such weak interactions decrease the accessible volume for polymer chains in the solution, thus leading to higher apparent polymer concentrations in flocculated NCC-free regions (Fig. 4c–4d).

2. Co-Dispersion

One of drawbacks of surfactant coupling is that the thermal and mechanical properties of the composites produced deteriorate when a high amount of surfactant is used to keep the nanofillers stable and ensure they remain in the final polymer composites [24, 25]. The introduction of extra nanomaterials not only improves the dispersion of nanocellulose but also endows it with other functionalities, such as thermal, mechanical, and barrier properties, etc. Zhu *et al.* [26] reported that 2-D boron nitride (BN) nanosheets can be stabilized by cellulose nanofibers. The nanofilm prepared shows superior thermal and mechanical properties, caused by

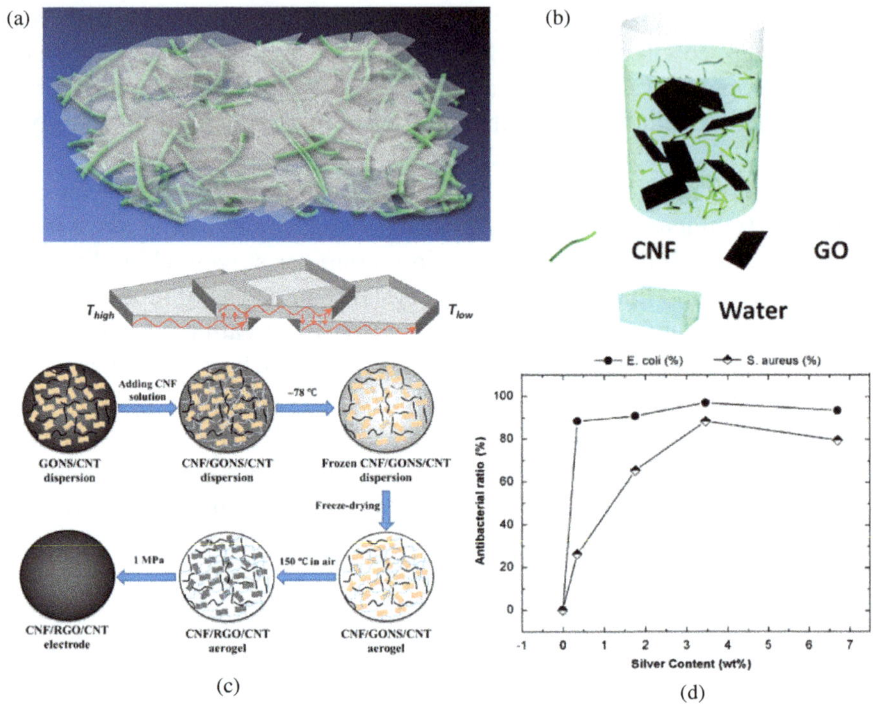

Fig. 5. (a) Schematic to show the heat transfer along the horizontal plane direction in a hybrid of cellulose nanofiber/boron nitride nanosheets [26]; (b) illustration of the RGO/CNF suspension [27]; (c) fabrication process of a CNF/RGO/CNT electrode [28]; (d) antibacterial property of PVA/CNs/AgNPs composites to *E. coli* and *S. aureus* [29].

two-dimensional (2D) nanosheets forming a thermally conductive network and NFC providing mechanical strength (Fig. 5a). Yang *et al.* [27] used graphene oxide (GO) to aid the dispersion of cellulose nanofibrils *via* hydrogen interaction (Fig. 5b). The reduced GO/cellulose nanofiber composite obtained had excellent EMI shielding performance and strong anisotropy of thermal conductivity. Zheng *et al.* [28] reported that highly flexible cellulose nanofibril (CNF)/reduced graphene oxide (RGO)/carbon nanotube (CNT) electrodes showed high specific capacitance (i.e., 252 Fg^{-1} at a discharge current density of 0.5 Ag^{-1}) and remarkable cycle stability, with the cellulose nanofibril not only serving as a binder but also providing the diffusion channels for electrolytes (Fig. 5c). Xu *et al.* [29] incorporated Ag nanoparticles (NPs) into a cellulose

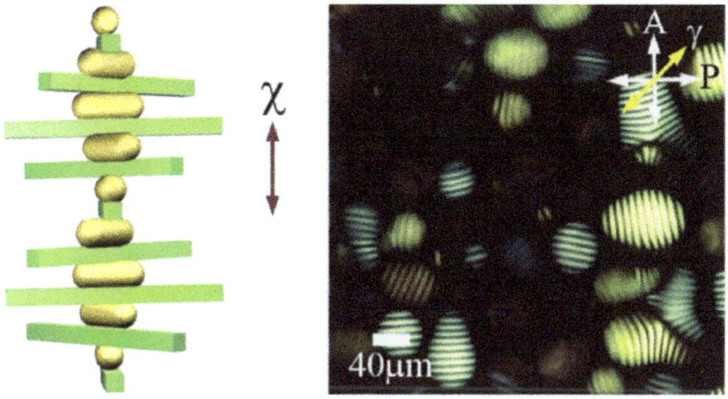

(a) (b)

Fig. 6. (a) Schematic of helicoidal assembly of GNRs in a cholesteric dispersion of CNCs; (b) POM image of continuous domains of GNRs in cholesteric CNCs with a full-wavelength (530 nm) retardation plate with a slow axis (γ) marked by yellow arrow [32].

nanocrystals aqueous system to produce homogenous PVA composite with enhanced mechanical property and antibacterial activities (Fig. 5d).

More interestingly, nanocellulose, including cellulose nanowhiskers, and other stable dispersed nanocellulose forms, can self-organize into nematic and cholesteric structures [30]. The nematic or cholesteric order can be preserved even after the removal of water, allowing the long-range orientational order of well-dispersed nanomaterials compared with their surfactant-based dispersion [31]. Liu *et al.* [32] demonstrated the nematic-like and helicoidal orientational self-assemblies of gold nanorods co-dispersed with cellulose nanocrystal form into liquid crystalline (LC) phases (Fig. 6). The phase behavior of LC aqueous solution can be controlled by varying sample preparation conditions, such as surface charge and concentration.

3. Chemical Modification

There is one drawback for the physical modification of nanocellulose, which it may release the adsorbed surfactant or deteriorate the functioning of the resulting materials. In contrast, surface modification *via* chemical bonding is thought to be a versatile approach to functionalize nanocellulose due to the presence of enriched hydroxyls and other active groups on the surface. In this section, chemical reactions, including oxidation, small molecule substitution, polymer grafting, and surface coating are addressed.

3.1 *Oxidation*

The polyhydric alcohol structure makes cellulose very sensitive to various oxidizing reagents. Usually, the oxidation of cellulose involves the primary and secondary hydroxyl groups of the glucopyranoside ring, giving rise to carboxylic acid or aldehyde functionalities. There are mainly three types of oxidation that occur in nanocellulose, including periodate oxidation, 2,2,6,6-tetramethyl-1-piperidinyloxy (TEMPO) oxidation, and ammonium persulfate (APS) oxidation.

3.1.1 *Periodate oxidation*

Periodate oxidation is a highly selective reaction to cleaves the C2–C3 bond of the glucopyranoside ring and introduces two aldehyde groups, which is beneficial for the further modification (Fig. 7) [34, 35]. Sun *et al.* [36] introduced an aldehyde group to cellulose nanocrystal *via* oxidation by sodium periodate. Reaction conditions, including oxidant dose, pH, temperature and reaction time, play significant roles in the oxidized nanocellulose yield and the aldehyde content. The efficiency of the periodate oxidation reaction can be further improved by using cellulose activators (e.g., LiCl, NaCl, $MgCl_2$ and other metal chlorides) that disrupt the inter and intra molecular hydrogen bonds between the cellulose chains, making the periodate oxidation more efficient and more environmentally friendly [37, 38]. Kim *et al.* [39] suggested that the crystallinity of cellulose oxidized by periodate oxidation decreased with increasing oxidation level and the dialdehyde groups were vert unevenly distributed on longitudinal

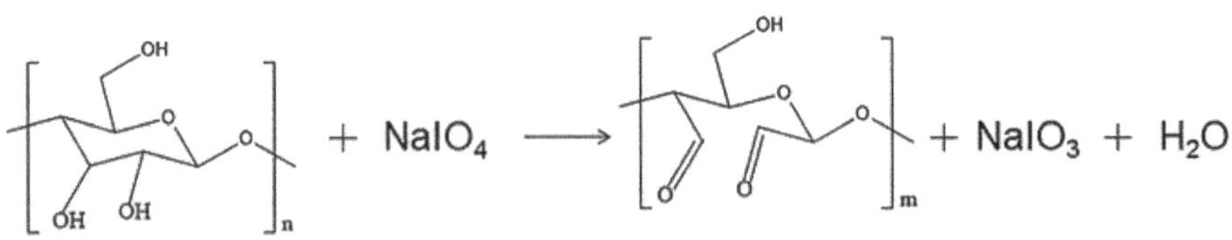

Fig. 7. Periodate oxidation of cellulose to produce dialdehyde modified cellulose [33].

axis of cellulose fibers. The periodate oxidized aldehyde group of nanocellulose can evolve into a carboxyl group by a subsequent oxidation with sodium chlorite at pH 5.0 [40].

3.1.2 *TEMPO oxidation*

The utilization of TEMPO to introduce carboxyl groups onto cellulose has been extensively studied. Nooy *et al.* [44, 45], first applied the TEMPO/NaBr/NaClO system to convert the C6 primary hydroxyl groups of cellulose to a carboxyl group and synthesized water-soluble polysaccharides, with hypochlorite being the primary oxidant and bromide acting as a cocatalyst. The reaction exhibited much higher selectivity to primary alcohols at $9 < pH < 11$ than at $pH < 8$ [46]. However, the oxidized products from native cellulose are not completely soluble in water, even extending the oxidation time, because of the high crystallinity and low accessibility of cellulose to the reagents [47, 48]. Therefore, nanocelluloses, including BC, NCC, and NFC, were subjected to TEMPO oxidation to produce carboxylated nanocellulose (Fig. 8a) [41, 49]. The DP of nanocellulose suffers a significant decrease from more than 1000 to approximately 200 within a short time when the TEMPO/NaBr/NaClO system is maintained at a pH 9–11 [50]. To avoid substantial degradation during the oxidation and discoloration during heating, the TEMPO/NaClO/$NaClO_2$ oxidation system was developed [43]. As shown in Fig. 8b, NaClO oxidizes 4-acetamide-TEMPO to the corresponding *N*-oxoammonium ion and results in the rapid oxidization of primary hydroxyl to aldehyde under acidic conditions, forming the hydroxylamine. Furthermore, the aldehyde is oxidized to carboxyl by the primary oxidant $NaClO_2$, forming NaClO, and then the hydroxylamine is oxidized to the *N*-oxoammonium ion again by the NaClO thus generated [42]. Still, the pH of the reaction medium is a vital factor for the oxidation of the primary hydroxyls to carboxyl groups. The oxidation efficiently proceeded at pH 6.8, while nearly no carboxyl groups were formed at pH 3.5 (Fig. 8c).

3.1.3 *APS oxidation*

The oxidation by APS can also transform hydroxyls of cellulose into carboxyls. ReAlfred *et al.* [51] prepared the carboxylated CNCs through heating cellulosic materials at 60°C in ammonium persulfate solution under acidic condition with vigorous stirring, by which free radicals

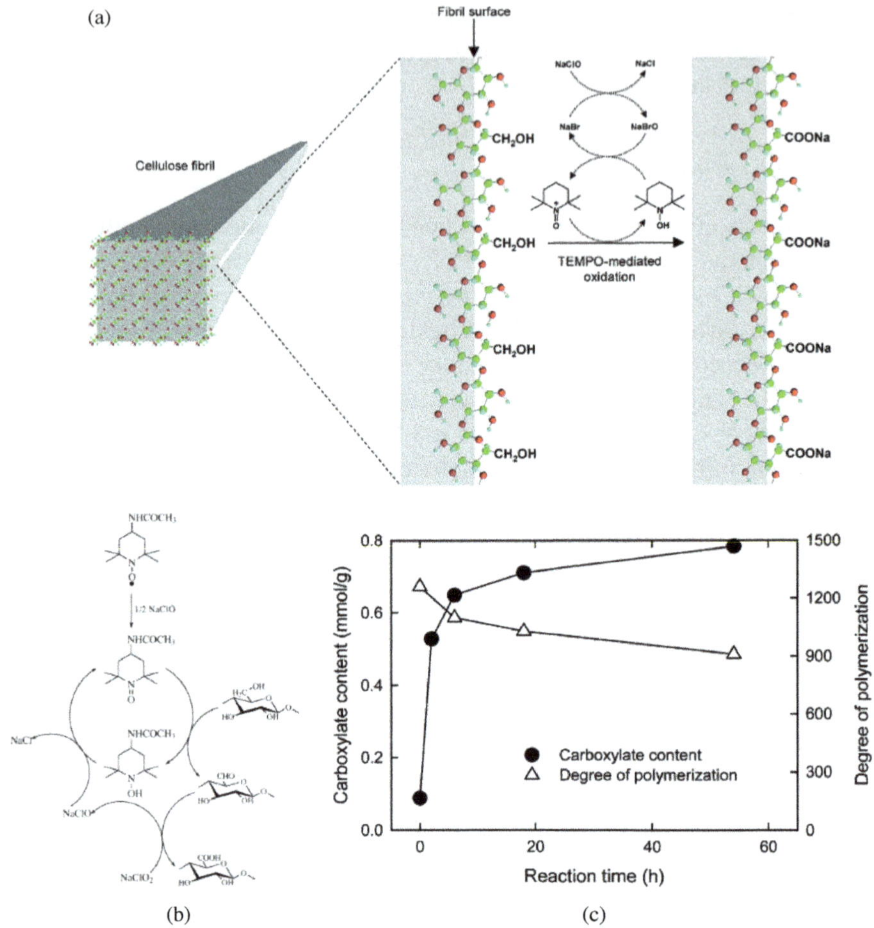

Fig. 8. (a) Schematic representation of the oxidation of C6 primary hydroxyl on cellulose microfibril by TEMPO/NaClO/NaBr system [41]; (b) catalytic cycle of the TEMPO/ NaClO/$NaClO_2$ system for oxidation of cellulose [42]; (c) carboxylate content of TEMPO-oxidized celluloses prepared by the TEMPO/NaClO/$NaClO_2$ system under different conditions [43].

formed due to heating ($S_2O_8^{2-} + \text{heat} \rightarrow 2SO_4^{\bullet -}$) and H_2O_2 formed under acidic conditions ($S_2O_8^{2-} + 2H_2O \rightarrow 2HSO_4^- + H_2O_2$), capable of destroying the amorphous cellulose to release nanocellulose. The reaction time varied depending on the source of cellulose materials, ranging from 3 hours for bacterial cellulose to 16 hours for complex substrates such as

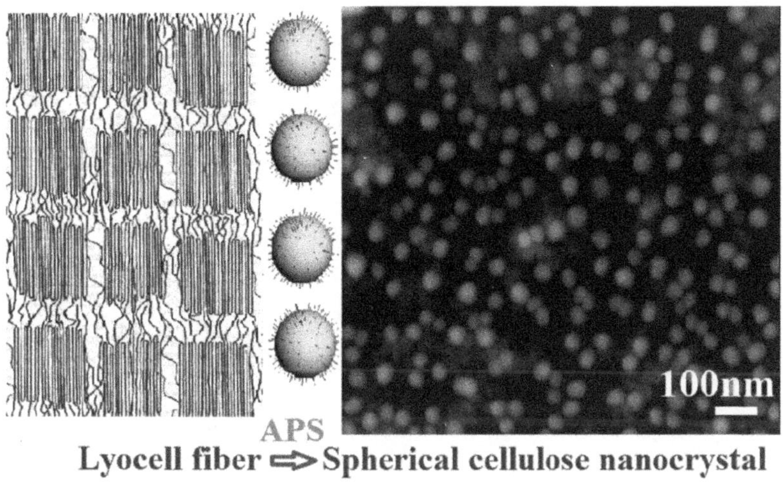

Fig. 9. Formation of spherical cellulose nanoparticles by APS oxidation [52].

hemp and flax. With the increase of reaction time, the carboxylated nanocellulose evolves into spherical nanoparticles with a more uniform diameter distribution Fig. 9 [52]. The oxidation level, viscosity, and transmittance of cellulose nanofibrils are comparable with TEMPO-oxidized nanocellulose [53]. This method avoids the difficulties of using and recycling concentrated sulfuric acid and other highly toxic agents.

3.2 *Esterification*

The most common esterification process for nanocellulose is sulfuric acid hydrolysis, which was first introduced by Rånby [54]. In this process, sulfuric acid breaks down the amorphous region and leaves rod-like crystallized fragments. At the same time, the esterification of hydroxyl groups by sulfate ions leads to negatively charged sulfate groups, thus leading to a stable dispersion of cellulose nanoparticles in water. The esterification levels of nanocellulose highly depend on H_2SO_4 treatment conditions [55, 56]. However, the introduction of sulfate groups causes a significant decrease in degradation temperatures by lowering the activation energy, which in turn increases the strength of the desulfation methods (such as acid and solvolytic desulfation), leading to a reduction in the sulfate esterification levels [57].

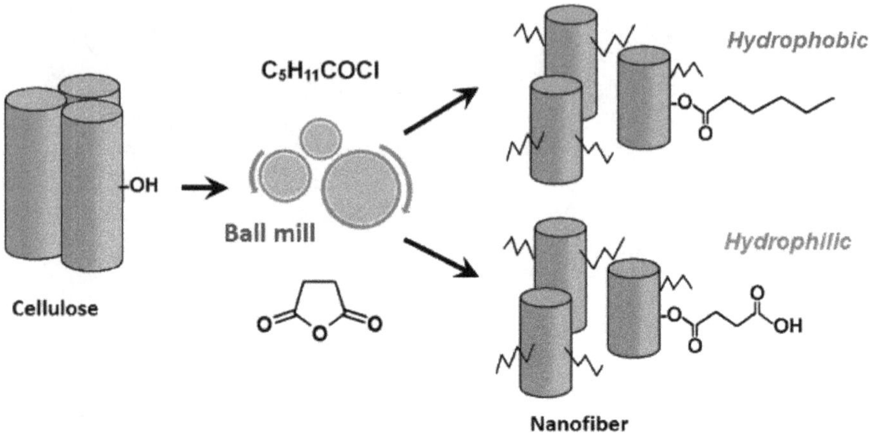

Fig. 10. Surface manipulation of cellulose nanofiber by anhydride and chloride [58].

To delicately manipulate the surface property of nanocellulose, esterifying modifiers, such as anhydride, organic acid, and isocyanate, have been used to carry out esterification on the surface (Fig. 10). Lin *et al.* [59] introduced a surface acetylation of cellulose nanocrystals by reacting with acetic anhydride. The acetylated NCC produced exhibited improved dispersion in various organic solvents and reduced polarity compared with unmodified cellulose nanocrystals, and therefore showed uniform dispersion of the acetylated NCC in poly(lactic acid) matrix. However, due to their lower reactivity, catalysts (e.g., perchloric acid, propionic acids, dimethylaminopyridine (DMAP), iodine and pyridine) and special treatments (e.g., ultrasonic, high temperature and microwave) have been applied to activate the esterification between esterifying agents and nanocellulose [59–67]. Other than the hydrophobic surface, the negatively or positively charged surface can be obtained by using succinic anhydride, dicarboxylic acids and so on [63, 67, 68]. Although chloride can directly react with the hydroxyl of nanocellulose at room temperature under anhydrous conditions, its high toxicity and the production of HCl limit its application in the surface modification of nanocellulose [58, 69]. Moreover, the enzyme has employed to catalyze the phosphorylation of CNFs with adenosine-5-triphosphate in the presence of Mg-ions, resulting in the predomination of phosphate groups at the C-6-O position hydroxyl groups of the cellulose monomer rings [70].

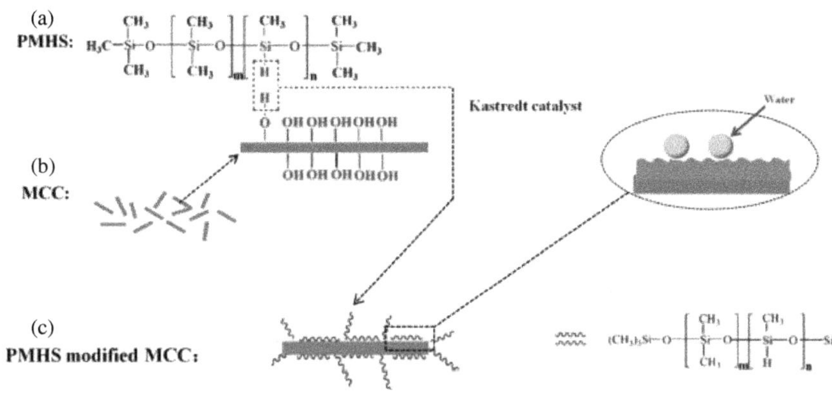

Fig. 11. (a) Schematic representation of PMHS, (b) MCC and (c) PMHS-modified MCC [72].

Silane-coupling agents are very reactive to hydroxyl groups, and thus readily react with hydroxyl groups presenting on surface of the nanocellulose, undergoing stages of hydrolysis, condensation, and bond formation. Grunert *et al.* [71] conducted a trimethylsilylation reaction onto the acid-hydrolyzed bacterial cellulose with hexamethyldisilazane in formamide, achieving an average substitution of 0.49. The silylated BC prepared had an improved compatibility with cellulose acetate butyrate, and therefore enhanced the mechanical characteristics of the composites. Yang *et al.* [72] presented a method for hydrophobic modification of MFC surface using poly(methylhydro)siloxane (PMHS) in the presence of catalyst (Fig. 11). Under the optimum reaction conditions, the water contact angle of modified MFC was 141.5°. Also, isocyanatepropyltriethoxysilane, alkyldimethylchlorosilanes, and γ-methacryloxypropyltrimethoxysilane have been employed to modify nanocellulose because of the advantages of rapid reaction at room temperature and low cost of the modifiers [73–75]. However, there is one disadvantage that the harsh silylation conditions lead to the partial solubilization of nanocellulose and a loss of the nanofiber character [76].

The amidation of nanocellulose involves amino and hydroxyl or carboxyl groups, and thus could be applied for further functionalization of nanocellulose. Huang *et al.* [77] reported the fluorescent labeling of nanocrystalline cellulose with 7-hydrazino-4-methylcoumarin (HMC) and 7-amino-4-methylcoumarin (AMC). The fluorescent tagged CNC is expected to have potential applications in determining the location of CNC in sensor systems. Barazzouk *et al.* [78] chemically coupled carboxylated CNC with *N*-ethyl-*N′*-(3-dimethylaminopropyl) carbodiimide hydrochloride to form a stable active ester in the presence of *N*-hydroxysuccinimide. Then the active ester was reacted with the amino group of the amino acid or peptide, forming an amide bond between CNC and the grafted molecule, achieving uniform coupling of these molecules on the CNC.

3.3 *Polymer Grafting*

Polymer grafting can result in steric hindrances that prevent the aggregation of nanomaterials by the polymer chains extending into the medium. In general, steric stabilization has several advantages over electrostatic stabilization, such as the following: (a) resistance to ionic strength, (b) higher solid contents achieved, and (c) surface property adjustment. Araki *et al.* [79], grafted the carboxylation of cellulose microcrystals with a single terminally aminated PEG (PEG-NH_2) by using a water-soluble carbodiimide, 1-ethyl-3-(3-dimethylaminopropyl)carbodiimide hydrochloride (EDC) (Fig. 12a). The PEG-grafted CNC showed drastically enhanced dispersion stability, that is, resistance to addition of 2 M sodium chloride, and ability to re-disperse into either water or chloroform from the freeze-dried state. Meanwhile, the concentrated aqueous suspension of PEG-grafted CNC formed a chiral nematic mesophase through a phase separation similar to that of the ungrafted sample. Ring-opening polymerisation of ε-CL from MFC in the presence of benzyl alcohol produces MFC-PCL dispersible in nonpolar organic solvents (Fig. 12b) [80]. Surface-grafted waterborne polyurethane (WPU) can form crystalline domains on the surface of NCCs, which induces the formation of a co-continuous phase between the matrix and filler, and therefore improves the thermal stability and mechanical strength of the composites [81]. The introduction of thermo-sensitive amine-terminated oligomers of poly(*N*-Isopropylacrylamide) (Pnipam_{2500}) onto the surface of cellulose nanocrystals leading to the steric stabilization of the NCC-g-Pnipam_{2500} in high ionic strength, and also enabled the realization of a thermo-reversible suspension [82]. Chitosan was grafted onto NCC *via* the carbodiimide reaction using *N*-hydroxysuccinimide and

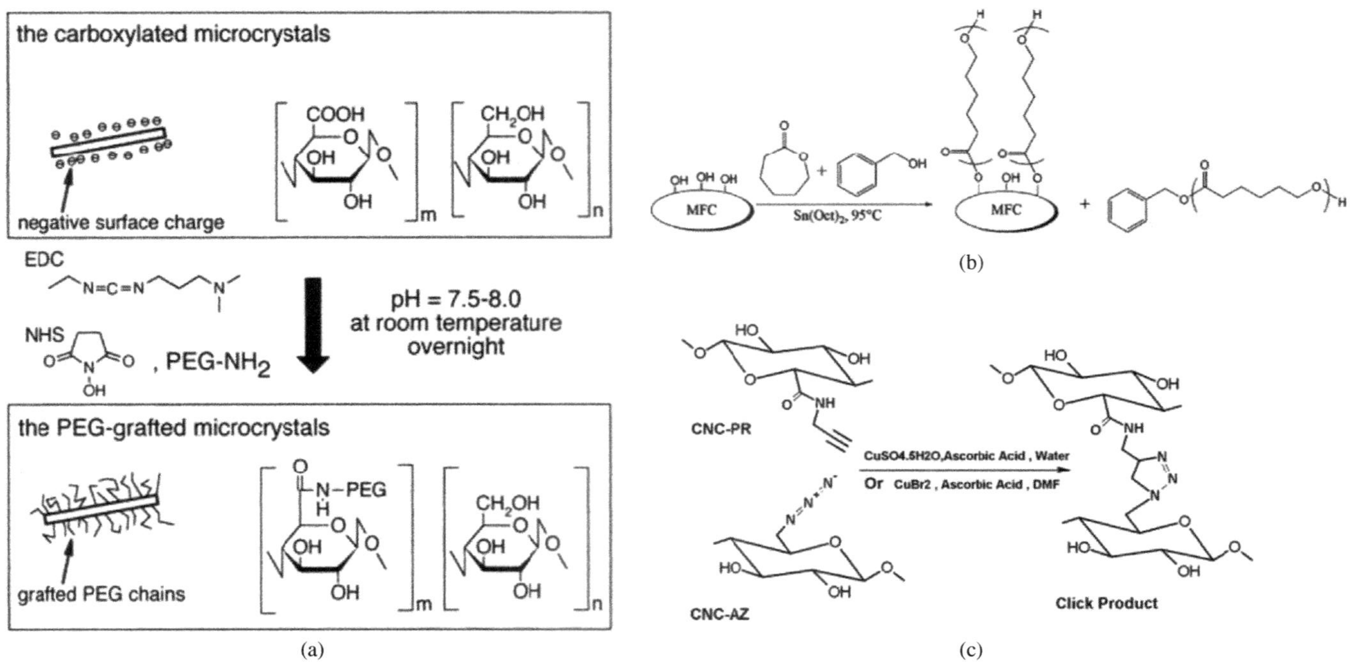

Fig. 12. (a) Illustration of the grafting of poly(ethylene glycol) with a terminal amino group by using water-soluble carbodiimide onto NCCs [79]; (b) ring-opening polymerization of ε-CL from MFC in the presence of benzyl alcohol as co-initiator [80]; (c) schematic of the click reaction between CNC-AZ and CNC-PR [84].

1-ethyl-3-(3-dimethylaminopropyl)-carbodiimide as coupling agents, donating the NCC a biocompatible and biodegradable drug carrier for transdermal drug delivery applications [83]. Moreover, click chemistry was employed to functionalize NCC by alkyne and azide precursors, creating a highly crystallized nanoplatelet with transverse dimensions ranging between 7–8 nm and lengths between 100 and 200 nm (Fig. 12c) [84].

3.4 *Nano-Deposition*

Possessing hydroxyls or other functionalities, cellulose can react with inorganic metal groups (Ti^{4+}, Zn^{2+}, Au^{2+}, Ag^{+}, Sn^{2+}, $Fe^{2+/3+}$, etc.) or even

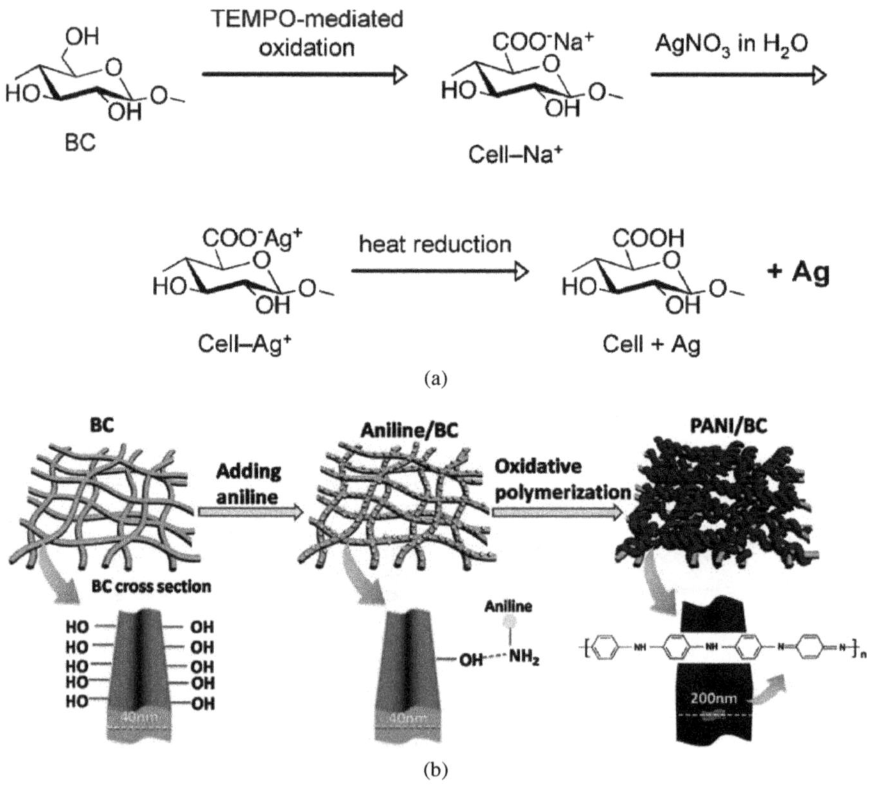

Fig. 13. (a) Synthesis of silver nanoparticles from bacterial cellulose nanofibers [85]; (b) schematic diagram of the formation of PANI/BC nanocomposites [86].

non-metal cations such as Si^{4+} (Fig. 13a). Besides low cost and high flexibility, additional specific properties such as high surface area, high chemical or biological reactivity, and, in some cases, high porosity promotes the use of nanocellulose as a template to deposit various inorganic nanomaterials for the purpose of catalytic and sensing applications (Table 1).

Table 1. Deposition of nanoparticles on the surface of nanocellulose.

Materials	Nanocellulose	Diameter	Method	Application	Refs.
Se	H_2SO_4-hydrolyzed cellulose nanowhisker	10–20 nm	Hydrothermal reduction		[88]
Au	Poly(diallyldimethyl ammonium chloride) -coated carboxylated nanocellulose	2.95 ± 0.06 nm	Physical absorption	Catalyst	[87]
Ag	Bacterial cellulose	10–30 nm	*In situ* reduction	Drug release, sensor	[90, 91]
Pb	HCl-hydrolyzed cellulose nanowhisker	3.6 ± 0.8	Reduction	Catalyst	[92]
Ni	H_2SO_4-hydrolyzed cellulose nanowhisker	5–12 nm	Thermal reduction		[93]
Au–Ag alloy	H_2SO_4-hydrolyzed cellulose nanowhisker	3–7 nm	Hydrothermal reduction		[94]
Pt	H_2SO_4-hydrolyzed cellulose nanowhisker	1–2 nm	Hydrothermal reduction	Catalyst	[95]
CdS	Bacterial cellulose	10–20 nm	Hydrothermal reduction	Photocatalyst	[96]
$Fe(OH)_3$	TEMPO-mediated nanocellulose	50–200 nm	Precipitation reaction	Water treatment	[97]
TiO_2	Bacterial cellulose	4.3–8.5 nm	Hydrothermal reduction or Surface sol–gel	Photocatalyst	[98, 99]
SiO_2	H_2SO_4-hydrolyzed cellulose nanowhisker	13 ± 3 nm	Sol–gel mineralization		[100]

General synthesis routes are based on a multistep chemical process involving hydrolysis, polycondensation, and perhaps nucleation and growth of inorganic species. Lam *et al.* [87] indicated that the Au nanoparticles deposited onto poly(diallyldimethyl ammonium chloride) (PDDA)-coated carboxylated nanocellulose exhibited comparable reaction rate to the benchmark literature value obtained using AuNPs decorated on a network of crystalline cellulose fibers. Shin *et al.* [88] reported that Selenium nanoparticles of 10–20 nm in diameter have been prepared using nanocellulose as a reducing and structure-directing agent under hydrothermal conditions. The embedding of magnetic nanoparticles (Fe_3O_4NPs) and Au nanoparticles on cellulose nanocrystals (CNCs) was used as a magnetic support for the covalent conjugation of papain and facilitated recovery of this immobilized enzyme [89]. Hu *et al.* [86] synthesized conductive polyaniline/bacterial cellulose (PANI/BC) nanocomposite by *in situ* polymerization of aniline onto BC (Fig. 13b). Results indicated that the composites exhibited excellent electrical conductivity (the highest value was 5.0×10^{-2} S/cm) and good mechanical properties (Young's modulus was 5.6 GPa and tensile strength was 95.7 MPa), which could be applied in sensors, flexible electrodes, and flexible displays.

4. Conclusions

Nanocelluloses, including nanofibrillated cellulose, cellulose nanocrystals, and bacterial cellulose, have been extensively studied for various applications, such as in the pharmaceutical, biomedical, paper, cosmetic and food industries. However, the inherent hydrophilic surface property makes nanocellulose tend to aggregate, especially in an apolar medium, hence requiring the surface modification of nanocellulose to adapt to the surrounding medium. The current chapter provides a general overview of physical modifications, including surfactant coupling and co-dispersion, *via* entropic effects, charge–charge interactions, hydrophobic interaction, solvency, and hydrogen bonding. In comparison, chemical modifications, such as oxidation, esterification, polymer grafting, and nano-deposition, cause the stable and permanent dispersion of nanocellulose in matrix or solvent, which inevitably increase the cost and lead to environmental problems. Further attention focusing on the development of innovative surface manipulation of nanocellulose in a simplified and low-cost way is now imperative to amplify their use in other fields.

References

[1] H.P.S. Abdul Khalil, Y. Davoudpour, M.N. Islam, A. Mustapha, K. Sudesh, R. Dungani, and M. Jawaid, Production and modification of nano fibrillated cellulose using various mechanical processes: A review. *Carbohydrate Polymers*, 99 (2014) 649–665.

[2] D. Klemm, F. Kramer, S. Moritz, T. Lindström, M. Ankerfors, D. Gray, and A. Dorris, Nanocelluloses: A new family of nature-based materials. *Angewandte Chemie International Edition*, 50(24) (2011) 5438–5466.

[3] K.S. Gordeyeva, A.B. Fall, S. Hall, B. Wicklein, and L. Bergström, Stabilizing nanocellulose-nonionic surfactant composite foams by delayed Ca-induced gelation. *Journal of Colloid and Interface Science*, 472 (2016) 44–51.

[4] L.H. Torn, L.K. Koopal, A. de Keizer, and J. Lyklema, Adsorption of nonionic surfactants on cellulose surfaces: Adsorbed amounts and kinetics. *Langmuir*, 21(17) (2005) 7768–7775.

[5] N. Quennouz, S.M. Hashmi, H.S. Choi, J.W. Kim, and C.O. Osuji, Rheology of cellulose nanofibrils in the presence of surfactants. *Soft Matter*, 12(1) (2016) 157–164.

[6] M. Salajková, L.A. Berglund, and Q. Zhou, Hydrophobic cellulose nanocrystals modified with quaternary ammonium salts. *Journal of Materials Chemistry*, 22(37) (2012) 19798–19805.

[7] K. Syverud, K. Xhanari, G. Chinga-Carrasco, Y. Yu, and P. Stenius, Films made of cellulose nanofibrils: Surface modification by adsorption of a cationic surfactant and characterization by computer-assisted electron microscopy. *Journal of Nanoparticle Research*, 13(2) (2011) 773–782.

[8] K. Xhanari, K. Syverud, G. Chinga-Carrasco, K. Paso, and P. Stenius, Reduction of water wettability of nanofibrillated cellulose by adsorption of cationic surfactants. *Cellulose*, 18(2) (2011) 257–270.

[9] N. Dhar, D. Au, R.C. Berry, and K.C. Tam, Interactions of nanocrystalline cellulose with an oppositely charged surfactant in aqueous medium, *Colloids and Surfaces A: Physicochemical and Engineering Aspects*, 415 (2012) 310–319.

[10] Y. Lapitsky, M. Parikh, and E.W. Kaler, Calorimetric determination of surfactant/polyelectrolyte binding isotherms. *The Journal of Physical Chemistry B*, 111(29) (2007) 8379–8387.

[11] D. Bondeson and K. Oksman, Dispersion and characteristics of surfactant modified cellulose whiskers nanocomposites. *Composite Interfaces*, 14(7–9) (2007) 617–630.

[12] R.J. Crawford, K.J. Edler, S. Lindhoud, J.L. Scott, and G. Unali, Formation of shear thinning gels from partially oxidised cellulose nanofibrils. *Green Chemistry*, 14(2) (2012) 300–303.

[13] L. Heux, G. Chauve, and C. Bonini, Nonflocculating and chiral-nematic self-ordering of cellulose microcrystals suspensions in nonpolar solvents. *Langmuir*, 16(21) (2000) 8210–8212.
[14] J. Kim, G. Montero, Y. Habibi, J.P. Hinestroza, J. Genzer, D.S. Argyropoulos, and O.J. Rojas, Dispersion of cellulose crystallites by nonionic surfactants in a hydrophobic polymer matrix. *Polymer Engineering and Science*, 49(10) (2009) 2054–2061.
[15] S.K. Obendorf, H. Liu, K. Tan, M.J. Leonard, T.J. Young, and M.J. Incorvia, Adsorption of aroma chemicals on cotton fabric in different aqueous environments. *Journal of Surfactants and Detergents*, 12(1) (2009) 43–58.
[16] M. Shimizu, T. Saito, and A. Isogai, Water-resistant and high oxygen-barrier nanocellulose films with interfibrillar cross-linkages formed through multivalent metal ions. *Journal of Membrane Science*, 500 (2016) 1–7.
[17] K. Sakakibara, H. Yano, and Y. Tsujii, Surface engineering of cellulose nanofiber by adsorption of diblock copolymer dispersant for green nanocomposite materials. *ACS Applied Materials & Interfaces*, 8(37) (2016) 24893–24900.
[18] M. Wang, A. Olszewska, A. Walther, J.-M. Malho, F.H. Schacher, J. Ruokolainen, M. Ankerfors, J. Laine, L.A. Berglund, M. Österberg, and O. Ikkala, Colloidal ionic assembly between anionic native cellulose nanofibrils and cationic block copolymer micelles into biomimetic nanocomposites. *Biomacromolecules*, 12(6) (2011) 2074–2081.
[19] Q. Zhou, H. Brumer, and T.T. Teeri, Self-Organization of cellulose nanocrystals adsorbed with xyloglucan oligosaccharide–Poly(ethylene glycol)–Polystyrene triblock copolymer. *Macromolecules*, 42(15) (2009) 5430–5432.
[20] S. Ahola, M. Österberg, and J. Laine, Cellulose nanofibrils — adsorption with poly(amideamine) epichlorohydrin studied by QCM-D and application as a paper strength additive. *Cellulose*, 15(2) (2008) 303–314.
[21] F. Aloulou, S. Boufi, and D. Beneventi, Adsorption of organic compounds onto polyelectrolyte immobilized-surfactant aggregates on cellulosic fibers. *Journal of Colloid and Interface Science*, 280(2) (2004) 350–358.
[22] Y. Boluk, L. Zhao, and V. Incani, Dispersions of nanocrystalline cellulose in aqueous polymer solutions: Structure formation of colloidal rods. *Langmuir*, 28(14) (2012) 6114–6123.
[23] P. Jenkins and M. Snowden, Depletion flocculation in colloidal dispersions. *Advances in Colloid and Interface Science*, 68 (1996) 57–96.
[24] E.E. Tkalya, M. Ghislandi, G. de With, and C.E. Koning, The use of surfactants for dispersing carbon nanotubes and graphene to make conductive nanocomposites. *Current Opinion in Colloid & Interface Science*, 17(4) (2012) 225–232.

[25] K. Oksman, A.P. Mathew, D. Bondeson, and I. Kvien, Manufacturing process of cellulose whiskers/polylactic acid nanocomposites. *Composites Science and Technology*, 66(15) (2006) 2776–2784.
[26] H. Zhu, Y. Li, Z. Fang, J. Xu, F. Cao, J. Wan, C. Preston, B. Yang, and L. Hu, Highly thermally conductive papers with percolative layered boron nitride nanosheets. *ACS Nano*, 8(4) (2014) 3606–3613.
[27] W. Yang, Z. Zhao, K. Wu, R. Huang, T. Liu, H. Jiang, F. Chen, and Q. Fu, Ultrathin flexible reduced graphene oxide/cellulose nanofiber composite films with strongly anisotropic thermal conductivity and efficient electromagnetic interference shielding. *Journal of Materials Chemistry C*, 5(15) (2017) 3748–3756.
[28] Q. Zheng, Z. Cai, Z. Ma, and S. Gong, Cellulose nanofibril/reduced graphene oxide/carbon nanotube hybrid aerogels for highly flexible and all-solid-state supercapacitors. *ACS Applied Materials & Interfaces*, 7(5) (2015) 3263–3271.
[29] X. Xu, Y.-Q. Yang, Y.-Y. Xing, J.-F. Yang, and S.-F. Wang, Properties of novel polyvinyl alcohol/cellulose nanocrystals/silver nanoparticles blend membranes. *Carbohydrate Polymers*, 98(2) (2013) 1573–1577.
[30] Y. Habibi, L.A. Lucia, and O.J. Rojas, Cellulose nanocrystals: Chemistry, self-assembly, and applications. *Chemical Reviews*, 110(6) (2010) 3479–3500.
[31] Q. Liu, Y. Cui, D. Gardner, X. Li, S. He, and I.I. Smalyukh, Self-Alignment of plasmonic gold nanorods in reconfigurable anisotropic fluids for tunable bulk metamaterial applications. *Nano Letters*, 10(4) (2010) 1347–1353.
[32] Q. Liu, M.G. Campbell, J.S. Evans, and I.I. Smalyukh, Orientationally ordered colloidal co-dispersions of gold nanorods and cellulose nanocrystals. *Advanced Materials*, 26(42) (2014) 7178–7184.
[33] H. Yang and T.G.M. van de Ven, Preparation of hairy cationic nanocrystalline cellulose. *Cellulose*, 23(3) (2016) 1791–1801.
[34] S. Vicini, E. Princi, G. Luciano, E. Franceschi, E. Pedemonte, D. Oldak, H. Kaczmarek, and A. Sionkowska, Thermal analysis and characterisation of cellulose oxidised with sodium methaperiodate. *Thermochimica Acta*, 418(1) (2004) 123–130.
[35] E. Maekawa and T. Koshijima, Properties of 2,3-dicarboxy cellulose combined with various metallic ions. *Journal of Applied Polymer Science*, 29(7) (1984) 2289–2297.
[36] B. Sun, Q. Hou, Z. Liu, and Y. Ni, Sodium periodate oxidation of cellulose nanocrystal and its application as a paper wet strength additive. *Cellulose*, 22(2) (2015) 1135–1146.
[37] J. Sirvio, U. Hyvakko, H. Liimatainen, J. Niinimaki, and O. Hormi, Periodate oxidation of cellulose at elevated temperatures using metal salts as cellulose activators. *Carbohydrate Polymers*, 83(3) (2011) 1293–1297.

[38] H. Yang, D. Chen, and T.G.M. van de Ven, Preparation and characterization of sterically stabilized nanocrystalline cellulose obtained by periodate oxidation of cellulose fibers. *Cellulose*, 22(3) (2015) 1743–1752.

[39] U.-J. Kim, S. Kuga, M. Wada, T. Okano, and T. Kondo, Periodate oxidation of crystalline cellulose. *Biomacromolecules*, 1(3) (2000) 488–492.

[40] H. Yang, M.N. Alam, and T.G.M.J.C. van de Ven, Highly charged nanocrystalline cellulose and dicarboxylated cellulose from periodate and chlorite oxidized cellulose fibers. *Cellulose*, 20(4) (2013) 1865–1875.

[41] Y. Okita, T. Saito, and A. Isogai, Entire surface oxidation of various cellulose microfibrils by TEMPO-mediated oxidation. *Biomacromolecules*, 11(6) (2010) 1696–1700.

[42] S. Coseri, G. Biliuta, B.C. Simionescu, K. Stana-Kleinschek, V. Ribitsch, and V. Harabagiu, Oxidized cellulose — Survey of the most recent achievements. *Carbohydrate Polymers*, 93(1) (2013) 207–215.

[43] T. Saito, M. Hirota, N. Tamura, S. Kimura, H. Fukuzumi, L. Heux, and A. Isogai, Individualization of nano-sized plant cellulose fibrils by direct surface carboxylation using TEMPO catalyst under neutral conditions. *Biomacromolecules*, 10(7) (2009) 1992–1996.

[44] A.E.J. de Nooy, A.C. Besemer, and H. van Bekkum, Highly selective nitroxyl radical-mediated oxidation of primary alcohol groups in water-soluble glucans. *Carbohydrate Research*, 269(1) (1995) 89–98.

[45] A.E.J. de Nooy, A.C. Besemer, and H. van Bekkum, Selective oxidation of primary alcohols mediated by nitroxyl radical in aqueous solution. Kinetics and mechanism. *Tetrahedron*, 51(29) (1995) 8023–8032.

[46] P.L. Bragd, H. van Bekkum, and A.C. Besemer, TEMPO-Mediated oxidation of polysaccharides: Survey of methods and applications. *Topics in Catalysis*, 27(1) (2004) 49–66.

[47] A. Isogai and Y. Kato, Preparation of polyuronic acid from cellulose by TEMPO-mediated oxidation. *Cellulose*, 5(3) (1998) 153–164.

[48] C. Tahiri and M.R. Vignon, TEMPO-oxidation of cellulose: Synthesis and characterisation of polyglucuronans. *Cellulose*, 7(2) (2000) 177–188.

[49] S. Montanari, M. Rountani, L. Heux, and M.R. Vignon, Topochemistry of carboxylated cellulose nanocrystals resulting from TEMPO-mediated oxidation. *Macromolecules*, 38(5) (2005) 1665–1671.

[50] T. Saito and A. Isogai, TEMPO-mediated oxidation of native cellulose. The effect of oxidation conditions on chemical and crystal structures of the water-insoluble fractions. *Biomacromolecules*, 5(5) (2004) 1983–1989.

[51] A.C.W. Leung, S. Hrapovic, E. Lam, Y. Liu, K.B. Male, K.A. Mahmoud, and J.H.T. Luong, Characteristics and properties of carboxylated cellulose nanocrystals prepared from a novel one-step procedure. *Small*, 7(3) (2011) 302–305.

[52] M. Cheng, Z. Qin, Y. Liu, Y. Qin, T. Li, L. Chen, and M. Zhu, Efficient extraction of carboxylated spherical cellulose nanocrystals with narrow distribution through hydrolysis of lyocell fibers by using ammonium persulfate as an oxidant. *Journal of Materials Chemistry A*, 2(1) (2014) 251–258.
[53] H. Liimatainen, M. Visanko, J.A. Sirviö, O.E.O. Hormi, and J. Niinimaki, Enhancement of the nanofibrillation of wood cellulose through sequential Periodate–Chlorite oxidation. *Biomacromolecules*, 13(5) (2012) 1592–1597.
[54] B.G. Ranby, Aqueous colloidal solutions of cellulose micelles. *Acta Chemica Scandinavica*, (1949) 3(5), 649–650.
[55] X.M. Dong, J.-F. Revol, and D.G. Gray, Effect of microcrystallite preparation conditions on the formation of colloid crystals of cellulose. *Cellulose*, 5(1) (1998) 19–32.
[56] J. Araki, M. Wada, S. Kuga, and T.J.J.o.W.S. Okano, Influence of surface charge on viscosity behavior of cellulose microcrystal suspension. *Journal of Wood Science*, 45(3) (1999) 258–261.
[57] J. Gu, J.M. Catchmark, E.Q. Kaiser, and D.D. Archibald, Quantification of cellulose nanowhiskers sulfate esterification levels. *Carbohydrate Polymers*, 92(2) (2013) 1809–1816.
[58] P. Huang, M. Wu, S. Kuga, D. Wang, D. Wu, and Y. Huang, One-Step dispersion of cellulose nanofibers by mechanochemical esterification in an organic solvent. *Chem Sus Chem*, 5(12) (2012) 2319–2322.
[59] N. Lin, J. Huang, P.R. Chang, J. Feng, and J. Yu, Surface acetylation of cellulose nanocrystal and its reinforcing function in poly(lactic acid). *Carbohydrate Polymers*, 83(4) (2011) 1834–1842.
[60] J.A. Ávila Ramírez, C.J. Suriano, P. Cerrutti, and M.L. Foresti, Surface esterification of cellulose nanofibers by a simple organocatalytic methodology. *Carbohydrate Polymers*, 114 (2014) 416–423.
[61] A. Ashori, M. Babaee, M. Jonoobi, and Y. Hamzeh, Solvent-free acetylation of cellulose nanofibers for improving compatibility and dispersion. *Carbohydrate Polymers*, 102 (2014) 369–375.
[62] W. Hu, S. Chen, Q. Xu, and H. Wang, Solvent-free acetylation of bacterial cellulose under moderate conditions. *Carbohydrate Polymers*, 83(4) (2011) 1575–1581.
[63] P. Huang, Y. Zhao, S. Kuga, M. Wu, and Y. Huang, A versatile method for producing functionalized cellulose nanofibers and their application. *Nanoscale*, 8(6) (2016) 3753–3759.
[64] A. Sato, D. Kabusaki, H. Okumura, T. Nakatani, F. Nakatsubo, and H. Yano, Surface modification of cellulose nanofibers with alkenyl succinic anhydride for high-density polyethylene reinforcement. *Composites Part A: Applied Science and Manufacturing*, 83 (2016) 72–79.

[65] D.-Y. Kim, Y. Nishiyama, and S. Kuga, Surface acetylation of bacterial cellulose. *Cellulose*, 9(3) (2002) 361–367.
[66] G. Siqueira, J. Bras, and A. Dufresne, Cellulose whiskers versus microfibrils: Influence of the nature of the nanoparticle and its surface functionalization on the thermal and mechanical properties of nanocomposites. *Biomacromolecules*, 10(2) (2009) 425–432.
[67] L. Chen, J.Y. Zhu, C. Baez, P. Kitin, and T. Elder, Highly thermal-stable and functional cellulose nanocrystals and nanofibrils produced using fully recyclable organic acids. *Green Chemistry*, 18(13) (2016) 3835–3843.
[68] M. Zaman, H. Xiao, F. Chibante, and Y. Ni, Synthesis and characterization of cationically modified nanocrystalline cellulose. *Carbohydrate Polymers*, 89(1) (2012) 163–170.
[69] K.-Y. Lee, F. Quero, J.J. Blaker, C.A.S. Hill, S.J. Eichhorn, and A. Bismarck, Surface only modification of bacterial cellulose nanofibres with organic acids. *Cellulose*, 18(3) (2011) 595–605.
[70] M. Božič, P. Liu, A.P. Mathew, and V. Kokol, Enzymatic phosphorylation of cellulose nanofibers to new highly-ions adsorbing, flame-retardant and hydroxyapatite-growth induced natural nanoparticles. *Cellulose*, 21(4) (2014) 2713–2726.
[71] M. Grunert and W.T. Winter, Nanocomposites of cellulose acetate butyrate reinforced with cellulose nanocrystals. *Journal of Polymers and the Environment*, 10(1) (2002) 27–30.
[72] Y. Xie, S. Cai, Z. Hou, W. Li, Y. Wang, X. Zhang, and W. Yang, Surface hydrophobic modification of microcrystalline cellulose by Poly(methylhydro)siloxane using response surface methodology. *Polymers*, 10(12) (2018) 1335.
[73] M. de Oliveira Taipina, M.M.F. Ferrarezi, I.V.P. Yoshida, and M.D.C. Gonçalves, Surface modification of cotton nanocrystals with a silane agent. *Cellulose*, 20(1) (2013) 217–226.
[74] C. Goussé, H. Chanzy, G. Excoffier, L. Soubeyrand, and E. Fleury, Stable suspensions of partially silylated cellulose whiskers dispersed in organic solvents. *Polymers*, 43(9) (2002) 2645–2651.
[75] A. Ben Mabrouk, H. Kaddami, A. Magnin, M.N. Belgacem, A. Dufresne, and S. Boufi, Preparation of nanocomposite dispersions based on cellulose whiskers and acrylic copolymer by miniemulsion polymerization: Effect of the silane content. *Polymer Engineering & Science*, 51(1) (2011) 62–70.
[76] M. Andresen, L.-S. Johansson, B.S. Tanem, and P. Stenius, Properties and characterization of hydrophobized microfibrillated cellulose. *Cellulose*, 13(6) (2006) 665–677.
[77] J.-L. Huang, C.-J. Li, and D.G. Gray, Cellulose nanocrystals incorporating fluorescent methylcoumarin groups. *ACS Sustainable Chemistry & Engineering*, 1(9) (2013) 1160–1164.

[78] S. Barazzouk and C.J.N. Daneault, Amino acid and peptide immobilization on oxidized nanocellulose: Spectroscopic characterization. *Nanomaterials*, 2(2) (2012) 187–205.

[79] J. Araki, M. Wada, and S. Kuga, Steric stabilization of a cellulose microcrystal suspension by poly(ethylene glycol) grafting. *Langmuir*, 17(1) (2001) 21–27.

[80] H. Lönnberg, L. Fogelström, L. Berglund, E. Malmström, and A. Hult, Surface grafting of microfibrillated cellulose with poly(ε-caprolactone) — Synthesis and characterization. *European Polymer Journal*, 44(9) (2008) 2991–2997.

[81] X. Cao, Y. Habibi, and L.A. Lucia, One-pot polymerization, surface grafting, and processing of waterborne polyurethane-cellulose nanocrystal nanocomposites. *Journal of Materials Chemistry*, 19(38) (2009) 7137–7145.

[82] E. Gicquel, C. Martin, L. Heux, B. Jean, and J. Bras, Adsorption versus grafting of poly(N-Isopropylacrylamide) in aqueous conditions on the surface of cellulose nanocrystals. *Carbohydrate Polymers*, 210 (2019) 100–109.

[83] S.P. Akhlaghi, R.C. Berry, and K.C. Tam, Surface modification of cellulose nanocrystal with chitosan oligosaccharide for drug delivery applications. *Cellulose*, 20(4) (2013) 1747–1764.

[84] H. Sadeghifar, I. Filpponen, S.P. Clarke, D.F. Brougham, and D.S. Argyropoulos, Production of cellulose nanocrystals using hydrobromic acid and click reactions on their surface. *Journal of Materials Science*, 46(22) (2011) 7344–7355.

[85] S. Ifuku, M. Tsuji, M. Morimoto, H. Saimoto, and H. Yano, Synthesis of silver nanoparticles templated by TEMPO-mediated oxidized bacterial cellulose nanofibers. *Biomacromolecules*, 10(9) (2009) 2714–2717.

[86] W. Hu, S. Chen, Z. Yang, L. Liu, and H. Wang, Flexible electrically conductive nanocomposite membrane based on bacterial cellulose and polyaniline. *The Journal of Physical Chemistry B*, 115(26) (2011) 8453–8457.

[87] E. Lam, S. Hrapovic, E. Majid, J.H. Chong, and J.H.T. Luong, Catalysis using gold nanoparticles decorated on nanocrystalline cellulose. *Nanoscale*, 4(3) (2012) 997–1002.

[88] Y. Shin, J.M. Blackwood, I.-T. Bae, B.W. Arey, and G.J. Exarhos, Synthesis and stabilization of selenium nanoparticles on cellulose nanocrystal. *Materials Letters*, 61(21) (2007) 4297–4300.

[89] K.A. Mahmoud, E. Lam, S. Hrapovic, and J.H.T. Luong, Preparation of well-dispersed gold/magnetite nanoparticles embedded on cellulose nanocrystals for efficient immobilization of papain enzyme *ACS Applied Materials & Interfaces*, 5(11) (2013) 4978–4985.

[90] J. Wu, Y. Zheng, W. Song, J. Luan, X. Wen, Z. Wu, X. Chen, Q. Wang, and S. Guo, *In situ* synthesis of silver-nanoparticles/bacterial cellulose composites for slow-released antimicrobial wound dressing. *Carbohydrate Polymers*, 102 (2014) 762–771.

[91] P. Liou, F.X. Nayigiziki, F. Kong, A. Mustapha, and M. Lin, Cellulose nanofibers coated with silver nanoparticles as a SERS platform for detection of pesticides in apples. *Carbohydrate Polymers*, 157 (2017) 643–650.

[92] C.M. Cirtiu, A.F. Dunlop-Brière, and A. Moores, Cellulose nanocrystallites as an efficient support for nanoparticles of palladium: Application for catalytic hydrogenation and Heck coupling under mild conditions. *Green Chemistry*, 13(2) (2011) 288–291.

[93] Y. Shin, I.-T. Bae, B.W. Arey, and G.J. Exarhos, Simple preparation and stabilization of nickel nanocrystals on cellulose nanocrystal. *Materials Letters*, 61(14) (2007) 3215–3217.

[94] Y. Shin, I.-T. Bae, B.W. Arey, and G.J. Exarhos, Facile stabilization of gold-silver alloy nanoparticles on cellulose nanocrystal. *The Journal of Physical Chemistry C*, 112(13) (2008) 4844–4848.

[95] X. Wu, Z. Shi, S. Fu, J. Chen, R.M. Berry, and K.C. Tam, Strategy for synthesizing porous cellulose nanocrystal supported metal nanocatalysts. *ACS Sustainable Chemistry & Engineering*, 4(11) (2016) 5929–5935.

[96] J. Yang, J. Yu, J. Fan, D. Sun, W. Tang, and X. Yang, Biotemplated preparation of CdS nanoparticles/bacterial cellulose hybrid nanofibers for photocatalysis application. *Journal of Hazardous Materials*, 189(1) (2011) 377–383.

[97] G. Cui, M. Liu, Y. Chen, W. Zhang, and J. Zhao, Synthesis of a ferric hydroxide-coated cellulose nanofiber hybrid for effective removal of phosphate from wastewater. *Carbohydrate Polymers*, 154 (2016) 40–47.

[98] D. Sun, J. Yang, and X. Wang, Bacterial cellulose/TiO_2 hybrid nanofibers prepared by the surface hydrolysis method with molecular precision. *Nanoscale*, 2(2) (2010) 287–292.

[99] J. Huang and T. Kunitake, Nano-Precision replication of natural cellulosic substances by metal oxides. *Journal of the American Chemical Society*, 125(39) (2003) 11834–11835.

[100] E. Dujardin, M. Blaseby, and S. Mann, Synthesis of mesoporous silica by sol–gel mineralisation of cellulose nanorod nematic suspensions. *Journal of Materials Chemistry*, 13(4) (2003) 696–699.

https://doi.org/10.1142/9781786349477_0004

Chapter 4

Nanocellulose-Reinforced Starch Nanocomposites

Preetha Balakrishnan[*], Sreerag Gopi[†], Muhammad Wajid Ullah[‡], Guang Yang[‡], and Sabu Thomas[*,§,¶]

[*]*International and Inter University Centre for Nanoscience and Nanotechnology, Mahatma Gandhi University, Kottayam, Kerala, India*

[†]*ADSO Naturals Private Limited Banglore, India*

[‡]*Department of Biomedical Engineering, College of Life Science and Technology, Huazhong University of Science and Technology, Wuhan, PR China*

[§]*School of Chemical Sciences, Mahatma Gandhi University, Kottayam, Kerala, India*

[¶]*sabuthomas@mgu.ac.in*

Abstract

With the advancement of science and technology, people of the modern civilized era are becoming more dependent on advanced materials. One of the major gifts that chemistry has ever conferred to the human society is the so-called "polymer" or "polymeric materials", without which the world have been in a totally different position. This chapter deals with recent developments in starch and nanocellulose-reinforced starch nanocomposites.

Keywords: Nanocellulose; starch; tailored properties; nanomaterials; applications

1. Polymers

A polymer is a large molecule or macromolecule composed of repeated subunits. Due to its wide range of applicability, both synthetic and natural polymers play a vital role in everyday life. Polymers, both natural and synthetic, are created via polymerization of many small molecules known as monomers. With growing environmental and health concerns, there has been significant focus in several sectors like scientific, industrial, and environment communities on the use of eco-friendly green material.

Figure 1 shows a classification of biodegradable polymers in four families. Except the fourth family, which is of fossil origin, most polymers are obtained from renewable resources (biomass). The first family are agro-based polymers (e.g., polysaccharides) obtained from biomass. The second and third families are polyesters, obtained, respectively, by fermentation from biomass or from genetically modified plants (e.g., polyhydroxyalkanoate: PHA) and by synthesis from monomers obtained from biomass (e.g., polylactic acid: PLA). The fourth family are poly esters e.g., Polycaprolactone: PCL. A large number of these

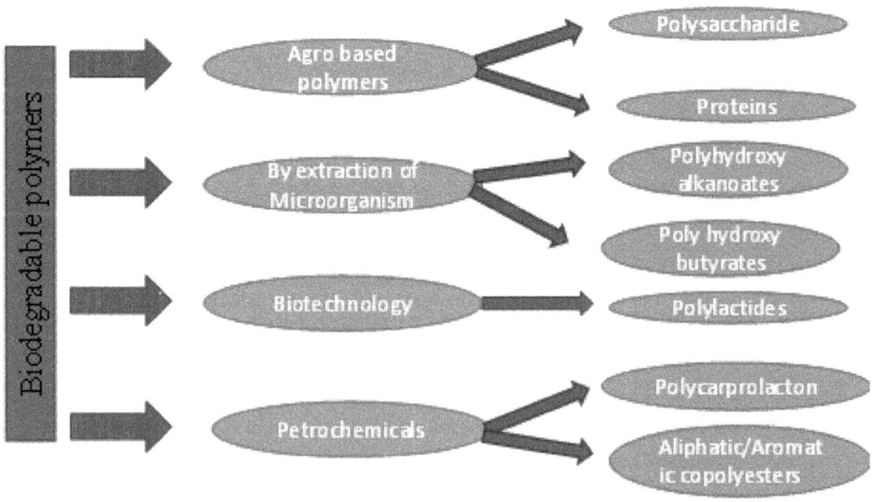

Fig. 1. Classification of biodegradable polymers.

biodegradable polymers (biopolymers) are commercially available. They show a wide range of properties and can compete with non-biodegradable polymers in different industrial fields.[1]

2. Starch

2.1 *Structure and Properties*

Among the natural polymers, starch has been considered as a promising candidate in this regard because of its attractive combination of price, abundance, and thermoplastic behavior in addition to biodegradability.[2,3] But starch also faces some drawbacks over synthetic polymers. For example, renewable starch-based systems, traditionally, are low cost but suffer from poor processability and final product quality whereas non-renewable synthetic polymers, traditionally, are easy to process and have excellent properties but are typically too cost prohibitive to sell.[4] So properly furnishing starch will lead to the generation of superior-quality products. Starch is a low-cost material when compared with most of the other synthetic polymers, and it is also a naturally available carbohydrate obtained from a wide variety of crops. Starch has been investigated widely for the potential manufacture of products such as water-soluble pouches for detergents and insecticide flushable lines and bags and medical delivery systems and devices.[5,6] Native starch commonly exists in a granular structure, which can be processed into thermoplastic starch at high temperature and shear forces by melting. Starch-based biodegradable polymers have great commercial potential for bio-plastics, but some of the properties like brittleness, low heat distortion temperature, high gas permeability, low melt viscosity for further processing, etc. restrict their use in a wide range of applications. For the above-mentioned reasons modification of starch-based materials for innovative applications is still a challenge for the material scientist. In order to solve these issues, various methods such as blending, reinforcing, and plasticization are usually performed. Adding nano-reinforcement to pristine polymers when preparing nanocomposite has already proven to be an effective way to improve these properties concurrently.[7,8] The major crops used for starch production include corn, potato, and rice. In all of these above-mentioned plants, starch is produced as granules, and these granules vary in size and somewhat in composition from plant to plant. Because of increasing prices and decreasing availability of conventional film-forming plastics, starch has been widely used as a raw material in film production. Starch films are thus attractive materials for food packaging, owing to their low permeability. Starch is also useful for making agricultural mulch films because it degrades into harmless products when in contact with the soil.

The major macromolecular components of starch are amylose and amylopectin (Fig. 2). Amylose is essentially a linear molecule of (1–4) linked *α-D*-glucopyranosyl units with some slight branches formed by (1–6)-*α*-linkages. Typically, amylose molecules have molecular weights ranging from 10^5 to 10^6 $gmol^{-1}$. Amylopectin is a highly branched molecule composed of chains of *α-D*-glucopyranosyl residues linked together mainly by (1–4) linkages but with (1–6) linkages at branch points. Amylopectin consists of hundreds of short chains of (1–4) linked *α-D*-glucopyranosyl interlinked by (1–6) *α*-linkages. It is an extremely large and highly branched molecule (molecular weight ranging from 10^6 to 10^8 $gmol^{-1}$), and thus it is unique compared to synthetic polymer.[9] Moreover, the branching within amylopectin molecules is not random.

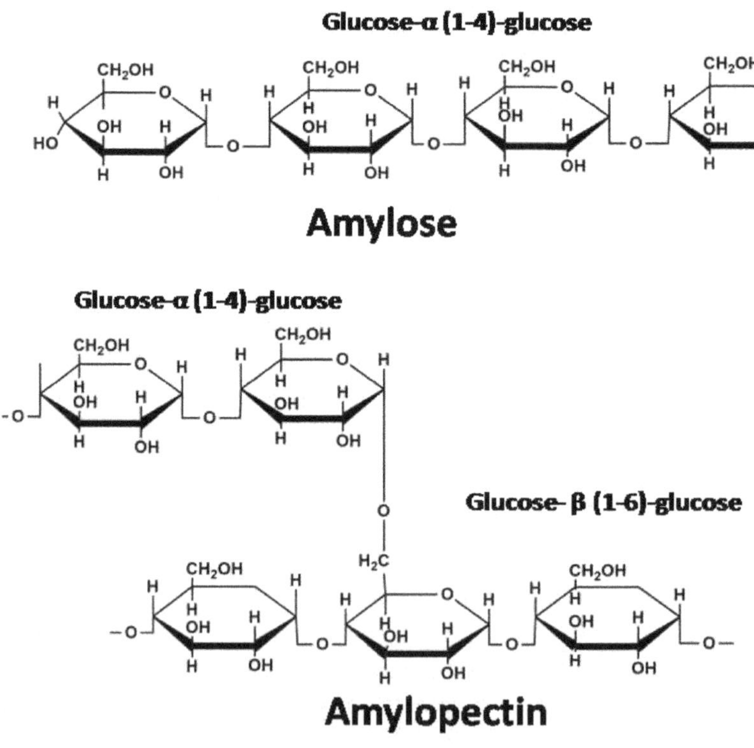

Fig. 2. Structure of amylose and amylopectin.

Depending on the plant, starch generally contains 20–25% amylose and 75–80% amylopectin by weight. This polysaccharide is produced by all green plants as an energy store.

Amylose and amylopectin in starch are organized into granules as alternating semi-crystalline and amorphous layers. As revealed by our SEM studies, starch granules are smooth, with a spherical or ellipsoid shape. The granule surface is smooth, with only minor ripples and bumps at the nanometer scale. SEM image of potato starch granules is shown in Fig. 3.

2.2 *Crystallinity and Multiscale Structure of Starch*

The starch granule displays a multiscale structure as shown in Fig. 4. From the figure, we can see that the starch granule consists of (a) starch granule (2–100 μm) and growth rings (b) composed of blocklets of (d) crystalline and amorphous lamellae (c) containing (g) amylopectin, and (h) amylose chains (0.1–1 nm). When observed under a microscope and polarized light, starch shows birefringence due to the semi-crystalline nature of starch. The so-called onion-like structure of starch granules with more or less concentric growth rings is composed of alternating hard crystalline and soft, less-ordered shells growing by opposition from the hilum (center of the granule). The thickness of the combined repeated crystalline and amorphous lamellae is 9 nm regardless the botanic origin. Native starches contain between 15% and 45% of crystalline material. Depending on their X-ray diffraction pattern, starches are categorized in three crystalline

Fig. 3. SEM image of potato starch.

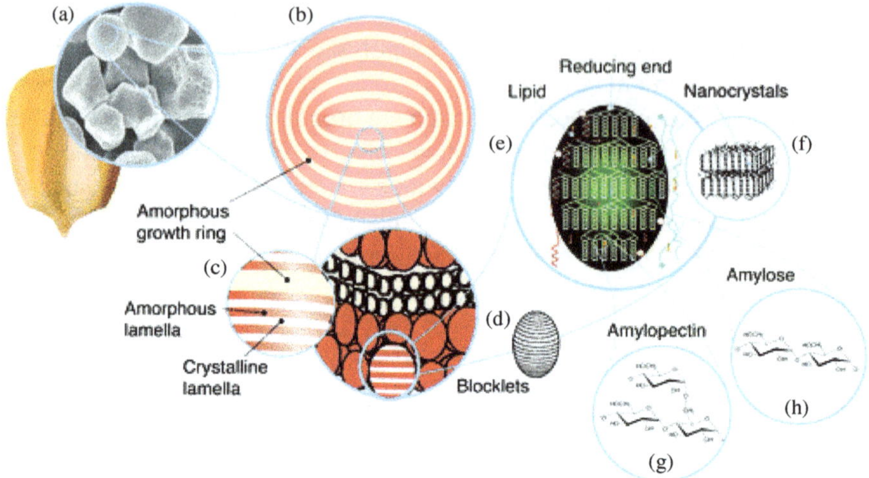

Fig. 4. Multiscale structure of starch.[10] Starch multiscale structure: (a) starch granules from normal maize (30 μm), (b) amorphous and semicrystalline growth rings (120–500 nm), (c) amorphous and crystalline lamellae (9 nm), magnified details of the semicrystalline growth ring, (d) blocklets (20–50 nm) constituting a unit of the growth rings, (e) amylopectin double helixes forming the crystalline lamellae of the blocklets, (f) nanocrystals: other representation of the crystalline lamellae called starch nanocrystals when separated by acid hydrolysis, (g) amylopectin's molecular structure, and (h) amylose's molecular structure (0.1–1 nm).

types, referred to as A, B, and C. A-type is characteristic of cereal starches (wheat and maize starch). B-type is typical of tuber and amylose-rich cereal starches. C-type is characteristic of leguminous starches and is made up of to a mixture of A and B crystalline types. V-type is observed during the formation of complexes between amylose and a complexing molecule (iodine, alcohols, cyclohexane, fatty acids, etc.). The appearance of a starch X-ray diffraction pattern depends on the water content of the granules during measurement. The more starch is hydrated, the thinner the diffraction pattern rings up to a given limit. Water is, therefore, one of the components of the crystalline organization of starch.[10,11]

3. Starch as a Natural Biopolymer

Biopolymer involves materials consisting of units with a biological origin. Accordingly, a wide range of materials can be considered as biopolymers, such as for example, Poly (Lactic Acid) (PLA), in which the monomer (repeating units) derived from natural resources by fermentation of starch is included in the definition even though the polymer itself is a product of a chemical or enzymatic synthesis. It is therefore necessary to distinguish natural polymeric materials synthesized as such in nature, e.g., cellulose, starch, proteins etc. In this case, all natural polymers can be considered as biopolymers, but not all biopolymers are natural polymers.[12] However, these resources need to be processed to some extent to be able to be used as a raw material for the preparation of biocomposites; even natural polymers such as starch, cellulose, chitin need to first be extracted from their natural origin and then modified accordingly to get the desired properties. For biocomposite design, modification to the reinforcements and/or to the matrices constitutes an important step in improving the interfacial adhesion between the components, thus enhancing macroscopic properties such as mechanical properties, barrier properties and resistance towards moisture absorption.[13] Betolini and co-workers studied starch irradiated with a mercury vapor lamp. They suggested that the starch photo-oxidation starts to break the C^2–C^3 bond of glucopyranose, producing dialdehyde followed by the formation of formaldehyde, formic acid, and CO_2.[14] In addition, starch-based materials have poor mechanical properties and particularly poor elongation (around 6%) at ambient conditions. Thus, the incorporation of a plasticizer is required to overcome the brittleness of the materials. Plasticizers reduce the intermolecular forces and increase the mobility of polymer chains, decreasing the glass transition temperature (Tg) and increasing the permeability. Other approaches to improve the final properties of the starch films is to blend starch with other polymers. However, most of the synthetic polymers are hydrophobic and thermodynamically immiscible with hydrophilic starch.

4. Modification of Starch

Among many kinds of candidates, including aliphatic polyesters, natural polymers, and their derivatives, starch, a polysaccharide produced by many plants as a storage polymer, is one of the most promising materials for biodegradable plastics because it is a versatile biopolymer with immense potential and low price for use in the non-food industries.[15] Incorporating plastifying agents, such as water and/or poly-alcohols, starch can be made into thermoplastic called thermoplastic starch (TPS) or plasticized starch (PS) through destructurization by the use of mechanical and heat energy.[16] Thermoplastic starch (TPS) has received considerable attention during the past two decades and is an interesting alternative for

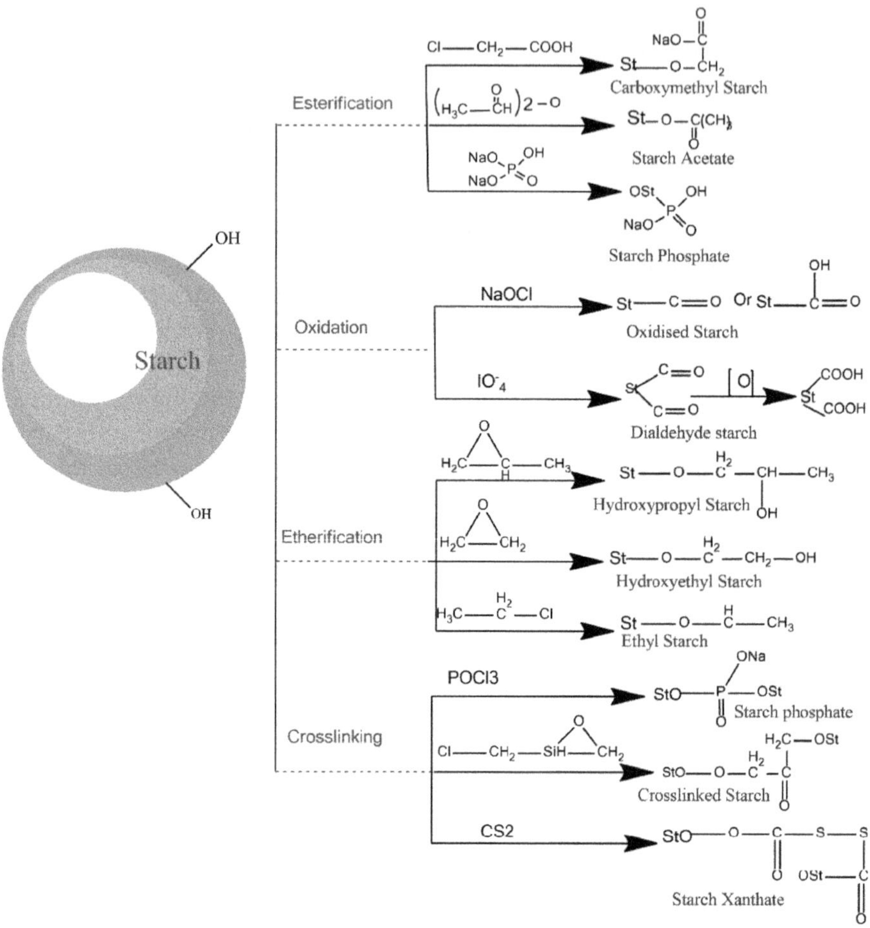

Fig. 5. Modification of starch.

synthetic polymers where long-term durability is not needed and rapid degradation is an advantage. Chemical modification, e.g., cross-linking, has long been studied as a way to reduce these problems and to produce low water-sensitive and high-strength materials (Fig. 5). Examples of cross-linking reagents include glutaraldehyde,[17] boric acid[18] and epichlorohydrin.[19] However, some of these cross-linking agents always display toxicity, and thus their potential applications as biomaterials are limited. To overcome these disadvantages, certain non-toxic functional additives and simple modification techniques are required to improve the mechanical properties and water resistibility of the starch films.[20] Godbillot, Dole, and co-workers reported that water content and interactions in starch films are dependent not only on plasticizer concentration but also on the relative humidity (RH) to which these materials are exposed. They reported that the interactions between the components of the system, formed by starch, plasticizer, and water, vary according to the relative quantities of these components. It is possible that, under some conditions, phase separation occurs.[21] Starch may be modified via many different chemical reactions, as shown in Fig. 5. Modified starches are the products of glucosidic bond cleavage (acid modification to dextrins), forming new functional groups (carbonyl group formation during oxidation), substitution of free available hydroxyl groups (by etherification or esterification), and bridging of molecular chains by cross-linking reactions.[18] By replacing –OH groups or cross-linking starch molecules, chemical modification reduces the diffusion coefficient and hence water absorption of TPS.

5. Thermoplastic Starch

Starch owes much of its functionality to two major high molecular weight carbohydrate components, amylose and amylopectin, as well as to the physical organization of these macromolecules into granular structure. Amylose is responsible for the film forming capacity of starches. However, the final organoleptic nutritional and mechanical properties of an edible film can be modified by the addition of various chemicals in minor amounts. Glycerol is a low molecular weight volatile substance that, even in small quantities, can bring drastic changes in processability, utility and cost during the processing of starch-based plastics. Plasticizers such as glycerol, sorbitol, and polyethylene glycol are defined as essentially high-boiling, non-separating substances that, when added to another material, change the physical and/or mechanical properties of that material. However, films formed from starch are brittle and difficult to handle. Plasticizers are normally added to the film-forming solution before casting and drying procedures, as a way to overcome the brittleness of the films. The plasticizers must be compatible with starch, and these plasticizers are used to decrease the intermolecular attractions between adjacent chains, thereby increasing chain flexibility. Incorporation of these additives may, however, cause significant changes in the barrier properties of film. Various plasticizers are added to improve the processability and final property of starch-based biocomposites, and various polyols (e.g., glycerol,[22,23] glycol,[24,25] and sorbitol,[26–28] urea,[29–31] sugars,[32] formamide,[33–35] and citric acid[36,37] have been widely used. Combination of various plasticizers in the right proportion can bring about drastic changes in the final properties of starch films as reported earlier.[38,39] Sometimes, cross-linking agents and plasticizing agents have adverse effect on tensile properties. Generally, the tensile strength increases and the strain at break decreases as the percentage of cross-linking agent is increased. Adding plasticizers to starch can help overcome strong inter and intra molecular hydrogen bond within the starch molecule. Thus, thermoplastic material can be achieved under high temperatures and shear, which is usually called thermoplastic starch (TPS).[40,41] TPS can not only improve the compatibility in water, but it is recognized as a much more efficient plasticizer than glycerol, and starch chain mobility is also greatly dependent on the water content of the premix.[42,43] Therefore, it is difficult to achieve a completely plasticized thermoplastic blend after excluding water. However, formamide in small volumes and with high polarity is a much more effective plasticizer than glycerol, which can form stable hydrogen bonds with starch and improve the plasticization of starch. The incorporation of plasticizers is necessary to reduce polymer intermolecular forces, increasing the mobility of the polymeric chains and improving the mechanical characteristics of the film, such as film extensibility.[29] Also, plasticizers affect the water barrier property of the films, since they have a great affinity for water. Starch plasticization is a three-stage process during which the following events take place[44] (Fig. 6).

Plasticizers are absorbed by starch granules and form hydrogen bonds with amylose and non-ordered amylopectin. This processes increase

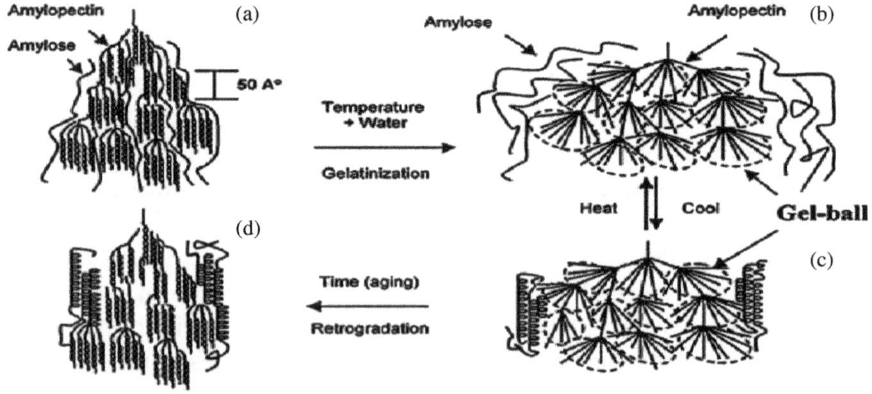

Fig. 6. Schematic representation of the phase transitions of starch during thermal processing and aging.[44]

amylose and amylopectin mobility in the amorphous regions. Amylose and amylopectin rearrange, forming new intermolecular interactions.

1. Amylose and amylopectin become more mobile and lose their intermolecular interactions and granular structure when heat and shearing forces are applied. The energy absorbed by the granules melts their crystallite structures and facilitates the formation of new bonds between the starch and the plasticizers. SEM images of this process are shown below in Fig. 6. Starch restructured in this way acquires flowing properties similar to that of the thermoplastics such as polyethylene and polypropylene.
2. A thermoplastic material is formed as starch is cooled back below the melting temperature of the starch crystallites. Starch chains re-associate and a new crystalline order is created, different from the original granule structure. The re-organizing of amylose molecules is rapid due to the greater mobility of amylose molecules compared to amylopectin molecules. On the other hand, the restructuring of amylopectin molecules proceeds slowly over several days. The final material contains amorphous regions as well as crystalline regions.

Starch is typically plasticized, restructured, and/or blended with other materials to provide useful mechanical properties. Such TPS compounds can be processed on existing plastics fabrication equipment. Plastics with high starch content are highly hydrophilic and readily disintegrate on contact with water. This can be overcome through blending as the starch has free hydroxyl groups, which readily undergo a number of reactions such as acetylation, esterification, and etherification. When native starch granules are heated in water, the granules exhibit irreversible transition in which the granules swell and finally burst out. They are soluble in water, and this process is called starch gelatinization. Gelatinization is an irreversible process that includes, in a broad sense and going by the time/temperature sequence, granular swelling, native crystalline melting (loss of birefringence which is due to the disruption of granular structure), and molecular solubilisation.[38] Starch gelatinization is a difficult pattern to explain, and the phenomenon was thought to explain the loss of crystallinity of starch granules, notwithstanding the process conditions applied. Two supporting views in this regard are as follows: Gelatinization occurs in two stages. The first stage occurs at a temperature of around 60–70°C, and then the second stage of gelation happens at above 90°C. The first stage is characterized by swelling of granules; the second stage is characterized by the disappearance of granular integrity due to excessive swelling of starch granules and solubilization. To make starch thermoplastics, its crystalline structure has to be destroyed by pressure, heat, mechanical work, and use of plasticizers.[45] The well-accepted concept of "gelatinization" refers to the destruction of the crystalline structure in starch granules,[46] which is an irreversible process that includes, in a broad sense and going by the time sequence of events, granular swelling, native crystalline melting (loss of birefringence), and molecular solubilization. Starch gelatinization is a multi-stage process that can be detected by Differential Scanning Calorimetry (DSC).[47] Gelatinized amylopectin initially remains in an amorphous state, even though there is still a small amount of ungelatinized amylopectin. Although the double-helical crystalline structure formed by the short-branched chains in amylopectin is torn apart during gelatinization, the chains remain in a regular pattern and retain a certain "memory". It has been considered that these short-branched chains form gel-balls, which mainly contain chains from the same sub-main chain (Fig. 6b). One amylopectin molecule may form a relatively separated super-globe. The molecular entanglements between gel-balls and super-globes are much less than those between linear polymer chains due to the size and length of the chains (only 4–6 glucose). V-type single-helix crystals (Fig. 6c) have been found to form directly after extrusion, which initially results in higher modulus and yield stress for amylose-rich materials. The crystallinity of the V-type crystals increases with time (Fig. 6d).

6. Starch as a Matrix for Biocomposite

Starch is one of the most studied natural polymers in biopolymer research for the creation of, for example, biodegradable food wraps, single-use trays and plates, food utensils, and non-food packaging, because of its availability, biodegradability, renewable nature, and low cost. However, the major drawbacks associated with starch are its poor mechanical strength and water holding capacity, which limits its industrial use. Thermoplastic starch or plasticized starch offers an interesting alternative for synthetic polymers in specific applications and is used, for example, in starch-based composites. Starch-based materials generally absorb moisture and undergo environmental stress cracking, causing significant problems with shelf life storage. Recently, however, starch-based nanocomposites, with the addition of low quantities of nanoscale reinforcements, have shown large-scale improvement in mechanical, thermal, physical, and barrier properties without high viscosity during processing.[48] In order to increase the water resistance and strength, various biodegradable polymers are blended with starch, e.g., Poly(lactic acid) (PLA),[49] Poly(e-caprolactone)(PCL),[50] Poly(Vinyl alcohol)(PVA),[51] and chitosan.[52] Addition of reinforcing agents to the starch matrix has proved to be an effective approach to obtaining high-performance starch-based biocomposites. To improve the properties of starch-based biocomposites, the matrix is reinforced with suitable fillers like biofiber, cellulose nanofiber, bacterial cellulose, inorganic clays, and so on. They consist of biodegradable plastics with reinforcement of biodegradable natural fibers. Natural plant fibers exist as hairs (cotton, kapok, and bamboo),[53,54] bast fibers(also called soft fibers; kenaf, ramie, hemp, jute, papyrus, cordial, and Indian mellow),[55–57] hard fibers (sisal, abaca (manila hemp), raffia, pineapple,[58] and flax), stem (bamboo, bagasse,[59] banana stalk, and corn stalk), fruit(coconut[60]), straw (rice, corn, and wheat), and others (seaweeds, palm).[61] Interfacial adhesion between the natural fiber and polymer matrix is a key factor in a biocomposites system using natural fiber reinforcement.[62] The improvement of interfacial adhesion between hydrophilic natural fibers and hydrophobic polymer matrix through surface treatment of natural fibers before composite processing contributes, to a greater or less extent, to an increase in the properties of the biocomposites. Poor wetting and bonding of fiber with the polymer results in a negative effect on the mechanical and thermal properties as well as on the water resistance of the biocomposites, limiting their potential and probable uses. Cellulose fibers, which are strongly polarized, are inherently incompatible with hydrophobic polymers due to their hydrophilic nature. Glycerol plasticized–starch (GPS)/cellulose nanoparticle (CN) composites were prepared by Chang *et al.*[30] Starch-based nanocomposites reinforced with flax cellulose nanocrystals were prepared by Cao *et al.*[63] Remarkable improvement of the tensile strength, Young's modulus, and water resistivity was observed. Wheat straw nanofiber-reinforced biocomposites were prepared by Sain *et al.*[64] by mixing the nanofiber with gelatinized starch solution followed by solution casting. Remarkable improvements in thermal and mechanical properties were also observed by them. Poly(vinyl alcohol) nanocomposites reinforced with cellulose microfibrils isolated by high-intensity ultrasonication were prepared by Cheng *et al.*[65] Increment of tensile modulus and tensile strength was observed by them.

6.1 *Effect of Reinforcing Filler on Improving Properties of Biocomposites*

Das *et al.* studied the physico-mechanical properties of jute fiber-reinforced starch composites. Jute micro/nanofibrils (JNF) were prepared from jute felt by acid hydrolysis. JNF reinforced starch/PVA-based biocomposite films were prepared by solution casting method.[56] It was observed that the moisture uptake decreased in the entire reinforced biocomposite film compared to the unreinforced one. The moisture uptake of the unreinforced film was 32% after 1 day, whereas it decreased significantly in the JNF loaded composites. There is a decrease in Tg value, and this deviation was due to the efficient blending of the highly compatible starch/PVA matrix and good inter component bonding between the matrix and the JNF. Dagang *et al.*[66] studied the effect of bamboo cellulose fibers on starch composites and reported that the reinforcing effect on the starch composites was indeed due to size and the reduction of water uptake. The dispersion and polymorph of cellulose crystals were severely influenced by different treatments and the surrounding matrix, which ultimately affected the reinforcing effect on the plasticized starch-based biocomposites. Gironès *et al.*[67] studied the effect of sisal fibers on thermoplastic starch matrix obtained by melt processing. Addition of natural latex has been proved to decrease the kinetics of water absorption of the composites, although its effect on water uptake, crystallinity, and mechanical properties was negligible. The results demonstrated that both tensile and flexural strength improved with the percentage of reinforcement fibers. Despite having similar mechanical properties, hemp strands provided composites with better mechanical properties that those obtained with sisal. This effect is assigned to the better fibrillation obtained during the mixing of the TPS with the hemp strands, and points out the mechanical anchoring as the main cause of the enhanced mechanical resistance.[68]

Lopez *et al.*[67] investigated the thermal, mechanical, and water sorption characteristics of newspaper fiber-reinforced starch composites prepared by melt processing. Lignocellulosic fiber obtained from newspaper was reinforced with glycerol plasticized corn starch. It was observed that the effects caused by the addition of natural strands (sisal, abaca, hemp, etc.) are similar to those observed here, when cellulosic fibers from recovered newspaper have been included in the formulation of the materials. Fibers from recovered newspaper effectively acted as reinforcing agent, increasing the tensile strength and the Young's modulus of TPS up to 260%. The bacterial cellulose (BC) nanofiber–starch biocomposites have been manufactured by solution impregnation method. The BC/starch biocomposites possess much higher tensile strength and modulus than the unreinforced starch but show lower elongation at break. The tensile strength of the BC/starch biocomposite containing 15.1 wt.% BC nanofibers and starch decreases drastically upon exposure to microorganism attacks. The BC/starch biocomposites shows a higher tensile strength retention when compared to the starch owing to the higher resistance to microorganism attacks of the BC fibers in comparison to starch. The BC/starch biocomposites consisting of biodegradable BC and starch are fully biodegradable, which renders them advantageous in terms of environmental protection.[69] Claudia *et al.*[70] studied the physico-chemical characteristic of non-conventional sources of starch obtained by cold and thermal gelatinization method. Starch-based films prepared with the thermal gelatinization method had higher thickness than those prepared with the cold gelatinization one, with the starch source playing an important role in these values. Starch films prepared with thermal gelatinization showed homogeneous surfaces without pores or cracks compared with starch films made with cold gelatinization; this integrity is important in the mechanical properties of the films. Elastic modulus, puncture, and elongation values were higher in films obtained by thermal gelatinization.

Marques *et al.*[24] studied the thermal property of cassava starch cross-linked with tetra ethylene glycol and found the thermal stability of starch cross-linked with tetra ethylene glycol diacrylate, analyzed using thermogravimetry and infrared spectroscopy. Although the thermo gravimetric curves for non-cross-linked and cross-linked starches seemed similar, with only one main stage of degradation, different activation energy values were observed. The increase in the E values after cross-linking also indicates changes in the mechanism of degradation, which is also dependent of the products formed during the degradation process. Although E values higher than 100 kJ mol^{-1} suggest a random chain scission, it seems clear that for the cross-linked starch, the main volatile products associated with hydrocarbons and carbonyl are associated with the scission of weak bonds in the chain (probably branched groups) and then the scission of stronger bonds (glycosidic linkages).

7. Cellulose Nanomaterials

Cellulose is the major part of lignocellulosic fibers. It is a linear polymer consisting of β-D-glucopyranose units joined together by glycosidic *β*-(1,4) bonds. Cellulose is characterized by the presence of six hydroxyl group. These hydroxyl groups and their hydrogen-bonding ability play a major role in crystalline packing and in governing important physical properties of this highly cohesive material. The intermolecular hydrogen bonds confer to cellulose a crystalline structure. The amorphous regions are susceptible to acid attack, which promotes the transverse cleavage of the microfibrils into short nanocrystals. Cellulose nanoparticles (fibers/crystals) (Fig. 7) can be extracted from cellulose fibers by top-down approach by mechanical or chemical treatment.[71] Aqueous suspension of nanofibrillated cellulose or nanofibers (CNF) can be prepared by mechanical treatment and acid hydrolysis using mild acids like oxalic acid; on the

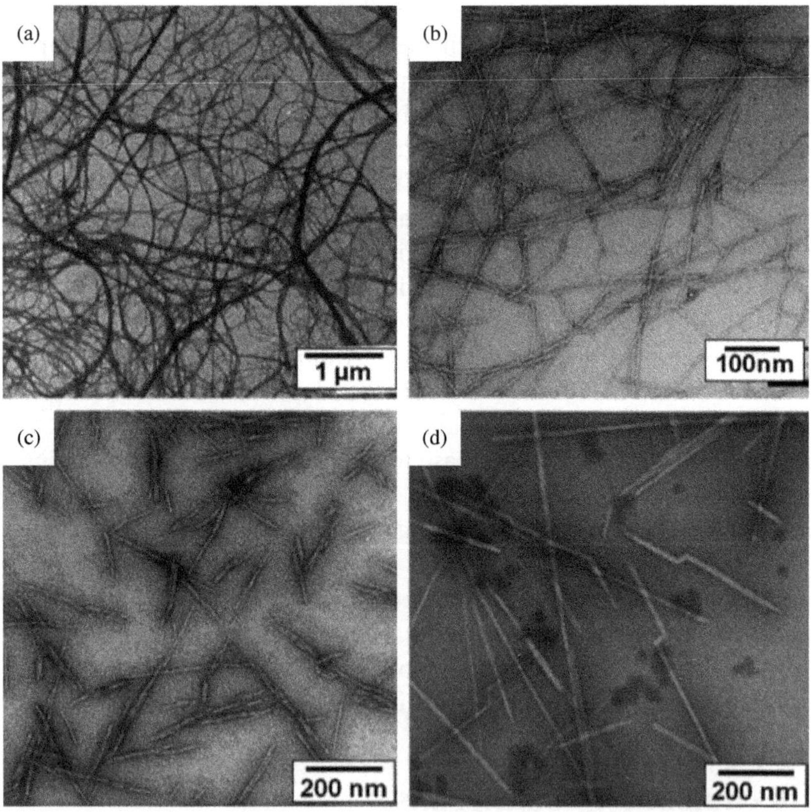

Fig. 7. (a) TEM image of microfibrillated cellulose,[75] (b) TEM image of TEMPO-oxidized cellulose,[76] (c) TEM image of wood CNCs,[77] (d) TEM image of t-CNC.[78]

other hand hydrolysis, strong acid leads to the formation of nanocrystals.[72] Cellulose nanomaterials a high value material since it can change the performance of existing products as well as to create new, unique and improved products.[73,74]

Cellulose nanomaterials can be isolated from a variety of biomass materials like hemp,[67] ramie,[79] pineapple leaf,[58] cassava bagasse,[80] cotton,[81] tunicate,[82] etc. by acid hydrolysis with mineral acid by strictly controlling the temperature and conditions of time and temperature. Acid hydrolysis results in elimination of the amorphous region, thus resulting in materials with high crystallinity. Tunicin[83] (animal cellulose) nanowhiskers have comparatively high aspect ratio (L/D: 50–200), which could enhance the nanofiller–matrix interfacial phenomena. Though direct acid hydrolysis is most frequently used to prepare cellulose nanowhiskers (CNWs), other methods have also been employed, such as enzymatic hydrolysis followed by combined shear and pressure treatment.[84]

In an interesting study, Grande *et al.*[85] prepared bacterial cellulose (BC)/starch self-assembled nanocomposites by allowing the bottom-up formation of a coherent network of cellulose nanofibers in the presence of partially gelatinized starch granules. The starch suspension was autoclaved at 121°C. During this process, the starch phase had a "first gelatinization" when the starch granules swelled, leached amylose, and lost their crystalline structure because of the melting of their ordered regions. At this stage, bacteria were inoculated and the nanofiber network was formed in the presence of a partially gelatinized starch structure. Starch was further gelatinized, and it formed a layer that covered the cellulose nanofibrils. The volume fraction of the strong phase in the nanocomposites produced by this technique was around 90%, which was similar to the proportions present in some biological nanocomposites, such as nacre. BC/starch gels were collected from the top, washed, and pressed in a hydraulic plate to produce sheets.

8. Applications of Starch Nanocomposites

One of the major problems that our world faces today is the problems due to plastic wastes. In this scenario, biodegradable polymers gain importance. Biopolymer-based packaging materials obtained from naturally derived renewable materials such as polysaccharides, proteins, and lipids have become the focus of worldwide attention in recent years. Such biopolymers offer favorable environmental advantages of recyclability and reutilization, compared to conventional petroleum-based synthetic polymers. One of the major fields in which these packaging applications of biodegradable polymers can be applied is in the food packaging industry. Because of the hydrophilic nature of starch and cellulose, packaging materials based on these materials have a low water vapor barrier, which leads to limited long-term stability and poor mechanical properties (sensitive to moisture content). Other drawbacks are bad processability, brittleness, and vulnerability to degradation.

8.1 *In the Food Industry*

Food packaging and edible films are two major applications of the starch-based biodegradable polymers in the food industry. The requirements for food packaging includes reducing the food losses, keeping food fresh, enhancing organoleptic characteristics of food such as appearance, odor, and flavor, and providing food safety. Traditional food packaging polymers like low-density poly ethylene (LDPE) face one major drawback, that is, they are not biodegradable. The starch-based biodegradable polymers can be a possible alternative for food packaging to overcome these disadvantages and keep the advantages of traditional packaging materials. However, the components in the conventional starch-based polymer packaging materials are not completely inert. The problems associated with starch-based composites for food packaging can be solved to a great extent by adding some naturally available filler, which can enhance the mechanical, barrier, and thermal properties, as well as shelf life. They display very low permeability to oxygen at low relative humidity and are proposed for food product protection to improve quality and shelf life without impairing consumer acceptability. In addition, starch can be transformed into a foamed material by using water steam to replace the polystyrene foam as packaging material. It can be pressed into trays or disposable dishes, which are able to dissolve in water and leave a non-toxic solution that can then be consumed by microbes. Ultimately, the starch-based biodegradable polymers are attractive for use in the food industry and will be of great help in the future. Figure 8 illustrates various application of starch-based materials.

8.2 *In Agriculture*

Starch-based biodegradable polymers have found three major applications in agriculture: the covering of greenhouses, mulch film, and fertilizer-controlled release materials. Generally, the methods of disposal of

Fig. 8. Applications of starch-based materials.

agricultural films are landfill, recycling, or burning. They are time-consuming, expensive, and lead to environmental pollution. On the other hand, the utilization efficiency of fertilizers is the key element in the development and growth of agricultural products. However, due to surface runoff, leaching, and vaporization, the fertilizers escape to the environment and cause diseconomy and environmental problems. The development of starch-based biodegradable polymers offers a possibility to overcome the mentioned problems. They can be used as the fertilizer-controlled release matrices to release the fertilizers slowly or in a controlled way. As a result, the loss of fertilizers and environment pollution can be avoided or reduced. After use, starch-based films can be ploughed into the soil and disposed directly. Moreover, no toxic residues are formed after the degradation of starch-based biodegradable polymers. Thus, the development of starch-based materials for agricultural applications is being continued. For example, to enhance the mechanical properties and solvent or gas resistance, starch-based biodegradable materials are mixed with some nano-grade additives, such as TiO_2, layered silicate and MMT to form bio-nanocomposites.

8.3 *In the Medical Field*

Starch-based biodegradable polymers have some advantages for use as medical polymer materials, and these include the following:

(a) Good biocompatibility.
(b) Biodegradability, with its degradation products being non-toxic.
(c) Proper mechanical properties.
(d) Degradation as requirement.

Starch-based biodegradable polymers have been widely investigated in various biomedical applications including bone tissue engineering. Starch-based materials are biodegradable, and so can be used as bone cements that provide immediate structural support to the body and can be degraded from the site of application. Moreover, they can be pinned together with bioactive particles, which allow for new bone growth to be induced in both the interface of cement–bone and the volume left empty by polymer degradation. Starch-based biodegradable materials can also be used as bone tissue engineering scaffolds and are suitable for drug delivery in the form of microspheres or hydrogels. There is no need for surgical removal of the device after drug depletion. These unique properties, such as hydrophilicity, permeability, biocompatibility, and to some extent similarity to soft biological systems, of starch-based hydrogels make them useful for various biomedical applications. The 3D structure of starch-based hydrogels enables them to absorb and store plenty of water and retain a good-enough mechanical property at the same time. Starch-based hydrogels have thus been of increasing interest for biomedical applications.

Acknowledgment

The authors wish to acknowledge DST for awarding the INSPIRE Fellowship to Preetha Balakrishnan.

References

1. B. Sharma, P. Malik, and P. Jain, Biopolymer reinforced nanocomposites: A comprehensive review. *Mater. Today Commun.*, 2018, **16**, 353–363.
2. J. Wróblewska-Krepsztul, T. Rydzkowski, G. Borowski, M. Szczypiński, T. Klepka, and V. K. Thakur, Recent progress in biodegradable polymers and nanocomposite-based packaging materials for sustainable environment. *Int. J. Polym. Anal. Charact.*, 2018, **23**, 383–395.
3. H. Almasi, B. Ghanbarzadeh, and A. A. Entezami, Physicochemical properties of starch-CMC-nanoclay biodegradable films. *Int. J. Biol. Macromol.*, 2010, **46**, 1–5.
4. P. J. Halley, Thermoplastic starch biodegradable polymers. *Biodegrad. Polym. Ind. Appl.*, 2005, 140–162.
5. G. Madhumitha, J. Fowsiya, S. Mohana Roopan, and V. K. Thakur, Recent advances in starch–clay nanocomposites. *Int. J. Polym. Anal. Charact.*, 2018, **23**, 331–345.
6. X. Ma, P. R. Chang, and J. Yu, Properties of biodegradable thermoplastic pea starch/carboxymethyl cellulose and pea starch/microcrystalline cellulose composites. *Carbohydr. Polym.*, 2008, **72**, 369–375.
7. S. Sinha Ray and M. Okamoto, Polymer/layered silicate nanocomposites: A review from preparation to processing. *Prog. Polym. Sci.*, 2003, **28**, 1539–1641.
8. J. Kaur, G. Kaur, S. Sharma, and K. Jeet, Cereal starch nanoparticles — A prospective food additive: A review. *Crit. Rev. Food Sci. Nutr.*, 2018, **58**, 1097–1107.
9. S. Palanikumar, P. Siva, B. Meenarathi, L. Kannammal, and R. Anbarasan, Effect of Fe(3)O(4) on the sedimentation and structure-property relationship of starch under different pHs., *Int. J. Biol. Macromol.*, 2014, **67**, 91–98.
10. D. Le Corre, J. Bras, and A. Dufresne, Starch nanoparticles: A review. *Biomacromolecules*, 2010, **11**, 1139–1153.
11. W. Fu and T. Nakamura, Effects of starches on the mechanical properties and microstructure of processed cheeses with different types of casein network structures. *Food Hydrocoll.*, 2018, **79**, 587–595.
12. F. Vilaplana, E. Strömberg, and S. Karlsson, Environmental and resource aspects of sustainable biocomposites. *Polym. Degrad. Stab.*, 2010, **95**, 2147–2161.
13. X. Li, L. G. Tabil, and S. Panigrahi, Chemical treatments of natural fiber for use in natural fiber-reinforced composites: A review. *J. Polym. Environ.*, 2007, **15**, 25–33.
14. A. C. Bertolini, C. Mestres, J. Raffi, A. Buléon, D. Lerner, and P. Colonna, Photodegradation of cassava and corn starches. *J. Agric. Food Chem.*, 2001, **49**, 675–682.

15. L. Averous and N. Boquillon, Biocomposites based on plasticized starch: Thermal and mechanical behaviours. *Carbohydr. Polym.*, 2004, **56**, 111–122.
16. N. Soykeabkaew, N. Laosat, A. Ngaokla, N. Yodsuwan, and T. Tunkasiri, Reinforcing potential of micro- and nano-sized fibers in the starch-based biocomposites. *Compos. Sci. Technol.*, 2012, **72**, 845–852.
17. B. Priya, V. K. Gupta, D. Pathania, and A. S. Singha, Synthesis, characterization and antibacterial activity of biodegradable starch/PVA composite films reinforced with cellulosic fibre. *Carbohydr. Polym.*, 2014, **109**, 171–9.
18. Y. Yin, J. Li, Y. Liu, and Z. Li, Starch crosslinked with poly(vinyl alcohol) by boric acid. *J. Appl. Polym. Sci.*, 2005, **96**, 1394–1397.
19. K. Das, D. Ray, N. R. Bandyopadhyay, A. Gupta, S. Sengupta, S. Sahoo, A. Mohanty, and M. Misra, Preparation and characterization of cross-linked starch/poly(vinyl alcohol) green films with low moisture absorption. *Ind. Eng. Chem. Res.*, 2010, **49**, 2176–2185.
20. I. Kelnar, L. Kaprálková, L. Brožová, J. Hromádková, and J. Kotek, Effect of chitosan on the behaviour of the wheat B-starch nanocomposite. *Ind. Crops Prod.*, 2013, **46**, 186–190.
21. L. Godbillot, P. Dole, C. Joly, B. Roge, and M. Mathlouthi, Analysis of water binding in starch plasticized films. *Food Chem.*, 2006, **96**, 380–386.
22. H. Staroszczyk, K. Gottfried, E. Malinowska-Pańczyk, and I. Kołodziejska, Clay-filled starch films. Part I: Effect of clay kind and glycerol concentration on functional properties of composites. *Starch — Stärke*, 2017, **69**, 1500325.
23. W. A. W. A. Rahman, L. T. Sin, A. R. Rahmat, and A. A. Samad, Thermal behaviour and interactions of cassava starch filled with glycerol plasticized polyvinyl alcohol blends. *Carbohydr. Polym.*, 2010, **81**, 805–810.
24. P. T. Marques, a. M. F. Lima, G. Bianco, J. B. Laurindo, R. Borsali, J. F. Le Meins, and V. Soldi, Thermal properties and stability of cassava starch films cross-linked with tetraethylene glycol diacrylate. *Polym. Degrad. Stab.*, 2006, **91**, 726–732.
25. J.-W. Rhim, H.-M. Park, and C.-S. Ha, Bio-nanocomposites for food packaging applications. *Prog. Polym. Sci.*, 2013, **38**, 1629–1652.
26. A. P. Mathew and A. Dufresne, Morphological investigation of nanocomposites from sorbitol plasticized starch and tunicin whiskers. *Biomacromolecules*, 2002, **3**, 609–617.
27. J. Viguié, S. Molina-Boisseau, and A. Dufresne, Processing and characterization of waxy maize starch films plasticized by sorbitol and reinforced with starch nanocrystals. *Macromol. Biosci.*, 2007, **7**, 1206–1216.
28. C. M. O. Müller, F. Yamashita, and J. B. Laurindo, Evaluation of the effects of glycerol and sorbitol concentration and water activity on the water barrier properties of cassava starch films through a solubility approach. *Carbohydr. Polym.*, 2008, **72**, 82–87.

29. X. F. Ma, J. G. Yu, and Y. B. Ma, Urea and formamide as a mixed plasticizer for thermoplastic wheat flour. *Carbohydr. Polym.*, 2005, **60**(1), 111–116.
30. P. R. Chang, R. Jian, P. Zheng, J. Yu, and X. Ma, Preparation and properties of glycerol plasticized-starch (GPS)/cellulose nanoparticle (CN) composites. *Carbohydr. Polym.*, 2010, **79**, 301–305.
31. K. Xie, Y. Zhang, and Y. Yu, Preparation and characterization of cellulose hybrids grafted with the polyhedral oligomeric silsesquioxanes (POSS). *Carbohydr. Polym.*, 2009, **77**, 858–862.
32. P. Veiga-Santos, L. M. Oliveira, M. P. Cereda, and A. R. P. Scamparini. Sucrose and inverted sugar as plasticizer. Effect on cassava starch–gelatin film mechanical properties, hydrophilicity and water activity. *Food Chem.*, 2007, **103**(2), 255–262.
33. N. Wang, J. Yu, P. R. Chang, and X. Ma, Influence of formamide and water on the properties of thermoplastic starch/poly(lactic acid) blends. *Carbohydr. Polym.*, 2008, **71**, 109–118.
34. L., Ping, Y. Wu, Y. Zhou, and Y. Zuo. Preparation and characterization of resorcinol-dialdehyde starch-formaldehyde copolycondensation resin adhesive. *Int. J. Biol. Macromol.*, 2019, **127**, 12–17.
35. X. Ma, J. Yu, and J. F. Kennedy, Studies on the properties of natural fibers-reinforced thermoplastic starch composites. *Carbohydr. Polym.*, 2005, **62**, 19–24.
36. X. Ma, R. Jian, P. R. Chang, and J. Yu, Fabrication and characterization of citric acid-modified starch nanoparticles/plasticized-starch composites. *Biomacromolecules*, 2008, **9**, 3314–3320.
37. B. Ghanbarzadeh, H. Almasi, and A. A. Entezami, Improving the barrier and mechanical properties of corn starch-based edible films: Effect of citric acid and carboxymethyl cellulose. *Ind. Crops Prod.*, 2011, **33**, 229–235.
38. J. L. Willett, M. M. Millard, and B. K. Jasberg, Extrusion of waxy maize starch: Melt rheology and molecular weight degradation of amylopectin. *Polymer (Guildf).*, 1997, **38**, 5983–5989.
39. B. S. Miller, R. I. Derby, and H. B. Trimbo, A pictorial explanation for the increase in viscosity of a heated wheat starch — water suspension. *Cereal Chem.*, 1973, **50**, 271–280.
40. M. Iman, A. K. Manhar, T. K. Maji, C. Jia, H. Bian, T. Gao, F. Jiang, I. M. Kierzewski, Y. Yao, L. Chen, Z. Shao, J. Y. Zhu, L. Hu, Y. Liu, B. Guo, Q. Xia, J. Meng, W. Chen, S. Liu, Q. Wang, K. Song, H. Xu, K. Xie, and Y. Yang, Keratin-Based biocomposites reinforced and cross-linked with dual-functional cellulose nanocrystals. *RSC Adv.*, 2017, **4**, 33826–33839.
41. H. Xu, H. Canisag, B. Mu, and Y. Yang, Robust and flexible films from 100% starch cross-linked by biobased disaccharide derivative. *ACS Sustain. Chem. Eng.*, 2015, **3**(11), 2631–2639.

42. R. L. Shogren, Effect of moisture content on the melting and subsequent physical aging of cornstarch. *Carbohydr. Polym.*, 1992, **19**, 83–90.
43. S. H. D. Hulleman, F. H. P. Janssen, and H. Feil, The role of water during plasticization of native starches. *Polymer*, 1998, **39**(10), 2043–2048.
44. H. Liu, F. Xie, L. Yu, L. Chen, and L. Li, Thermal processing of starch-based polymers. *Prog. Polym. Sci.*, 2009, **34**, 1348–1368.
45. L. Liu, M. Tian, W. Zhang, L. Zhang, and J. E. Mark, Crystallization and morphology study of polyhedral oligomeric silsesquioxane (POSS)/polysiloxane elastomer composites prepared by melt blending. *Polymer (Guildf)*., 2007, **48**, 3201–3212.
46. A. L. Da Róz, A. J. F. Carvalho, A. Gandini, and A. A. S. Curvelo, The effect of plasticizers on thermoplastic starch compositions obtained by melt processing. *Carbohydr. Polym.*, 2006, **63**, 417–424.
47. L. Yu and G. Christie, Measurement of starch thermal transitions using differential scanning calorimetry. *Carbohydr. Polym.*, 2001, **46**, 179–184.
48. R. Smith, *Biodegradable Polymers for Industrial Applications*, Woodhead, 2005.
49. S. H. Sung, Y. Chang and J. Han, Development of polylactic acid nanocomposite films reinforced with cellulose nanocrystals derived from coffee silverskin, *Carbohydr. Polym.*, DOI:10.1016/j.carbpol.2017.04.037.
50. L. Vertuccio, G. Gorrasi, A. Sorrentino, and V. Vittoria, Nano clay reinforced PCL/starch blends obtained by high energy ball milling. *Carbohydr. Polym.*, 2009, **75**, 172–179.
51. S. Do Yoon, M. H. Park, and H. S. Byun, Mechanical and water barrier properties of starch/PVA composite films by adding nano-sized poly(methyl methacrylate-co-acrylamide) particles. *Carbohydr. Polym.*, 2012, **87**, 676–686.
52. I. Kelnar, L. Kaprálková, L. Brožová, J. Hromádková, and J. Kotek, Effect of chitosan on the behaviour of the wheat B-starch nanocomposite. *Ind. Crops Prod.*, 2013, **46**, 186–190.
53. H. P. S. Abdul Khalil, I. U. H. Bhat, M. Jawaid, a. Zaidon, D. Hermawan, and Y. S. Hadi, Bamboo fibre reinforced biocomposites: A review. *Mater. Des.*, 2012, **42**, 353–368.
54. J. Prachayawarakorn, P. Sangnitidej, and P. Boonpasith, Properties of thermoplastic rice starch composites reinforced by cotton fiber or low-density polyethylene. *Carbohydr. Polym.*, 2010, **81**, 425–433.
55. S. Y. Z. Zainuddin, I. Ahmad, H. Kargarzadeh, I. Abdullah, and A. Dufresne, Potential of using multiscale kenaf fibers as reinforcing filler in cassava starch-kenaf biocomposites. *Carbohydr. Polym.*, 2013, **92**, 2299–2305.
56. K. Das, D. Ray, N. R. Bandyopadhyay, S. Sahoo, A. K. Mohanty, and M. Misra, Physico–mechanical properties of the jute micro/nanofibril reinforced starch/polyvinyl alcohol biocomposite films. *Compos. Part B Eng.*, 2011, **42**, 376–381.

57. J. Gironès, J. P. López, P. Mutjé, A. J. F. Carvalho, A. A S. Curvelo, and F. Vilaseca, Natural fiber-reinforced thermoplastic starch composites obtained by melt processing. *Compos. Sci. Technol.*, 2012, **72**, 858–863.
58. P. Balakrishnan, M. S. Sreekala, M. Kunaver, M. Huskić, and S. Thomas, Morphology, transport characteristics and viscoelastic polymer chain confinement in nanocomposites based on thermoplastic potato starch and cellulose nanofibers from pineapple leaf. *Carbohydr. Polym.*, DOI:10.1016/j.carbpol.2017.04.017.
59. E. de M. Teixeira, D. Pasquini, A. A. S. Curvelo, E. Corradini, M. N. Belgacem, and A. Dufresne, Cassava bagasse cellulose nanofibrils reinforced thermoplastic cassava starch. *Carbohydr. Polym.*, 2009, **78**, 422–431.
60. M. G. Lomelí-Ramírez, S. G. Kestur, R. Manríquez-González, S. Iwakiri, G. B. De Muniz, and T. S. Flores-Sahagun, Bio-composites of cassava starch-green coconut fiber: Part II — Structure and properties. *Carbohydr. Polym.*, 2014, **102**, 576–583.
61. M. H. Salehudin, E. Salleh, S. N. H. Mamat, and I. I. Muhamad, Starch based active packaging film reinforced with Empty Fruit Bunch (EFB) cellulose nanofiber. *Procedia Chem.*, 2014, **9**, 23–33.
62. M. E. Mngomezulu, M. J. John, V. Jacobs, and A. S. Luyt, Review on flammability of biofibres and biocomposites. *Carbohydr. Polym.*, 2014, **111**, 149–182.
63. S. Cao, D. Song, Y. Deng, and A. Ragauskas, Preparation of starch-fatty acid modified clay and its application in packaging papers. *Ind. Eng. Chem. Res.*, 2011, **50**, 5628–5633.
64. A. Alemdar and M. Sain, Biocomposites from wheat straw nanofibers: Morphology, thermal and mechanical properties. *Compos. Sci. Technol.*, 2008, **68**, 557–565.
65. J. Cheng, P. Zheng, F. Zhao, and X. Ma, The composites based on plasticized starch and carbon nanotubes. *Int. J. Biol. Macromol.*, 2013, **59**, 13–19.
66. D. Liu, T. Zhong, P. R. Chang, K. Li, and Q. Wu, Starch composites reinforced by bamboo cellulosic crystals. *Bioresour. Technol.*, 2010, **101**, 2529–2536.
67. J. Gironès, J. P. López, P. Mutjé, A. J. F. Carvalho, A. A. S. Curvelo, and F. Vilaseca, Natural fiber-reinforced thermoplastic starch composites obtained by melt processing. *Compos. Sci. Technol.*, 2012, **72**, 858–863.
68. M. F. Rosa, B. S. Chiou, E. S. Medeiros, D. F. Wood, T. G. Williams, L. H. C. Mattoso, W. J. Orts, and S. H. Imam, Effect of fiber treatments on tensile and thermal properties of starch/ethylene vinyl alcohol copolymers/coir biocomposites. *Bioresour. Technol.*, 2009, **100**, 5196–5202.
69. C. M. O. Müller, J. B. Laurindo, and F. Yamashita, Effect of cellulose fibers addition on the mechanical properties and water vapor barrier of starch-based films. *Food Hydrocoll.*, 2009, **23**, 1328–1333.

70. C. A. Romero-Bastida, L. A. Bello-Pérez, M. A. García, M. N. Martino, J. Solorza-Feria, and N. E. Zaritzky, Physicochemical and microstructural characterization of films prepared by thermal and cold gelatinization from non-conventional sources of starches. *Carbohydr. Polym.*, 2005, **60**, 235–244.
71. S. Gopi, P. Balakrishnan, C. Divya, S. Valic, E. Govorcin Bajsic, A. Pius, and S. Thomas, Facile synthesis of chitin nanocrystals decorated on 3D cellulose aerogels as a new multi-functional material for waste water treatment with enhanced anti-bacterial and anti-oxidant properties. *New J. Chem.*, 2017, **41**, 12746–12755.
72. E. J. Foster, R. J. Moon, U. P. Agarwal, M. J. Bortner, J. Bras, S. Camarero-Espinosa, K. J. Chan, M. J. D. Clift, E. D. Cranston, S. J. Eichhorn, D. M. Fox, W. Y. Hamad, L. Heux, B. Jean, M. Korey, W. Nieh, K. J. Ong, M. S. Reid, S. Renneckar, R. Roberts, J. A. Shatkin, J. Simonsen, K. Stinson-Bagby, N. Wanasekara, and J. Youngblood, Current characterization methods for cellulose nanomaterials. *Chem. Soc. Rev.*, 2018, **47**, 2609–2679.
73. Y. Habibi, L. A. Lucia, and O. J. Rojas, Cellulose nanocrystals: Chemistry, self-assembly, and applications. *Chem. Rev.*, 2010, **110**(6), 3479–3500.
74. S. Malmir, B. Montero, M. Rico, L. Barral, and R. Bouza, Morphology, thermal and barrier properties of biodegradable films of poly (3-hydroxybutyrate-co-3-hydroxyvalerate) containing cellulose nanocrystals. *Compos. Part A Appl. Sci. Manuf.*, 2017, **93**, 41–48.
75. M. Mondragón, K. Arroyo, and J. Romero-García, Biocomposites of thermoplastic starch with surfactant. *Carbohydr. Polym.*, 2008, **74**, 201–208.
76. T. Saito, S. Kimura, Y. Nishiyama, and A. Isogai, Cellulose nanofibers prepared by TEMPO-mediated oxidation of native cellulose. *Biomacromolecules*, 2007, **8**, 2485–2491.
77 N. Lin, J. Huang, and A. Dufresne, Preparation, properties and applications of polysaccharide nanocrystals in advanced functional nanomaterials: A review. *Nanoscale*, 2012, **4**, 3274.
78. R. J. Moon, A. Martini, J. Nairn, J. Simonsen, and J. Youngblood, Cellulose nanomaterials review: Structure, properties and nanocomposites. *Chem. Soc. Rev.*, 2011, **40**, 3941.
79. M. Wollerdorfer and H. Bader, Influence of natural fibres on the mechanical properties of biodegradable polymers. *Ind. Crops Prod.*, 1998, **8**, 105–112.
80. A. Rani, S. Monga, M. Bansal, and A. Sharma, Bionanocomposites reinforced with cellulose nanofibers derived from sugarcane bagasse. *Polym. Compos.*, DOI:10.1002/pc.24112.
81. D. Chen, D. Lawton, M. R. Thompson, and Q. Liu, Biocomposites reinforced with cellulose nanocrystals derived from potato peel waste. *Carbohydr. Polym.*, 2012, **90**, 709–716.

82. A. P. Mathew and A. Dufresne, Morphological investigation of nanocomposites from sorbitol plasticized starch and tunicin whiskers. *Biomacromolecules*, 2002, **3**, 609–617.
83. A. Dufresne, Plasticized starch/tunicin whiskers nanocomposites. 1. *Structural Analysis*, 2000, 8344–8353.
84. E. Neter, O. Westphal, O. Luderitz, E. A. Gorzynski, and E. Eichenberger, Studies of enterobacterial lipopolysaccharides: Effects of heat and chemicals on erythrocyte-modifying, antigenic, toxic and pyrogenic properties. *J. Immunol.*, 1956, **76**, 377–85.
85. C. J. Grande, F. G. Torres, C. M. Gomez, O. P. Troncoso, J. Canet-Ferrer, and J. Martínez-Pastor, Development of self-assembled bacterial cellulose–starch nanocomposites. *Mater. Sci. Eng. C*, 2009, **29**, 1098–1104.

https://doi.org/10.1142/9781786349477_0005

Chapter 5

Recent Developments in Synthesis, Properties, and Biomedical Applications of Cellulose-Based Hydrogels

Zhijun Shi, Muhammad Wajid Ullah, Xinshuo Liang, and Guang Yang*

Department of Biomedical Engineering, College of Life Science and Technology, Huazhong University of Science and Technology, Wuhan, PR China

yang_sunny@yahoo.com

Abstract

The cellulose-based hydrogels have received immense interest for various biomedical applications owing to their unique structural, physicochemical, mechanical, and biological properties. This chapter provides an overview of different cellulose-dissolving systems, cellulose derivatives, cellulose nanostructures, and cellulose-producing microorganisms, with emphasis on the fabrication, properties, and biomedical applications of cellulose-based hydrogels. Further, it highlights the current developments of the latest synthesis methods and technologies, such as 3D and 4D printing, inspired from the naturally occurring biomaterials or soft/hard tissues with numerous excellent properties, to fabricate cellulose-based hydrogels with a highly ordered microstructure, to further improve their properties and applications.

Keywords: Nanocellulose; cellulose derivatives; hydrogels; regeneration; bioprinting; biomedical applications

1. Introduction

Hydrophilic polymers form a three-dimensional (3D) network that retains a large amount of interstitial water (free/freezable or bound/non-freezable water), which results in the formation of hydrogels. Hydrogels are composites that contain two phases, a solid phase and a liquid phase. The 3D polymeric network in a hydrogel provides the solid phase, which provides the hydrogels with soft elastomeric mechanical properties while the liquid phase (water) is incorporated into the 3D network [1–3]. Moreover, the amount of liquid phase can be changed by external stimuli such as temperature, pH, or ionic gradient, which results in volume or mass swelling change seen in the hydrogel [4]. Further, the liquid phase may also possess reactive, monomeric, and potentially polymerizable species in the interstices of the hydrogel matrix and occupy the empty spaces. Cellulose is the most abundant natural polymer, and its hydrogels have been of substantial interest in various research fields owing to their highly safety level and unique properties, such as high biocompatibility, biodegradability, and hydrophilic nature. To date, various methods have been developed to fabricate cellulose-based hydrogels. Natural cellulose is difficult to dissolve in common organic and inorganic solvents; therefore, several solvent systems including LiCl/dimethylacetamide (DMAc) [5], *N*-methylmorpholine-*N*-oxide (NMMO) [6, 7], ionic liquids (ILs) [8, 9], and alkali/urea (or thiourea) aqueous systems [10], have been developed to dissolve cellulose, which provide the basis for designing and fabrication of regenerated cellulose-based hydrogels. In addition, the water-soluble cellulose derivatives of cellulose, such as methylcellulose (MC), hydroxypropyl methylcellulose (HPMC), ethyl cellulose (EC), hydroxyethyl cellulose (HEC), carboxymethylcellulose (CMC), and carboxyethyl cellulose (CEC), are also used to form hydrogels by using appropriate cross-linking approaches [11]. Owing to the several advantages such as renewability, biodegradability, low density, and high tensile strength, cellulose nanomaterials such as such as cellulose nanocrystals (CNC), nanofibrillar cellulose (NFC), and TEMPO-oxidized cellulose nanofibers (TOCN) are widely used both for the preparation of hydrogels or as reinforcement materials to improve the properties of composites hydrogels [12].

In addition, bacterial cellulose (BC), a natural cellulose hydrogel produced by different microorganisms such as bacteria, algae, and fungi, has also received immense consideration for various biomedical and other applications owing to its unique structural, physico-chemical, mechanical, and biological features [13]. These cellulose-based hydrogels are environmentally-friendly and low cost, and possess several useful properties such as hydrophilicity, biodegradability, biocompatibility, high mechanical

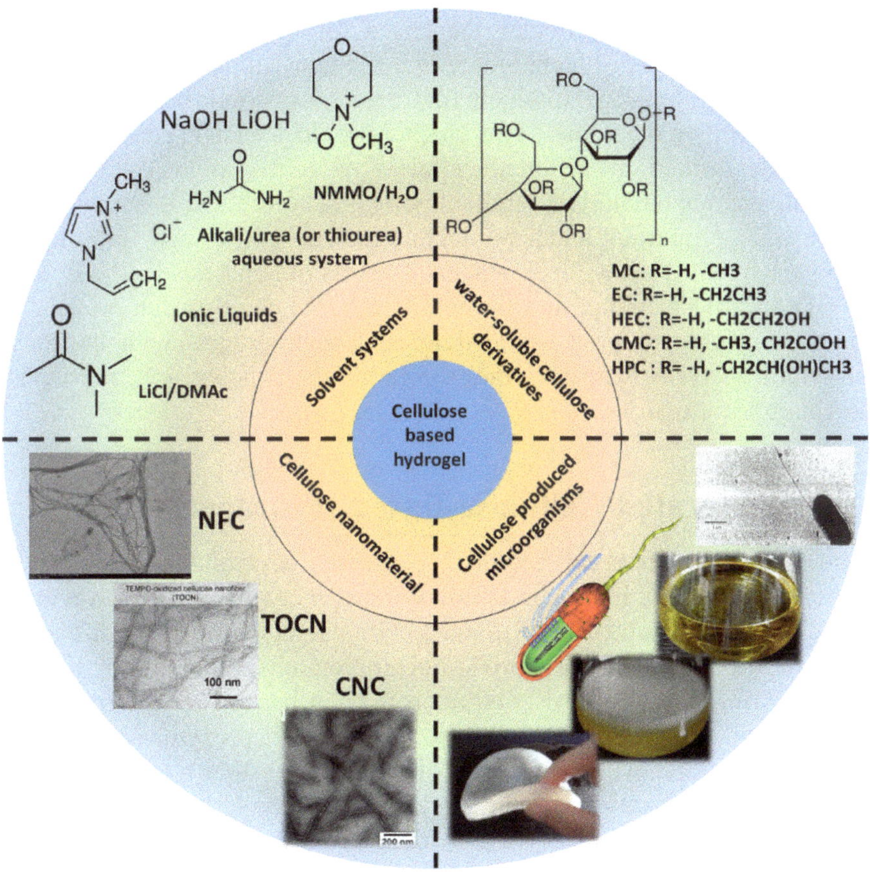

Fig. 1. Different cellulose-dissolving systems, cellulose derivatives, cellulose nanostructures, and cellulose-producing microorganisms with emphasis on fabrication of cellulose-based hydrogels.

Source: Reproduced with permission from Refs. [12, 13, 27].

strength, low weight, flexibility, and optical transparency, which are important from the medical perspective [14–16]. In general, these nanocellulose-based materials have received immense consideration in different fields such as agriculture (water reservoir and fertilizer release) [17, 18], environment treatment (heavy metal adsorption, oil/water separation) [19, 20], energy (transfer, storage, and capture) [21, 22], and biomedicine (tissue engineering and drug delivery) [23–26].

Inspired by nature, the highly ordered self-assembled structures provide the naturally occurring biomaterials and soft/hard tissues with excellent properties. This provides the based for the development of new methods and advanced technologies for the fabrication of highly ordered microstructure hydrogels, which find promising applications in bionics and regenerative medicines. This chapter provides an overview of various hydrogels formed by different cellulose solvent systems (see Fig. 1), derivatives, nanomaterials, and microbial cellulose, with emphasis on fabrication, properties, and potential biomedical applications. Importantly, it highlights the recent developments in terms of synthesis methods, such as 3D and 4D printing, and the latest technologies that can be used to fabricate highly ordered nanocellulose-based hydrogels with improved structural features, with the aim to improve the existing uses and devise novel applications.

2. Regenerated Cellulose-Based Hydrogels

The molecular chain of cellulose possesses several hydroxyl (−OH) groups that form a hydrogen bonding-linked network [28]. However, as the skeletal component in all plants, cellulose is organized in a distinct fiber morphology, and the morphological hierarchy is defined by the elementary fibrils, microfibrils, and microfibrillar bands [6]. The fiber morphology allows the cellulose to possess a highly extended hydrogen bonded structure, which makes it difficult to be dissolved in common solvents [29]. Recently, several solvent systems have been developed to dissolve cellulose, which have provided the base to design and fabricate regenerated cellulose hydrogels. These solvent systems include LiCl/dimethylacetamide (DMAc) [5], *N*-methylmorpholine-*N*-oxide (NMMO) [6, 7], ionic liquids (ILs) [8, 9], and alkali/urea (or thiourea) aqueous systems [10] (Table 1). These solvent systems disrupt the hydrogen bonding, which results in almost complete loss of the supramolecular structure of cellulose [6]. In contrast to the commonly used solvent systems, the

Table 1. Summary of various cellulose-dissolving solvent systems, cross-linking methods, improved features of hydrogels, and their applications.

Solvent systems	Celluloses origin	Combine	Coagulated/Cross linking method	Property	Applications	Refs.
N-methylmorpholine-*N*-oxide (NMMO)	Cellulose	—	Deionized water	—	—	[30]
	Sugarcane bagasse	—	Ethanol vapor	—	—	[47]
DMAc/LiCl system	Wood pulp cotton and BC	—	Water	High mechanical strength and optical transparency	—	[48]
	Softwood dissolving pulp	—	Ion exchange resins	Colorless, transparent, and optical anisotropic behavior	—	[49]
	Cotton pulp	CNC	Isopropanol (IPA)	pH-responsiveness and dielectric behavior	—	[50]
	Defatted cotton	Mimosa	Ethanol atmosphere	—	Ultrasound (US)-responded drug release	[51]
LiBr aqueous solution	Filter paper pulp	Silk fibroin	Methanol	Highly porous, 3D networks composed of long and interconnected fibrils	Improved adhesion and growth of hASCS	[37]
Ionic liquids: [EMIm][OAc]: DMSO	BC	Carbohydrate-binding modules	Methanol/water	Bioaffinity	Bioaffinity applications	[52]
Ionic liquids: BMIMCl	MCC	PPy	*N*,*N*-methylene bisarcylamide (BIS)	Improved mechanical and conductive properties	Biological and semiconductive materials	[33]
Ionic liquids: BMIMAC BMIMCL	MCC	—	Ethanol/water	Enhanced adsorption ability	Adsorbent of dyes	[8]

(Continued)

Table 1. *(Continued)*

Solvent systems	Celluloses origin	Combine	Coagulated/Cross linking method	Property	Applications	Refs.
N,N-dimethylacetamide (DMA), triethylamine (TEA)	Urea-modified MCC	—	Methanol	Improved swelling behavior	Controlled fertilizer release	[18]
Tetrabutylammonium fluoride/dimethyl sulfoxide (TBAF/DMSO)	MCC	—	Water	High mechanical strength	—	[53]
NaOH/urea system	MCC	—	Epichlorohydrin (ECH)	Hydrogel-coated nylon mesh	High-efficiency oil–water separation	[54]
	Cellulose derived from kenaf core powder	—	ECH	—	—	[55]
	MCC	Cyclosophoraose	ECH	High loading and sustained release of galangin	Efficient galangin delivery system	[56]
	Whatman CF-11 fibrous cellulose powder	Sodium alginate	ECH	Macroporous structure	—	[57]
	Cellulose powder	Black phosphorus	ECH	Microporous structure with excellent photothermal response, mechanical stability, flexibility, and biocompatibility, *in vitro* and *in vivo*	Photothermal agents in effective cancer therapy	[36]

LiOH/urea system	Filter paper pulp	—	Methanol	No cytotoxicity	Successful synthesis of hASCS-spheroids on hydrogels	[58]
	TEMPO-oxidized cellulose	—	Methanol	High adsorption capacity of heavy metal ion	Heavy metal ion adsorbent	[59]
	Cellulose carbamate	—	ECH	Highly porous structure	—	[60]
	Cotton linter pulps	—	ECH	Microstructural replication ability	Transparent microchip material	[61]
	Cotton linter pulp	—	ECH and ethanol	High mechanical strength and toughness	—	[46]
	Cellulose powder	Halloysite nanotube	ECH and hydrogen bonding interactions between HNT and cellulose	Improved mechanical properties	Curcumin delivery	[35]

cellulose/NMMO/water solvent system is of considerable significance because of its safe nature and economically effective production of cellulose fibers by the aminoxide process. Owing to its strong N–O dipole, NMMO in combination with water can dissolve cellulose without prior activation or derivatization. Thus, regenerated cellulose hydrogels with improved thermal stability, high equilibrium swelling, and high retention ratios were prepared using the NMMO aqueous solution. These improved features of cellulose hydrogels could be attributed to the high content of regenerated cellulose [30]. The LiCl/dimethylacetamide (DMAc) is an efficient solvent system that dissolves cellulose from different sources such as wood pulp, cotton, BC, and pre-treated bagasse, [5] each possessing different molecular weight ($>10^6$), at an ambient temperature without noticeable degradation [31]. It finds wide applications in organic synthesis and for other analytical uses. Besides, ionic liquids are also used for the preparation of regenerated cellulose-based hydrogels. The ionic liquids are solely composed of ions and exist in the form of liquid, and they are known for their promising and environmentally-friendly features such as non-toxicity, no vapor pressure, strong polarity, stability, and reusability. A variety of ionic liquids such as 1-butyl-3-methylimidazolium chloride (BMIMCl) [8], 1-allyl-3-methylimidazolium chloride (AMIMCL), 1-ethyl-3-methylimidazolium chloride (EMIMCL) [9], 1-ethyl-3-methylimidazolium acetate (EMIMAC) [32], etc. can effectively dissolve cellulose, and thus these can be applied in the preparation of cellulose-based hydrogels. Zhang and co-workers developed an alkali/urea aqueous system for dissolution of cellulose involving a dynamic self-assembly process among the solvent molecules (NaOH, urea, and water) and cellulose macromolecules. The self-assembled clusters created a possible new pathway for the "dissolution" of natural polymers through the hydrogen-bonded arrangement of macromolecules and some special small molecules in an aqueous system [10]. This alkali/urea aqueous system can be used for the preparation of regenerated cellulose hydrogels.

The regenerated cellulose hydrogels provide a matrix that can be used to fabricate functional materials such as electroconductive hydrogels. The monomers of functional polymers, such as pyrrole [33] and aniline, [34] can be polymerized within the matrix to enhance the electroactivity of the hydrogel, which can then find potential applications in actuation, electrochemical biosensors, electro-stimulated controlled drug release systems, and neural prosthetics. Similarly, functional nanomaterials such as halloysite nanotubes (HNTs) [35] and black phosphorus (BP) [36] can also be added to cellulose solution to prepare composite hydrogels for their applications in drug delivery systems, wound dressings, and photothermal therapy. Furthermore, different natural polymers such as silk fibroin [37], pectin [38], and cyclodextrin [39] can be mixed with cellulose solution to prepare biocompatible hydrogels for various biomedical applications.

Cellulose in the form of solutions can be molded into different shape hydrogels for diverse biomedical applications. The feasibility of using cellulose-based hydrogels could be further increased by improving their

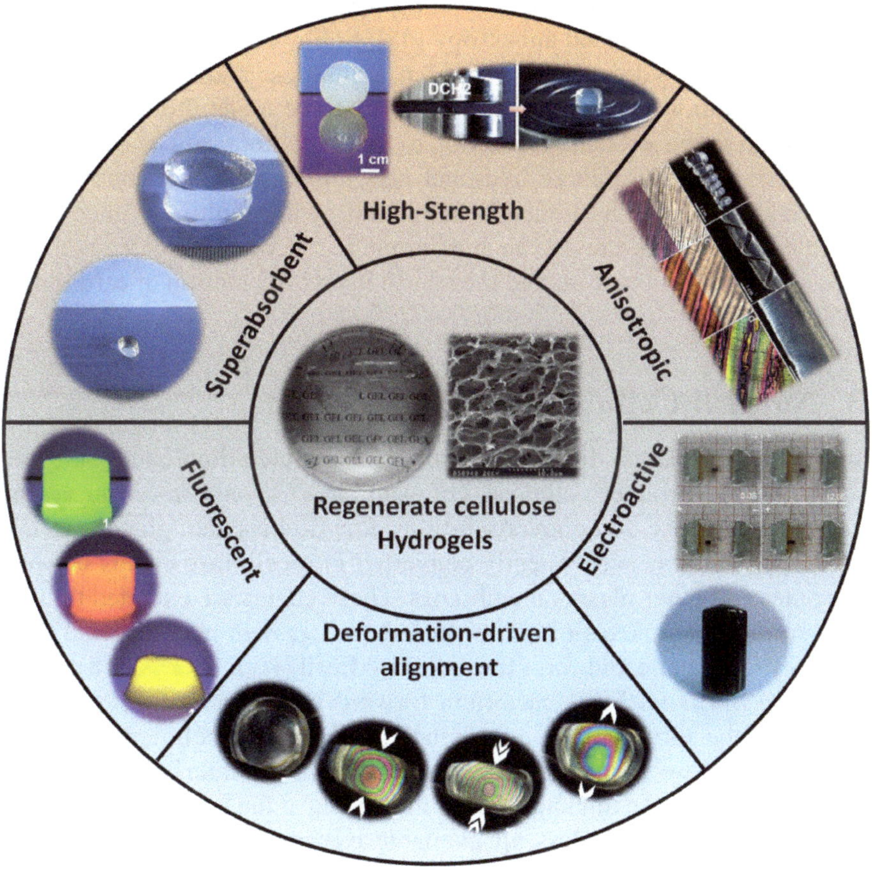

Fig. 2. Regenerative cellulose hydrogels can be formed with special properties, such as superabsorbence, high strength, anisotropicity, electroactivity, fluorescence, deformation alignment, and so on.

Source: Reproduced with permission from Refs. [34, 40–45].

mechanical features, cell biocompatibility, and structural bionics requirements for practical applications. To improve the mechanical features of regenerated cellulose hydrogels (see Fig. 2), Zhao *et al.* combined a sequential process of chemical cross-linking with epichlorohydrin and physical cross-linking from an aqueous ethanol solution using the cellulose/alkali hydroxide/urea aqueous system, resulting in a high-strength and high-toughness double-cross-linked cellulose hydrogel [46]. In another study, the toughness of cellulose hydrogels was improved by reacting cellulose with a small amount of epichlorohydrin (EPI) in LiOH/urea solution and subsequently treating it with dilute acid, which exhibited deformation-induced anisotropy [45]. Following this research, regenerate cellulose hydrogels formed from a cellulose/alkali hydroxide/urea aqueous solution pre-stretched by external force made the hydrogel network align in the stretching direction to form a temporarily oriented structure. This slightly oriented hydrogel was immersed in aqueous sulfuric acid to improve the hydrogen bonding and obtain highly oriented structured hydrogels. As a result, an anisotropic regenerated cellulose hydrogel was obtained, which promoted the adhesion and orientation of cardiomyocytes [44].

3. Natural Cellulose-Based Hydrogels

In addition to plants, different microorganisms including bacteria, algae, and fungi also produce cellulose. For example, *Gluconacetobacter xylinus* consumes glucose and converts it to uridine diphosphate glucose (UDP-glucose), which is subsequently converted into cellulose chains between the outer and inner plasma membranes. These chains are excreted through the terminal complexes (TCs) to the external growth medium, and these then self-aggregate and form the cellulose fibril (Kim *et al.*, 2018). These cellulose fibrils push the bacterium forward in a random fashion, which results in the formation of a reticulated web-shaped cellulose structure. Finally, cellulose is accumulated at the air–medium interface in the form of a hydrogel whose thickness keeps increasing at the bottom face towards the growth medium unless all available oxygen and nutrients are consumed and the cells die due to depletion of oxygen or nutrients. This kind of cellulose hydrogel helps to colonize the bacteria and maintain a high oxygen content near the surface and serves as a protective barrier against drying, natural enemies, and radiation. Bacterial cellulose (BC) is a natural hydrogel possessing 3D polymeric networks and adsorbs large amounts of water (99 vol%). The degree of polymerization (DP) of BC is considerable, and its value ranges between 2000–8000. Besides, it also possesses high crystallinity (60–90%) [62]. It possesses good mechanical properties and shows mechanical anisotropy [63]. Precisely, the mechanical properties of the BC are weaker in the stratified direction as compared to those along the fiber layer direction as bacteria synthesize cellulose in a layer-by-layer manner. Due to this behavior, 2D network layers are formed on top of each other, which are comprized of several entangled fibers. Consequently, no fibers are entangled in the stratified direction, and the strength of the mechanical property lies in the entangled fibers. Previous studies have reported various mechanical properties of BC fibers along the layer direction, stratified direction, and that of a single fiber [64, 65]. The values of the elastic modulus of BC vary over a wide range due to varying bacterial strains under different culture conditions, cultivation methods, incubation times, and post-synthesis processing. The elastic modulus of a BC hydrogel along the fiber-layer direction ranges from 1 to 15 MPa in wet state, whereas its elongation at break is about 10–30% [63, 66–68]. Specifically, the fiber-layer direction consists of interwoven fibers that can potentially form anisotropic aligned structure under external loading, consequently enhancing stiffness [69–71].

BC hydrogels have received immense consideration in the biomedical field, where these are mostly used as wound dressing materials because of their high water holding capacity, slow water release rate, high mechanical strength, fibrous structure, and excellent biocompatibility [72]. In the research of Fu *et al.* [23], a low cytotoxicity of the BC film and good proliferation of human adipose-derived stem cells on it were observed, indicating its high biocompatibility, thus finding potential clinical applications. However, the innate biocompatibility of BC to support different types of cells is still below the mark. Therefore, extensive efforts have been made to improve its biocompatibility, by making its composites with different natural polymers such as hyaluronan [73], gelatin [74], silk sericin [25], and chitosan [75]. The 3D fibrous network supports the loading of different growth factors [76] and drugs [77] and supports their controlled release. Further, the BC network also interacts with different monomers, which react and polymerize along its fibers and occupy the empty spaces in the network [78]. The 3D fibrous structure of BC also supports the impregnation of different inorganic nanoparticles, metal ions and oxides, carbon nanotubes, graphene and grapheme oxide, and conducting polymers, thus improving the existing properties and imparting additional features to BC, such as antimicrobial activity, magnetic functionality, antioxidant properties, and electroactivity [41, 79]. For example, BC decorated with 4,6-diamino-2-pyrimidinethiol-modified gold nanoparticles showed better efficacy than most of the antibiotics (cefazolin/sulfamethoxazole) against Gram-negative bacteria, while maintaining excellent physico-chemical properties, including better water uptake capability, mechanical strength, and biocompatibility. The developed BC-Au composite hydrogel can be a promising platform for treating superinfected wounds [80]. In another study, a super-soft multichannel electrode was fabricated by depositing gold layers on BC. The Au-BC electrodes showed high electrical conductivity and durability (conductivity > 95% after 100 cycles of 180° bending). *In vivo* recording of brain electrical activity demonstrated a great potential of the developed Au-BC electrodes for neural interfacing applications [81]. Using electrochemical polymerization method, double network BC-based electroactive hydrogels can be developed by depositing conductive polymers (CPs) on the BC fibers, where one layer is comprised of electroactive polymer and the other layer is BC. The electroconductive polymer confers the voltage and current response, while BC provides biocompatibility, bioadhesion, and mass transport properties to the composites. These double network BC–CP hydrogels are biphasic Janus hydrogels that integrate electroactivity with biocompatibility and can serve as biology–device interfaces to develop implantable devices for personalized and regenerative medicines [82].

BC, as a natural cellulose hydrogel (see Fig. 3), can form multi-shaped pulps, filaments, spheres, and sheets during the fermentation process [85]. In addition, using the specialized molds, the BC hydrogels with macro-order structures such as tubes [86, 87], cartilages [88], or organ shapes such as ear [89] can be fabricated. This feature of BC is of interest in vascular and cartilage tissue engineering applications. Further, the ordered structure of BC hydrogel in micro size can also be achieved [90–93]. For example, micropatterned BC (5–45 μm) was bioprinted during the fermentation by *Acetobacter xylinus* using a slab of micropatterned polydimethylsiloxane (PDMS) as a template. The micropatterned BC can potentially guide the aligned cell growth, mimicking the aligned cell organization of the native tissue, which can be used in annulus fibrosus [94] and nerve tissue repair and regeneration [95].

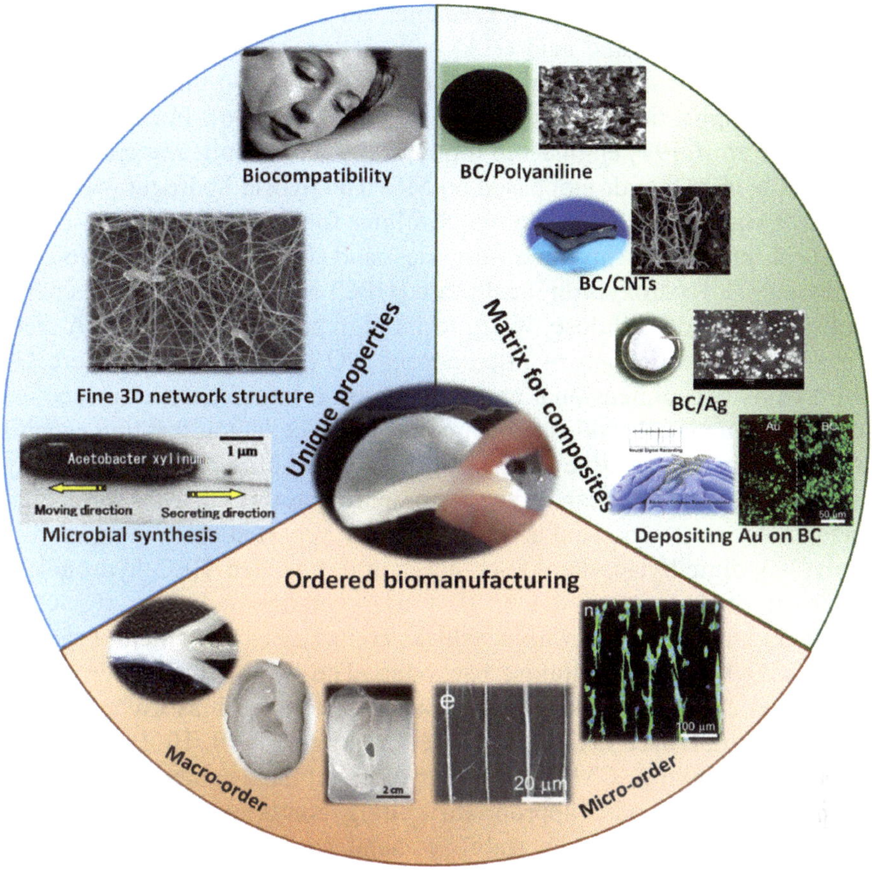

Fig. 3. Bacterial cellulose hydrogel.

Source: Reproduced with permission from Refs. [41, 81, 83–85].

4. Cellulose-Derivatives-Based Hydrogels

In addition to using different solvent systems, the cellulose hydrogels can also be formed through cross-linking of aqueous solutions of water-soluble cellulose derivatives, such as methylcellulose (MC), hydroxypropyl methylcellulose (HPMC), ethyl cellulose (EC), hydroxyethyl cellulose (HEC), carboxymethylcellulose (CMC), and carboxyethyl cellulose (CEC).

Methylcellulose (MC) is a thermosensitive hydrophilic compound derived from pure cellulose. Using polyethylene glycol as a cross-linker, a fast gelling injectable, degradable, and biocompatible composite hydrogel of hyaluronic acid (HA) and methylcellulose was prepared to meet the special shape requirement of injured tissues; thus, this can find potential applications in neural tissue engineering [96]. In another study, Pakulska *et al.* [97] engineered a physically and chemically cross-linked injectable hybrid hydrogel based on MC. The hybrid hydrogel was developed *in situ* and showed minimal swelling for long time; thus, it can be used in minimally invasive surgeries and for sustained biomolecule release [97]. Hydroxypropyl cellulose (HPC) is another important chemical derivative of cellulose owing to its high solubility in water. A study reported the fabrication of microporous GO/HPC hydrogels where GO worked as two-dimensional nanofillers with a large specific surface and multiple functional groups, which conferred excellent adsorption capability to the hydrogel in the removal of methylene blue [98]. The hydroxyethyl cellulose (HEC) can be used to prepare porous hydrogels using citric acid (CA) as a cross-linker [11]. A study reported the preparation of poly(*N,N*-dimethylacrylamide-co-acrylamide)-grafted HEC hydrogel for removal of congo red dye [99]. Carboxymethyl cellulose (CMC) is another important derivative which contains carboxymethyl groups ($-CH_2COO^-$) on the cellulose backbone. The carboxymethyl groups on CMC can coordinate with multivalent metal ions such as Ca^{2+}, Al^{3+}, and Fe^{3+}, etc. to prepare ionic cross-linked hydrogels [100, 101]. Apart from ionic cross-linking, CMC can also be cross-linked under irradiation to form hydrogels under appropriate gelling conditions [102, 103]. The CMC-based hydrogels can stabilize metal nanoparticles, such as ZnO, [100, 104] to form bionanocomposite hydrogels, which can be used as wound dressing and drug carriers. In addition, CMC can combine with synthetic polymers such as polyacrylic acid [105] and polyvinyl alcohol [106] through photo-cross-linking and chemical-cross-linking to form hydrogels that can be used as drug delivery systems and as scaffolds in tissue engineering. Natural polymers such as starch [107] and β-cyclodextrin [108] can also be combined with CMC through photo-cross-linking and chemical-cross-linking to form hybrid hydrogels, which can be then used as superabsorbent materials and in drug delivery applications (see Table 2).

Self-healing cellulose-based hydrogels are rarely reported because cellulose is difficult to dissolve in common solvents and lacks the bonding sites (e.g., amino or carboxyl) to form dynamic covalent bonds. Fortunately, the cellulose derivatives are water-soluble and can be grafted by various

Table 2. Summary reinforcement material, cross-linking methods, improved properties, and applications of various cellulose derivatives-based hydrogels.

Cellulose derivatives	Reinforcement materials	Cross linking method	Improved features	Applications	Refs.
Carboxymethyl cellulose	—	Fe^{2+}, Ca^{2+}	—	Plant root-targeted delivery of fertilizers	[101]
	—	Radiation technique (^{60}Co γ source with dose rate 3.15 Gy/s)	—	—	[103]
	Oxidized regenerated cellulose (ORC)	—	Reduced synthesis of intraabdominal adhesion	Adhesion barriers	[109]
	HEC	Citric acid, fumaric acid, malic acid, and 1-ethyl-3-(3-dimethylaminopropyl) carbodiimide hydrochloride	Superabsorbent	Water reservoir in agriculture	[17, 110, 111]
	ZnO nanoparticle	Fe^{3+}	pH sensitive swelling behavior	Sustained and controlled drug releases	[100]
	ZnO mesoporous silica	Citric acid	Improved tensile strength, swelling, erosion, and gas permeability	Wound dressing and sustained drug delivery	[104]
Hydroxyethyl cellulose	—	Citric acid	Fast swelling property	—	[11]
Hydroxypropyl cellulose	Graphene oxide	Divinyl sulfone	Excellent adsorption performance	Removal of dyes	[98]
Hydroxypropyl methylcellulose	Silicium	NaOH	Injectable, self-setting	Repair of articular cartilage defects	[112]
Methylcellulose	—	Physical cross-linking: hydrophobic interactions between MC chains and chemical cross-linking: reaction of thiol-modified MC with PEG-bismaleimide cross-linker	Injectable, *in situ* gelling, minimal swelling, and durability	A predictable and tunable platform for local drug delivery	[97]

functional radical groups, which can be used to develop self-healing hydrogels derived from cellulose. A self-healing CMC hydrogel based on physical bonding can be achieved by acidifying the CMC in citric acid solution. In one study, after immersing the Na-CMC into citric acid solution, the H^+ ions gradually diffused into the Na-CMC and bonded to the CMC polymer chains *via* hydrogen bonding, which resulted in a hydrogel with 2.5 MPa compressive strength and excellent self-healing behavior, with over 80% healing efficiency [113]. In another study, the self-healing and photoluminescent CMC-based hydrogel was fabricated by doping Al^{3+} and photoluminescent citric acid derivatives (PCAD) into the CMC polymer networks. The $-COO^-$ groups of PCAD appropriately attenuated the strong ionic interaction between the Al^{3+} ions and $-COO^-$ groups of the CMC, which facilitated dynamic ionic coordinating interactions to achieve self-healing capability of the hydrogels [114]. The dynamic covalent hydrogels can be easily prepared from the carboxylate-containing polysaccharides through the formation of phenylboronate ester [115] or acylhydrazone [116] cross-links. Yang *et al.* [116] used CEC and poly(ethylene glycol) (PEG) to synthesize CEC-graft-dithiodipropionate dihydrazide and dibenzaldehyde-terminated PEG. The two grafted precursor polymers formed cellulose-based self-healing hydrogels *via* covalent acylhydrazone linkages. The resulting hydrogels exhibited excellent self-healing ability with a high healing efficiency (≈96%), good mechanical properties, and excellent biocompatibility for 3D cell culture.

5. Cellulose as Reinforcement Material in Composite Hydrogels

The well-dispersed nano-size materials can effectively disperse in polymer hydrogel matrices to improve the various features of hydrogels such as mechanical properties, smart properties, and so on. The cellulose nanomaterials such as CNC, NFC, and TEMPO-oxidized cellulose nanofibers (TOCN) are widely used to reinforce various polymers (Table 3).

Cellulose nanocrystals. Cellulose nanocrystals are a renewable nanomaterial obtained from wood pulp through acid hydrolysis. Morphologically, CNCs exist in a rod/needle-like form with dimensions less than 100 nm in length with a high degree of crystallinity. These can be viewed as functional and robust fillers for different applications owing to their

Table 3. Summary of various methods, improved properties, and structural features of composites hydrogels prepared by using different forms of celluloses as reinforcement materials.

Cellulose form	Methods	Property	Morphology and size	Refs.
Cellulose nanocrystals	Acid hydrolysis	High surface area (~250 m^2/g), high tensile strength (7500 MPa), high stiffness (E > 140 GPa), and high aspect ratio (~72)	Rod/needle-shaped nanocrystals with length less than 500 nm	[27]
Nanofibrillar cellulose	Homogenization process	High aspect ratio (~100) and great stiffness (E > 100 GPa)	A mixture of amorphous and crystalline cellulose chains with diameter 5–50 nm and length few micrometer	[117]
TEMPO-oxidized cellulose nanofibers	TEMPO (2,2,6,6-tetramethylpiperidine-1-oxyl radical)-mediated oxidation	Aspect ratios > 100	Nanofibers with diameter 3–4 nm wide and length few microns	[12]

renewability, biodegradability, low density, and high tensile strength. Therefore, CNCs as reinforcing material have been blended with a wide range of polymeric hydrogel matrices such as polyacrylamide [118–120], poly(*N,N*-dimethylacrylamide) [121], poly(acrylic acid) [122], poly(acrylic acid-coacrylamide) [123], poly(vinyl alcohol) [124], alginate [125], gelatin [126, 127], etc., to improve their stiffness, strength, and swelling properties. Several other functional composites hydrogel such as double-membrane hydrogel [128], self-healing hydrogel [116, 129], and macroporous hydrogel [130] are also developed through reinforcement of CNCs. CNC suspensions can form hydrogels upon addition of salt (such as Na^+, K^+, Mg^{2+}, Ca^{2+}). A high salt concentration in CNC suspension enables the aggregation of CNCs to allow the formation of hydrogels that usually exhibit nematic ordered capacity [131]. The CNCs-reinforced composite hydrogels possess long-range chiral nematic structure and photonic properties, which can be useful in applications that exploit their photonic properties, such as sensors or tunable filters, and also as soft templates for new materials [132].

Cellulose nanofibrils. Cellulose nanofibrils consist of long, flexible, and interconnected fibrils, 5–50 nm wide and few micrometers in length, and a mixture of amorphous and crystalline cellulose chains, and they are commonly obtained by a homogenization process. CNFs-based hydrogels can be produced by mechanically treating the disintegrated CNF aqueous suspensions with alkali followed by neutralization [133, 134]. NFC has a high aspect ratio, specific surface area, great stiffness, and crystallinity as well as low percolation threshold. Owing to these features, NFCs can form a rigid network in different matrices, subsequently benefiting their mechanical properties. It has been reported that NFC leads to a significant improvement of the compactness of the 3D network structure and the properties of polyacrylamide [135] and gelatin [127] hydrogels.

TEMPO-oxidized cellulose nanofibers. In de Nooy *et al.* [136] developed 2,2,6,6-tetramethylpiperidinyl-1-oxyl radical (TEMPO)-mediated oxidation with hypochlorite/bromide and reported that it selectively oxidized the primary alcohol of polysaccharides [136]. Later on, Isogai and co-workers applied the TEMPO-mediated oxidation to cellulose and established the optimal conditions for TEMPO-mediated oxidation of cellulose [12]. The TEMPO system, consisting of TEMPO/NaBr/NaClO/H_2O, is an effective and environmentally-friendly process to introduce carboxylic groups onto the C6 position of the cellulose molecule, and the cellulose can be converted into individual nanofibers of 3–4 nm diameter that are at least several microns in length and possess an aspect ratio >100. Hydrogels from TEMPO-oxidized cellulose nanofiber suspensions can be produced by the protonization of the carboxylate groups at low pH [137] or through cross-linking using metal cations [138, 139]. TEMPO-oxidized cellulose nanofibers as a reinforcing candidate have been successfully reinforced into poly(*N*-isopropylacrylamide), polyacrylamide, gelatin, and so on, which simultaneously improved the swelling and mechanical properties of the composite hydrogel [140–142].

6. 3D/4D Printing of Cellulose-Based Hydrogels

3D printing, also known as additive manufacturing, is a term used to describe a process of fabrication of a 3D object layer by layer, where each new layer is deposited on the previously formed one. This technology enables the fabrication of multidimensional, ordered micro-structured materials. Nowadays, it is widely used in consumer and industrial applications, including motor vehicle and aerospace manufacturing, as well as in medical applications. In the medical field, 3D printing technology enables the fabrication of artificial scaffolds based on tissue images scientifically designed or captured with common medical imaging techniques such as computed tomography (CT) and magnetic resonance imaging (MRI). Hydrogels are particularly promising 3D-printable materials for fabrication of artificial scaffolds because they recapitulate several features of the natural extracellular matrix and allow cell encapsulation in a highly hydrated, mechanically supportive 3D environment. However, when using hydrogels as 3D-printable materials, collapse and shape fidelity are common challenges encountered during the printing process. The incorporation of nanofillers has been a reliable approach to improve and tune the different properties of 3D printing inks. As discussed earlier, the different forms of nanocellulose, such as CNC, CNF, and TOCN are used as reinforcement materials in composite hydrogel; besides this, other forms of cellulose such as CMC [143] and CNF [144, 145] are also used as reinforcement materials in alginate composite hydrogel and as a component in 3D-printable inks. The collapsing of 3D printed layers can be avoided and the print pattern can have a significant effect on the shape fidelity and stability if CMC or CNF is used. Another kind of nanocellulose, CNCs, are also used as stiff fillers for 3D printing, showing significant improvement in mechanical and surface properties [146]. Furthermore, during the printing process, CNCs undergo shear-induced alignment as the ink flows through the deposition nozzle, which leads to the provision of favorable structural orientation and rheological and mechanical properties. Hausmann *et al.* studied the dynamics of CNC alignment under the shear stresses applied to the concentrated inks during direct ink writing, and this offers powerful quantitative guidelines for the digital manufacturing of composite materials with programmed particle orientations and properties [147]. CNC-based inks for 3D printing can be used to create microstructures with gradient porosity [148] and textured cellular architectures [149].

The highly ordered self-assembled structures of naturally occurring biomaterials and soft/hard tissues provide them with numerous excellent properties. For example, *Mimosa pudica*, Venus flytrap, responds to environmental stimuli (such as humidity, light, or touch) by varying its internal turgor, which leads to dynamic conformations governed by the tissue composition and microstructural anisotropy of cell walls. This phenomenon inspires the fabrication of multidimensionally ordered micro-structured materials to achieve functional modulation. 4D printing, a recently developed field originated from 3D printing, is an emerging pathway for mimicking the natural dynamic architectures by incorporating materials that can respond to external stimuli [150]. As discussed above, CNCs or CNFs containing inks usually exhibit shear thinning behavior, which orient preferably in the printing direction. Gladman *et al.* [151] fabricated a composite hydrogel ink which mimicked the structure of plant cell walls, where CNFs in inks formed printed filaments with anisotropic stiffness. Meanwhile, the swelling behavior of printed filaments in the longitudinal direction is different from that in the transverse direction. Using this composite hydrogel, plant-inspired architectures can be programmably fabricated, and these can change their shape upon immersion in water and yield complex 3D morphologies [151]. In this research, the cellulose content was low and was a hard component in the 4D printing process. However, another study showed the formation of an ink-containing CMC hydrocolloid with incorporated cellulose pulp fibers, with a high total cellulose content (fiber volume fraction≈50% for the dehydrated composite) and good dispersion of fibers within the hydrogel matrix. This high cellulose content ink can be used to fabricate complex structure by 4D printing (see Fig. 4), and this is capable of morphing according to the predetermined design rules in response to hydration/dehydration [152].

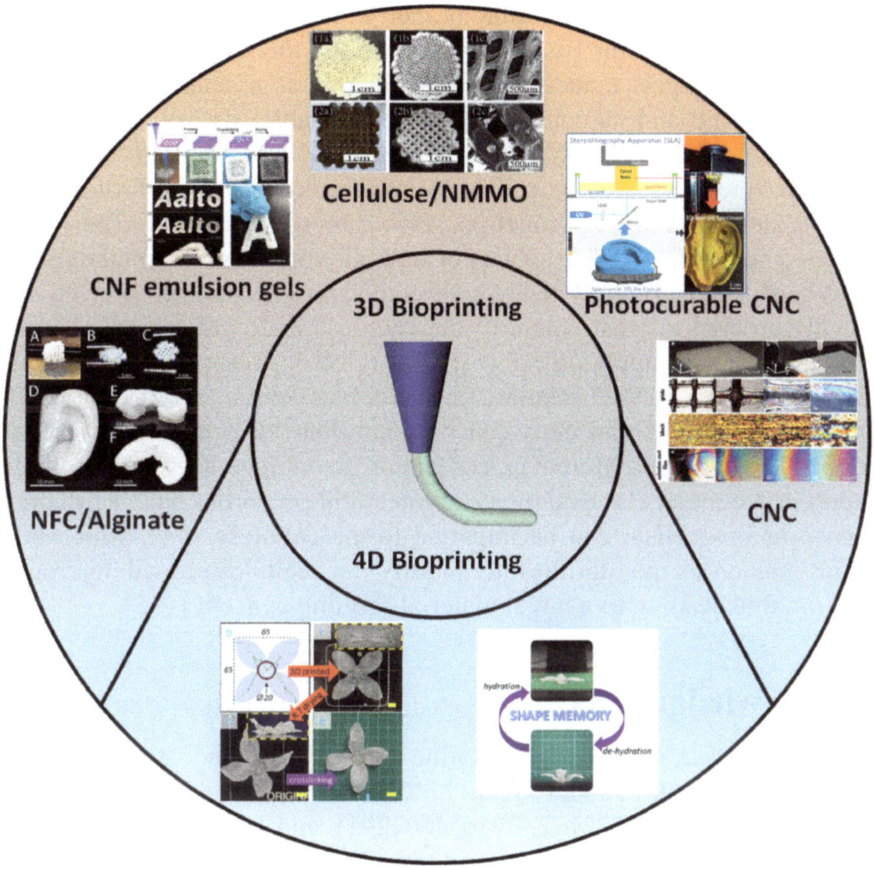

Fig. 4. 3D/4D printing of cellulose-based hydrogel.
Source: Reproduced with permission from Refs. [145, 146, 152–154].

7. Conclusions and Future Outlook

Cellulose is the most abundant biopolymer on earth. This chapter provided an overview of the recent research progress in fabrication, properties, and possible applications of cellulose-based hydrogels. Different methods are used to fabricate cellulose-based hydrogels to impart them with different properties for diverse applications. For example, the superabsorbent property of cellulose hydrogels confers it with wide applications in agriculture, horticulture, water treatment, and medical fields; the highly porous and biocompatible property of cellulose hydrogels enables its use as scaffolds for tissue engineering; the thermosensitive nature of cellulose hydrogels makes it capable for use as a smart interventional embolic agent; and its conductive property enables its use in biosensors and bioconductors.

The future research on cellulose-based hydrogels needs further study on the design of the hydrogel-possessing bionics structure. The highly ordered self-assembled structures of naturally occurring biomaterials and soft/hard tissues provide them with numerous excellent properties. The highly ordered microstructure of the cellulose hydrogel can be designed and fabricated by applying external force (electric, magnetic, and mechanical force, etc.), 3D/4D printing, or *in situ* biosynthesis. For example, the movement of bacterial cells can be controlled to synthesize micropatterned BC hydrogel. Moreover, combining the biological systems with the appropriate molecular design, some unnatural properties such as fluorescence or magnetism can be imparted to the cellulose fiber, which may offer numerous possibilities to modify the cellulose-based hydrogels *in situ,* thus leading to a novel material-farming concept [155].

Acknowledgments

This work was supported by National Natural Science Foundation of China (31270150, 51603079, 21774039), China Postdoctoral Science Foundation (2015M572132, 2016M602291), and Fundamental Research Funds for Central Universities, Open Research Fund of State Key Laboratory of Polymer Physics and Chemistry, Changchun Institute of Applied Chemistry, Chinese Academy of Sciences.

References

[1] Annabi N, Tamayol A, Uquillas J A, *et al.*, 25th Anniversary article: Rational design and applications of hydrogels in regenerative medicine. *Adv. Mater.*, 2014, 26(1): 85–124.

[2] Thiele J, Ma Y, Bruekers S M C, *et al.*, 25th anniversary article: Designer hydrogels for cell cultures: A materials selection guide. *Adv. Mater.*, 2014, 26(1): 125–148.

[3] Malda J, Visser J, Melchels F P W, *et al.*, 25th Anniversary article: Engineering hydrogels for biofabrication. *Adv. Mater.*, 2013, 25(36): 5011–5028.

[4] Döring A, Birnbaum W, Kuckling D, Responsive hydrogels — Structurally and dimensionally optimized smart frameworks for applications in catalysis, micro-system technology and material science. *Chem. Soc. Rev.*, 2013, 42(17): 7391–7420.

[5] Potthast A, Rosenau T, Buchner R, *et al.*, The cellulose solvent system *N,N*-dimethylacetamide/lithium chloride revisited: The effect of water on physicochemical properties and chemical stability. *Cellulose*, 2002.

[6] Klemm D, Heublein B, Fink H, *et al.*, Cellulose: Fascinating biopolymer and sustainable raw material. *Angewandte Chemie*, 2005, 44(22): 3358–3393.

[7] Ulislam M, Khattak W A, Ullah M W, *et al.*, Synthesis of regenerated bacterial cellulose-zinc oxide nanocomposite films for biomedical applications. *Cellulose*, 2014, 21(1): 433–447.

[8] Muginova S V, Myasnikova D, Kazarian S G, *et al.*, Evaluation of novel applications of cellulose hydrogel films reconstituted from acetate and chloride of 1-butyl-3-methylimidazolium by comparing their optical, mechanical, and adsorption properties. *Mater. Today. Commun.*, 2016, 8(3): 108–117.

[9] Pang J, Liu X, Wu M, *et al.*, Fabrication and characterization of regenerated cellulose films using different ionic liquids. *Spectroscopy*, 2014, 2014: 1–8.

[10] Cai J, Zhang L, Liu S, *et al.*, Dynamic self-assembly induced rapid dissolution of cellulose at low temperatures. *Macromolecules*, 2008, 41(23): 9345–9351.

[11] Peng Z, Chen F. Hydroxyethyl cellulose-based hydrogels with various pore sizes prepared by freeze-drying. *J. Macromol. Sci. B.*, 2011, 50(2): 340–349.

[12] Isogai A, Saito T, Fukuzumi H. TEMPO-oxidized cellulose nanofibers. *Nanoscale*, 2011, 3(1): 71–85.

[13] Klemm D, Schumann D, Udhardt U, *et al.*, Bacterial synthesized cellulose — Artificial blood vessels for microsurgery. *Prog. in Polym. Sci.*, 2001, 26(9): 1561–1603.

[14] Czaja W, Krystynowicz A, Bielecki S, *et al.*, Microbial cellulose — The natural power to heal wounds. *Biomaterials*, 2006, 27(2): 145–151.

[15] Li H X, Kim S, Lee Y, *et al.*, Determination of the stoichiometry and critical oxygen tension in the production culture of bacterial cellulose using saccharified food wastes. Korean *J. Chem. Eng.*, 2011, 28(12): 2306–2311.

[16] Di Z, Shi Z, Ullah M W, *et al.*, A transparent wound dressing based on bacterial cellulose whisker and poly(2-hydroxyethyl methacrylate). *Int. J. Biolog. Macromol.*, 2017, 105: 638–644.

[17] Demitri C, Scalera F, Madaghiele M, *et al.*, Potential of cellulose-based superabsorbent hydrogels as water reservoir in agriculture. *Int. J. Polym. Sci.*, 2013, 2013: 1–6.

[18] Mohammadikhoo S, Moghadam P N, Fareghi A R, *et al.*, Synthesis of a cellulose-based hydrogel network: Characterization and study of urea fertilizer slow release. *J. Appl. Polym. Sci.*, 2016, 133(5).

[19] Nascimento D M, Nunes Y L, De Figueiredo M C B, *et al.*, Nanocellulose nanocomposite hydrogels: Technological and environmental issues. *Green Chem.*, 2018, 20(11): 2428–2448.

[20] Phanthong P, Reubroycharoen P, Kongparakul S, *et al.*, Fabrication and evaluation of nanocellulose sponge for oil/water separation. *Carbohydr. Polym.*, 2018, 190: 184–189.

[21] Li S, Huang D, Zhang B, *et al.*, Flexible supercapacitors based on bacterial cellulose paper electrodes. *Adv. Ener. Mater.*, 2014, 4(10): 1301655.

[22] Hanninen A, Sarlin E, Lyyra I, *et al.*, Nanocellulose and chitosan based films as low cost, green piezoelectric materials. *Carbohydr. Polym.*, 2018, 202: 418–424.

[23] Fu L, Zhang Y, Li C, *et al.*, Skin tissue repair materials from bacterial cellulose by a multilayer fermentation method. *J. Mater. Chem.*, 2012, 22(24): 12349–12357.

[24] Li Y, Jiang K, Feng J, *et al.*, Construction of small-diameter vascular graft by shape-memory and self-rolling bacterial cellulose membrane. *Adv. Health. Mater.*, 2017, 6(11): 1601343.

[25] Lamboni L, Li Y, Liu J, *et al.*, Silk sericin-functionalized bacterial cellulose as a potential wound-healing biomaterial. *Biomacromolecules*, 2016, 17(9): 3076–3084.

[26] Yang J, Du M, Wang L, *et al.*, Bacterial cellulose as a supersoft neural interfacing substrate. *ACS Appl. Mater. & Int.*, 2018, 10(39): 33049–33059.

[27] Grishkewich N, Mohammed N, Tang J, *et al.*, Recent advances in the application of cellulose nanocrystals. C*urr. Opin. in Collo. & Int. Sci.*, 2017, 29: 32–45.

[28] Chang C, Zhang L. Cellulose-based hydrogels: Present status and application prospects. *Carbohydr. Polym.*, 2011, 84(1): 40–53.

[29] Edgar K J, Buchanan C M, Debenham J S, *et al.*, Advances in cellulose ester performance and application. *Prog. in Polym. Sci.*, 2001, 26(9): 1605–1688.

[30] L Z A F X. Preparation and properties of regenerated cellulose hydrogels. *IOP Conference Series: Mater. Sci. Eng.*, 2017, 170(1): 12038.

[31] Matsumoto T, Tatsumi D, Tamai N, *et al.*, Solution properties of celluloses from different biological origins in LiCl · DMAc. *Cellulose*, 2001, 8(4): 275–282.

[32] Alfassi G, Rein D M, Cohen Y. Enhanced hydrolysis of cellulose hydrogels by morphological modification. *Biopro. and Biosyst. Eng.*, 2017, 40(11): 1635–1641.

[33] Liang X, Qu B, Li J, *et al.*, Preparation of cellulose-based conductive hydrogels with ionic liquid. *Reac. & Fun. Polym.*, 2015, 86: 1–6.
[34] Shi X, Hu Y, Tu K, *et al.*, Electromechanical polyaniline–cellulose hydrogels with high compressive strength. *Soft Matter.*, 2013, 9(42): 10129–10134.
[35] Huang B, Liu M, Zhou C, Cellulose–halloysite nanotube composite hydrogels for curcumin delivery. *Cellulose*, 2017, 24(7): 2861–2875.
[36] Xing C, Chen S, Qiu M, *et al.*, Conceptually novel black phosphorus/cellulose hydrogels as promising photothermal agents for effective cancer therapy. *Adv. Health. Mater.*, 2018, 1701510.
[37] Kim H J, Yang Y J, Oh H J, *et al.*, Cellulose–silk fibroin hydrogels prepared in a lithium bromide aqueous solution. *Cellulose*, 2017, 24(11): 5079–5088.
[38] Lopez-Sanchez P, Martinez-Sanz M, Bonilla M R, *et al.*, Cellulose-pectin composite hydrogels: Intermolecular interactions and material properties depend on order of assembly. *Carbohydr. Polym.*, 2017, 162:71–81.
[39] Pinho E, Henriques M, Soares G M B, Cyclodextrin/cellulose hydrogel with gallic acid to prevent wound infection. *Cellulose*, 2014, 21(6): 4519–4530.
[40] Ye D, Chang C, Zhang L, High-strength and tough cellulose hydrogels chemically dual cross-linked by using low- and high-molecular-weight cross-linkers. *Biomacromolecules*, 2019, 20(5): 1989–1995.
[41] Shi Z, Gao X, Ullah M W, *et al.*, Electroconductive natural polymer-based hydrogels. *Biomaterials*, 2016, 111: 40–54.
[42] Chang C, Duan B, Cai J, *et al.*, Superabsorbent hydrogels based on cellulose for smart swelling and controllable delivery. *European Polym. J.*, 2010, 46(1): 92–100.
[43] Chang C, Peng J, Zhang L, *et al.*, Strongly fluorescent hydrogels with quantum dots embedded in cellulose matrices. *J. Mater. Chem.*, 2009, 19(41): 7771–7776.
[44] Ye D, Yang P, Lei X, *et al.*, Robust anisotropic cellulose hydrogels fabricated *via* strong self-aggregation forces for cardiomyocytes unidirectional growth. *Chem. Mater.*, 2018, 30(15): 5175–5183.
[45] Ye D, Cheng Q, Zhang Q, *et al.*, Deformation drives alignment of nanofibers in framework for inducing anisotropic cellulose hydrogels with high toughness. *ACS Appl. Mater. & Int.*, 2017, 9(49): 43154–43162.
[46] Zhao D, Huang J, Zhong Y, *et al.*, High-strength and high-toughness double-cross-linked cellulose hydrogels: A new strategy using sequential chemical and physical cross-linking. *Adv. Funct. Mater.*, 2016, 26(34): 6279–6287.
[47] Nakasone K, Kobayashi T, Effect of pre-treatment of sugarcane bagasse on the cellulose solution and application for the cellulose hydrogel films. *Polym. for Adv. Tech.*, 2016, 27(7): 973–980.

[48] Patchan M, Graham J L, Xia Z, *et al.*, Synthesis and properties of regenerated cellulose-based hydrogels with high strength and transparency for potential use as an ocular bandage. *Mater. Sci. Eng. C.*, 2013, 33(5): 3069–3076.

[49] Ishii D, Tatsumi D, Matsumoto T, *et al.*, Investigation of the structure of cellulose in LiCl/DMAc solution and its gelation behavior by small-angle X-ray scattering measurements. *Macromol. Biosci.*, 2006, 6(4): 293–300.

[50] Gao X, Sadasivuni K K, Kim H, *et al.*, Designing pH-responsive and dielectric hydrogels from cellulose nanocrystals. *J. Chem. Sci.*, 2015, 127(6): 1119–1125.

[51] Jiang H, Tovarcarrillo K L, Kobayashi T, Ultrasound stimulated release of mimosa medicine from cellulose hydrogel matrix. *Ultra. Sonochem.*, 2016, 32: 398–406.

[52] Johns M A, Bernardes A, De Azevedo E R, *et al.*, On the subtle tuneability of cellulose hydrogels: Implications for binding of biomolecules demonstrated for CBM 1. *J. Mater. Chem. B*, 2017, 5(21): 3879–3887.

[53] Choe D, Kim Y M, Nam J E, *et al.*, Synthesis of high-strength microcrystalline cellulose hydrogel by viscosity adjustment. *Carbohydr. Polym.*, 2018, 180: 231–237.

[54] Lu F, Chen Y, Liu N, *et al.*, A fast and convenient cellulose hydrogel-coated colander for high-efficiency oil–water separation. *RSC Adv.*, 2014, 4(61): 32544–32548.

[55] Kaco H, Zakaria S, Razali N F, *et al.*, Properties of cellulose hydrogel from kenaf core prepared *via* pre-cooled dissolving method. *Sains Malaysiana*, 2014, 43(8): 1221–1229.

[56] Jeong D, Kim H K, Jeong J, *et al.*, Cyclosophoraose/cellulose hydrogels as an efficient delivery system for galangin, a hydrophobic antibacterial drug. *Cellulose*, 2016, 23(4): 2609–2625.

[57] Chang C, Duan B, Zhang L, Fabrication and characterization of novel macroporous cellulose–alginate hydrogels. *Polymer*, 2009, 50(23): 5467–5473.

[58] Kim H J, Castaneda R, Kang T H, *et al.*, Cellulose hydrogel film for spheroid formation of human adipose-derived stemcells. *Cellulose*, 2018, 25(4): 2589–2598.

[59] Isobe N, Chen X, Kim U, *et al.*, TEMPO-oxidized cellulose hydrogel as a high-capacity and reusable heavy metal ion adsorbent. *J. Hazard. Mater.*, 2013, 260: 195–201.

[60] Gan S, Zakaria S, Chia C H, *et al.*, Highly porous regenerated cellulose hydrogel and aerogel prepared from hydrothermal synthesized cellulose carbamate. *PLOS ONE*, 2017, 12(3).

[61] Pei Y, Wang X, Huang W, *et al.*, Cellulose-based hydrogels with excellent microstructural replication ability and cytocompatibility for microfluidic devices. *Cellulose*, 2013, 20(4): 1897–1909.

[62] Shi Z, Zhang Y, Phillips G O, *et al.*, Utilization of bacterial cellulose in food. *Food Hydrocoll.*, 2014, 35(0): 539–545.
[63] Nakayama A, Kakugo A, Gong J P, *et al.*, High mechanical strength doublenetwork hydrogel with bacterial cellulose. *Adv. Funct. Mater.*, 2004, 14(11): 1124–1128.
[64] Hsieh Y, Yano H, Nogi M, *et al.*, An estimation of the Young's modulus of bacterial cellulose filaments. *Cellulose*, 2008, 15(4): 507–513.
[65] Guhados G, Wan W, Hutter J L, Measurement of the elastic modulus of single bacterial cellulose fibers using atomic force microscopy. *Langmuir*, 2005, 21(14): 6642–6646.
[66] Fu L, Zhou P, Zhang S, *et al.*, Evaluation of bacterial nanocellulose-based uniform wound dressing for large area skin transplantation. *Mater. Sci. Eng. C.*, 2013, 33(5): 2995–3000.
[67] Astley O M, Chanliaud E, Donald A M, *et al.*, Tensile deformation of bacterial cellulose composites. *Int. J. Biolog. Macromol.*, 2003, 32(1): 28–35.
[68] Mckenna B A, Mikkelsen D, Wehr J B, *et al.*, Mechanical and structural properties of native and alkali-treated bacterial cellulose produced by Gluconacetobacter xylinus strain ATCC 53524. *Cellulose*, 2009, 16(6): 1047–1055.
[69] Gao X, Shi Z, Liu C, *et al.*, Inelastic behaviour of bacterial cellulose hydrogel: In aqua cyclic tests. *Polym. Test.*, 2015, 44: 82–92.
[70] Wang S, Jiang F, Xu X, *et al.*, Super-strong, super-stiff macrofibers with aligned, Long Bacterial Cellulose Nanofibers. *Adv. Mater.*, 2017, 29(35): 1702498.
[71] Wang S, Li T, Chen C, *et al.*, Transparent, anisotropic biofilm with aligned bacterial cellulose nanofibers. *Adv. Func. Mater.*, 2018, 28(24) 1707491.
[72] Fu L, Zhang J, Yang G. Present status and applications of bacterial cellulose-based materials for skin tissue repair. *Carbohydr. Polym.*, 2013, 92(2): 1432–1442.
[73] Li Y, Jiang H, Zheng W, *et al.*, Bacterial cellulose–hyaluronan nanocomposite biomaterials as wound dressings for severe skin injury repair. *J. Mater. Chem. B*, 2015, 3(17): 3498–3507.
[74] Wang B, Lv X, Chen S, *et al.*, Bacterial cellulose/gelatin scaffold loaded with VEGF-silk fibroin nanoparticles for improving angiogenesis in tissue regeneration. *Cellulose*, 2017, 24(11): 5013–5024.
[75] Lin W, Lien C, Yeh H, *et al.*, Bacterial cellulose and bacterial cellulose-chitosan membranes for wound dressing applications. *Carbohydr. Polym.*, 2013, 94(1): 603–611.
[76] Lin Y K, Chen K H, Ou K L, *et al.*, Effects of different extracellular matrices and growth factor immobilization on biodegradability and biocompatibility of macroporous bacterial cellulose. *J. Bioact. Compa. Polym.*, 2011, 26(5): 508–518.

[77] Huang L, Chen X, Nguyen T X, *et al.*, Nano-cellulose 3D-networks as controlled-release drug carriers. *J. Mate. Chem. B.*, 2013, 1(23): 2976–2984.
[78] Shi Z, Zang S, Jiang F, *et al.*, *In situ* nano-assembly of bacterial cellulose-polyaniline composites. *RSC Adv.*, 2012, 2(3): 1040–1046.
[79] Shi Z, Phillips G O, Yang, Nanocellulose electroconductive composites. *Nanoscale*, 2013, 5(8): 3194–3201.
[80] Li Y, Tian Y, Zheng W, *et al.*, Composites of bacterial cellulose and small molecule-decorated gold nanoparticles for treating gram-negative bacteria-infected wounds. *Small*, 2017, 13(27): 1700130.
[81] Yang J, Du M, Wang L, *et al.*, Bacterial cellulose as a supersoft neural interfacing substrate. *ACS Appl. Mater. & Int.*, 2018, 10(39): 33049–33059.
[82] Shi Z, Li Y, Chen X, *et al.*, Double network bacterial cellulose hydrogel to build a biology–device interface. *Nanoscale*, 2014, 6(2): 970–977.
[83] Yang J, Wang L, Zhang W, *et al.*, Reverse reconstruction and bioprinting of bacterial cellulose-based functional total intervertebral disc for therapeutic implantation. *Small*, 2018, 14(7): 1702582.
[84] Greca L G, Lehtonen J, Tardy B L, *et al.*, Biofabrication of multifunctional nanocellulosic 3D structures: a facile and customizable route. *Mater. Horizons*, 2018, 5(3): 408–415.
[85] Shi Z, Shi X, Ullah M W, *et al.*, Fabrication of nanocomposites and hybrid materials using microbial biotemplates. *Adv. Compo. and Hybr. Mater.*, 2018, 1(1): 79–93.
[86] Klemm D, Schumann D, Udhardt, *et al.*, Bacterial synthesized cellulose — artificial blood vessels for microsurgery. *Prog. in Polym. Sci.*, 2001, 26(9): 1561–1603.
[87] Zang S, Zhang R, Chen H, *et al.*, Investigation on artificial blood vessels prepared from bacterial cellulose. *Mater. Sci. Eng. C.*, 2015, 46: 111–117.
[88] Svensson A, Nicklasson E, Harrah T, *et al.*, Bacterial cellulose as a potential scaffold for tissue engineering of cartilage. *Biomaterials*, 2005, 26(4): 419–431.
[89] Nimeskern L, Martínezávila H, Sundberg J, *et al.*, Mechanical evaluation of bacterial nanocellulose as an implant material for ear cartilage replacement. *J. Mech. Behav. Biomed. Mater.*, 2013, 22: 12–21.
[90] Sano M B, Rojas A D, Gatenholm P, *et al.*, Electromagnetically controlled biological assembly of aligned bacterial cellulose nanofibers. *Annals of Biomed. Eng.*, 2010, 38(8): 2475–2484.
[91] Putra A, Kakugo A, Furukawa H, *et al.*, Production of bacterial cellulose with well oriented fibril on PDMS substrate. *Polym. J.*, 2008, 40(2): 137–142.
[92] Uraki Y, Nemoto J, Otsuka H, *et al.*, Honeycomb-like architecture produced by living bacteria, *Gluconacetobacter xylinus*. *Carbohydr. Polym.*, 2007, 69(1): 1–6.

[93] Zang S, Sun Z, Liu K, *et al.*, Ordered manufactured bacterial cellulose as biomaterial of tissue engineering. *Mater. Lett.*, 2014, 128: 314–318.
[94] Yang J, Wang L, Zhang W, *et al.*, Reverse reconstruction and bioprinting of bacterial cellulose-based functional total intervertebral disc for therapeutic implantation. *Small*, 2017, 14(7): 1702582.
[95] Geisel N, Clasohm J, Shi X, *et al.*, Microstructured multilevel bacterial cellulose allows the guided growth of neural stem cells. *Small*, 2016, 12(39): 5407–5413.
[96] Zhuo F, Liu X, Gao Q, *et al.*, Injectable hyaluronan-methylcellulose composite hydrogel crosslinked by polyethylene glycol for central nervous system tissue engineering. *Mater. Sci. Eng. C.*, 2017, 81: 1–7.
[97] Pakulska M M, Vulic K, Tam R Y, *et al.*, Hybrid crosslinked methylcellulose hydrogel: A predictable and tunable platform for local drug delivery. *Adv. Mater.*, 2015, 27(34): 5002–5008.
[98] Liu X, Zhou Y, Nie W, *et al.*, Fabrication of hydrogel of hydroxypropyl cellulose (HPC) composited with graphene oxide and its application for methylene blue removal. *J. Mater. Sci.*, 2015, 50(18): 6113–6123.
[99] Jana S, Pradhan S S, Tripathy T. Poly(*N*,*N*-dimethylacrylamide-co-acrylamide) grafted hydroxyethyl cellulose hydrogel: A useful congo red dye remover. *J. Polym. and the Envir.*, 2017, 1–18.
[100] Zare-Akbari Z, Farhadnejad H, Furughi-Nia B, *et al.*, PH-sensitive bionanocomposite hydrogel beads based on carboxymethyl cellulose/ZnO nanoparticle as drug carrier. *Int. J. Biolo. Macromol.*, 2016, 93: 1317–1327.
[101] Davidson D W, Verma M S, Gu F X, Controlled root targeted delivery of fertilizer using an ionically crosslinked carboxymethyl cellulose hydrogel matrix. *SpringerPlus*, 2013, 2(1): 318.
[102] Fei B, Wach R A, Mitomo H, *et al.*, Hydrogel of biodegradable cellulose derivatives. I. Radiation-induced crosslinking of CMC. *J. Appl. Polym. Sci.*, 2000, 78(2): 278–283.
[103] Liu P, Zhai M, Li J, *et al.*, Radiation preparation and swelling behavior of sodium carboxymethyl cellulose hydrogels. *Radiation Phy. and Chem.*, 2002, 63(3): 525–528.
[104] Rakhshaei R, Namazi H, A potential bioactive wound dressing based on carboxymethyl cellulose/ZnO impregnated MCM-41 nanocomposite hydrogel. *Mater. Sci. and Eng. C.*, 2017, 73: 456–464.
[105] Zhu L, Qiu J, Sakai E, *et al.*, Design of a rubbery carboxymethyl cellulose/polyacrylic acid hydrogel *via* visible-light-triggered polymerization. *Macromol. Mater. and Eng.*, 2017, 302(6): 1600509.
[106] Dahlan N A, Pushpamalar J, Veeramachineni A K, *et al.*, Smart hydrogel of carboxymethyl cellulose grafted carboxymethyl polyvinyl alcohol and properties studied for future material applications. *J. Polym. Envir.*, 2018, 26(5): 2061–2071.

[107] Fekete T, Borsa J, Takacs E, *et al.*, Synthesis of carboxymethylcellulose/starch superabsorbent hydrogels by gamma-irradiation. *Chem. Central J.*, 2017, 11(1): 46.

[108] Malik N S, Ahmad M, Minhas M U, Cross-linked β-cyclodextrin and carboxymethyl cellulose hydrogels for controlled drug delivery of acyclovir. *PLOS ONE*, 2017, 12(2): 0.

[109] Aktekin A, Sahin I, Sezer U A, *et al.*, Carboxymethyl cellulose/oxidized regenerated cellulose hydrogels as adhesion barriers: Comparative study with different molecular weights and substitution degrees. *Cellulose*, 2016, 23(5): 3145–3156.

[110] Demitri C, Sole R D, Scalera F, *et al.*, Novel superabsorbent cellulose-based hydrogels crosslinked with citric acid. *J. Appl. Polym. Sci.*, 2008, 110(4): 2453–2460.

[111] Seki Y, Altinisik A, Demircioglu B, *et al.*, Carboxymethylcellulose (CMC)–hydroxyethylcellulose (HEC) based hydrogels: Synthesis and characterization. *Cellulose*, 2014, 21(3): 1689–1698.

[112] Vinatier C, Gauthier O, Fatimi A, *et al.*, An injectable cellulose-based hydrogel for the transfer of autologous nasal chondrocytes in articular cartilage defects. *Biotech. Bioengin.*, 2009, 102(4): 1259–1267.

[113] Zheng W J, Gao J, Wei Z, *et al.*, Facile fabrication of self-healing carboxymethyl cellulose hydrogels. *Eur. Polym. J.*, 2015, 72: 514–522.

[114] Chen Y M, Sun L, Yang S A, *et al.*, Self-healing and photoluminescent carboxymethyl cellulose-based hydrogels. *Eur. Polym. J.*, 2017, 94 501–510.

[115] Tarus D, Hachet E, Messager L, *et al.*, Readily prepared dynamic hydrogels by combining phenyl boronic acid- and maltose-modified anionic polysaccharides at neutral pH. *Macromol. Rapid Commun.*, 2014, 35(24): 2089–2095.

[116] Yang X, Liu G, Peng L, *et al.*, Highly efficient self-healable and dual responsive cellulose-Based hydrogels for controlled release and 3D cell culture. *Adv. Func. Mater.*, 2017, 27(40): 1703174.

[117] Nechyporchuk O, Belgacem M N, Bras J, Production of cellulose nanofibrils: A review of recent advances. *Ind. Crops and Prod.*, 2016, 93: 2–25.

[118] Yang J, Zhao J, Zhang X, Modification of cellulose nanocrystal-reinforced composite hydrogels: Effects of co-crosslinked and drying treatment. *Cellulose*, 2014, 21(5): 3487–3496.

[119] Yang J, Han C, Duan J, *et al.*, Synthesis and characterization of mechanically flexible and tough cellulose nanocrystals–polyacrylamide nanocomposite hydrogels. *Cellulose*, 2013, 20(1): 227–237.

[120] Yang J, Zhao J, Han C, *et al.*, Tough nanocomposite hydrogels from cellulose nanocrystals/poly(acrylamide) clusters: Influence of the charge

density, aspect ratio and surface coating with PEG. *Cellulose*, 2014, 21(1): 541–551.

[121] Yang J, Han C, Xu F, *et al.*, Simple approach to reinforce hydrogels with cellulose nanocrystals. *Nanoscale*, 2014, 6(11): 5934–5943.

[122] Zhang T, Cheng Q, Ye D, *et al.*, Tunicate cellulose nanocrystals reinforced nanocomposite hydrogels comprised by hybrid cross-linked networks. *Carbohydr. Polym.*, 2017, 169: 139–148.

[123] Zhang T, Zuo T, Hu D, *et al.*, Dual physically cross-linked nanocomposite hydrogels reinforced by tunicate cellulose nanocrystals with high toughness and good self-recoverability. *ACS Appl. Mater. & Int.*, 2017, 9(28): 24230–24237.

[124] Abitbol T, Johnstone T, Quinn T M, *et al.*, Reinforcement with cellulose nanocrystals of poly(vinyl alcohol) hydrogels prepared by cyclic freezing and thawing. *Soft Matter.*, 2011, 7(6): 2373–2379.

[125] Wang K, Nune K C, Misra R D K, The functional response of alginate-gelatin-nanocrystalline cellulose injectable hydrogels toward delivery of cells and bioactive molecules. *Acta Biomater.*, 2016, 36: 143–151.

[126] Ooi S Y, Ahmad I, Amin M C I M. Cellulose nanocrystals extracted from rice husks as a reinforcing material in gelatin hydrogels for use in controlled drug delivery systems. *Ind. Crops Prod.*, 2016, 93: 227–234.

[127] Wang W, Zhang X, Teng A, *et al.*, Mechanical reinforcement of gelatin hydrogel with nanofiber cellulose as a function of percolation concentration. *Int. J. Biol. Macromol.*, 2017, 103: 226–233.

[128] Lin N, Geze A, Wouessidjewe D, *et al.*, Biocompatible double-membrane hydrogels from cationic cellulose nanocrystals and anionic alginate as complexing drugs codelivery. *ACS Appl. Mater. & Int.*, 2016, 8(11): 6880–6889.

[129] Shao C, Wang M, Chang H, *et al.*, A self-healing cellulose nanocrystal-poly(ethylene glycol) nanocomposite hydrogel *via* diels–alder click reaction. *ACS Sust. Chem. & Eng.*, 2017, 5(7): 6167–6174.

[130] Liu S, Jin M, Chen Y, *et al.*, High internal phase emulsions stabilised by supramolecular cellulose nanocrystals and their application as cell-adhesive macroporous hydrogel monoliths. *J. Mater. Chem. B.*, 2017, 5(14): 2671–2678.

[131] Bertsch P, Isabettini S, Fischer P, Ion-induced hydrogel formation and nematic ordering of nanocrystalline cellulose suspensions. *Biomacromol.*, 2017, 18(12): 4060–4066.

[132] Kelly J A, Shukaliak A M, Cheung C C, *et al.*, Responsive photonic hydrogels based on nanocrystalline cellulose. *Angewandte Chemie.*, 2013, 125(34): 8912–8916.

[133] Abe K, Yano H, Formation of hydrogels from cellulose nanofibers. *Carbohydr. Polym.*, 2011, 85(4): 733–737.

[134] Abe K, Yano H, Cellulose nanofiber-based hydrogels with high mechanical strength. *Cellulose*, 2012, 19(6): 1907–1912.
[135] Niu J, Wang J, Dai X, *et al.*, Dual physically crosslinked healable polyacrylamide/cellulose nanofibers nanocomposite hydrogels with excellent mechanical properties. *Carbohydr. Polym.*, 2018, 193: 73–81.
[136] de Nooy A E J, Besemer A C, van Bekkum H, Highly selective nitroxyl radical-mediated oxidation of primary alcohol groups in water-soluble glucans. *Carbohydr. Res.*, 1995, 269(1): 89–98.
[137] Saito T, Uematsu T, Kimura S, *et al.*, Self-aligned integration of native cellulose nanofibrils towards producing diverse bulk materials. *Soft Matter.*, 2011, 7(19): 8804–8809.
[138] Dong H, Snyder J F, Williams K S, *et al.*, Cation-induced hydrogels of cellulose nanofibrils with tunable moduli. *Biomacromol.*, 2013, 14(9): 3338–3345.
[139] Dong H, Snyder J F, Tran D T, *et al.*, Hydrogel, aerogel and film of cellulose nanofibrils functionalized with silver nanoparticles. *Carbohydr. Polym.*, 2013, 95(2): 760–767.
[140] Li N, Chen W, Chen G, *et al.*, Rapid shape memory TEMPO-oxidized cellulose nanofibers/polyacrylamide/gelatin hydrogels with enhanced mechanical strength. *Carbohydr. Polym.*, 2017, 171: 77-84.
[141] Masruchin N, Park B, Causin V, Dual-responsive composite hydrogels based on TEMPO-oxidized cellulose nanofibril and poly(N-isopropylacrylamide) for model drug release. *Cellulose*, 2018, 25(1): 485–502.
[142] Wei J, Chen Y, Liu H, *et al.*, Effect of surface charge content in the TEMPO-oxidized cellulose nanofibers on morphologies and properties of poly(N-isopropylacrylamide)-based composite hydrogels. *Ind. Crops and Prod.*, 2016, 92: 227-235.
[143] Habib A, Sathish V, Mallik S, *et al.*, 3D printability of alginate-carboxymethyl cellulose hydrogel. *Materials*, 2018, 11(3): 454.
[144] Leppiniemi J, Lahtinen P, Paajanen A, *et al.*, 3D-printable bioactivated nanocellulose–alginate hydrogels. *ACS Appl. Mater. & Int.*, 2017, 9(26): 21959–21970.
[145] Markstedt K, Mantas A, Tournier I, *et al.*, 3D bioprinting human chondrocytes with nanocellulose–alginate bioink for cartilage tissue engineering applications. *Biomacromol.*, 2015, 16(5): 1489–1496.
[146] Palaganas N, Mangadlao J D, De Leon A, *et al.*, 3D Printing of photocurable cellulose nanocrystal composite for fabrication of complex architectures *via* stereolithography. *ACS Appl. Mater. & Int.*, 2017, 9(39): 34314–34324.
[147] Hausmann M K, Rühs P A, Siqueira G, *et al.*, Dynamics of cellulose nanocrystal alignment during 3D printing. *ACS Nano*, 2018, 12(7): 6926–6937.

[148] Sultan S, Mathew A P, 3D printed scaffolds with gradient porosity based on a cellulose nanocrystal hydrogel. *Nanoscale*, 2018, 10(9): 4421–4431.
[149] Siqueira G, Kokkinis D, Libanori R, *et al.*, Cellulose nanocrystal inks for 3D printing of textured cellular architectures. *Advan. Func. Mater.*, 2017, 27(12): 1604619.
[150] Momeni F, M. Mehdi Hassani. N S, Liu X, *et al.*, A review of 4D printing. *Mater. & Design*, 2017, 122: 42-79.
[151] Gladman A S, Matsumoto E A, Nuzzo R G, *et al.*, Biomimetic 4D printing. *Nat. Mater.*, 2016, 15(4): 413–418.
[152] Mulakkal M C, Trask R S, Ting V P, *et al.*, Responsive cellulose-hydrogel composite ink for 4D printing. *Mater. & Design*, 2018, 160: 108–118.
[153] Huan S, Ajdary R, Bai L, *et al.*, Low solids emulsion gels based on nano-cellulose for 3D-printing. *Biomacromol.*, 2019, 20(2): 635–644.
[154] Li L, Zhu Y, Yang J, 3D bioprinting of cellulose with controlled porous structures from NMMO. *Mater. Lett.*, 2018, 210: 136–138.
[155] Natalio F, Fuchs R, Cohen S R, *et al.*, Biological fabrication of cellulose fibers with tailored properties. *Science*, 2017, 357(6356): 1118–1122.

https://doi.org/10.1142/9781786349477_0006

Chapter 6

Nanocellulose-Based Biomaterials for Skin Tissue Repair

Lallepak Lamboni[*,†,¶], **Amarachi Rosemary Osi**[‡], **Biaou O. Ode Boni**[*], **Bianza Moise Bakadia**[*], **and Guang Yang**[*,§]

[*]*National Engineering Research Center for Nano-Medicine, Department of Biomedical Engineering, College of Life Science and Technology, Huazhong University of Science and Technology, Wuhan, PR China*

[†]*Laboratoire de Biologie Moléculaire et Virologie, Institut National d'Hygiène-Togo, 26 Rue Nangbéto, Quartier Administratif- PO. Box 1396, Lomé-Togo*

[‡]*Ningbo Institute of Materials Technology and Engineering, Chinese Academy of Sciences, Ningbo, China*

[¶]*lallepak@yahoo.fr*

[§]*yang_sunny@yahoo.com*

Abstract

As a virtuous material for biomaterial fabrication, nanocellulose has been extensively explored for biomedical applications, and specifically for the repair and regeneration of skin tissue, due to its unique physico-chemical properties particularly interesting for wound healing. Owing to its high hydrophilicity, collagen-like fibrous structure, biocompatibility, and mechanical stability, it satisfies the main requirements for wound dressing application, with a chemical structure allowing for further functionalization and composite formation for the development of improved and adaptable materials. Following a brief description of skin physiology and wound healing biology, the potential of this biopolymer for the repair of skin tissue is outlined throughout this chapter, emphasizing its advantages and beneficial properties for wound healing application. Modification modalities for endowing the material with novel properties are also discussed, before addressing the various nanocellulose-incorporating biomaterials that are being proposed for wound healing application.

Keywords: Nanocellulose; bacterial nanocellulose; plant nanocellulose; wound healing; biomaterials; wound dressings; skin

1. Introduction

The skin is the main barrier that protects the body against external harmful media. It is also an organ of perception, while it is responsible for preserving the balance of electrolytes and water and, at the same time, regulating the body temperature. Hence, it is counted among the most vital organs of the body, ensuring various essential functions [1]. However, besides iatrogenic factors, as the outermost tissue of the body, the skin constantly incurs high risks of damage due to repeated exposure to potentially harmful microbial, thermal, chemical, or mechanical assaults, which disrupt the body's homeostasis and compromises life in certain cases. In fact, while altered skin tissue generally initiates a complex and continuous molecular wound healing cascade to repair itself, acute but major trauma such as burns, or chronic wounds and surgery-related sequelae, often involve extensive debilitating skin losses that degrade the quality of life, with the potential of causing death [2]. In such deteriorated skin, the barrier function is interrupted, subjecting the body to bacterial infection and eventual sepsis, one of the main causes of fatality in burns. Moreover, following considerable skin loss extending to the dermis, the regenerative ability of the skin is jeopardized due to complete loss of progenitor cells [3]; such injuries are classically treated using organ transplantation, which frequently leads to rarely mild or irrelevant scars [3], and additionally to the multiple limitations inherent to such practices, particularly the lack of donor organs and the risk of infection [4]. Hence, tissue engineering proposes bioengineered skin substitutes that primarily aim at restoring the barrier function of the skin [5]. Simulating physiological skin, these materials initiate and assist repair in wounds deprived of self-healing ability, promoting effective healing with restoration of skin structure and function, aiming at reduced scar formation [1].

Among tissue engineered products for skin repair, wound dressings serve not only as alternative cellular niches towards the formation of new tissue but also as supportive microenvironments for cell migration, adhesion, and proliferation [6, 7]. Used as temporary implants, they modulate the binding and release of signaling molecules which regulate cell response and behavior [8, 9]. In fact, as biomaterials, they constitute artificial extracellular matrices (ECM) that provide necessary cues and components for tissue regeneration by mimicking the biophysical, biochemical, and biological properties of targeted tissue [10]. Hence, they should be biocompatible and bio-inert to prevent antigenic reaction, and present bioactivity to instruct cellular behavior [11]. Moreover, importantly, they should mimic the nanodimensional organization and mechanical and physical properties of the ECM [10, 11]. While these materials should adopt a topography that allows easy adhesion to the wound so as to enable interaction with the covered tissue, they should also be readily and painlessly removable without causing damage to the underlying neotissue. Other parameters include pH maintenance in the wound bed [12], the ability to retain wound hydration, and permit oxygen permeability, thereby facilitating biological processes, absorbing wound exudates to retain the bioactive factors released in the wound bed and prevent infection, and protecting the wound from microbial infiltration [8]. To address these requirements, natural biopolymers and, particularly, polysaccharides are found to be suitable materials based on their remarkable structural features and valuable physical, chemical, and mechanical properties, as well as bioactivity, biodegradability, and biocompatibility [13, 14].

As a natural polysaccharide, cellulose is a biocompatible material that has a classical use in wound care in forms of cotton gauze or carboxymethyl cellulose bandages. Exhibiting good hemostatic properties with overall minimal adverse effects on wound healing [15], cellulose is a highly hydrophilic, biodegradable, and renewable fibrous polymer. Widely available, it is the main constituent of woody and non-woody plants and agricultural residues, while also produced by some algae and certain bacteria [16]. Nanocellulose corresponds to cellulose fibers with nanoscopic dimensions. While bacterial nanocellulose (BNC) is generated from bacteria in the form of nanofibrous networks, plant-derived nanocelluloses are obtained following mechanical, enzymatic, and/or chemical treatments of plant or wood components [17] to remove hemicellulose, pectin and lignin contaminants, yielding cellulose nanofibrils (CNF) on one hand, and shorter rode-like cellulose nanocrystals or crystalline cellulose (CNC) on the other [18]. Due to its versatile and tailorable properties, nanocellulose has emerged as a novel material with various applications and huge potential in the biomedical field [19]. In the current chapter, nanocellulose is discussed with regard to skin repair and wound healing applications. Following a brief description of skin physiology and wound healing, the advantages of different nanocellulose types are explained in terms of properties and suitability for the targeted application. Moreover, reportedly proposed biomaterials for skin care are presented.

2. Physiology of Skin Tissue Repair and Wound Healing

2.1 *Structure and Biology of Skin Tissue*

The skin is an organ of multiple layers whose main function is to protect the body against pathogens and other exogenous substances [20, 21]. It also maintains body homeostasis by controlling water loss and regulating temperature and salt concentrations, additionally to contributing to sensation [22].

The skin is subdivided into three main layers that are the epidermis, the dermis, and the hypodermis (Fig. 1) [23]. These layers consist of

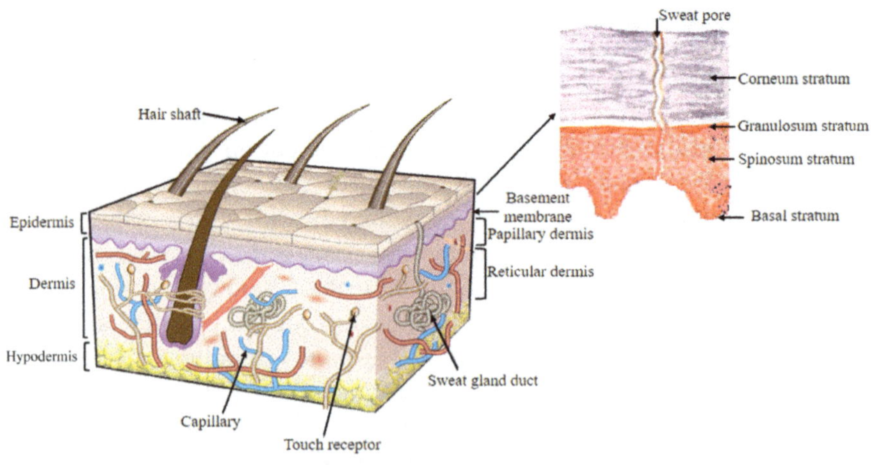

Fig. 1. Structural composition of the human skin.

Source: Adapted from Ref. [26] with permission from Nature and Ref. [141].

extracellular matrix (ECM) and different cells which have determining roles in skin functions, with the latter directly dependent on the normal tissue structure. In fact, following partial or deep tears of the skin, after cleaning the wound, biomaterials commonly referred to as wound dressing or skin substitutes are used to restore the structure and functions of the skin. These biomaterials must mimic the native tissue structure in order to support the proliferation and differentiation of different cells. Therefore, understanding skin structure and the role of each of its constituents is essential for the engineering of such required biomaterials.

2.1.1 *Epidermis*

The epidermis, the outer skin layer, is itself multilayered and is responsible for the protective and hydration functions of the skin [24]. From the inside-out, it contains the basal stratum, spinosum stratum, granulosum stratum, and corneum stratum (Fig. 1) [25]. Cellular components include keratinocytes as the main cell type, along with melanocytes, Langerhans cells, Merkel cells, and lymphocytes [25, 26]. Melanocytes produce melanin that generates skin pigmentation and protects against ultraviolet radiation, [22], while Langerhans cells and lymphocytes are immune cells involved in the immune response to microorganisms [27].

Epidermogenesis occurs in a biological process known as keratinization where stem cells located in the basal stratum differentiate into keratinocytes, which move towards the interface, synthesize keratin, differentiate, become enucleated, and die [22, 28, 29]. In the spinosum stratum, keratinocytes synthesize lipids that are extruded in the granulosum and corneum strata to form the ECM [25]. The dead keratinocytes, called corneocytes, rich in proteins (keratin), are threaded into the lipid matrix (ceramides, cholesterol, and free fatty acids) to form the corneum stratum responsible for the physical impermeability to environmental toxic agents [25, 30]. Further, the lipids therein prevent the loss of water and salts, retaining water and skin moisture. In order to maintain its protective role and assure restoration after partial destruction of the skin, the stratum corneum is constantly renewed *via* keratinization. The process consists of regulating the proliferation and apoptosis of keratinocytes. It has been shown that regeneration of the epidermal tissue depends strictly on the dermal tissue through the mutual interaction between keratinocytes and fibroblasts, which are the main cells of the dermis [20, 31]. The latter, in fact, controls the proliferation of keratinocytes. Hence, engineering biomaterials for the restoration of the epidermis should take into consideration its interaction with the dermis.

2.1.2 *Dermis*

The dermis, the middle layer of the skin, is composed mainly of connective tissue, which supports blood and lymphatic vessels, nerve fibers, hair follicles, and secretory glands [32, 33]. The connective tissue mostly consists of collagens and elastin fibers that provide flexibility, elasticity, and tensile strength to the skin [22, 24]. Blood vessels supply nutrients and oxygen, while the lymphatic vessels are specialized in supplying the lymphatic cells which are responsible for the immune response; nerve endings provide sensation [34].

The dermis divides into the papillary layer in contact with the epidermis and the reticular layer in contact with the hypodermis (Fig. 1), [35, 36], made distinct by their cell and ECM densities [37]. While the papillary dermis is rich in cells and poor in connective tissue, the reticular dermis is poor in cells but rich in well-arranged collagen [35]. Accordingly, fibroblasts distribute heterogeneously among the layers, ensuring different but complementary functions [35, 38]. It has been reported that papillary fibroblasts better support the proliferation and differentiation of keratinocytes as compared with reticular fibroblasts [39]. Papillary fibroblasts synthetize less collagen I and chondroitin proteoglycan than reticular fibroblasts, while the latter produce lower levels of keratinocyte growth-promoting factors, from GM-CSF to KGF [37]. However, it has been shown that variation in growth factor production is not the cause of the different effects on keratinocyte proliferation, which is rather explained by the ECM composition [35, 39]. Overall, papillary fibroblasts induce dense basement membrane formation, which supports the growth and differentiation of keratinocytes [35]. However, the mechanism by which papillary fibroblasts affect epidermogenesis remains unclear. From the foregoing, it would be judicious to take into account the characteristics of each type of fibroblast in the engineering of biomaterials for skin substitute or wound dressing applications for optimized tissue repair.

2.1.3 *Hypodermis*

Known as the subcutaneous fat layer, the hypodermis is the innermost region of the skin (Fig. 1). It connects the skin to other organs (muscles and bones). It is essentially made of adipose and connective (mainly collagen) tissues that support nerves and blood and lymphatic vessels [23]. Functionally, the fatty acids contained in the adipose tissue are responsible for thermoregulation and protect deeper tissues from injuries [33].

2.2 *Physiological Background of Skin Repair*

A wound is a disruption (resulting from physical, chemical, or thermal damage) of the normal structure of skin, [9] which alters the functions of the latter. This is ensued by healing, the name given to the biological process resulting in the structural and functional restoration of the skin [40]. It takes place in the following stages: homeostasis, inflammation, proliferation, and remodeling (Fig. 2) [21]. Homeostasis takes place immediately after injury and consists in plugging the latter with a fibrin clot derived from blood coagulation thanks to a cascade of reactions involving coagulation factors and leading to the conversion of fibrinogen to fibrin and the activation of platelets [9]. Hence, homeostasis aims to stop bleeding, temporarily restoring the physical barrier of the skin. It also induces the inflammatory phase, *via* cytokine secretion by activated platelets [41]. Thus, the inflammatory phase initiates during homeostasis [42]. The former consists of the recruitment of immunological cells, mainly neutrophils and pro-inflammatory macrophages [43]. These cells recognize and remove microbes and cell debris by phagocytosis, cleaning the site of the wound [40]. They then contribute to the resolution of inflammation by their reverse migration; moreover, neutrophil death (by apoptosis) and efferocytosis induce the conversion of pro-inflammatory macrophages

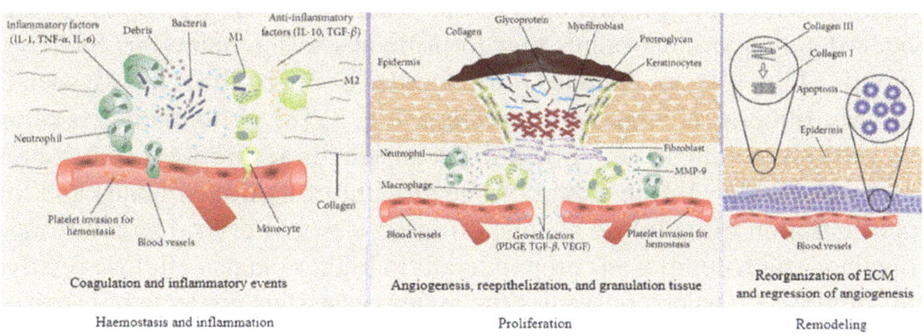

Fig. 2. Descriptive stages of normal adult wound healing Ref. [142].

into anti-inflammatory cells [21, 43]. These are responsible for the secretion of growth factors such as transforming growth factor beta and vascular endothelial growth factor, which induce the migration and proliferation of fibroblasts and endothelial cells in the proliferation and new tissue formation phase [42, 44]. Fibroblasts, the main cells in the genesis of the dermis, also secrete cytokines responsible for the proliferation and differentiation of epithelial cells, thus playing a determining role in the morphogenesis of the epidermis [31]. On the other hand, endothelial cells are responsible for the genesis of blood vessels that supply oxygen and nutrients to the newly formed tissue [45]. Remodeling is the last step of wound healing; it consists of the reorganization of ECMs and the regression of angiogenesis [21].

The different healing steps contribute to the restoration of the structure and functions of the skin. However, alteration of any of these steps in time or by any event is likely to delay the recovery of the skin structure, with increased chances of generating fibrosis and scar. Hence, to assure optimal wound healing and promote uniform regeneration of the skin, application of biomaterials such as wound dressing is recommended.

3. Potential of Nanocellulose for Wound Healing and Skin Tissue Engineering

Selected on their ability to simulate *in vivo* conditions or to be manufactured with such features, biomaterials for tissue regeneration satisfy not only biological but also physico-chemical properties, generally originating from their structural backbone. Cellulose consists of fibrillated chains of glucose units connected one to the other by β-1,4 links, and further organized into microfibrils by Van der Waals forces as well as intra- and inter-chains hydrogen bonds that form thanks to the numerous hydroxyl groups of the glucose residues [46]; bundles of microfibrils then associate to form ribbons of cellulose nanofibers (Fig. 3) [8]. With diameters of up to 100 nm, cellulose nanofibers differ in length according to the source material (Table 1). Depending on the amount of hydrogen bonds and the configuration of cellulose chains, different types of crystal structures are obtained, with cellulose I_β preponderant in higher plants cellulose relative to cellulose I_α, which constitutes the major crystalline contents of bacterial cellulose [11]. Moreover, plant cellulose presents increased amorphous regions and reduced crystallinity as compared to bacterial cellulose,

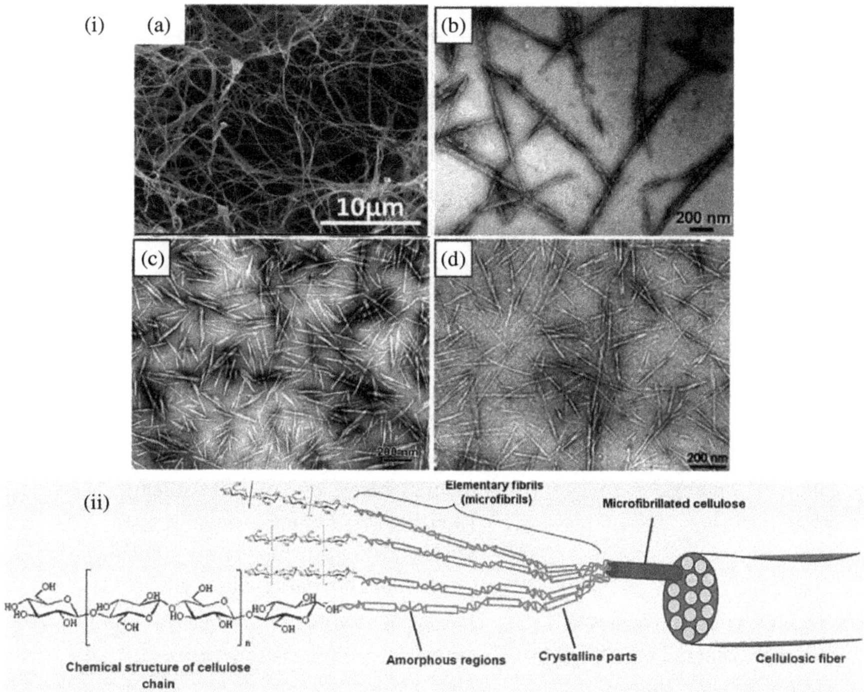

Fig. 3. Morphology (i) and structural organization (ii) of nanocellulose. (a) SEM image of bacterial cellulose fibers; (b), (c), and (d) TEM images of dried dispersion of cellulose nanocrystals derived from bacteria, ramie, and sisal, respectively.

Source: (i): (a) Reproduced from Ref. [6] with the permission of the American Chemical Society; (b), (c), and (d) Reproduced from Ref. [143] with the permission of the American Chemical Society; (ii): Reproduced from Ref. [144] with permission from Elsevier.

although CNC with greater crystallinity can be prepared following chemical degradation of the amorphous regions. While pure BNC is produced from bacteria in the form of meshed hydrogel containing up to 90% water, CNF forms hydrocolloid suspensions in water and can be utilized in various formulations including gels and films; CNC however are rod-like structures hardly deformable due to their rigid nature [47] (Fig. 3). Hence, contrary to BNC and CNF, CNC is not frequently utilized for the fabrication of stable 3D materials due to its minimal fiber entanglement, but is rather mostly employed as reinforcement [18].

Despite its non-degradability in humans that somewhat limits its usefulness as an implant in tissue engineering, cellulose conveys a firmly

Table 1. Types of nanocelluloses and properties [47].

Cellulose type	Properties
Cellulose Nanofibers (CNF) Synonyms: Microfibrillated cellulose, cellulose microfibrils, nanocellulose fibers, nanowiskers	– Produced by mechanical (high pressure homogenizer), chemical (TEMPO-mediated oxidation), enzymatic treatment – Plant and algae as sources – 5–60 nm diameter and several microns in length (high aspect ratio) – Flexible and long fibers able of entanglement – Gel-like characteristics in water – Contains both crystalline and amorphous regions
Cellulose Nanocrystals (CNC) Synonyms: Nanocellulose crystals, nanocrystalline cellulose, cellulose crystallites	– Produced by strong acid hydrolysis (H_2SO_4, HCl) followed by ultrasonic treatment – Plant source – 5 nm diameter and 20–100 nm length – Elongated crystalline rod-like shapes – Rigid rods (elastic modulus ~ 100 GPa) – Crystalline: no amorphous regions
Bacterial Nanocellulose (BNC) Synonyms: Bacterial cellulose, microbial nanocellulose	– Produced as polymer by bacteria from low molecular weight carbon sources – 20–100 nm diameter and several microns in length in a continuous 3D network of CNF – Form stable hydrogels with up to 99% water – High crystallinity and pure cellulose (no hemicellulose, lignin) – Good mechanical stability

acknowledged biocompatibility and a range of properties that make it suitable for biomaterial applications, particularly in wound healing. It is a xeno-free, cheap, and sustainable macromolecule with available and widespread source material. Mimicking the ECM morphology by its fibrous structure, nanocellulose further presents the nanodimension of the latter, thus offering the potential of regulating cellular microenvironments at such levels [10]. Due to the small fiber size, nanocelluloses exhibit high specific surface area supportive of cell adhesion [48]. Hence, it has been shown to promote fibroblast and keratinocyte cell attachment and growth [6, 49]. In nanocellulose-based materials, the fiber entanglement and the degree of crystallinity can be controlled to optimize the porosity within the material, thereby directing cell infiltration and migration as well as gas and fluid circulation; besides, the pores can be exploited for lodging of useful small molecules [50–52]. Similarly, nanocelluloses possess tailorable and high mechanical properties attractive for biomaterial fabrication. Indeed, nanocellulose has been utilized as reinforcing material to achieve desired biomaterial properties [49, 53, 54]. While nanocelluloses can be fabricated into various formulations (films, gels, foams), these materials display good moldability and flexibility, facilitating adaptation to the shape of the wound and efficient covering. Attributes for wound dressing applications further include the high hydrophilicity arising from the high porosity, the numerous hydroxyl groups of the component glucose molecules, and the hydrogen bonds that lock water molecules, imparting high swelling abilities necessary to absorb wound exudates [6]. Because of this behavior, nanocellulose-based materials form a hydrated layer on the wound bed, accelerating tissue regeneration and decreasing adhesiveness while relieving the patient's pain [12]. Thus, nanocellulose outperformed carboxymethylcellulose in the treatment of full-thickness skin injuries [12].

Moreover, various functions and bioactivities can be endowed to the materials through chemical functionalization or composite preparation, as demonstrated in many studies [55, 56]. Among the nanocellulose types, BNC stands out as the most promising for wound healing applications [57] and has thus been intensively researched in this respect over the years, due to its highly porous and attractive network [55, 58]. As it possesses higher crystallinity in comparison with plant cellulose, it is also more stable, with higher swelling ability deriving from the higher porosity due to smaller fiber size [11, 47]. Despite its debated production costs, its purity remains quite an advantage. Nevertheless, nanocellulose is devoid of antimicrobial activity, one of the wound dressing essentials. Fortunately, efforts in modifying the macromolecule to this end have been amply successful, as detailed below [52, 59, 60]. Overall, with similar chemistries despite the diversity of sources, nanocelluloses present similar types of properties but differ mainly in their degrees of crystallinity and dependent features.

4. Tuning the Properties of Nanocellulose for Optimal Application

Optimized biomaterials are obtained from raw polymer materials and macromolecules possessing attributes that address the needs of specified applications. However, due to the large array of criteria, singular primary materials rarely present all the required features for a successful biomaterial in a given application, encouraging material modification and the use of composite materials for achieving adequate biomaterial behavior. To attain suitable wound healing, various, often co-dependent, wound dressing properties can be leveraged, including the biocompatibility, antimicrobial activity, as well as water absorbency and retention along with the ability of delivering bioactive substances to the wound bed [8]. According to the wound type (infected wound, chronic non-healing wound, low or high exuding wound, different degrees of burn wounds, etc), fitting properties are assembled by tailoring the structure and network of main matrix components. Interestingly, as indicated above, nanocellulose presents a hierarchical structure enabling multilevel modification, both physically and chemically (summarized in Fig. 4 with the illustration of bacterial cellulose). In fact, the numerous hydroxyl groups of cellulose, while enabling physical interactions with other charged or functionalized molecules (mostly by hydrogen bonding) also allow for covalent bonding with other functional groups by their chemical reactivity [61], leading to the versatile application of this polysaccharide.

4.1 *Enhancing the Wound Healing Ability*

To improve biocompatibility and promote the active participation of the dressing to the healing process, cell–material interactions can be enhanced by mimicking the various regulating factors of *in vivo* ECM (Fig. 5). Hence, as for tissue engineering biomaterials in general, dressings may incorporate adhesion ligands or natural ECM proteins for facilitating cell anchorage. For example, nanocomposites were prepared from BNC with keratin to enhance the attachment of dermal fibroblast cells [62]; similarly, the physical incorporation of hyaluronan in BNC yielded improved growth of primary human fibroblast cells [63]. In the same scope, composites may be prepared with bioactive natural compounds that stimulate and support healing, including silk sericin with its reputed wound healing effects [64–66], silk fibroin [66–68], and curcumin [69] among others. Moreover, intrinsic physical characteristics of the nanocellulose network such as the porosity and pore size as well as the fibrous structure and size may be controlled to facilitate cell compatibility in terms of attachment, migration, and growth [70]. In this prospect, the BNC matrix can be adapted *in situ* during production by the bacteria by use of porogens,

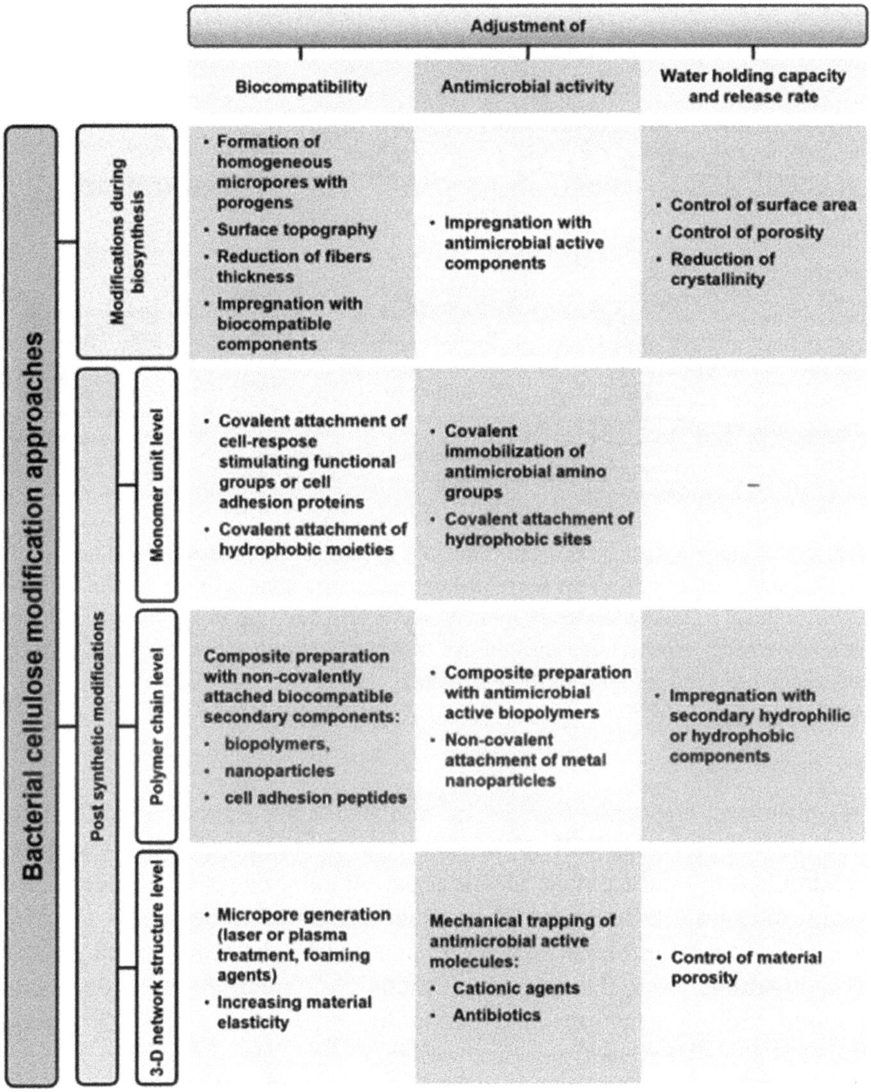

Fig. 4. Tailoring the properties of nanocellulose for wound healing application: The illustration of bacterial nanocellulose.

Source: Reproduced from Ref. [8] with the permission of Elsevier.

lithography techniques, and/or by post-processing methods such as phase separation. As for CNF and CNC suspensions and hydrogels, they can generally be subjected to fabrication methods such as electrospinning or

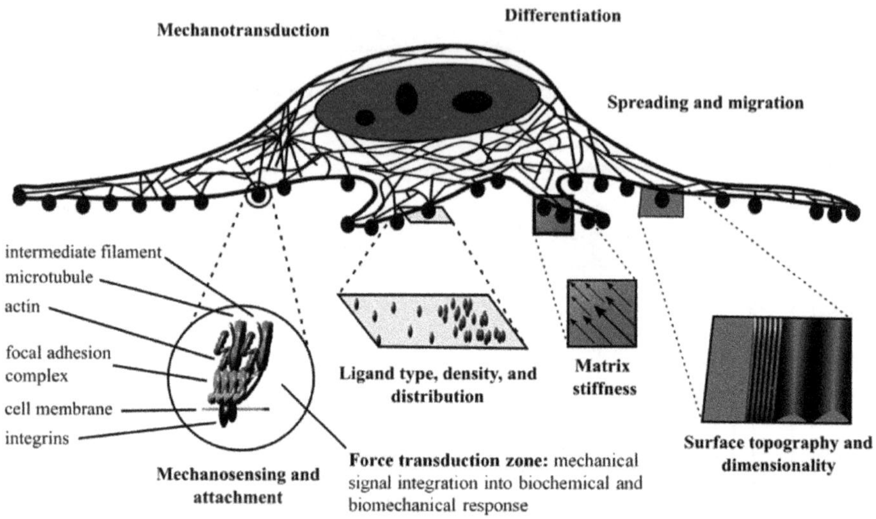

Fig. 5. Factors regulating cell fate through biomaterial properties. In addition to the biochemical composition of the microenvironment surrounding cells (such as in biomaterials), features including topography and elasticity also have the ability of regulating cell behavior through mechanotransduction that leads to chemical response and adaptive cell migration, proliferation, and/or differentiation.

Source: Reproduced from Ref. [11] with permission.

3D printing [71], with lithography and phase separation as well as the inclusion of porogens [72] applicable when forming materials by solution casting. Additional to these physical parameters, material surface topography may be altered by molding or printing techniques so as to foster normal and efficient healing as demonstrated by the reduced scar formation resulting from the use of meshed BNC membranes upon which crossed groove structures had been imprinted using photolithography, which induced native tissue-like collagen structure in rat skin wounds [73]. When stripe structures were formed on the BNC membrane during biosynthesis, inflammation, which is also a hallmark of scars, considerably reduced in animal wound healing, leading to reduced scar size [74, 75]. Actually, recent findings revealed that fiber alignment enhanced cell migration and vascularization as compared to non-aligned materials, efficiently replicating in rat's healed skin the collagen organization found in non-wounded tissue, thus improving the overall wound healing [67]. Other important physical parameters in cell–material interactions involve mechanical properties such as material stiffness and viscoelasticity, explaining the success of hydrogels as wound dressings as they reproduce the texture of natural ECM [76]. Thus, CNF, which readily forms hydrogels by its high viscosity in water is often applied in such formulations. The fiber structure of CNF and CNC can be varied by optimizing the method of purification from the source material, which further affects material properties such as the swelling ability as well as the mechanical properties, all reflecting on the biological performance of fabricated biomaterials [77]. Moreover, the water absorbency, which also depends on material chemistry, may be regulated by composite preparation, such as in the case of addition of plasticizers like polyethylene glycol [78] or by addition of other molecules (e.g., collagen) presenting different hydrophilic profiles relatively to the to-be-modified nanocellulose material [79].

Besides physical modulation, chemical functionalization may also be applied to aid biocompatibility and enable the active involvement of the dressing in wound healing. In a recent study, the grafting of specific fluorescent peptides onto the nanocellulosic interface of a cellulose-based dressing promoted the inactivation by capture and sequestration of lytic elastase that is released in the wound by neutrophils; the proteolytic enzyme is responsible for the growth factor and ECM degradation that delays healing in chronic wounds [80]. Moreover, chemical treatments are often helpful in the incorporation of active functional sites onto the polymer structure, in view of facilitating further reaction with other molecules for the fabrication of composite materials [61, 79].

In summary, active wound dressings are those biomaterials designed so as to interact with cells and direct their behavior, ultimately controlling the outcome of the wound healing. This relies on various material parameters, including the biochemical, physical, and mechanical properties of the final biomaterial. Additional to these factors discussed above, wound dressing biomaterials may serve as delivery vehicles, bringing regulatory growth factors or stem cells and populating cells to the wound [7, 81].

4.2 *Addition of Antimicrobial Property*

Imparting antimicrobial abilities to nanocellulose also constitutes an important aspect of its optimization for wound dressing applications, given that cellulose itself is intrinsically devoid of such properties. This is achieved by associating cellulose (in composite dressings) with other materials with potential interest, through physical interactions or by use of chemical cross-linking. In this context, chitosan is a natural antimicrobial biopolymer often utilized, enabling the development of promising wound dressing biomaterials [72, 82–84]. Other natural compounds with effective antibacterial activity may be used, as illustrated by the incorporation of *C. colocynths* or *Aloe vera* extracts in BNC-based dressings [60, 85]. Moreover, nanocellulose can be endowed with antimicrobial activity by use of inorganic compounds and metal ions with such effects. Thus, ion-cross-linked montmorillonite-BNC membranes were reported that exhibited adequate antimicrobial properties in addition to successfully acting as tissue regenerative dressings in the animal setting [86]. Similar to gold nanoparticles [59], silver nanoparticles have also been used on several occasions, displaying antibacterial abilities while contributing to wound healing by reducing inflammation [60, 87, 88]. Further, antimicrobial drugs such as polyhexamethylene biguanide, tetracycline hydrochloride, octenidine, or others may be integrated in dressing materials for preventing wound infection [89–92]. Besides these modification approaches, CNC and CNF have occasionally been reported to inhibit microbial growth to some extent [16, 93], though this does not represent a constituent property of the cellulose macromolecule and thus is not consistent; however, following activation by ion-cross-linking that enables immobilization of bivalent cations (Cu^{2+}, Ca^{2+}) on the nanocellulose matrix, strong antibacterial and antibiofilm activities are obtained [94]. Similarly, composites with cationic polymers endowed with antimicrobial abilities or the simple addition of cationic groups have the potential of imparting these properties to nanocellulose [95, 96]. Engraftment of anionic groups have also been reported, leading to antiviral activity in CNCs [97].

5. Nanofibrillar Cellulose Materials in Skin Repair

As a bioinert material, cellulose, and thus CNF, possesses no inherent property capable of directly impacting tissue regeneration or repair; however, the latter provides the non-toxic and biocompatible platform necessary for sustaining and protecting the involved actors, namely growth factors and cells, which will activate and/or accelerate the process of tissue repair and regeneration. Indeed, calcium ion-cross-linked CNF hydrogels supported the growth of primary human dermal fibroblasts without affecting the cytokine production of inflammatory cells, while enabling absorptive behaviors suitable for dressing applications in variable wound types [98]. Besides, by tunability of physical properties, CNF enables the ready replication of the rheological properties of the ECM, essential for tissue engineering applications [99]. Further, for wound healing and skin repair applications, CNF-based biomaterials (wound dressings) offer the ability to absorb wound exudates, and low adhesiveness enabling removal of the material from the wound surface after recovery, both important criteria for wound dressing materials. Hence, ultrapure wood CNF materials were produced by TEMPO-mediated oxidation that demonstrated good biocompatibility with human primary skin cells, in addition to presenting outstanding swelling abilities (superior to the control commercial dressing), indicating potential for the dressing of exuding chronic wounds [100]. Likewise, in a study by Hakkarainen *et al.* the CNF-based wound dressing outperformed the commercial Suprathel® dressing in human skin graft donor site treatment, with faster epithelialization and wound healing thanks to the hydration ability of CNF [12]. Further, in a porcine model, it was demonstrated that CNF inhibits wound contraction without compromising proper wound healing, which would reduce the formation of debilitating excessive fibrotic scars [101]. However, most materials involving CNF are composites that take advantage of its physical properties while incorporating components with beneficial bioactivities for required applications. Gao *et al.* recently proposed a wound dressing combining highly porous CNF sponges with quaternized β-chitin/organic rectorite possessing antimicrobial effects, with overall improved wound healing ability [102]. A sericin-releasing composite with CNF as supporting material has also been proposed, with the reputed positive effects of the silk protein expected to ensure effective wound healing [64]. Elsewhere, a transparent wound dressing was proposed based on CNF and chitosan, which displayed mechanical properties comparable to that of human skin, anticipating potential in skin repair as artificial skin or wound dressing [103]. When nanocellulose was added to a mixture of chitosan-poly (vinyl pyrrolidone) solution for composite preparation, improved mechanical properties and swelling abilities were obtained from the produced materials which displayed blood compatibility, antibacterial activity, and *in vitro* biocompatibility to NIH3T3 cells, substantiating their potential for wound dressing application [83]. With the addition of NaCl, the porosity could be controlled, which was further enhanced with the increasing content of nanocellulose; moreover, as stearic acid was utilized to form a hydrophobic surface on one side of the samples thus yielding asymmetric materials, a good biocompatibility was demonstrated in cell culture with L929 cells whereas higher wound healing rate was observed in rats relative to the povidone-iodine control ointment [72]. By carboxymethylating periodate oxidized cellulose nanofibrils, Rees *et al.* obtained a printable bioink with adequate rheological properties; interestingly, the printed materials exhibited antibacterial properties, useful for efficient wound healing [104]. Other CNF-based printed hydrogels incorporating gelatin methacrylate were recently reported that showed promise for wound healing application promoting *in vitro* the growth of fibroblast cells. Thanks to the *in situ* cross-linking of the different components and the strong interaction between the latter, printable ink was obtained at low polymer concentrations [71]. Similarly, Ca^{2+}-cross-linking enabled the stabilization of CNF-alginate 3D-printed grid matrices and helped correct their size and shape that were negatively affected by the presence of alginate, which also diminished the print fidelity though substantially contributing to the overall mechanical properties of the composites [105]. Presented as potential candidates for wound dressing application, these successfully supported the growth of L929 mouse fibroblasts [105].

6. Bacterial Nanocellulose-Based Biomaterials for Skin Tissue Repair

With regard to wound treatment application, BNC has long been recognized as a satisfactory biomaterial since the 1980s and indeed possesses qualities ideal for wound dressing, while providing a relatively cost-effective alternative to other wound dressings [48]. BNC is a moisture-retentive dressing that offers a warm and moist environment favorable to healing processes such as granulation and removal of dead tissues [106]. In addition, BNC provides a soothing sensation to the wound that relieves pain and encourages accelerated recovery. Moreover, while BNC reduces maceration of the wound edges by trapping and retaining exudates from the wound site, microbial invasion is inhibited as this moist-based wound dressing can easily adhere to the contour of any wound bed, creating a protective barrier between the wound and the surrounding environment, with infection further limited by the asymmetric pore size distribution of the BNC when the latter is applied such that the membrane side with smaller pores is the one in contact with the external environment [107, 108]. As a biocompatible, non-toxic, transparent, and mechanically strong material, BNC is in fact a very practical wound dressing with attractive clinical values [109]. The clinical investigation of a commercial BNC-based wound dressing (XCell®) revealed that its use in the treatment of venous leg ulcer generated a protective, moist, and hypoxic environment, thus facilitating the formation of healthy granulation tissue and accelerating re-epithelialization [110]. These advantageous effects have been repeatedly observed in similar studies following the application of BNC-based wound dressings [6]. Thus, BNC is applicable for the treatment of various wounds, including skin tears, venous stasis, ischemic and diabetic wounds, pressure sores, and third-degree burns [110–113], as illustrated in Table 2 which presents a list of available commercial BNC-based wound dressings [114].

Table 3 summarizes the different types of BNC-based dressing materials described herein.

6.1 *Bacterial Nanocellulose as a Biocompatible Biomaterial*

Besides controlling scar tissue formation, one critical goal of wound healing is to reestablish the biological and structural function of the skin. In this regard, the ability of a material to stay in contact with living tissues without provoking any adverse reaction is considered critically important when creating novel biomaterials. BNC serves as a good cell-supporting artificial ECM that provides for cell adhesion and growth of connective tissue. Great biocompatibility was observed in a study on skin lesion treatment in BALB/c mice, with BNC film securing a suitable niche for cell proliferation and good wound protection while providing proper mechanical properties and ease of dressing removal [115]. Furthermore, BNC films exerted skin tolerance by integrating into the full-thickness skin wounds of Wistar rats upon implantation [116]. In the latter study, the porosity of the BNC membrane was beneficial for the attachment and proliferation of fibroblasts and epithelial cells. Apparently, the high surface area and pore volume of BNC greatly contributed to cellular activities and abilities to infiltrate the BNC films.

6.2 *Bacterial Nanocellulose-Based Antimicrobial Biomaterials*

Fundamental to wound healing is the use of materials that can overcome infectious complications in wound treatment. Although BNC itself lacks antimicrobial qualities, efforts at imparting this feature to BNC have been

Table 2. List of commercial bacterial nanocellulose-based wound healing biomaterials. From Ref. [114].

Product name	Medical application	Company (country)
BioFill®	Temporary substitute for human skin	Fibrocel Produtos Biotechnologicos[a] (Brazil)
BioProcess®	Skin transplant, treatment of third degree burns, ulcers, and decubitus	Fibrocel Produtos Biotechnologicos[a] (Brazil)
NexFill®	Dry bandage for burns and wounds	Fibrocel Produtos Biotechnologicos[a] (Brazil)
Gengiflex®	Bandage for dental wounds	Fibrocel Produtos Biotechnologicos[a] (Brazil)
Suprasorb® X[b]	HydroBalance wound dressing for lightly to moderately exuding, non-infected wounds	Lohmann & Rauscher (Germany)
Suprasorb® X[b] + PHMB	Antimicrobial hydroBalance wound dressing containing 0.3% polyhexamethylene biguanide, for lightly to moderately exuding wounds at risk of infection or infected	Lohmann & Rauscher (Germany)
DermaFill™	Translucent wound dressing	Cellulose Solutions (USA)
CELMAT® Wound/Eye/Face	Hydrated wound dressing	Bowil Biotech (Poland)
CELMAT® Wound/Eye/Face H	Hydrated wound dressing containing sodium hyaluronate	Bowil Biotech (Poland)
CELMAT® Wound/Eye/Face P	Hydrated, antiseptic wound dressing containing polyhexanide	Bowil Biotech (Poland)

Notes: [a]Earlier Biofill Industrias, Brazil.
[b]Earlier XCell® delivered by Xylos, USA.

Table 3. BNC-based biomaterials for wound healing and skin repair applications.

BNC-based biomaterial	Applications	Therapeutic effects	Refs.
BNC	Treatment of chronic leg venous ulcer, diabetic ulcer	Enhanced epithelialization and reduced wound pain, enhanced rate of wound healing	[110, 140]
BNC	Skin tissue repair	Suitable niche for proliferation and adhesion of human adipose-derived stem cells	[115]
BNC/Silver nanoparticles	Antibacterial wound dressing	Increased biocompatibility and antibacterial activity	[120, 133]
BNC/Chitosan	Antibacterial wound dressing	Improved mechanical and antimicrobial properties	[134]
BNC/Silk sericin	Wound dressing	Increased skin cell viability	[115]
BNC with silk sericin and polyhexamethylene biguanide	Antimicrobial wound dressing	Improved antimicrobial activity and collagen production	[117]
BNC-Carboxymethyl cellulose/ε-Poly *L*-lysine	Antimicrobial wound dressing	Contact inhibition of *Staphylococcus epidermis*	[127]
BNC/Lidocaine	Burn injuries	Healing of deep burn lesions with mild inflammation, no maceration, and no fibrosis	[113]
BNC/Antimicrobial agents	Wound dressing with drug controlled release	Improved antibacterial activity and biocompatibility	[123–125]
BNC/Antibiotics	Antibiotic-loaded wound dressing	Improved antibacterial activity	[126]

(*Continued*)

Table 3. (*Continued*)

BNC-based biomaterial	**Applications**	**Therapeutic effects**	**Refs.**
BNC/Acrylic acid	Cell carrying wound dressing	Enhanced cell transfer and survival, hastened wound healing	[106]
BNC/Poly(lactic acid)	Antibacterial dressing with drug loading capacity	Improved antibacterial and mechanical properties, promoted fibroblast adhesion and growth	[118]
BNC/Poloxamer		Increased absorption capacity of BNC and controlled release of octenidine with increased mechanical and antimicrobial properties	[130]
BNC/Agarose		Improved wound healing performance of BNC, enhanced mechanical properties and water retaining ability	[135]
BNC/Solidum alginate		Improved functional properties of BNC with enhanced mechanical properties	[136]
BNC/Gelgum/AMSCs[a]	Skin tissue regeneration	Infiltration of AMSCs in the dressing, high regenerative potential, increased cell attachment and proliferation	[137]

Note: [a] AMSCs: adipose derived mesenchymal stem cells.

successful. Besides biological and synthetic polymers rich in antimicrobial activity [117, 118], inorganic compounds (e.g., metal or graphene nanoparticles and metal oxide) [119, 120] and cationic (e.g., quaternary ammonium compounds) antiseptics [121] have been used in composites with BNC to inhibit the growth of pathogens. It is quite clear that in addition to the bacterial source, several conditions such as the physicochemical properties of the active compound as well as the formulation and configuration of the BNC-network determine the choice of incorporation route. Of particular interest is the significant association of the time-dependent release rate of the active compounds with the functionalization strategy. Solution impregnation is the more preferred and simple route for incorporating foreign elements into the BNC network. Indeed, the evaluation of the wound healing performance of BNC impregnated with a commercial cationic antimicrobial surfactant demonstrated bioinhibitory activity against variety of microorganisms, while providing optimal moisture needed for healing chronic and burn wounds [122]. Studies on BNC functionalized by immersion in octenidine antiseptic solutions seeded with human keratinocytes confirmed the advantages of its use as wound dressing including high biocompatibility, slow release of active compounds, and antimicrobial effect on *Staphylococcus aureus,* even after 6 months of storage [123]. These features together with unaltered mechanical properties are in agreement with those of polyhexanide-loaded BNC but inconsistent with those of BNC modified with povidone–iodine (PI) that showed a lower antimicrobial activity due to delayed PI release deriving from the high molar mass of this component which actually stimulated significant structural changes in BNC, thereby increasing its compressive strength [124]. Elsewhere, antibiotic conjugation with BNC by submersion procured antimicrobial activity by inhibiting protein translation through binding of the antibiotics with 50S ribosomal subunits; the resulting BNC/tetracycline [125] and BNC/chloramphenicol [126] conjugates revealed antimicrobial effects against tested bacteria and beneficial biocompatibility following slow release of the antibiotics. Significant antimicrobial efficiency, biodegradability, and cytocompatibility to human fibroblasts were obtained from BNC membranes incorporating the natural antimicrobial peptide ε-poly-*L*-lysine (ε-PLL) and this served as a barrier to microbial growth in skin wounds [127]. Besides, laccase immobilization on BNC by immersion presented high antimicrobial activity for Gram-positive bacteria with essential cytocompatibility, as required for wound dressing application [116]. With regard to conjugation strategies, increased collagen formation and antimicrobial activity were obtained from simultaneous blending of BNC with silk sericin and polyhexanide, highlighting the potential of introducing more than one functionality into BNC wound dressing [117].

6.3 *Bacterial Nanocellulose-Based Composites and Conjugates*

Based on the chemical composition of BNC which is equivalent to vegetal cellulose, the surface properties of BNC are readily tunable by chemical modification used for cellulose derivatization. This approach in turn presents the advantage of loading the BNC matrix with various molecules that can be control-released, with the possibility for various compounds potentializing the effects one of the other. In fact, the adsorption capacity of proteins onto BNC was increased by functionalizing the BNC structure with various groups such as acetyl groups, carboxylmethyl groups, quaternary ammonium groups, and phosphoric groups, allowing for the introduction of additional adsorption capacity for antimicrobial agents to BNC surface [121, 128, 129]. Besides the increased mechanical and antibacterial properties, octenidine's release time frame approached 96 hours in a control release investigation aimed at prolonging the release of the antiseptic from the dressing, using poloxamers in micellar structures and gels as drug carriers to modify the BNC matrix [130]. Using carbodiimide chemistry, ε-PLL was covalently cross-linked to carboxymethyl-modified BNC (CMC-BNC) membrane and directly cross-linked to pure BNC membranes [127]. The mechanical properties and nanofibrous structure of both pure BNC and CMC-BNC membranes remained unaffected after modification, with the growth of *Staphylococcus epidermidis* inhibited upon contact with the composites. Also, inclusion of amine and methyl groups on the BNC surface promoted the attachment and proliferation of normal dermal fibroblast cells by improving the hydrophobic and electrostatic interactions between cells and the BNC surface [131]. When used as an antimicrobial agent, various BNC/metal nanocomposites displayed increased levels of antibacterial activity. Rather than the usual physical interaction approach to incorporate silver nanoparticles along with strong reducing agents into the BNC network, a modified approach using mild, chemical dimethyl sulfoxide as reducing agent was employed to activate BNC membranes where AgNPs were chemically immobilized [132]. The generated BNC/AgNP exerted effective antimicrobial action against *E. coli* while exhibiting elevated retention time. Another effective

BNC/AgNP developed by ion exchange following soaking of TEMPO-oxidized BNC in $AgNO_3$ solution resulted in low cytotoxicity for NIH3T3 fibroblast cells and strong antibacterial activities of 100% and 99.2% against *E. coli* and *S. aureus,* respectively [133].

Further, altering material properties such as porosity, fiber structure, and wettability by blending BNCs with other materials significantly widens its applications. Blends of BNC with different natural polymers such as chitosan, agarose, silk sericin, and alginate significantly ameliorated BNC's functional properties as wound dressing material [6, 134–136]. BNC/chitosan composites generated by impregnation of BNC in chitosan solution were also more efficient in enhancing BNC's wound healing efficacy [134]. Chitosan penetrated and integrated into the cellulose chain, exhibiting greater water-retaining properties, mechanical properties, and high antibacterial activity against Gram-positive and Gram-negative bacteria. Furthermore, coating of poly(lactic acid) (PLA) over dry BNC sheet generated a microenvironment with an improved mechanical properties to accelerate wound repair [118]. Layers of PLA on the BNC influenced the morphology of the biocomposite sheet, thereby revealing a porous-fiber like structure which allows low moisture uptake and prolonged swelling in simulated wound fluid. Other approaches have involved the blending of BNC with sodium alginate possessing a semi-interpenetrating polymer network structure [136]; the improved swelling ratios and mechanical properties of the composite arose from the interpenetrating structure, which is promising for biomedical purposes such as wound dressing and skin tissue engineering.

6.4 *Bacterial Nanocellulose-Based Biomaterials for Cell Delivery Applications*

One of the trending concepts towards improving the qualities of BNC as a prospective skin substitute is the incorporation of cells into the BNC matrix. In particular, the incorporation of human epidermal keratinocytes and dermal fibroblasts into BNC membranes is expanding and features essential benefits to accelerate wound healing and stimulate wound re-epithelialization. Considering their roles in stimulating the production of regulatory cytokines and ECM proteins essential in healing, these cells seeded on BNC/acrylic acid hydrogels promoted the wound healing action of BNC *in vitro* and in mice models *in vivo* [106]. Stem cells, aside from possessing self-renewal ability, can differentiate into various specialized cell lineages and are effective in inducing wound epithelialization, thus presenting high regenerative potential for burn wounds. When seeded within the BNC/gel gum matrix and tested on rats' burn wounds, adipose mesenchymal stem cells (AMSCs) demonstrated higher regenerative potential than the commercial membrane; the ability of AMSCs to infiltrate into the BNC membrane and migrate into the injured area to improve healing along with the adhesion and proliferation of the cells on the membranes present BNC/AMSCs membranes with desirable dressing abilities for burns and sores [137].

7. Conclusions and Future Prospects

As the nanodimensional product of cellulose, nanocellulose has gained tremendous attention in materials science thanks to its availability and unique natural virtues, namely a chemical structure and physical properties that enable wide and versatile application in various fields. In biomaterials, the biopolymer is attractive owing to its biocompatibility, strong mechanical properties and ECM-like fibrous structure readily amenable for the fabrication of matrices supportive of cellular activity, tissue repair, and regeneration. For skin tissue repair specifically, nanocellulose is further attractive owing to its high hydrophilicity and hydration ability, propitious for wound healing. In fact, contrary to its somewhat limited application as a regenerative implant due to slow *in vivo* degradation, nanocellulose constitutes an advantageous and convenient material for wound dressing applications. Easily formulated in hydrogels which represent optimal niches for skin wound healing, nanocellulose further enables the delivery of bioactive components beneficial for healing, or modification with the latter. BNC, which is considered best suited for skin tissue repair among all types of nanocelluloses, is also the most studied, with CNC and CNF only beginning to come into focus due to their more renewable source and cheaper production costs. Hence, there is still work to be carried out before these biomaterials are fully accessible in the market, though few BNC-based dressings are already available in clinics (Table 2). For instance, the biocompatibility of CNC- and CNF-based materials require deeper optimization studies. While BNC is excreted in pure form by bacteria and possesses an established biocompatibility, CNF and CNC are mostly extracted from plant source materials after purification; therefore, any remaining impurities may compromise biocompatibility. Moreover, optimization needs to be performed to minimize potentially toxic leachables or extractables, as illustrated by the study of Liu *et al.* [138]. Moreover, these compounds have been reported to present some extent of cytotoxicity at certain concentrations, as clearly summarized by Jorfi and Foster in their review article [139]. In a very recent study, though CNF did not affect the inflammatory reaction, it induced a foreign body response and was encapsulated due to slow *in vivo* degradation, inciting further modification studies to prevent the latter response and enhance the degradation of the nanocellulose. Simultaneously, alternative production methods of BNC are needed to reduce its costs and facilitate the scale-up production, in order to render this material more accessible in the market.

Acknowledgments

The authors would like to acknowledge Dr. Lina Fu for advising on the contents of the current chapter.

This work was supported by the National Natural Science Foundation of China [grant numbers 21774039, 21574050].

References

[1] Nourian Dehkordi, A., F. Mirahmadi Babaheydari, M. Chehelgerdi, and S. Raeisi Dehkordi, Skin tissue engineering: Wound healing based on stem-cell-based therapeutic strategies. *Stem Cell Res. & Therapy*, 2019, 10(1): p. 111.

[2] Wells, A., A. Nuschke, and C.C. Yates, Skin tissue repair: Matrix microenvironmental influences. *Mat. Biol.*, 2016. 49: pp. 25–36.

[3] Böttcher-Haberzeth, S., T. Biedermann, and E. Reichmann, Tissue engineering of skin. *Burns*, 2010, 36(4): pp. 450–460.

[4] Kundu, B. and S.C. Kundu, Silk sericin/polyacrylamide *in situ* forming hydrogels for dermal reconstruction. *Biomaterials*, 2012, 33(30): pp. 7456–7467.

[5] Chua, A.W.C., Y.C. Khoo, B.K. Tan, K.C. Tan, C.L. Foo, and S.J. Chong, Skin tissue engineering advances in severe burns: Review and therapeutic applications. *Burns & Trauma*, 2016, 4: 3.

[6] Lamboni, L., Y. Li, J. Liu, and G. Yang, Silk sericin-functionalized bacterial cellulose as a potential wound-healing biomaterial. *Biomacromolecules*, 2016, 17(9): pp. 3076–3084.

[7] Mertaniemi, H., C. Escobedo-Lucea, A. Sanz-Garcia, C. Gandía, A. Mäkitie, J. Partanen, O. Ikkala, and M. Yliperttula, Human stem cell decorated nanocellulose threads for biomedical applications. *Biomaterials*, 2016, 82: pp. 208–220.

[8] Sulaeva, I., U. Henniges, T. Rosenau, and A. Potthast, Bacterial cellulose as a material for wound treatment: Properties and modifications. A review. *Biotech. Adv.*, 2015, 33(8): pp. 1547–1571.

[9] Boateng, J.S., K.H. Matthews, H.N.E. Stevens, and G.M. Eccleston, Wound healing dressings and drug delivery systems: A review. *J. Pharma. Sci.*, 2008, 97(8): pp. 2892–2923.

[10] Milan, P.B., S. Kargozar, M.T. Joghataie, and A. Samadikuchaksaraei, Chapter 11 — Nanoengineered biomaterials for skin regeneration, in *Nanoengineered Biomaterials for Regenerative Medicine*, M. Mozafari, J. Rajadas, and D. Kaplan, Editors, 2019. Elsevier, pp. 265–283.

[11] Hickey, R.J. and A.E. Pelling, Cellulose biomaterials for tissue engineering. *Frontiers Bioeng. Biotech.*, 2019, 7(45).

[12] Hakkarainen, T., R. Koivuniemi, M. Kosonen, C. Escobedo-Lucea, A. Sanz-Garcia, J. Vuola, J. Valtonen, P. Tammela, A. Mäkitie, K. Luukko, M. Yliperttula, and H. Kavola, Nanofibrillar cellulose wound dressing in skin graft donor site treatment. *J. Controll. Rel.*, 2016, 244: pp. 292–301.

[13] Das, S. and A.B. Baker, Biomaterials and nanotherapeutics for enhancing skin wound healing. *Frontiers in Bioeng. Biotech.*, 2016, 4: 82.

[14] Rahimnejad, M., S. Derakhshanfar, and W. Zhong, Biomaterials and tissue engineering for scar management in wound care. *Burns & Trauma*, 2017, 5: p. 4.

[15] Kang, B.S., Y.C. Na, and Y.W. Jin, Comparison of the wound healing effect of cellulose and gelatin: An *in vivo* study. *Archives of Plastic Surgery*, 2012, 39(4): pp. 317–321.

[16] Jack, A.A., H.R. Nordli, L.C. Powell, K.A. Powell, H. Kishnani, P.O. Johnsen, B. Pukstad, D.W. Thomas, G. Chinga-Carrasco, and K.E. Hill, The interaction of wood nanocellulose dressings and the wound pathogen *P. aeruginosa. Carbohydr. Polym.*, 2017, 157: pp. 1955–1962.

[17] Kargarzadeh, H., M. Mariano, D. Gopakumar, I. Ahmad, S. Thomas, A. Dufresne, J. Huang, and N. Lin, Advances in cellulose nanomaterials. *Cellulose*, 2018, 25(4): pp. 2151–2189.

[18] Abitbol, T., A. Rivkin, Y. Cao, Y. Nevo, E. Abraham, T. Ben-Shalom, S. Lapidot, and O. Shoseyov, Nanocellulose, a tiny fiber with huge applications. *Curr. Opin. in Biotech.*, 2016, 39: pp. 76–88.

[19] Lu, Y., H.L. Tekinalp, W.H. Peter, C. Eberle, A.K. Naskar, and S. Ozcan, Nanocellulose in polymer composites and biomedical. *Res. and Appl.*, 2014.

[20] Groeber, F., M. Holeiter, M. Hampel, S. Hinderer, and K. Schenke-Layland, Skin tissue engineering — *In vivo* and *in vitro* applications. *Clinics in Plastic Surgery*, 2011, 39(1): pp. 33–58.

[21] Boni, B.O.O., L. Lamboni, T. Souho, M. Gauthier, and G. Yang, Immunomodulation and cellular response to biomaterials: The overriding role of neutrophils in healing. *Mater. Horizons*, 2019.
[22] Kolarsick, P.A.J., M.A. Kolarsick, and C. Goodwin, Anatomy and Physiology of the Skin. *J. of the Dermato. Nur. Asso.*, 2011, 3(4): pp. 203–213.
[23] Yildirimer, L., N.T.K. Thanh, and A.M. Seifalian, Skin regeneration scaffolds: A multimodal bottom-up approach. *Tren. Biotechnol.*, 2012, 30(12): pp. 638–648.
[24] D.J. Tobin, Introduction to skin aging. *J. Tissue Viab.*, 2017, 26(1): pp. 37–46.
[25] Baroni, A., E. Buommino, V. De Gregorio, E. Ruocco, V. Ruocco, and R. Wolf, Structure and function of the epidermis related to barrier properties. *Clinics in Dermatol.*, 2012, 30(3): pp. 257–262.
[26] Sheila, M.N. Progress and opportunities for tissue-engineered skin. *Nature*, 2007, 445(7130): pp. 874–880.
[27] Bandarchi, B., L. Ma, R. Navab, A. Seth, and G. Rasty, From melanocyte to metastatic malignant melanoma. *Dermatol. Res, Pract.*, 2010. 2010: pp. 583748.
[28] Bi, H. and Y. Jin, Current progress of skin tissue engineering: Seed cells, bioscaffolds, and construction strategies. *Burns & Trauma*, 2013, 1(2): p. 63.
[29] Blanpain, C., W.E. Lowry, A. Geoghegan, L. Polak, and E. Fuchs, Self-renewal, multipotency, and the existence of two cell populations within an epithelial stem cell niche. *Cell*, 2004, 118(5): pp. 635–648.
[30] Vig, K., A. Chaudhari, S. Tripathi, S. Dixit, R. Sahu, S. Pillai, V. Dennis, and S. Singh, Advances in skin regeneration using tissue engineering. *Int. J. of Mol. Sci.*, 2017. 18(4): p. 789.
[31] Boehnke, K., N. Mirancea, A. Pavesio, N.E. Fusenig, P. Boukamp, and H.-J. Stark, Effects of fibroblasts and microenvironment on epidermal regeneration and tissue function in long-term skin equivalents. *Eur. J. Cell Biol.*, 2007, 86(11–12): pp. 731–746.
[32] Yu, J.R., J. Navarro, J.C. Coburn, B. Mahadik, and J.P. Fisher, Current and future perspectives on skin tissue engineering: Key features of biomedical research, translational assessment, and clinical application. *Adv. Health. Mater.*, 2019, 8(5): p. e1801471.
[33] Graham, H.K., A. Eckersley, M. Ozols, K.T. Mellody, and M.J. Sherratt, Human Skin: Composition, structure and visualisation methods, in skin biophysics. 2019, Springer. pp. 1–18.
[34] Gilaberte, Y., L. Prieto-Torres, I. Pastushenko, and Á. Juarranz, Anatomy and function of the skin, ed. N.i. Dermatology. 2016, 1–14(2016), *Nanoscience in Dermatology*: Academic Press.
[35] Rippa, A.L., E.P. Kalabusheva, and E.A. Vorotelyak, Regeneration of dermis: Scarring and cells involved. *Cells*, 2019, 8(6): p. 607.

[36] Korosec, A., S. Frech, B. Gesslbauer, M. Vierhapper, C. Radtke, P. Petzelbauer, and B.M. Lichtenberger, Lineage identity and location within the dermis determine the function of papillary and reticular fibroblasts in human skin. *J. Inv. Dermatol.*, 2019, 139(2): pp. 342–351.

[37] Sriram, G., P.L. Bigliardi, and M. Bigliardi-Qi, Fibroblast heterogeneity and its implications for engineering organotypic skin models *in vitro*. *Eur. J. Cell Biol.*, 2015, 94(11): pp. 483–512.

[38] Pageon, H., H. Zucchi, and D. Asselineau, Distinct and complementary roles of papillary and reticular fibroblasts in skin morphogenesis and homeostasis. *Eur. J. Dermatol.*, 2012, 22(3): pp. 324–332.

[39] Janson, D., M. Rietveld, C. Mahé, G. Saintigny, and A. El Ghalbzouri, Differential effect of extracellular matrix derived from papillary and reticular fibroblasts on epidermal development *in vitro*. *Eur. J. Dermatol.*, 2017, 27(3): pp. 237–246.

[40] Vasconcelos, A. and A. Cavaco-Paulo, Wound dressings for a proteolytic-rich environment. *Appl. Microbiol. Biotechnol.*, 2011, 90(2): pp. 445–460.

[41] Enoch, S. and D.J. Leaper, Basic science of wound healing. *Surgery*, 2008, 26(2): pp. 31–37.

[42] Minutti, C.M., J.A. Knipper, J.E. Allen, and D.M.W. Zaiss, Tissue-specific contribution of macrophages to wound healing. *Seminars in Cell & Developmental Biol.*, 2016, 61: p. 3.

[43] Koh, T.J. and L.A. Dipietro, Inflammation and wound healing: The role of the macrophage. *Expert Rev. in Mol. Med.*, 2011, 13(31): p. e23.

[44] Scatena, M., K.V. Eaton, M.F. Jackson, S.A. Lund, and C.M. Giachelli, Macrophages: The bad, the ugly, and the good in the inflammatory response to biomaterials. *The immune response to implanted materials and devices: The impact of the immune system on the success of an implant*, ed. B. Corradetti. 2017, Cham: Springer International Publishing.

[45] Ning, X.L., D. Li, and M. Ståhle, Transition from inflammation to proliferation: A critical step during wound healing. *Cellular & Molecular Life Sciences Cmls.*, 2016, 73(20): pp. 3861–3885.

[46] Klemm, D., E.D. Cranston, D. Fischer, M. Gama, S.A. Kedzior, D. Kralisch, F. Kramer, T. Kondo, T. Lindström, S. Nietzsche, K. Petzold-Welcke, and F. Rauchfuß, Nanocellulose as a natural source for groundbreaking applications in materials science: Today's state. *Mater. Today*, 2018, 21(7): pp. 720–748.

[47] Curvello, R., V.S. Raghuwanshi, and G. Garnier, Engineering nanocellulose hydrogels for biomedical applications. *Adv. in Coll. Int. Sci.*, 2019, 267: pp. 47–61.

[48] Fu, L., J. Zhang, and G. Yang, Present status and applications of bacterial cellulose-based materials for skin tissue repair. *Carbohydr. Polym.*, 2013, 92(2): pp. 1432–1442.

[49] Bacakova, L., J. Pajorova, M. Bacakova, A. Skogberg, P. Kallio, K. Kolarova, and V. Svorcik, Versatile application of nanocellulose: From industry to skin tissue engineering and wound healing. *Nanomaterials* (Basel, Switzerland), 2019, 9(2): p. 164.
[50] Wen, X., Y. Zheng, J. Wu, L. Yue, C. Wang, J. Luan, Z. Wu, and K. Wang, *In vitro* and *in vivo* investigation of bacterial cellulose dressing containing uniform silver sulfadiazine nanoparticles for burn wound healing. *Progress in Natural Science: Materials International*, 2015, 25(3): pp. 197–203.
[51] Khalid, A., H. Ullah, M. Ul-Islam, R. Khan, S. Khan, F. Ahmad, T. Khan, and F. Wahid, Bacterial cellulose–TiO_2 nanocomposites promote healing and tissue regeneration in burn mice model. *RSC Adv.*, 2017, 7(75): pp. 47662–47668.
[52] de Mattos, I.B., J.C.J. Holzer, A.-C. Tuca, F. Groeber-Becker, M. Funk, D. Popp, S. Mautner, T. Birngruber, and L.-P. Kamolz, Uptake of PHMB in a bacterial nanocellulose-based wound dressing: A feasible clinical procedure. *Burns*, 2019, 45(4): pp. 898–904.
[53] Kargarzadeh, H., M. Mariano, J. Huang, N. Lin, I. Ahmad, A. Dufresne, and S. Thomas, Recent developments on nanocellulose reinforced polymer nanocomposites: A review. *Polymer*, 2017, 132: pp. 368–393.
[54] Wei, D., Q. Liu, Z. Liu, J. Liu, X. Zheng, Y. Pei, and K. Tang, Modified nano microfibrillated cellulose/carboxymethyl chitosan composite hydrogel with giant network structure and quick gelation formability. *Int. J. Biol. Macromol.*, 2019, 135: pp. 561–568.
[55] Stumpf, T.R., X. Yang, J. Zhang, and X. Cao, *In situ* and *ex situ* modifications of bacterial cellulose for applications in tissue engineering. *Mater. Sci. Eng. C*, 2018, 82: pp. 372–383.
[56] Li, Y., S. Qing, J. Zhou, and G. Yang, Evaluation of bacterial cellulose/hyaluronan nanocomposite biomaterials. *Carbohydr. Polym.*, 2014, 103: pp. 496–501.
[57] Jorfi, M. and E.J. Foster, Recent advances in nanocellulose for biomedical applications. *J. Appl. Polym. Sci.*, 2015, 132(14).
[58] Portela, R., C.R. Leal, P.L. Almeida, and R.G. Sobral, Bacterial cellulose: A versatile biopolymer for wound dressing applications. *Microb. Biotechnol.*, 2019, 12(4): pp. 586–610.
[59] Li, Y., Y. Tian, W. Zheng, Y. Feng, R. Huang, J. Shao, R. Tang, P. Wang, Y. Jia, J. Zhang, W. Zheng, G. Yang, and X. Jiang, Composites of bacterial cellulose and small molecule-decorated gold nanoparticles for treating gram-negative bacteria-infected wounds. *Small*, 2017, 13(27): p. 1700130.
[60] Moniri, M., A. Boroumand Moghaddam, S. Azizi, R. Abdul Rahim, S.W. Zuhainis, M. Navaderi, and R. Mohamad, *In vitro* molecular study of wound healing using biosynthesized bacteria nanocellulose/silver

nanocomposite assisted by bioinformatics databases. *Int. J. Nanomedicine*, 2018, 13: pp. 5097–5112.

[61] Sharma, A., M. Thakur, M. Bhattacharya, T. Mandal, and S. Goswami, Commercial application of cellulose nano-composites — A review. *Biotechnol. Rep.*, 2019, 21.

[62] Keskin, Z., S.A. Urkmez, and E.E. Hames, Novel keratin modified bacterial cellulose nanocomposite production and characterization for skin tissue engineering. *Mater. Sci. Eng. C*, 2017, 75: pp. 1144–1153.

[63] Li, Y., H. Jiang, W. Zheng, N. Gong, L. Chen, X. Jiang, and G. Yang, Bacterial cellulose-hyaluronan nanocomposite biomaterials as wound dressings for severe skin injury repair. *J. Mater. Chem. B.*, 2015, 3(17): pp. 3498–3507.

[64] Ang-atikarnkul, P., A. Watthanaphanit, and R. Rujiravanit, Fabrication of cellulose nanofiber/chitin whisker/silk sericin bionanocomposite sponges and characterizations of their physical and biological properties. *Compos. Sci. Technol.*, 2014, 96: pp. 88–96.

[65] Lamboni, L., M. Gauthier, G. Yang, and Q. Wang, Silk sericin: A versatile material for tissue engineering and drug delivery. *Biotechnol. Adv.*, 2015, 33(8): pp. 1855–1867.

[66] Lamboni, L., T. Souho, A.R. Osi, and G. Yang, Silk proteins: A natural resource for biomaterials, in Bioinspired Materials Science and Engineering, G. Yang, L. Xiao, and L. Lamboni, Editors, 2020, John Wiley & Sons, pp. 185–199.

[67] Lu, G., Z. Ding, Y. Wei, X. Lu, Q. Lu, and D.L. Kaplan, Anisotropic biomimetic silk scaffolds for improved cell migration and healing of skin wounds. *ACS Appl. Mater. & Int.*, 2018, 10(51): pp. 44314–44323.

[68] Ullah, M.W., L. Fu, L. Lamboni, Z. Shi, and G. Yang, Chapter 3 — Current trends and biomedical applications of resorbable polymers, in *Materials for Biomedical Engineering*, V. Grumezescu and A.M. Grumezescu, Editors, 2019, Elsevier, pp. 41–86.

[69] Tejada, S., A. Manayi, M. Daglia, F.S. Nabavi, A. Sureda, Z. Hajheydari, O. Gortzi, H. Pazoki-Toroudi, and M.S. Nabavi, Wound healing effects of curcumin: A short review. *Curr. Pharm. Biotechnol.*, 2016, 17(11): pp. 1002–1007.

[70] Li, Z., D. Chu, Y. Gao, L. Jin, X. Zhang, W. Cui, and J. Li, Biomimicry, biomineralization, and bioregeneration of bone using advanced three-dimensional fibrous hydroxyapatite scaffold. *Mater. Today Adv.*, 2019, 3: p. 100014.

[71] Xu, W., B.Z. Molino, F. Cheng, P.J. Molino, Z. Yue, D. Su, X. Wang, S. Willför, C. Xu, and G.G. Wallace, On low-concentration inks formulated by nanocellulose assisted with gelatin methacrylate (GelMA) for 3D Printing toward Wound Healing Application. *ACS Appl. Mater. Int.*, 2019, 11: pp. 8838–8848.

[72] Poonguzhali, R., S. Khaleel Basha, and V. Sugantha Kumari, Novel asymmetric chitosan/PVP/nanocellulose wound dressing: *In vitro* and *in vivo* evaluation. *Int. J. Biolo. Macromol.*, 2018, 112: pp. 1300–1309.
[73] Hu, Y., H. Liu, X. Zhou, H. Pan, X. Wu, N. Abidi, Y. Zhu, and J. Wang, Surface engineering of spongy bacterial cellulose *via* constructing crossed groove/column micropattern by low-energy CO_2 laser photolithography toward scar-free wound healing. *Mater. Sci. Eng. C*, 2019, 99: pp. 333–343.
[74] Bottan, S., F. Robotti, P. Jayathissa, A. Hegglin, N. Bahamonde, J.A. Heredia-Guerrero, I.S. Bayer, A. Scarpellini, H. Merker, N. Lindenblatt, D. Poulikakos, and A. Ferrari, Surface-Structured bacterial cellulose with guided assembly-based biolithography (GAB). *ACS Nano*, 2015, 9(1): pp. 206–219.
[75] Jin, M., W. Chen, Z. Li, Y. Zhang, M. Zhang, and S. Chen, Patterned bacterial cellulose wound dressing for hypertrophic scar inhibition behavior. *Cellulose*, 2018, 25(11): pp. 6705–6717.
[76] Gupta, A., M. Kowalczuk, W. Heaselgrave, T.S. Britland, C. Martin, and I. Radecka, The production and application of hydrogels for wound management: A review. *Eur. Polym. J.*, 2019, 111: pp. 134–151.
[77] Sun, F., H.R. Nordli, B. Pukstad, E.K. Gamstedt, and G. Chinga-Carrasco, Mechanical characteristics of nanocellulose-PEG bionanocomposite wound dressings in wet conditions. *J. Mechan. Behav. Biomed. Mater.*, 2017, 69: pp. 377–384.
[78] Tehrani, Z., H.R. Nordli, B. Pukstad, D.T. Gethin, and G. Chinga-Carrasco, Translucent and ductile nanocellulose-PEG bionanocomposites — A novel substrate with potential to be functionalized by printing for wound dressing applications. *Ind. Crops Prod.*, 2016, 93: pp. 193–202.
[79] Lu, T., Q. Li, W. Chen, and H. Yu, Composite aerogels based on dialdehyde nanocellulose and collagen for potential applications as wound dressing and tissue engineering scaffold. *Compos. Sci. Technol.,* 2014, 94: pp. 132–138.
[80] Fontenot, K.R.J., V. Edwards, D. Haldane, N. Pircher, F. Liebner, B.D. Condon, H. Qureshi, and D. Yager, Designing cellulosic and nanocellulosic sensors for interface with a protease sequestrant wound-dressing prototype: Implications of material selection for dressing and protease sensor design. *J. Biomater. Appl.*, 2017, 0: pp. 1–16.
[81] Laurén, P., P. Somersalo, I. Pitkänen, Y.-R. Lou, A. Urtti, J. Partanen, J. Seppälä, M. Madetoja, T. Laaksonen, A. Mäkitie, and M. Yliperttula, Nanofibrillar cellulose-alginate hydrogel coated surgical sutures as cell-carrier systems. *PLoS ONE* 2017, 12(8).
[82] Wei, D., Q. Liu, Z. Liu, J. Liu, X. Zheng, Y. Pei, and K. Tang, Modified nano microfibrillated cellulose/carboxymethyl chitosan composite hydrogel with giant network structure and quick gelation formability. *Int. J. Biol. Macromol.*, 2019, 135: pp. 561–568.

[83] Poonguzhali, R., S. Khaleel Basha, and V. Sugantha Kumari, Synthesis and characterization of chitosan-PVP-nanocellulose composites for *in vitro* wound dressing application. *Int. J. of Biolo. Macromol.*, 2017, 105: pp. 111–120.
[84] Qu, J., X. Zhao, Y. Liang, T. Zhang, P.X. Ma, and B. Guo, Antibacterial adhesive injectable hydrogels with rapid self-healing, extensibility and compressibility as wound dressing for joints skin wound healing. *Biomaterials* 2018, 183: pp. 185–199.
[85] Moniri, M., A.B. Moghaddam, S. Azizi, A.R. Raha, W.Z. Saad, M. Navaderi, P. Arulselvan, and R. Mohamad, Molecular study of wound healing after using biosynthesized BNc/Fe3O4 nanocomposites assisted with a bioinformatics approach. *Int. J. Nanomed.*, 2018, 13: pp. 2955–2971.
[86] Sajjad, W., T. Khan, M. Ul-Islam, R. Khan, Z. Hussain, A. Khalid, and F. Wahid, Development of modified montmorillonite-bacterial cellulose nanocomposites as a novel substitute for burn skin and tissue regeneration. *Carbohydr. Polym.*, 2019, 2016: pp. 548–556.
[87] Singla, R., S. Soni, P.M. Kulurkar, A. Kumari, S. Mahesh, V. Patial, Y.S. Padwad, and S.K. Yadav, *In situ* functionalized nanobiocomposites dressings of bamboo cellulose nanocrystals and silver nanoparticles for accelerated wound healing. *Carbohydr. Polym.*, 2017, 155: pp. 152–162.
[88] Tabaii, M.J., and G. Emtiazi, Transparent nontoxic antibacterial wound dressing based on silver nano particle/bacterial cellulose nano composite synthesized in the presence of tripolyphosphate. *J. Drug Deli. Sci. Technol.*, 2018, 44: pp. 244–253.
[89] Napavichayanun, S., P. Amornsudthiwat, P. Pienpinijtham, and P. Aramwit, Interaction and effectiveness of antimicrobials along with healing-promoting agents in a novel biocellulose wound dressing. *Mater. Sci. Eng. C*, 2015, 55: pp. 95–104.
[90] Bernardelli de Mattos, I., J.C.J. Holzer, A.-C. Tuca, F. Groeber-Becker, M. Funk, D. Popp, S. Mautner, T. Birngruber, and L.-P. Kamolz, Uptake of PHMB in a bacterial nanocellulose-based wound dressing: A feasible clinical procedure. *Burns*, 2019, 52: pp. 898–904.
[91] Alkhatib, Y., M. Dewaldt, S. Moritz, R. Nitzsche, D. Kralisch, and D. Fischer, Controlled extended octenidine release from a bacterial nanocellulose/Poloxamer hybrid system. *Eur. J. Pharma. Biopharma.*, 2017, 112: pp. 164–176.
[92] Liu, Y., Y. Sui, C. Liu, C. Liu, M. Wu, B. Li, and Y. Li, A physically cross-linked polydopamine/nanocellulose hydrogel as potential versatile vehicles for drug delivery and wound healing. *Carbohydr. Polym.*, 2018, 188: pp. 27–36.
[93] Powell, L.C., S. Khan, G. Chinga-Carrasco, C.J. Wright, K.E. Hill, and D.W. Thomas, An investigation of Pseudomonas aeruginosa biofilm

growth on novel nanocellulose fibre dressings. *Carbohydr. Polym.*, 2016, 137: pp. 191–197.

[94] Basu, A., K. Heitz, M. Stromme, K. Welch, and N. Ferraz, Ion-crosslinked wood-derived nanocellulose hydrogels with tunable antibacterial properties: Candidate materials for advanced wound care applications. *Carbohydr. Polym.*, 2018, 181: pp. 345–350.

[95] Vilela, C., H. Oliveira, A. Almeida, A.J.D. Silvestre, and C.S.R. Freire, Nanocellulose-based antifungal nanocomposites against the polymorphic fungus *Candida albicans. Carbohydr. Polym.*, 2019, 217: pp. 207–216.

[96] Chaker, A., and S. Boufi, Cationic nanofibrillar cellulose with high antibacterial properties. *Carbohydr. Polym.*, 2015, 131: pp. 224–232.

[97] Zoppe, J.O., V. Ruottinen, J. Ruotsalainen, S. Rönkkö, L.-S. Johansson, A. Hinkkanen, K. Jarvinen, and S. Jukka, Synthesis of cellulose nanocrystals carrying tyrosine sulfate mimetic ligands and inhibition of alphavirus infection. *Biomacromolecules*, 2014, 15: pp. 1534−1542.

[98] Basu, A., J. Lindh, E. Ålander, M. Strømme, and N. Ferraz, On the use of ion-crosslinked nanocellulose hydrogels for wound healing solutions: Physicochemical properties and application-oriented biocompatibility studies. *Carbohydr. Polym.* 2017, 174: pp. 299–308.

[99] Palacios Hinestroza, H., Hern, J.A. ndez Diaz, M. Esquivel Alfaro, G. Toriz, O.J. Rojas, Sulbar, and B.C. n-Rangel, Isolation and characterization of nanofibrillar cellulose from Agave tequilana Weber Bagasse. *Adv. Mater. Sci. Eng.*, 2019, 2019: p. 7.

[100] Nordli, H.R., G. Chinga-Carrasco, A.M. Rokstad, and B. Pukstad, Producing ultrapure wood cellulose nanofibrils and evaluating the cytotoxicity using human skin cells. *Carbohydr. Polym.*, 2016, 150: pp. 65–73.

[101] Nuutila, K., A. Laukkanen, A. Lindford, S. Juteau, M. Nuopponen, J. Vuola, and E. Kankuri, Inhibition of skin wound contraction by nanofibrillar cellulose hydrogel. *Plastic and Reconstructive Surgery*, 2018, 141: pp. 357e–366e.

[102] Gao, H., Z. Zhong, H. Xia, Q. Hu, Q. Ye, Y. Wang, L. Chen, Y. Du, X. Shi, and L. Zhang, Construction of cellulose nanofibers/quaternized chitin/organic rectorite composites and their application as wound dressing materials. *Biomater. Sci.*, 2019, 7(6): pp. 2571–2581.

[103] Wu, T., R. Farnood, K. O'Kelly, and B. Chen, Mechanical behavior of transparent nanofibrillar cellulose–chitosan nanocomposite films in dry and wet conditions. *J. Mechani. Behav. Biomed. Mater.*, 2014, 32: pp. 279–286.

[104] Rees, A., L.C. Powell, G. Chinga-Carrasco, D.T. Gethin, K. Syverud, K.E. Hill, and D.W. Thomas, 3D Bioprinting of carboxymethylated-periodate oxidized nanocellulose constructs for wound dressing applications. *Bio Med Res. Int.*, 2015, 2015: pp. 925757–925757.

[105] Chinga-Carrasco, G., N.V. Ehman, D. Filgueira, J. Johansson, M.E. Vallejos, F.E. Felissia, J. Håkansson, and M.C. Area, Bagasse — A major agro-industrial residue as potential resource for nanocellulose inks for 3D printing of wound dressing devices. *Add. Manuf.*, 2019, 28: pp. 267–274.

[106] Loh, E.Y.X., N. Mohamad, M.B. Fauzi, M.H. Ng, S.F. Ng, and M.C.I. Mohd Amin, Development of a bacterial cellulose-based hydrogel cell carrier containing keratinocytes and fibroblasts for full-thickness wound healing. *Scienti. Rep.*, 2018, 8(1): p. 2875.

[107] Rajwade, J.M., K.M. Paknikar, and J.V. Kumbhar, Applications of bacterial cellulose and its composites in biomedicine. *Appl. Microbiol. Biotechnol.*, 2015, 99(6): pp. 2491–2511.

[108] de Olyveira, G.M., L.M. Manzine Costa, P. Basmaji, and L. Xavier Filho, Bacterial nanocellulose for medicine regenerative. *J. Nanotechnol. Eng. Med.*, 2011, 2(3).

[109] Fu, L.H., C. Qi, M.G. Ma, and P. Wan, Multifunctional cellulose-based hydrogels for biomedical applications. *J. Mater. Chem. B*, 2019, 7(10): pp. 1541–1562.

[110] Alvarez, O.M., M. Patel, J. Booker, and L. Markowitz, Effectiveness of a biocellulose wound dressing for the treatment of chronic venous leg ulcers: Results of a single center randomized study involving 24 patients. *Wounds*, 2004, 16(7): pp. 224–233.

[111] Levinson, D.R.S.M.C.D.J. Microbial cellulose wound dressings in the treatment of skin tears in the frail elderly. *Wounds*, 2010, 22(1).

[112] Muangman, P., S. Opasanon, S. Suwanchot, and O. Thangthed, Efficiency of microbial cellulose dressing in partial-thickness burn wounds. *J. Am. Col. Certif. Wound Spec.*, 2011, 3(1): pp. 16–19.

[113] Brassolatti, P., H.W. Kido, P.S. Bossini, P.R. Gabbai-Armelin, A.N. Otterco, L. Almeida-Lopes, L.M. Zanardi, M.A. Napolitano, L. de Avo, L.A. Forato, F.M. Araujo-Moreira, and N.A. Parizotto, Bacterial cellulose membrane used as biological dressings on third-degree burns in rats. *Biomed. Mater. Eng.*, 2018, 29(1): pp. 29–42.

[114] Ludwicka, K., M. Jedrzejczak-Krzepkowska, K. Kubiak, M. Kolodziejczyk, T. Pankiewicz, and S. Bielecki, Chapter 9 — Medical and cosmetic applications of bacterial nanocellulose, in bacterial nanocellulose, M. Gama, F. Dourado, and S. Bielecki, Editors, 2016, Elsevier: Amsterdam, pp. 145–165.

[115] Fu, L., Y. Zhang, C. Li, Z. Wu, Q. Zhuo, X. Huang, G. Qiu, P. Zhou, and G. Yang, Skin tissue repair materials from bacterial cellulose by a multilayer fermentation method. *J. Mater. Chem.*, 2012, 22(24): p. 12349.

[116] Li, Y., S. Wang, R. Huang, Z. Huang, B. Hu, W. Zheng, G. Yang, and X. Jiang, Evaluation of the effect of the structure of bacterial cellulose on

full thickness skin wound repair on a microfluidic chip. *Biomacromolecules*, 2015, 16(3): pp. 780–789.

[117] Napavichayanun, S., R. Yamdech, and P. Aramwit, The safety and efficacy of bacterial nanocellulose wound dressing incorporating sericin and polyhexamethylene biguanide: *In vitro, in vivo* and clinical studies. *Arch. Dermatol. Res.*, 2016, 308(2): pp. 123–132.

[118] Foong, C.Y., M.S.A. Hamzah, S.I.A. Razak, S. Saidin, and N.H.M. Nayan, Influence of poly(lactic acid) layer on the physical and antibacterial properties of dry bacterial cellulose sheet for potential acute wound healing materials. *Fibers Polym.*, 2018, 19(2): pp. 263–271.

[119] Galateanu, B., M.-C. Bunea, P. Stanescu, E. Vasile, A. Casarica, H. Iovu, A. Hermenean, C. Zaharia, and M. Costache, *In vitro* Studies of bacterial cellulose and magnetic nanoparticles smart nanocomposites for efficient chronic wounds healing. *Stem Cells Int.*, 2015, 2015: pp. 1–10.

[120] Pal, S., R. Nisi, M. Stoppa, and A. Licciulli, Silver-functionalized bacterial cellulose as antibacterial membrane for wound-healing applications. *ACS Omega*, 2017, 2(7): pp. 3632–3639.

[121] Förch, R., L. Duque, and A. Lotz, Antimicrobial bioactive polymer coatings. *Compreh. Mater. Proc.*, 2014, pp. 449–461.

[122] G. Serafica, R. Mormino, G. A. Oster, K. E. Lentz, K. P. Koehler, Microbial cellulose wound dressings for the treatment of chronic wounds. United State Patents 7704523, 2010.

[123] Moritz, S., C. Wiegand, F. Wesarg, N. Hessler, F.A. Müller, D. Kralisch, U.-C. Hipler, and D. Fischer, Active wound dressings based on bacterial nanocellulose as drug delivery system for octenidine. *Int. J. of Pharmac.*, 2014, 471(1–2): pp. 45–55.

[124] Wiegand, C., S. Moritz, N. Hessler, D. Kralisch, F. Wesarg, F.A. Müller, D. Fischer, and U.-C. Hipler, Antimicrobial functionalization of bacterial nanocellulose by loading with polihexanide and povidone-iodine. *J. Mater. Sci. Mater. Med.*, 2015, 26(10).

[125] Shao, W., H. Liu, S. Wang, J. Wu, M. Huang, H. Min, and X. Liu, Controlled release and antibacterial activity of tetracycline hydrochloride-loaded bacterial cellulose composite membranes. *Carbohydr. Polym.*, 2016, 145: pp. 114–120.

[126] Laçin, N.T., Development of biodegradable antibacterial cellulose based hydrogel membranes for wound healing. *Int. J. Biol. Macromol.*, 2014, 67: pp. 22–27.

[127] Fürsatz, M., M. Skog, P. Sivlér, E. Palm, C. Aronsson, A. Skallberg, G. Greczynski, H. Khalaf, T. Bengtsson, and D. Aili, Functionalization of bacterial cellulose wound dressings with the antimicrobial peptide ε-poly-*L*-Lysine. *Biomed. Mater.*, 2018, 13(2): pp. 025014.

[128] Lin, Q., Y. Zheng, G. Wang, X. Shi, T. Zhang, J. Yu, and J. Sun, Protein adsorption behaviors of carboxymethylated bacterial cellulose membranes. *Int. J. Biol. Macromol.*, 2015, 73: pp. 264–269.

[129] Oshima, T., S. Taguchi, K. Ohe, and Y. Baba, Phosphorylated bacterial cellulose for adsorption of proteinsCarbohydrate polymers. *Carbohydr. Polym.*, 2011, 83(2): pp. 953–958.

[130] Alkhatib, Y., M. Dewaldt, S. Moritz, R. Nitzsche, D. Kralisch, and D. Fischer, Controlled extended octenidine release from a bacterial nanocellulose/Poloxamer hybrid system. *Eur. J. Pharm. Biopharm.*, 2017, 112: pp. 164–176.

[131] Taokaew, S., M. Phisalaphong, and B.-m.Z. Newby, modification of bacterial cellulose with organosilanes to improve attachment and spreading of human fibroblasts. *Cellulose*, 2015, 22(4): pp. 2311–2324.

[132] Berndt, S., F. Wesarg, C. Wiegand, D. Kralisch, and F.A. Müller, Antimicrobial porous hybrids consisting of bacterial nanocellulose and silver nanoparticles. *Cellulose*, 2013, 20(2): pp. 771–783.

[133] Wu, C.-N., S.-C. Fuh, S.-P. Lin, Y.-Y. Lin, H.-Y. Chen, J.-M. Liu, and K.-C. Cheng, TEMPO-oxidized bacterial cellulose pellicle with silver nanoparticles for wound dressing. *Biomacromolecules*, 2018, 19(2): pp. 544–554.

[134] Savitskaya, I., A. Kistaubayeva, I. Digel, and D. Shokatayeva, Physicochemical and antibacterial properties of composite films based on bacterial cellulose and chitosan for wound dressing materials. *Eur. Chem. Technol. J.*, 2017, 19(3): p. 255.

[135] Awadhiya, A., D. Kumar, K. Rathore, B. Fatma, and V. Verma, Synthesis and characterization of agarose–bacterial cellulose biodegradable composites. *Polym. Bulletin*, 2017, 74(7): pp. 2887–2903.

[136] Lin, Q., Y. Zheng, L. Ren, J. Wu, H. Wang, J. An, and W. Fan, Preparation and characteristic of a sodium alginate/carboxymethylated bacterial cellulose composite with a crosslinking semi-interpenetrating network. *J. Appl. Polym. Sci.*, 2014, 131(3): pp. 1–9.

[137] Souza, C.M.C.O., L.A.F. Mesquita, D. Souza, A.C. Irioda, J.C. Francisco, C.F. Souza, L.C. Guarita-Souza, M.R. Sierakowski, and K.A.T. Carvalho, Regeneration of skin tissue promoted by mesenchymal stem cells seeded in nanostructured membrane. *Transplan. Proc.*, 2014, 46(6): pp. 1882–1886.

[138] Liu, J., S. Willför, and A. Mihranyan, On importance of impurities, potential leachables and extractables in algal nanocellulose for biomedical use. *Carbohydr. Polym.*, 2017, 172: pp. 11–19.

[139] Jorfi, M., and E.J. Foster, Recent advances in nanocellulose for biomedical applications. *J. Appl. Polym. Sci.*, 2015, 132.

[140] Solway, D.R., W.A. Clark, and D.J. Levinson, A parallel open-label trial to evaluate microbial cellulose wound dressing in the treatment of diabetic foot ulcers. *Int. Wound J.*, 2011, 8: pp. 69–73.
[141] Varkey, M., D.O. Visscher, P.P.M. van Zuijlen, A. Atala, and J.J. Yoo, Skin bioprinting: The future of burn wound reconstruction? *Burns & Trauma.*, 2019, 7(1): p. 4.
[142] Serra, M.B., W.A. Barroso, N.N. da Silva, S.D.N. Silva, A.C.R. Borges, I.C. Abreu, and M.O.D.R. Borges, From inflammation to current and alternative therapies involved in wound healing. *Int. J. Inflam.*, 2017, 2017.
[143] Habibi, Y., A.L. Lucia, and J.O. Rojas, Cellulose nanocrystals: Chemistry, self-assembly, and applications. *Chem. Rev.*, 2010, 110: pp. 3479–3500.
[144] Lavoine, N., I. Desloges, A. Dufresne, and J. Bras, Microfibrillated cellulose — Its barrier properties and applications in cellulosic materials: A review. *Carbohydr. Polym.*, 2012, 90(2): pp. 735–764.

https://doi.org/10.1142/9781786349477_0007

Chapter 7

Applications of Nanofibrillar Celluloses in Drug Delivery: From Conventional Tablet Excipient to Novel Drug Carrier

Munair Badshah[*], Taous Khan[*,§], Hanif Ullah[*], Fazli Wahid[†,¶], and Muhammad Wajid Ullah[‡]

[*]*Department of Pharmacy, COMSATS University Islamabad, Abbottabad Campus, Pakistan*

[†]*Department of Biomedical Sciences, Pak-Austria Fachhochschule: Institute of Applied Sciences and Technology, Mang, Khanpur Road, Haripur, Pakistan*

[‡]*Department of Biomedical Engineering, College of Life Science and Technology, Huazhong University of Science and Technology, Wuhan, PR China*

[§]*taouskhan@cuiatd.edu.pk*

[¶]*fazliwahid@cuiatd.edu.pk*

Abstract

The discovery of new drugs, poor patient compliance, limitations of the conventional dosage forms, and switching to more suitable routes of administration have led to the development of new dosage forms, which in turn demands innovation/discovery of pharmaceutical excipients. In addition to number of excipients, the addition of polymer(s) in drug formulation(s) is unavoidable, as this plays a vital role in the stability, solubility, release, and bioavailability of the drug. The selection of an excipient with ideal properties such as non-toxicity, stability, excess availability, biodegradability, biocompatibility, ease of processing, contaminant-free, and non-allergic response are the major concerns. These issues can be addressed by using cellulose, its derivatives, and different forms as drug delivery systems. To this end, nanocellulose (NC) has received great interest due to its outstanding mechanical, rheological, chemical, and optical properties. It serves as an eco-friendly and sustainable material with vast pharmaceutical applications and is obtained from numerous natural sources of cellulose. NC is hydrophilic in nature; however, the alteration of its surface properties may enhance its applications to meet the challenges of the future. The cellulose produced by microbes, also known as bacterial/microbial cellulose, is a biomaterial with fascinating properties including extremely high purity, high porosity, ultrafine fiber network, high water holding capacity, great tensile strength, and permeability to diffusion of liquids, gases, and small molecules. In the field of drug delivery, cellulose plays a pivotal role in dosage formulations in the form of filler, stabilizer, thickening agent, coating agent, in the preparation of transdermal patches, controlled drug delivery, protein delivery, and macromolecular pro-drug delivery. The current chapter gives an overview of the applications of nanocellulose in drug delivery based on the available literature. An explorative look to NFC applications in the pharmaceutical field and various approaches adopted in the field of drug delivery are also discussed. The approaches made for the safe, effective, and convenient delivery of drugs into body are discussed in detail. Based on current literature, it is suggested that further experimental work is required for the exploration of NFC applications in the stimuli-responsive delivery of drugs.

Keywords: Bacterial cellulose; cellulose nanocrystals; cellulose nanofibrils; drug delivery; nanocrystalline cellulose; protein delivery; tissue engineering

1. Introduction

Biomaterials, particularly natural polymers obtained from plant sources have had a great impact and relationship with human life for decades (Czaja *et al.*, 2007; Hubbell, 1995; Ratner and Bryant, 2004; Shoichet, 2009;

Ullah *et al.*, 2016a, b). This relationship ranges from the use of daily-life, household items to clothing and personal care products as well as to agriculture residue utilization for food items and their preservation, utilization in conventional paper and e-paper making, use in electronics and biomedical devices, and applications in traditional drugs formulations and for the design of novel drug delivery systems (Panday *et al.*, 2018; Shi *et al.*, 2016; Adel *et al.*, 2016; Boufi *et al.*, 2016; Duran *et al.*, 2012; Ullah *et al.*, 2016b). In general, the applications and importance of biopolymers can never be ruled out from human life (Badshah *et al.*, 2017; Ellis and Smith, 2008; Murphy, 2001; Jay *et al.*, 2008).

Cellulose, a natural biopolymer is produced as a semi-crystalline polysaccharide by plants with fiber length varying from 0.5 mm to several mm. The polymeric backbone of cellulose basically consists of smaller glucose units, linked by β-1,4 glycosidic linkage. The small units aggregate and form elementary fibrils with diameters up to 3 mm. These elementary fibrils further aggregate and form macrofibrils with diameters of 10–25 mm. Nanofibrillar cellulose (NFC) production is usually carried out using different chemical and mechanical processes. Initially, the fibers are broken down by mechanical shearing, followed by complete dissolution *via* chemical processes and regeneration (Boufi *et al.*, 2016; Voisin *et al.*, 2017). The features like mechanical strength, chemical inertness, large surface area, and multipurpose chemical structure make it a material of interest for research purpose in parallel with conventional applications. The higher utilization of cellulose is possible due to its annual production of 1.5×10^{12} tons from various plant sources (Czaja *et al.*, 2004; Klemm *et al.*, 2005; Siró and Plackett, 2010). The sources of cellulose are not only limited to plants, it is also produced by microorganisms, i.e., different species of fungus and green algae (Klemm *et al.*, 2005), non-pathogenic bacterial species including *Acetobacter*, *Sarcina*, *Rhizobium,* and *Agrobacterium* (Dufresne, 2013; Khan *et al.*, 2007; Shezad *et al.*, 2010; Ullah *et al.*, 2016b; Ullah *et al.*, 2017), and cell-free systems (Ullah *et al.*, 2015; Kim *et al.*, 2019), these represent the purest form of cellulose. NFC is biochemically processed nano-structured cellulose, which is also called bacterial nanocellulose (BNC), bacterial cellulose (BC), cellulose nanocrystals (CNC), microfibrils, microfibrillated cellulose (MFC), nanofibrous cellulose or nanocrystalline cellulose (NCC), and nanowhiskers (George *et al.*, 2015; Kolakovic *et al.*, 2012b; Peng *et al.*, 2011). NFC is a biopolymer that is of light weight, has a low production cost, is extremely safe in handling, is non-toxic in nature, and has versatile mechanical strengths. It has high surface area, ability to be shaped into different materials with promising properties, and act as a starting material for different copolymerization processes. These properties make it a unique bio-based material for potential applications in different fields related to human life (Mahfoudhi *et al.*, 2017). From the structural aspect, nanocellulose is of two types: amorphous and crystalline. On exposure to acidic conditions for hydrolysis, cellulose forms nanowhiskers, which are composed of short rods having highly crystalline structure. The whiskers diameter ranges from 2 to 20 nm while its width and length distribution reach 100–600 nm (Kolakovic *et al.*, 2012b). General applications of NFC are summarized in Fig. 1.

Cellulose of microbial origin is extensively obtained from the culturing of different microbial strains such as *Gluconacetobacter xylinus*

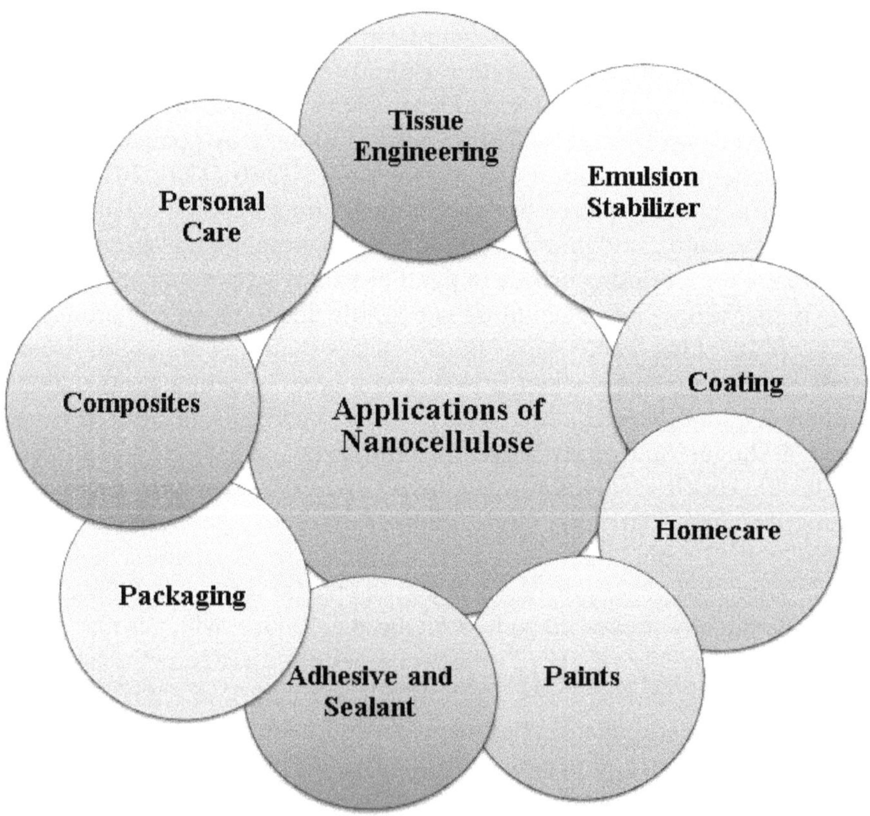

Fig. 1. An overview of general applications of nanofibrillar cellulose.

(*G. xylinus*) and *Gluconacetobacter hansenii* (*G. hansenii*) due to their non-pathogenic nature and capabilities to ingest various carbon-rich compounds and waste materials (Ul-Islam *et al.*, 2017, 2020) and excrete cellulose membrane as exudate (Ross *et al.*, 1991; Saxena and Brown Jr, 2012; Ullah *et al.*, 2016b; Ullah *et al.*, 2017; Ullah *et al.*, 2017; Ul-Islam *et al.*, 2017). The cellulose that is produced is odorless, tasteless, and transparent membrane, from the presence of any contamination after washing (Lin *et al.*, 2013). Nanocellulose, especially BC, has received immense consideration in different fields due to its outstanding mechanical, rheological, chemical, and optical properties (Farooq *et al.*, 2020; Khan *et al.*, 2018a, b; Aljohani *et al.*, 2018; Shoukat *et al.*, 2019; McCarthy *et al.*, 2019a, b). Cellulose is generally hydrophilic in nature; however, the alteration of its surface properties can potentially enhance its applications to meet the requirements for future applications (George *et al.*, 2015). The general features of NFC are summarized in Fig. 2. The NFCs have numerous applications in drug delivery for humans (Lin *et al.*, 2013; Ullah *et al.*, 2016a) and these are summarized in Fig. 3.

The promising properties of NFC in biomedical and tissue engineering applications include the stability, large surface area, hydrophilicity, biocompatibility, non-immunogenic nature, and ability to support cell adhesions and proliferation. In addition, the non-toxic and inert nature make NFC a suitable candidate for applications in drug delivery (Badshah *et al.*, 2017, 2018; Kolakovic *et al.*, 2012a; Helenius *et al.*, 2006; Millon & Wan, 2006; Qiu & Netravali, 2014; Svensson *et al.*, 2005; Ullah *et al.*, 2016a). This chapter provides a detailed description on the applications associated with NFC in the field of drug delivery, ranging from conventional formulations to the novel and advanced techniques. It also provides an insight to the role of NFC as diluent, conventional formulations stabilizer, inert polymer for coating of drugs, and drug carrier in sustained and controlled drug delivery. In connection with the above discussion, it is expected that the researchers in academia, industry, and research and development organizations would be benefitted from this chapter to design NFC-based, tailored drug delivery systems.

2. Properties of Nanofibrillated Cellulose and Its Potential Applications in Drug Delivery

Features like purity and absence of plant structural components such as lignin, hemicelluloses, and pectin; being inert and comparatively less

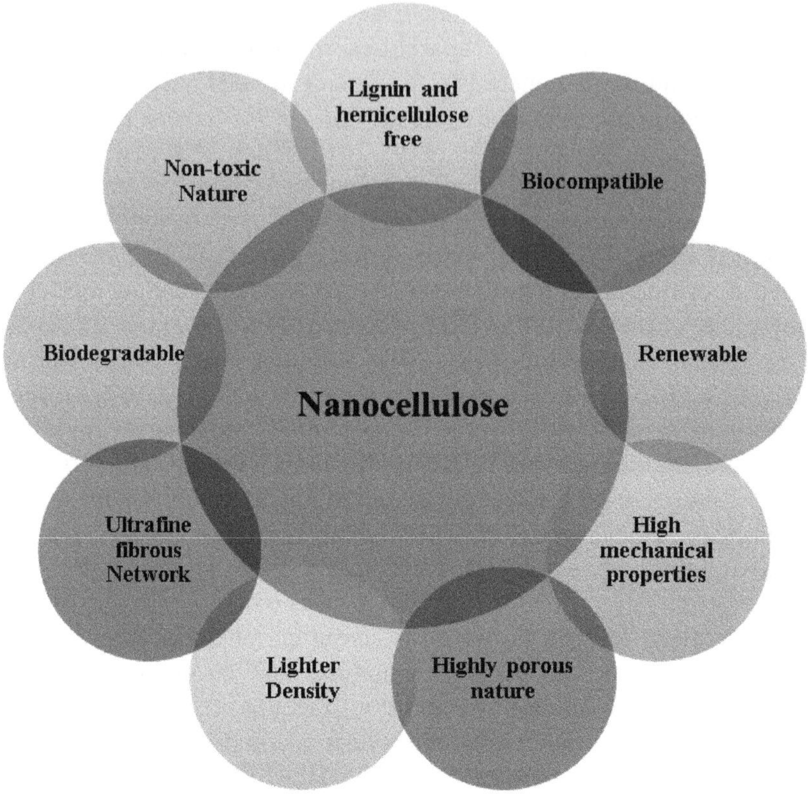

Fig. 2. An understanding of the general features of microfibrils.

Source: Adapted with permission from Ullah *et al.* (2016a). Copyright Springer Publishers.

immunogenic in nature; having biocompatibility with other polymers; and the ability to retain water make NFC a unique biomaterial for applications in drug delivery (Chawla *et al.*, 2009; Lin *et al.*, 2013; Shi *et al.*, 2014; Ullah *et al.*, 2016). The literature shows that NFC has been studied widely for application in conventional formulations as agents for thickening, coating, and stabilizing, and as additive for controlling the drug release, as excipients in transdermal drug delivery, topical drug delivery, intestinal drug delivery, and protectant of proteins by coating (Campano *et al.*, 2015; Lin *et al.*, 2015; Lin *et al.*, 2016; Shi *et al.*, 2014). Numerous factual features of NFC for the desired applications considering the scope of the current chapter are shown in Fig. 3.

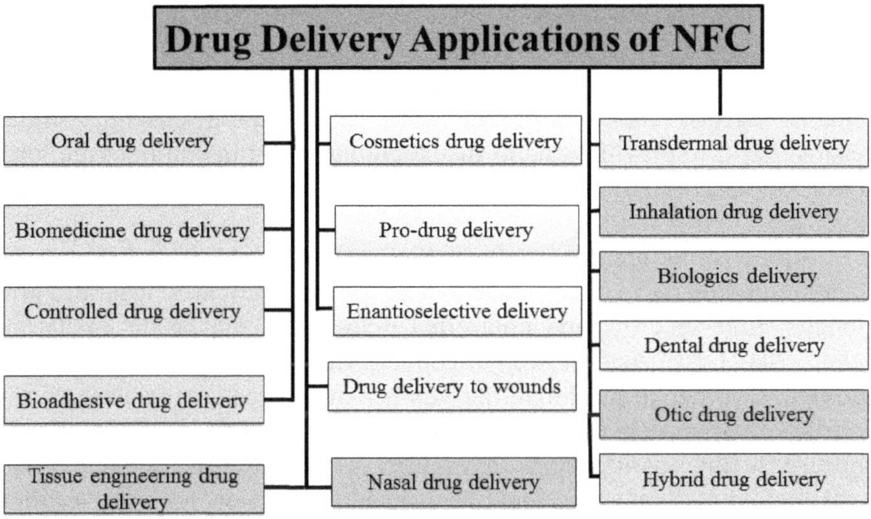

Fig. 3. Drug delivery applications of cellulose nanofibrils.
Source: Adapted with permission from Ullah *et al.* (2016a).

In addition to the existing properties, NFC can be provided with newfangled features using physical and chemical modification with polymers, solvents, specialized radiations, and specific techniques of drying (Lin *et al.*, 2013; Lin *et al.*, 2016; Olyveira *et al.*, 2013; Petersen and Gatenholm, 2011; Sulaeva *et al.*, 2015; Ullah *et al.*, 2016b; Yadav *et al.*, 2010). Besides application in foods and cosmetics, the superior features of NFC make it of use in a number of applications in drug delivery, which are discussed in the succeeding sections of this chapter.

2.1 *Applications in Conventional Dosage Forms*

Cellulose has a key role in the designing of conventional dosage forms as additives and fillers for the delivery of drugs at the target sites at the desired rate for immediate and controlled release. Cellulose-based excipients help in enhancing the drug solubility, coating of the dosage forms for taste masking or controlling drug release, enabling drug delivery in the desired site in acidic or basic environment, and helping in adhesion of drug to the target site (Debotton & Dahan, 2017; Shokri & Adibkia, 2013). In tablet form, cellulose and its derivatives such as MCC act as binder, filler, and disintegrant; hydroxypropyl methylcellulose (HPMC) and hydroxypropyl cellulose (HPC) act as binders and disintegrant; and cellulose acetate phthalate (CAP), ethylcellulose (EC), and HPMC act as coating agents. The hydrophilic nature and water absorbing abilities enable MCC to play the role of disintegrating agent in solid dosage form (Kolakovic *et al.*, 2011; Macuja *et al.*, 2015; Xin *et al.*, 2017; Yassin *et al.*, 2015).

In one of the approaches for achieving prolonged drug delivery, the coating process is of great interest. This process provides accurate, prolonged, time-specific, and controlled delivery of the products (Tiwari *et al.*, 2012). NFC has showed effective controlled drug release from the dosage. In order to minimize the number of excipients and observe the effects of variation in the concentration of nanocellulose, i.e., BC, on drug release in film coating, Amin *et al.* (2012a) prepared the powder of BC and applied it as a coating agent for paracetamol tablets *via* spray-coating method without the addition of plasticizer. This was compared with the coating carried out using aqueous dispersion of ethyl cellulose as a coating agent. The results of coating with nanocellulose were highly flexibility, uniformity in quality, and foldability. The *in vitro* drug release studies demonstrated that concentration of nanocellulose in the coating has high impact on the drug release behavior, and increase in the nanocellulose amount was inversely proportional to the drug release (Amin *et al.*, 2012a). Cellulose nanowhiskers have the ability to work as co-stabilizer for direct compression of beads into the tablet dosage form. The study by Villanova *et al.* (2011) showed that the presence of cellulose nanowhiskers in the tablet formulations resulted in reduced particles and uniform distribution, which facilitated the tablet's direct compression. The *in vitro* drug release studies of these tablets showed controlled drug release (Villanova *et al.*, 2011). Similarly, Kolakovic *et al.* (2011) also reported the spray-dried NFC for preparing tablets by direct compression. The studies were carried out using NFC and commercial grades of MCC. The results showed that NFC demonstrated far better powder flow and porosity than commercial grades of MCC. The studies also showed that NFC-based tablets released >90% drug in only 4 minutes in comparison to MCC-based tablets, which released 78% drug in 14 minutes (Kolakovic *et al.*, 2011). However, in another study, Kolakovic *et al.* (2012a) reported a matrix system prepared from spray-dried NFC for sustained delivery of hydrophobic drugs. In this case, the drug-loaded NFC microspheres with a size of up to 5 μm were prepared using spray-drying method. *In vitro* dissolution studies were carried out to assess the drug release from the microspheres. The release data showed that drug release from the microsphere was sustained for all the formulations. The possible reason for this could be the interaction between the drug and cellulose fibers, in addition to the hydrophobic nature of the drugs, which make it difficult to be dissolved in the dissolution medium (Kolakovic *et al.*, 2012a). Thus, NFC can provide a platform for immediate and sustained drug delivery depending upon the intended use of excipient, method of preparation, and nature of the drug.

Different applications of cellulose in drugs delivery are discussed in detail in the succeeding section and summarized in Fig. 3.

2.2 *Antimicrobial Drug Delivery Applications*

The non-toxic and biocompatible nature make microbial nanocellulose (MC) safe for different drug delivery applications (Shi *et al.*, 2014; Park *et al.*, 2009). MC possesses promising structural, physico-chemical, and mechanical properties and has the potential to be applied as an ideal dressing material for wounds; however, it still lacks the properties to be a realistic antimicrobial material in wound treatment. Composites of NFC with antimicrobial agents (polymeric and non-polymeric) have been limited in this regard. The effects of drug concentration and coating layers on the drug release were studied by Jipa *et al.* (2012) by mixing BC with sorbic acid (BC-SA). The composites were coated with single- and multilayer films and tested against *Escherichia coli* (*E. coli*) K12-MG1655. The results showed that multilayer formulation reduced the susceptibility of SA to water, which resulted in reduced solubility and subsequently controlled drug release; it improved the antimicrobial activities. In general, it was observed that BC-SA combination possesses promising biocompatible, non-toxic, and antimicrobial properties (Jipa *et al.*, 2012). In another study, CNF was loaded with different concentrations of fusidic acid and tested against *Staphylococcus aureus* (*S. aureus*). The drug was released in a controlled manner, and high antibacterial activities were observed (Liyaskina *et al.*, 2017). The polymer matrix of grinded BC and SA was prepared in combination with poly(vinyl alcohol) (PVA) by Dobre *et al.* (2012). The results revealed that SA was effective against *E. coli* and the matrix prepared showed potential ability for applications in delivery of antimicrobial agents (Dobre *et al.*, 2012). The application of BC in the pellicle form for drug loading and release properties was studied by

Janpetch *et al.* (2016). BC was successfully loaded with zinc oxide (ZnO) by immersion in ZnO solution. The nanostructured and permeable pores of BC facilitated the diffusion of ZnO into the pellicles. The BC-ZnO showed higher antibacterial properties against *S. aureus* and *E. coli* during *in vitro* studies. Such BC-ZnO pellicles were observed to have potential for applications in wound dressing and disinfection (Janpetch *et al.*, 2016). Other than *in vitro* studies, Khalid *et al.* (2017) reported the wound healing and antimicrobial effects of ZnO and BC nanocomposites in the *in vivo* study. The composites were applied on BALBc mice models, and they found that BC-ZnO composites showed high efficiency against *E. coli*, *Pseudomonas aeruginosa* (*P. aeruginosa*), *S. aureus*, and *Clostridium freundii* (*C. freundii*). The wound healing activities were found to be 66%. It was suggested that BC-ZnO composites have the potential for applications as wound dressing material (Khalid *et al.*, 2017). In addition to ZnO, Khalid *et al.* (2017b) also reported the BC and titanium dioxide (BC-TiO_2) composite for wound healing and antimicrobial applications. The composites showed faster wound healing, epithelial cell formation, and wound contraction. In contrast, the study conducted with pure BC showed no significant wound healing properties and inflammation prevention (Khalid *et al.*, 2017b). Similar studies were carried out by preparing composites of BC with different polymers by cross-linking and coating. BC-based hybrid materials with chitosan (CS) were prepared in the form of a film by using cross-linker, coated with poly(acrylic acid) (PLA), and finally loaded with model drug curcumin. The BC-CS film was tested at physiological pH for self-healing and antimicrobial effects. The drug release study revealed that BC-CS has non-toxic nature and delivered curcumin at a controlled rate. The BC-CS also showed good wound healing and antimicrobial effects (Khamrai *et al.*, 2017). Negatively charged NFC were prepared by modifying it with tetramethyl-1-piperidinyloxyl, which has the ability to adsorb the positively charged polymers, proteins, or other molecules. A surface-modified NFC was used to adsorb the positively charged nisin protein, which was released at different pH. The prepared composites were studied for antibacterial activity against *S. aureus* and *Bacillus subtilis* (*B. subtilis*). The antimicrobial NFC-nisin composites showed that the interactions between the modified NFC and nisin led to an increased fibril radius and reduced fiber flexibility with higher efficacy against bacteria as compared to pure nisin (Weishaupt *et al.*, 2018). The modification of NFC with benzyl penicillin showed good antibacterial properties. A film of NFC-penicillin was prepared and tested against *E. coli* and *S. aureus*. The results showed that NFC-penicillin film has good bacteriostatic effects as the drug was not diffused to the surrounding medium (Saini *et al.*, 2015). Similarly, NFC modified with quinoliniumsilane showed bacteriostatic and bactericidal activities against *S. aureus* and *E. coli* (Hassanpour *et al.*, 2017). The attachment of aminosilanes on the modified NFC showed inhibition of the bacterial strains for 24 hours with very low aminosilane (22.236 g/L) concentration (Saini *et al.*, 2017). The addition of functional group to CNC enables the polymer to adopt new properties. CNC modification with bis-2-chloroethylamine leads to the formation of hyperbranching cellulose, which is soluble in organic solvent. Further experiments conducted showed that hyperbranched CNC has higher efficacy to be used as an antimicrobial agent in future (Demircan & Zhang, 2017). In another approach, 2,2,6,6-tetramethylpiperidine-1-oxyl oxidized (TEMPO) CNC was grafted with PVA and loaded with polypyrrole (PPy), and this showed good antibacterial properties (Bideau *et al.*, 2016). Likewise, the esterification of CNC with rosin through the green route provided it with antibacterial features and inhibited the growth of selected species of Gram-positive and Gram-negative bacteria (de Castro *et al.*, 2016).

The above discussion shows that BC nanofibrils and plant-derived cellulose are interesting and thought-provoking biopolymers with abilities to develop dosage forms for the meaningful delivery of antimicrobial agents. Studies related to the applications of NFC in wound healing and antimicrobial drug delivery are summarized in Table 1.

In addition to the applications of BC nanofibrils in the delivery of antimicrobial agents, it can be potentially delivered as therapeutic agent *via* transdermal route in the form of a hydrogel. In this regard, other than being a moisturizing agent, BC nanofibrils could be applied for treating the conditions such as psoriasis, atopic dermatitis, and xerosis.

2.3 *Medicated Cosmetics*

The applications of BC alone or in combination with other polymeric ingredients and extracts have been studied for facial treatment. In a study, Lin *et al.* (2005) applied BC at a concentration of 0.05–1.0% w/w in cosmetics and found that the presence of BC nanofibrils not only enhanced the moisturizing action but also improved the permeability of the skin. This property can be potentially used for the delivery of topically applied therapeutic agents such as hyaluronic acid, salicylic acid, polypeptides,

Table 1. Antimicrobial drug delivery applications of nanofibrillar cellulose.

Active	Form	Purpose	Refs.
Zinc oxide	BC composites	Wound healing/ Antimicrobial	Khalid *et al.* (2017a)
Titanium dioxide	BC composites	Wound healing/ Antimicrobial	Khalid *et al.* (2017b)
Sorbic acid	BC composites	Antimicrobial	Jipa *et al.* (2012)
Sorbic acid	BC composite	Antimicrobial	Dobre *et al.* (2012)
Zinc oxide	Pellicle	Antimicrobial	Janpetch *et al.* (2016)
Curcumin	Coated composite	Antimicrobial	Khamrai *et al.* (2017)
Nisin	Composites	Antimicrobial	Weishaupt *et al.* (2018)
CNC	Hyperbranched polymer	Antimicrobial	Demircan & Zhang, 2017
Polypyrrole	CNC composites	Antibacterial	Bideau *et al.*, 2016
Rosin	CNC composites	Antimicrobial	de Castro *et al.*, 2016
Benzyl penicillin	NFC composites	Antimicrobial	Saini *et al.*, 2015
Quinoliniumsilane	NFC	Antimicrobial	Hassanpour *et al.*, 2017

growth factors, and enzymes, which are not stable *via* the oral route (Lin *et al.*, 2015). For the delivery of salicylic acid and other therapeutic agents, an approach was developed by Heath *et al.* (2012). BC nanofibrils were formulated with starch and cellulose derivatives. In these formulations, the particle combination was suspended in the liquid matrix system having pH value of 4.0 and 7.0. It was observed that the formulations did not show any unwanted reaction or cause skin injury. It was also suggested that the drugs which are stable at acidic pH, such as salicylic acid, can be used in the formulation with pH value of 4.0. Such formulations are more suitable for the delivery of drugs to treat sensitive skin (Heath *et al.*, 2012). The applications of BC as supporting material in preparing facial scrub in combination with glutinous rice, ascorbic acid, aloe vera extract, and olive oil was reported by Hasan *et al.* (2012). In their study, the scrub not only showed good stability but also demonstrated higher viscosity and longer shelf life as compared to the commercial scrub. Such formulations have the potential for application as natural scrubs for keeping facial skin safe (Hasan *et al.*, 2012). Similarly, BC membranes were loaded with cosmetic agents and masks were prepared. The studies on human volunteers showed that BC-masks showed good skin adhesion, improved the skin moisture content, and prevented skin dehydration (Pacheco *et al.*, 2017). Due to higher interest for the application of BC-based facial masks, gel containing silk sericin was prepared and tested as facial mask. The study showed that BC-sericin gel possessed good moisture-absorbing properties and mechanical strength. The sericin showed controlled release from the gel and was easily removed from the skin as compared to commercial gel of sericin. The gel showed cytocompatibility with human keratinocytes and L929 cells (Aramwit & Bang, 2014).

2.4 *Drug Delivery Applications*

2.4.1 *Protein delivery*

Proteins have various applications such as vaccines, in diagnosis, and in treatment of several diseases. These are more superior to small drug molecules for applications in pharmaceutical drug delivery (Malik, 2008; Vermonden *et al.*, 2012). However, there are certain limitations associated with proteins as drugs such as enzymatic degradation results in their destabilization in addition to changes in their structure. In addition, the physical unfolding and aggregate formation as well the immunological response also limit therapeutic applications of proteins (Bruno *et al.*, 2013; De Groot and Martin, 2009; Kaliyaperumal and Jing, 2009; Manning *et al.*, 1989; Manning *et al.*, 2010; Vermonden *et al.*, 2012; Yang, 2015). Furthermore, the hepatic first-pass effect, faster renal clearance, and shorter half-life lead to regular intravenous administration of proteins. These limitations result in inconvenience, patient discomfort, and patient drug refusal for applications with conventional delivery methods (Harris and Chess, 2003; Tang *et al.*, 2004). The above-mentioned limitations necessitated the exploration of alternative approaches for protein delivery. One such approach is the preparation of hydrogels loaded with proteins, which may have the potential for better proteins bioavailability and patient compliance (Peppas *et al.*, 2004). The application of CNC as potential candidate for proteins delivery was studied by the adsorption of human serum albumin (HSA) and bovine serum albumin (BSA) proteins using physical as well chemical methods. The composites were found to be non-cytotoxic and stable. The release of proteins from the composites was studied using acidic pH (4.0) and basic pH (7.4) for 24 hours. It was found that not only were the proteins released in a controlled manner, excellent biocompatibility was also observed (Singla *et al.*, 2017b).

The physical properties of BNC such as biocompatibility, hydrophilic nature, ultrafine fibrous network, swollen nature, and presence of a number of hydroxyl groups make it an ideal candidate for preparation of hydrogels to be applied for protein delivery (Pandey *et al.*, 2014; Sulaeva *et al.*, 2015). Keeping in view the above properties of BC, Ahmad *et al.* (2014) prepared a hydrogel consisting of BC and polyacrylamide (BC-γ-PAM) for protein delivery. The hydrogels was fabricated using electron beam irradiation instead of cross-linkers and loaded with (BSA). The *in vitro* analysis showed a 10% drug release in the simulated gastric environment. The analysis further showed that BSA release was stable and showed better penetration through the mucosal membrane in the simulated intestinal environment. In addition, no toxic effects were observed with fabricated hydrogels, which were biocompatible and safe for application in the *in vivo* studies (Ahmad *et al.*, 2014). The release of BSA from BC was also studied by Müller *et al.* (2013) using never-dried and freeze-dried membranes. It was observed that never-dried BC has better protein loading capabilities as compared to the freeze-dried version. The drug release studies showed that most of the drug was released *via* diffusion in a controlled manner. Interestingly, the bioactivities and integrity of proteins were not disturbed during protein loading and release (Müller *et al.*, 2013). Similarly, BC and poly(acrylic acid) (BC-PAA) hydrogels for protein loading and *in vitro* evaluations were investigated by Amin *et al.* (2012b). The hydrogels were loaded with BSA and release analysis was carried out at gastric and alkaline pH. The hydrogel swelling in the gastric environment was 10 times more at pH below 5, thus demonstrating a slow protein release (2 hours). In the alkaline environment, the swelling of hydrogels was higher and protein release was controlled (8 hours) as compared to the gastric environment. It was observed that protein release from hydrogels into the medium was pH-dependent (Amin *et al.*, 2012b). The superhydrogel based on the combination of CMCS and cellulose powder for BSA controlled delivery was studied by Chang *et al.* (2010). CMCS and cellulose were cross-linked using epichlorohydrin and loaded with BSA. The drug release studied at pH 7.4 showed that CMCS-cellulose hydrogels have better swelling capacity and highly controlled protein release (Chang *et al.*, 2010). Hydrogels of HPMC with alginate also have the potential for application in the delivery of BSA to colon. A study showed that HPMC-alginate beads have higher swelling ability, which enables the hydrogels to absorb BSA. The drug release studies showed that HPMC-alginate beads have excellent BSA release in the basic medium (Nochos *et al.*, 2008). The delivery of BSA from magnetic-cellulose microspheres were studied using regenerated cellulose microspheres, which were loaded with magnetic Fe_3O_4 particles. The study showed that cellulose-Fe_3O_4 microspheres have excellent protein loading and release properties, which can be used for the site-specific and controlled delivery of proteins (Luo *et al.*, 2009).

2.4.2 *Cellulose in wound healing*

Naturally, BC and other celluloses lack antimicrobial properties. Therefore, in order to incorporate antimicrobial property and to deliver therapeutic/wound healing agent(s) to the wound sites, it is necessary to add such agent(s) to cellulose-based wound dressing materials to achieve an optimal and ideal wound healing environment.

For this purpose, BC films are fabricated with various antimicrobial agents to treat the wounds. To this end, lyophilized BC film was tailored with antimicrobial properties by dipping in the benzalkonium chloride (BZK) solution and dried by lyophilization. It was observed that BC film attained water absorbing capacity and swelling ratio of 26.2 and 37.3 for deionized water and saline solution, respectively. The antimicrobial studies against *S. aureus* for 24 hours showed good bacterial growth inhibition. It was suggested that BC-BZK combination has the potential for use as a wound dressing material that can potentially promote extended drug delivery to treat acute traumatic conditions (Wei *et al.*, 2011). Similar studies were carried out by loading BC with PLA and BZK. The experimental studies showed that the composites were highly effective against selected bacterial species and have the potential for application in wound healing (Foong *et al.*, 2018). Kaplan *et al.* (2014) reported the BC-based delivery of antibiotics ampicillin (AMP) and gentamicin (GTM) for drug loading, release, and effects against selected species of bacteria during *in vitro* studies. Drug loading for AMP and GTM was 99 and 48 mg/cm^2, respectively. During drug release studies for the initial 24 hours, very small amounts of AMP (0.107%) and GTM (0.113%) were released. Thus, no burst release of the drugs was observed during experiments. The release of AMP and GTM was 28% and 17%, respectively, after 7 days. Furthermore, the extended release of drug showed antibacterial effects against *E. coli, S. aureus, Enterococcus feacalis* (*E. feacalis*),

Staphylococcus epidermidis (*S. epidermidis*) and *P. aeruginosa* (Kaplan *et al.*, 2014). The enhancement in the function of GTM *via* chemical attachment with BC was reported by Rouabhia *et al.* (2014). In this study, GTM was initially modified with group of peptides RGDC (glycine, arginine, cysteine, and aspartic acid), followed by attachment with BC. The new conjugate, i.e., BC-RGDC-GM was found to have good bactericidal properties in addition to non-toxic effects on fibroblast cells obtained from human skin (Rouabhia *et al.*, 2014). Similarly, irradiated BC was loaded with tetracycline and evaluated during *in vitro* studies. The drug release studies showed that irradiation has considerable effects on the drug release from BC and resulted in decreased drug release (Stoica-Guzun *et al.*, 2007). Furthermore, BC impregnated with silver sulfadiazine (SSD) particles was evaluated and showed better antimicrobial efficacy, biocompatibility with human epidermal cells, good epithelization, and faster burned skin wound healing in model rats (Luan *et al.*, 2012; Wen *et al.*, 2015). Serafica *et al.* (2010) carried out further studies for the evaluation of hydrated BC loaded with polyhexamethylenebiguanide (PHMB) as wound dressing material. The studies were carried out on guinea pig for hemolysis, cytotoxicity, and sensitization. The studies for irritation potential and systemic toxicity were carried out using model rabbits (Serafica *et al.*, 2010). The prepared BC-PHMB showed interesting antimicrobial effects. In case of wound healing, the BC-PHMB demonstrated 70% wound healing compared to 20% with hydrogel and 50% with hydrocolloids-treated models. The studies on humans for clinical effectiveness showed that BC-PHMB wound healing results were promising (Serafica *et al.*, 2010). Furthermore, when these studies were compared to commercial dressing, BC-PHMB dressings showed superior results in term of pain management and bacterial load reduction (Haemmerle *et al.*, 2012). The effect of controlled drug delivery from BC for wound healing was investigated by preparing BC-vaccarin (BC-Vcr) composites. The composites were prepared by immersion of BC in Vcr solution. Characterization technique showed that hydroxyl groups on BC surface have strong effects on drug attachment. The wound healing studies of BC-Vcr were conducted using rat skin and it was found that BC-Vcr has no cytotoxic effects. This study also showed that BC-Vcr composite has higher wound healing and faster cells regeneration potential as compared to pure BC (Qiu *et al.*, 2016).

In addition to the above studies, various nanotherapeutics (Abdalla *et al.*, 2016) such as chitosan (viral protective membranes) (Wanling *et al.*, 2012), silver nanoparticles (Dobre and Stoica-Guzun, 2013), copper nanoparticles (Pinto *et al.*, 2013), gold nanoparticles (Khan *et al.*, 2018a, b), montmorillonite (Ul-Islam *et al.*, 2013), titanium dioxide (Khan *et al.*, 2015), and deacetylated chitosan (Butchosa *et al.*, 2013) were loaded onto BC and used as wound dressing materials. However, there were reports of potential toxicities and health-related problems such as hyperplasia, neuroplasty, hepatotoxcity, phototoxicity, cytotoxicity, genortoxicity, and formation of edema and dermal toxicities (Koohi *et al.*, 2011; Lu *et al.*, 2010; Prabhu *et al.*, 2010; Ray *et al.*, 2009; Samberg *et al.*, 2010; Wang and Wang, 2014; Wang *et al.*, 2014). Furthermore, studies on the kidneys, hearts, and bones of guinea pigs have reported toxicities with applications of these agents for longer time, which result in limited applications of these agents (Korani *et al.*, 2013). The study reported by Singla *et al.* (2017a) showed that composites of CNC with sliver nanoparticles (Ag-NP) have the advantages to stop microbial growth at the healing site as well promote better cell proliferation, angiogenesis, and collagen deposition. Based on these results, it was suggested that CNC-Ag-NP composites can be ideal candidates as wound dressing materials (Singla *et al.*, 2017a). In another study, the BC was modified with poly hydroxyethyl methacrylate (PHEMA) and loaded with Ag-NP. The composite showed 90% colony inhibition properties during culturing (Di *et al.*, 2017). Surface functionalization of the NFC has strong impact on the adsorption and desorption of bioactive molecules. TEMPO oxidation of NFC results in negative charges on the surface, which has the ability to attract and adsorb small proteins and peptides. This feature can be used for the immobilization of enzymes. This was proved by the NFC surface modification *via* TEMPO-oxidation and enzymes cysteine-protease papain immobilization on NFC with the help of glutaraldehyde and ethyl dimethylamino propyl cabodiimde. The study showed that there was excellent adsorption of the enzyme on NFC, which has the ability to control the release of protein at a desired rate. It was suggested that surface-modified NFC can find potential application as wound dressing material and for protein delivery (Weishaupt *et al.*, 2015). The wound healing features of BC can be further improved by the addition of cytoprotective agents. The study carried out by Lamboni *et al.* (2016) showed that BC loaded with sericin has good stability and showed controlled drug release. It was observed that BC-sericin promoted fibroblast proliferation, accelerated wound healing, improved extracellular matrix formation, and did not influence the keratinocytes growth (Lamboni *et al.*, 2016). Similarly, sericin and polyhexmethylene biguanide combined with BC as supporting platform showed improved wound healing and antimicrobial effects on human volunteers and in rats. It was observed that BC composites have supported good collagen production and antimicrobial features. In addition, the applications of BC composites also showed lower inflammation and absence of irritation at the healing site (Napavichayanun *et al.*, 2016). The addition of glycerin to the abovementioned formulation further improved the rehydration effect of the BC-composite and reduced the adhesion to the skin (Napavichayanun *et al.*, 2017). The ability of bacterial species to resist the effects of antibiotics is getting severe with the passage of time. Therefore, alternate antimicrobial agents are to be used for treating infections. In this regard, Li *et al.* (2017) prepared a composite of BC loaded with modified gold nanoparticles using diaminopyrimidinethiol. The prepared BC composites showed excellent features such as mechanical strain, biocompatibility, and high water uptake. The studies on selected bacterial cultures showed good growth inhibition as well as wound repair of rat skin (Li *et al.*, 2017). Table 2 summarizes the applications of NFC- and cellulose-based products for transdermal drug delivery.

2.4.3 *Tissue engineering drug delivery*

The applications of nanocellulose-based materials for the delivery of proteins, antibiotics, and implants in tissue engineering and regeneration have been widely studied. The delivery of antibiotics was studied by Mori *et al.* (2011) by applying a combination of vancomycin and GM as BC bone cement. The release studies showed that the usage of BC in bone cement prevented the fracture brittleness and compression, while the elution of antibiotics and weariness life of the bone cement was boosted (Mori *et al.*, 2011). These studies suggested that the application of bone cement loaded with antibiotics has the potential for clinical significance. In this regard, another study was conducted by Shi *et al.* (2012) for investigation of local delivery of proteins using a combination of BC with bone morphogenetic protein-2. The *in vitro* results showed that the prepared composites were highly compatible with body tissues and showed good differentiation capabilities among the fibroblast cells such as C2C12 and osteoblasts of the mouse. The *in vivo* results of the same study indicated that scaffolds of BC were highly efficient in the formation of bone deposited with high concentration of calcium when compared to scaffolds of pristine BC. These results demonstrate that nanocellulose has the capability to be used as a therapeutic candidate for drug delivery in localized treatment (Shi *et al.*, 2012). The delivery of human adipose mesenchymal

Table 2. Applications of BC in wound healing drug delivery.

Drug	BC matrix	Properties	Refs.
Nanoparticles of silver, copper and titanium dioxide	BC composites	Antimicrobial effects	Dobre and Stoica-Guzun (2013); Khan *et al.* (2015); Pinto *et al.* (2013)
Deacetylated chitosan and chitosan montmorillonite	BC composite	Antimicrobial properties	Butchosa *et al.* (2013); Wanling *et al.* (2012); Ul-Islam *et al.* (2013)
Benzalkonium chloride	BC composites	Wound healing/antimicrobial	Wei *et al.* (2011)
Gentamycin	BC peptide composite	Antibacterial and wound healing	Rouabhia *et al.* (2014)
Sulfadiazine	BC composite	Antibacterial and wound healing	Luan *et al.* (2012); Wen *et al.* (2015)
Vaccarin	BC fragments	Wound healing	Qiu *et al.* (2016)
Polyhexamethylenebiguanide	BC fragments	Antibiotic and cytotoxic effects	Serafica *et al.* (2010); Haemmerle *et al.* (2012)

cells (HAMC) for preventing the infection and wound healing in an injectable form faced the challenges of cell deposition in other organs, lower cell retention, and availability of minor concentrations of HAMC for wound healing. In addition, HAMC delivery to the wounds *via* animal-derived biomaterials may induce allergic reactions and risk pathogens transmission. To overcome these discrepancies, Mertaniemi *et al.* (2016) prepared bionanocomposites of nanofibrillated cellulose (NFC) using cross-linker glutaraldehyde (NFC-G) under wet conditions and loaded with HAMC. It was observed that HAMC was compatible with NFC-G as observed from their proliferation, adherence, and migration into NFC-G in addition to their non-toxic effects. The results of *ex-vivo* studies showed that HAMC-loaded NFC-G sutures were highly compatible and that HAMC delivery into the wounds was higher, which led to improved wound healing. These studies revealed that application of NFC-G has revolutionized the delivery of HAMC to treat chronic wounds and post-operative inflammations (Mertaniemi *et al.*, 2016). BC is a biocompatible and natural polymer; however, in combination with other polymers it may affect the growth and proliferation of body cells, both positively and negatively. In this regard, BC and silk fibroin (SF) nanocomposites were evaluated for stability, solubility, and effects on tissue regeneration and proliferation. The prepared nanocomposites of BC with different concentrations of SF (25%, 50%, and 75%) were characterized, which confirmed its stability and cell attachment to the nanocomposites. Cytotoxicity studies were carried out *via in vitro* micronucleus assays and it was found the BC-SF was highly biocompatible. These studies also revealed that the presence of fibroin in the scaffold accounted for improved cells adhesion in comparison to pristine BC. It was concluded that such scaffolds could be used for safe delivery of drugs and proteins, and in required quantity for the desirable results (Barud *et al.*, 2015). In addition, BC loaded with bone morphogenetic protein-2 (BMP_2) was investigated for osteogenic potential in tissue engineering (Shi *et al.*, 2012). The BC-BMP_2 was found to be highly compatible and capable of differentiating into osteoblasts cells and C2C12 cells of mouse model. The subcutaneous implants in *in vivo* studies showed that the BC-BMP_2 scaffolds have the capability to form bone with greater calcium concentration in comparison to the pristine BC scaffold (Shi *et al.*, 2012).

The delivery of regenerative medicines requires biomaterials as carriers with versatile properties that protect the agents used in therapy till their delivery to the targeted tissues. De Carvalho *et al.* (2016) reported negatively charged macromolecules (MMOLs) beads prepared *via* ionic gelation process using oxidized NFC (ONFC). The prepared MMOLs were loaded with osteoblast-I bone cells obtained from the rat. The release results indicated that MMOLs released the cells in a controlled manner, while maintaining the cell viability during the experimental period (De Carvalho *et al.*, 2016). It was concluded that NFC has the potential ability to be used for the targeted and controlled delivery of regenerative medicines to the cells and tissues. These studies reveal that NFC has the capability to be applied as carrier for delivery of therapeutic agents in tissue engineering.

2.4.4. *Controlled drug delivery*

The conventional therapeutic delivery approaches to the targeted sites have certain limitations (Vashist *et al.*, 2013) such as variation in drug concentration in plasma, shorter half-life, multiple dosing, and poor patient compliance with the associated dosage form (Amin *et al.*, 2012a). These associated problems have raised the need to design and develop new formulations for drug delivery. Pharmaceutical industries and researchers have made collective efforts to tailor the properties of polymers *via* advanced technology by alteration of the structure, surface functionalization, and shape (Vashist *et al.*, 2013). The objectives of all these efforts are to control the drug release from the delivery system.

The properties of hydrogels such as high compatibility, extraordinary drug loading and releasing capabilities, and ease of administration in various dosage forms make them interesting drug delivery systems for researchers. Cellulose and its derivatives are reported to be ideal candidates for the designing and development of aerogels and hydrogels. Lewis *et al.* (2016) prepared hydrogels by utilizing the negatively charged cellulose nanocrystals (CNC) at different temperatures under moist conditions. The prepared CNC hydrogels were highly porous and showed high shear thinning and thixotropic recovery properties upon gelation. There were no salts or solvents added during the preparation process, so the hydrogels were highly biocompatible and able to be administered *via* injection and did not show any alteration in their properties within the body. It was suggested that these hydrogels could be used for the controlled drug delivery (Lewis *et al.*, 2016). Lin *et al.* (2013) applied the method of grafting for enhancing the controlled drug release properties of cellulose. They applied the β-cyclodextrin and pluronic polymer *via* annexation with CNC, which was facilitated by the application of poly(ethylene glycol) (PEG), and thus the hydrogels loading capacity was enhanced. The newly grafted hydrogel was found to be thermally and structurally stable. The hydrogels were loaded with doxorubicin, and the release of drug was tested at different pH. It was observed that these hydrogels not only adopted the ideal drug loading but also showed prolonged drug releasing properties (Lin *et al.*, 2013). Hydrogels prepared by the applications of polysaccharides are of great advantages; however, it has certain limitations such as stability and deprived mechanical properties (You *et al.*, 2016). To overcome these limitations, nanocomposites of CNC and quaternized cellulose (QC) were linked *via* β-glycerophosphate (β-GP) and loaded with doxorubicin. The characterization studies showed that CNC-QC hydrogels have ideal mechanical properties and ability to resist the degradation. The prepared hydrogels were evaluated for *in vitro* drug release and biocompatibility studies (You *et al.*, 2016). The results showed that the prepared hydrogels avoided the cytotoxicity and inflammatory reactions. The hydrogels were injected into mice with liver, and it was found that doxorubicin was released in a sustained manner. It was concluded from the studies that CNC-QC hydrogels have the potential for applications in sustained drug delivery with improved therapeutic efficacy and better patient compliance (You *et al.*, 2016).

The release of nanoparticles from nanocrystalline cellulose (NCC) varies with sources, modification methods, pH-dependent surface charges, swelling properties of the matrices, and the presence of fibrils. The applications of various sources of NCC and their drying methods have impact on the drug release behavior. Valo *et al.* (2013) worked with NCC obtained from microcrystalline cellulose (MCC), red pepper, quine seed, BC, and TEMPO-oxidized cellulose. All the celluloses were freeze-dried and coated with amphiphilichydrophobin and beclomethasonediprpionate (BDP) nanoparticles, which lead the aerogels to act as reservoir. The *in vitro* drug release studies showed that the aerogels prepared with red pepper and MCC released the drug immediately as it lacked the interconnected fibers. In the case of aerogels prepared with BC, TEMPO-oxidized cellulose, and quine seed, drug was released in a sustained manner. It was concluded that coating on the BDP nanoparticles produced pH-dependent charges on the surface of nanoparticles, which in turn facilitated the diffusion of solvents into the matrices. The presence of interconnected fibers has an effect on the solvent penetration and offers resistance to the drug diffusion from matrices, and thus drug was release in a sustained manner.

Furthermore, it was observed that the application of CNF from various sources despite the same chemical structure showed difference in physical properties, which may be utilized for the designing of delivery system to release drug in controlled, sustained, and immediate manner, or even as a combination of all (Valo *et al.*, 2013). Similar studies were reported by Pavaloiu *et al.* (2014c) using CMC in combination with BC (BC-CMC) and epicholorohydrin (as cross-linker) to prepare hydrogels with model drug ibuprofen sodium. The results showed that the concentration of cross-linker and CMC has strong influence on the water holding and release properties of drug-loaded hydrogels. It was concluded that hydrogels like BC-CMC have the potential to be used for controlled drug delivery (Pavaloiu *et al.*, 2014c). The surface amine functionalization of NFC was carried out by Zhang *et al.* (2017) using dimethyl maleic anhydride, PEG, and cystamine. The prepared composites were loaded with doxorubicin as model drug. The composites were studied at pH 5.0 and 6.5 and it was found that bonds made by amine groups at the surface can be broken. The results showed that such composites can release the drug in a controlled manner at the target site (Zhang *et al.*, 2017).

To make it hydrophobic, BC in combination with polyethylene oxide caprolactone (CPT) and retinol (R) was fabricated by making the CPT-R mixture as a suspension, followed by immersion of BC gels in the suspension for 24 hours. The characterization data of the prepared BC-CPT-R confirmed the presence and stability of retinol. The *in vitro* release showed that retinol was sustained for 24 hours at pH 5.2, while the same was released by BC gel in 4 hours in a comparative study. These results led to the conclusion that BC-CPT bionanocomposites have the ability to be used for the delivery of hydrophilic as well hydrophobic drugs in a controlled pattern and has the potential for applications in transdermal drug delivery and wound healing. Furthermore, it has the capability to encapsulate inorganic, organic, and charged particles to be delivered in a controlled manner (Numata *et al.*, 2015). Similarly, multilayered hydrogels were prepared by Pavaloiu *et al.* (2014a) by using BC, chitosan, and PVA. To study the drug relaese properties and swelling behavior of hydrogels, ibuprofen sodium was loaded as a model drug. The drug release studies were carried out at different pH. The release studies showed a controlled and pH-dependent drug release. The applications of various mathematical models showed that the drug release from the hydrogels was following Fickian diffusion. It was the point of interest that the drug release from the matrices was inverse to the concentraiton of BC in the hydrogels (Pavaloiu *et al.*, 2014a). An advanced technique was developed by Domingues *et al.* (2015), who prepared injectable hydrogels of hyaluronic acid and CNC. They used adipic acid dihydrazide-modified hyaluronic acid and aldehyde-modified hyaluronic acid with different concentrations of aldehyde-functionalized nanocellulose. They studied the mechanical properties, degradation, swelling behavior, and morphology of the hydrogels using the enzyme hyaluronidase. The study finding suggested that the addition of the CNC has good impact on the structural organization and produced hydrogels with high loading capacity. In addition, it was observed that CNC had improved the hydrogels swelling properties and reduced the degradation. It was also observed with the studies on human adipose stem cells that the hydrogels exhibited high cell supportive and proliferation activities. In general, it was concluded that by applying the proposed technique, the biochemical, structural, and biomechanical properties of hydrogels can be improved (Domingues *et al.*, 2015).

The solubility issues related to the drugs has great impact on the drug release and bioavailability. In this regard, complex formation of the drugs with polymer is of great interest. Jeong *et al.* (2015) carried out a study to enhance the solubility of Galangin, used as antibacterial agent. They prepared hydrogels of cyclosophoraose (Cys) (a water-soluble complexation agent) with cellulose by molecular encapsulation and loaded it with a drug. The drug release studies indicated that as the concentration of Cys in the hydrogels was increased, the drug loading was enhanced and release efficiency was extended up to 48 hours. Comparative studies showed that the addition of Cys increased the drug release rate 1.65 fold than pure cellulose. The drug released from the cellulose/Cys hydrogels were also studied against *S. aureus* and showed growth inhibition up to 72 hours. In addition, the Cys/cellulose hydrogel did not show any cytotoxic effects on human fibroblast, thus showing the potential to be applied as antibacterial dressing material (Jeong *et al.*, 2015). Grafting of nanocellulose with other polymers for enhancement in the drug loading and release properties has becomes one of the most interesting and novel research areas. Aerogels of nanocellulose with polyethylenimin (PEI) and NFC were prepared *via* chemical modification. The prepared aerogels were found to be highly porous and have a number of functional groups. The PEI-NFC aerogels were compared for drug loading and release properties with pure NFC. The porous structure of aerogels was found to be highly efficient in drug loading. The drug releasing studies showed that PEI-NFC aerogels have made a remarkable difference in a pH-dependent manner by providing controlled and sustained release, when compared with pure NFC aerogels. The drug was released in pseudo-second order, which was not observed with pure NFC aerogels. It was also observed that PEI-NFC aerogels are highly compatible and biodegradable and have promising properties for application in drug delivery in future (Zhao *et al.*, 2015). It was concluded from the studies that BC in combination with different biomaterials has the potential for controlled and pH-responsive drug delivery, be they non-proteinaceous or proteinaceous. Such systems can be applied for the delivery of acid-labile therapeutic agents, proteins, peptides, and other oral drugs.

The swelling and water absorbing properties of BC makes it an ideal candidate for preparing hydrogels in combination with other polymers. The BC-gelatine composites were prepared with the help of crosslinker glutaraldehyde and loaded with the model drug methylene blue. The drug release from the hydrogels was carried out in aqueous solution for 48 hours. The release data showed that the drug-loaded hydrogels have good capabilities to continuously release the drug for 40 hours (Treesupphar *et al.*, 2017). Similarly, Păvăloiu *et al.* (2015) reported that BC-based hydrogels with gelatine (Gs) having different layers. The BC-Gs hydrogels were evaluated for the effect of gelatine concentration on the swelling behavior in acidic and alkaline environment. The hydrogels showed higher swelling capabilities in acidic conditions rather than alkaline ones due to the polyelectrolyte properties of gelatin. However, an increase in the concentration of BC for the coating of hydrogels showed inverse effect on the swelling rate of hydrogels (Păvăloiu *et al.*, 2015). The properties of these hydrogels in the acidic environment suggested the potential applications for localized drug delivery in the stomach. Cellulose derivative such as CMC in combination with cross-linkers has the potential to be applied in the preparation of hydrogels beads. Zare-Akbari *et al.* (2016) prepared nanocomposites using CMC and ZnO micro-particles and iron ions loaded with propranolol. It was observed that these hydrogel have high drug loading efficiency. The *in vitro* drug release at pH 7.4 showed that the prepared hydrogels have high swelling and absorbing behavior in basic and aqueous media as compared to neat hydrogels prepared without cross-linkers. It was also concluded that increase in the concentration of ZnO in bionanocomposite hydrogels has a high impact on the controlled release of drugs (Zare-Akbari *et al.*, 2016). Nanocellulose scaffold with properties such as ultrafine fiberous network, oleophilicty, lipophilicity and hydrophobicity, and diverse loading ability make it a potential biomaterial for smart and ideal drug delivery. The addition of any other materials to CNC will add new features in addition to the existing properties. In this regard, CNC and sodium dodecyl sulphate (SDS) were formulated together. The combination was loaded with alisartanisproxil, and drug loading up to 62% was observed. The drug release data showed that drug dissolution was increased four fold as compared to pure drug (Hou *et al.*, 2017). The research work reported by Masruchin *et al.* (2018) showed that the presence of surface hydroxyl groups of CNF had the ability to be attached with polymers to show thermal responsiveness. Experimental work was carried out by chemical attachment of poly(*N*-ispropyleacrylamide) with CNF and loaded with methylene blue. The studies showed that increase in the concentration of CNF may result in an increase in the thermal response. The drug release studies were carried out at various pH and temperatures. The drug release data showed that these hydrogels are not only thermal responsive but also have the ability to control the drug release at a desired rate (Masruchin *et al.*, 2018). In addition to pH and thermoresponsive hydrogels, electrically responsive hydrogels for controlled drug delivery are also reported. These hybrid hydrogels were prepared from BC and loaded with diclofenac sodium and coated with electrically responsive polymer poly(ethylene dioxythiophene). The drug release was checked using an electric field of –0.6 V to 0 V for 5 seconds. The release data showed an electric stimuli-dependent drug release. The prepared hydrogels were found to be biocompatible during studies using PC12 cell culture (Chen *et al.*, 2017). Surface-modified NFC has the ability to control the drug release loaded into the hydrogels. In this regard, anionic NFC was prepared and loaded with model drugs under freeze drying conditions. The drug release from the hydrogels was NFC concentration-dependent, while the drying process did not show any effect on the drug release (Paukkonen *et al.*, 2017a). From the above study, it was concluded that changes in the surface charges of BC using different chemical and physical techniques enhances its features. In addition, the applications of various polymers with different features also enhanced the drug retaining features of BC. It was observed that utilizing these procedures and techniques will facilitate the designing of drug delivery system with desired drug release at the target site.

Encapsulation *via* coating of drugs using layer-by-layer technique becomes interesting for the researchers (Kiprono *et al.*, 2018). In this way, alternation in the coating layer thickness changes the diffusion of water across the coat and thus results in a controlled drug release (Ai *et al.*, 2003).

Similarly, another approach of multilayered film was adopted by using chitosan and nanocellulose combination for loading of hydrophilic doxorubicin and hydrophobic coumarin (Mohanta *et al.*, 2014). The encapsulation of drug-loaded matrices was carried out by using nanocellulose and chitosan. The drug release from the matrices was evaluated at pH 6.4 and 7.4. The doxorubicin showed sustained release behavior at pH 6.4, while coumarin was observed to be released in a sustained manner at pH 7.4. It was concluded that the differences in nanocellulose and chitosan surface charges could be utilized for the sustained drug delivery of hydrophobic and hydrophilic drugs (Mohanta *et al.*, 2014). The formation of an efficient drug delivery system for the controlled drug delivery is a novel approach. In this regard, Galkina *et al.* (2015) assessed the formation of complexes/grafting of the penicillamin-D, diclofenac sodium, and phosphomycine with cellulose nanocomposites and titanium dioxide. The drug release from the prepared composites was studied using *in vitro* drug release for diclofenac sodium and penicillamin, while phosphomycine release was evaluated *via* radiolabeled analysis. The drug release from the nanocomposites was observed as a controlled system for extended time. The results of drug release showed that the interaction between the drugs and titanium dioxide can be used for controlling the drug release from the composites, thus having the potential for controlled and extended drug delivery of potent drugs. It was also found that such interactions can be applied for the transdermal drug delivery of antibiotics, analgesics, and anesthetics (Galkina *et al.*, 2015). Although BC has provided sustained drug delivery in many instances (Ullah *et al.*, 2016a; Ullah *et al.*, 2017), its properties such as highly porous fibrous network and permeability to solvents and small molecules limit its applications in controlled drug delivery. To overcome these limitations and improve the drug retaining properties, Luo *et al.* (2017) prepared BC-graphene oxide (BC-GO) nanocomposites/nanocarriers (NCs) loaded with ibuprofen. It was observed that such NCs not only have high surface area but also have better drug loading capacity. The drug release studies showed that BC-GO-based NCs have better drug retaining properties and prevented the premature drug release. The comparison studies showed that BC-GO has better viability with cells as compared to pure BC and has the potential for drug delivery as a carrier (Luo *et al.*, 2017). In addition to the existing conventional dosage form such as matrices, hydrogels, and tablets, capsules are one of the alternatives to these dosage forms, which is most preferred and patient friendly. Capsules have gained popularity due to use of minimum number of excipients for filling and is a suitable dosage form for drugs that may face problem(s) during tablet manufacturing. The applications of gelatine-based capsules faces limitations such as sensitivity, ethical, and stability issues. To find alternative materials for capsule preparation and applications in drug delivery, Ullah *et al.* (2017) studied the potential of BC. They reported as-synthesized BC capsule shells for controlled and immediate drug release. The BC capsules were studied using *in vitro* dissolution apparatus and found that irrespective of the variation in BC thickness, the drug was released in an immediate manner. The capsules were also loaded with cellulose-derived polymer such as carboxymethyl cellulose sodium and hydroxypropyl methyl cellulose, which showed controlled drug release up to 12 hours. It was concluded that BC could be used as an alternative to gelatine-based capsules for desired drug delivery (Ullah *et al.*, 2017).

Applications of non-conventional materials such as surfactant in combination with NFC for oral drug delivery have not been reported. These materials can play an important role in the designing of various dosage forms, particularly tablets. In one of such approaches, sodium salt of lauric acid was mixed with NFC by Svagan *et al.* (2016) and floating matrices were produced. Interestingly, such matrices can be designed in the desired shape and thickness (Svagan *et al.*, 2016). The matrices were prepared and loaded with different concentrations of riboflavin in wet form and were freeze-dried. It was observed that due to the presence of surfactants and air entrapment, the matrices achieved drug loading up to 50%. The drug release studies were carried out with the help of Franz cells, and it was found to occur in a controlled manner. In addition, the drug release was found to be dependent upon the dimension of foam thickness. Matrices with 8 mm thickness released 59% of the drug in 24 hours, while 0.6 mm released 78% within 8 hours, and matrices of 0.009 mm thickness released 72% drug content within 1 hours. It was concluded that this system has the potential for designing of gastroretentive drug delivery system (Svagan *et al.*, 2016). The concentration-dependent hydrophilic properties and hydrogen bond formation ability of NFC with hydrophobic polymers were studied by Cheng *et al.* (2017). A composite of NFC with poly(3-hydroxybutyrate-co-3-hydroxyvalerate) (PHBV) was prepared to enhance the NFC drug loading and retaining properties. NFC was mixed with PHBV in different concentrations and it was found that the addition of NFC enhanced the mechanical and water absorbing properties significantly. Tetracycline was used as model drug for studying the drug release properties. It was observed that NFC concentration up to 6% increased the composite tensile strength from 110 to 125%. In addition, higher drug loading (up to 98%) and prolonged drug release (up to 540 hours) were achieved (Cheng *et al.*, 2017). In addition to hydrogen bond formation with other polymers, NFC has the potential for applications in gastroretentive drug delivery due to its light weight and floating ability. Aerogels of NFC with bendamustine hydrochloride were prepared using physical adsorption. The aerogels were evaluated for drug release and bioavailability under acidic and basic conditions. It was observed that aerogels have good floating ability in acidic conditions (pH 1.2) with drug retention up to 24 hours, while under basic conditions (pH 7.4) drug was released in 7.5 hours. The most interesting result was the 3-fold increase in drug absorption during *in vivo* studies. It was observed that the aerogels with NFC have the potential for gastroretentive drug delivery (Bhandari *et al.*, 2017). Applications of BC in combination with other materials to prepare composites for drug delivery were studied by Kim *et al.* (2017). They prepared BC-alginate composites by culturing colonies of *G. xylinus* in alginate containing medium. It was observed that BC-alginate has higher crystallinity, surface area, and water holding ability as compared to pure alginate and BC. Lipase enzymes were used in the study for evaluation of the potential role of the composite in drug delivery. The enzymes released from the composite showed many fold increase in activities as compared to pure alginate and BC. It was suggested that such hybrid materials of BC have the potential for applications in drug delivery, tissue engineering, and enzymes immobilization (Kim *et al.*, 2017).

Introduction of charges on the surface of cellulose plays an important role in the absorption and delivery of charged drugs into and from the matrices. Various techniques have been applied for modification of cellulose surface and creation of charges. The reported method TEMPO-oxidation was applied to produce surface charges on the CNC without any major change in its basic backbone. The drug aspirin was loaded onto the surface *via* simple mixing techniques and incubated for various periods of time. The matrices were checked for drug stability using UV-spectroscopy, and it was observed that surface charges have significant effects on the absorption of water. The matrix with higher surface charges resisted the water adsorption. It was concluded that increase in the surface charges can be applied for controlled and extended drug delivery (Carlsson *et al.*, 2014). Similarly, Badshah *et al.* (2018) reported the surface-modified matrices of BC loaded with famotidine and tizanidine as model drugs. The *in vitro* drug release study showed that surface modification *via* acetylation has altered the surface properties of matrix, which resulted in retaining the drug for extended time periods as compared to pristine BC (Badshah *et al*., 2018). The crystalline nature of NFC was studied to investigate its application as single excipient or in combination for controlled drug delivery and cellular uptake. Jackson *et al*. (2011) prepared NFC crystals loaded with drugs of hydrophilic, hydrophobic, and ionizable nature. Hydrophilic drugs doxorubicin and tetracycline were used in this study. The *in vitro* drug release studies indicated that hydrophilic drugs were release within 1 day. On the second study, NFC was surface-modified using cetyltrimethylammonium bromide (CTAB) to increase surface charges on NFC. Surface-modified NFC was loaded with hydrophobic anticancer drugs such as etoposide, paclitaxel, and docetaxel. The drug release data showed that these drugs were bound closely to the modified nanocellulose and were released gradually for several days. In general, it was concluded that the modified nanocellulose has the potential for application as single ingredient in drug delivery, particularly controlled delivery system (Jackson *et al*., 2011). Abraham *et al*. (2017), prepared cellulosic biobased multifunctional material *via* esterification and freeze-drying process. The loading results of different hydrophilic solvents, oils, and proteins showed higher efficiency, and so it was proposed that esterified nanocellulose has the ability to be applied for the controlled delivery of drugs (Abraham *et al*., 2017).

The delivery of drugs in sustained manner for longer duration was achieved through the formation of NFC matrix. In this method, the matrix was prepared *via* filtration method and loaded with different concentrations (20–40%) of model drugs. The advantage of using filtration method was the higher drug loading efficiency into the matrix, which was found to be more than 90%, while the film thickness was found to be adjustable. It was also revealed that filtration and film formation methods have the potential to be applied for the loading and coating of heat-sensitive drugs and have the ease of preparation. Such formulations of NFC have the advantage to retain the drugs for longer duration and release it in the zero-order kinetic pattern. Thus, these films have the potential for application in the sustained drug delivery system (Kolakovic *et al*., 2012b). In addition to the ones described above, there are several other hydrophilic and hydrophobic cellulose derivatives meant for controlled drug delivery, which are beyond the scope of this chapter.

2.4.5 *Stimuli-responsive drug delivery*

Masruchin *et al.* (2018) reported that the presence of surface hydroxyl groups of CNF endow it with the ability to be attached with other polymers and show stimuli responsiveness. In this scenario, stimuli-responsive hydrogels of BC with G-poly(acrylic acid-co-acrylamide) were prepared by Amin *et al.* (2014) *via* co-polymerization, loaded with theophylline, and were evaluated for oral drug delivery. BC was irradiated prior to polymer grafting using microwave technique. The higher porosity of hydrogels was found more suitable for higher drug loading. *In vitro* evaluation was carried out at different pH and hydrogels were found to possess higher swelling ability at alkaline pH compared to acidic (Amin *et al.*, 2014). In this regard, the drug release from the hydrogels was slow in simulated gastric fluid (SGF) as compared to simulated intestinal fluids (SIF). It was hypothesized that such type of hydrogels have the potential for delivery of acid-sensitive therapeutic agents such as proteins, certain drugs, peptides, and drugs used for colonic cancer treatment. The application of NFC in thermal sensitive drug delivery systems was studied by Zhang *et al.* (2016). They prepared hybrid microspheres using temperature-sensitive polymer and NFC using cross-linker. The microspheres were grafted with *N*-isopropylacrylamide (NIPA) monomer using *in situ* radical polymerization technique. The NFC-NIPA microspheres were studied for drug release at various temperatures. During *in vitro* drug release study in water, it was observed that NFC-NIPA hybrid exhibited highest controlled release at 32°C. The drug release studies showed that variation in temperature had significant effect on the drug release from the microspheres (Zhang *et al.*, 2016). CNC-based hydrogels in combination with poly(N-isopropylacrylamide) without the addition of cross-linker were prepared and loaded with metronidazole. The hydrogels showed higher drug loading and controlled release at normal body temperature. It was suggested that these hydrogels have the potential for application as wound dressing material (Zubik *et al.*, 2017). In another study, Shi and co-workers (2014) reported the fabrication of hydrogels with dual response. The hydrogels were made up of BC and sodium alginate (SA), loaded with ibuprofen sodium as model drug. The *in vitro* drug release studies were carried out in the pH- and stimuli-responsive electric field. It was observed that these hydrogels have lower swelling at acidic pH, i.e., up to 8 times than alkaline. In case of the electric field with 0.5 V, the swelling rate of the hydrogels was 8 times higher than at 0 V. It was also observed that changing the drug release environment from acidic to alkaline affected the drug from the hydrogels and lower drug release was observed in the acidic environment as compared to alkaline. It was concluded that hybrid hydrogels of BC-SA are potential candidates for the controlled delivery of drugs (Shi *et al*., 2014). Dual stimuli-responsive hydrogels were also reported with polysaccharide salecan and poly dimethyl aminoethymethycrylate and loaded with model drug insulin. The hydrogels showed good stability and the ability to swell at pH 1.2 and 7.4. At temperature below 40°C, the hydrogels also showed good swelling ability. The drug release data revealed that it can be controlled from the hydrogel by changing the pH and ionic strength. In addition, the hydrogels did not show any cytotoxic effects during the studies (Wei *et al*., 2017). In contrast to temperature and pH dual stimuli-responsive hydrogels, pH-near infrared (NIR) responsive hydrogels were prepared by the application of oxidized CNF and polydiamide. The hydrogels were cross-linked by using Ca^{2+} ions and loaded with tetracycline as model drug. The hydrogels were able to release the desired concentration of drug when exposed to NIR as well acidic environment. It was further observed that these hydrogels have better wound healing capabilities, biocompatibility, and biodegradability (Liu *et al*., 2018). In another approach, cellulose was first converted to cellulose acetoacetate and then mixed with cystamine di-hydrochloride to prepare hydrogels, loaded with rhodamine B as model drug. The hydrogels prepared showed dual responsive (pH and redox) drug release and stability at physiologic conditions for 3 days. The drug release from the hydrogels showed that maximum drug retention time in acidic and basic environment was 10 hours and 25 hours, respectively (Liu *et al*., 2017). Dual pH and glucose responsive hydrogels were prepared by combination of cellulose and vinyl phenyl boronic acid and loaded with insulin. The glucose responsiveness of hydrogels was evaluated based on the release of insulin from the hydrogels, which make the hydrogel an ideal system for drug delivery with self-regulation (Peng *et al*., 2018).

The delivery of anticancer drugs to the desired site is always a great challenge. Due to the changes in pH and temperature of the cancerous cells, stimuli-responsive bionanogels are one possible approach to deliver drugs at the desired site. In this regard, dual responsive nanogels were prepared by using CMC as base material. The bionanogels were found to be stable for longer duration, and no interaction with body proteins was observed. The *in vitro* studies revealed that these dual responsive nanogels have the ability to release the drug only at the pH and temperature of the cancerous cells (Wen & Oh, 2015). The cross-linking *via* disulphide bonding of the hydroxpropyl cellulose (HPC) with other polymers to prepare carriers for hydrophobic anticancer drugs was reported by Rahimian *et al.* (2015). Nanocarriers were prepared by combining oligo ethylene oxide and triethylamin through disulfide bond and grafted on the HP. The nanogels were found to have better colloidal stability in addition to higher drug release at the tumor site as a result of redox response (Rahimian *et al.*, 2015).

The triple stimuli responsive hydrogels have the ability for drug release at different responses. Dutta *et al.* (2016) prepared hydrogels that were responsive to stimuli such as temperature, pH, and redox. The hydrogels were prepared by copolymerization of methyacylated CMC and *N*-isoprolyacrylamide, using *N,N′*-methylenebisacrylamide in addition to *N,N′*-bis-(acryloyl) cystamine and loaded with lysozyme as model protein. The hydrogels were found to have temperature-dependent water retention and swelling ability. The drug release from the hydrogels was dependent upon the presence of glutathione, lower pH, and lower temperature (Dutta *et al.*, 2016). Similarly, hydrogels with temperature, pH, and magnetic response were reported by Siangsanoh *et al.* (2018). These hydrogels were prepared using magnetic nanoparticles coated with gelatin and later on loaded on BC-gelatin matrices, cross-linked with the help of glutaraldehyde and loaded with methylene blue as model drug. The hydrogels showed higher swelling and drug release at pH 1.2 at 37°C under the effect of externally applied magnetic field. It was observed that the hydrogels have the potential as triple stimuli-responsive controlled drug delivery system (Siangsanoh *et al.*, 2018).

The application of stimuli-responsive polymers makes it possible to prepare composites that enable the delivery of drugs to the targeted site in a desired concentration. In this regard, polyurathanes and cellulose-based matrices were prepared with propionic acid, glycolic acid, and lactic acid. The matrices were loaded with felodipine, and the drug release was monitored in acidic and basic solutions. The release study showed that the matrix has low drug releasing properties in acidic environment as compared to basic environment. In addition, the matrices prepared were degraded under the basic environment. It was suggested that such stimuli-responsive systems are suitable for drug delivery to the colon and vagina (Solanki & Thakore, 2015).

2.4.6 *Transdermal drug delivery*

The most convenient alternative route to oral and hypodermal routes for drug delivery is the transdermal route (Prausnitz & Langer, 2008; Prausnitz *et al.*, 2004). In conventional formulations as well advanced approached, various strategies have been adopted for the delivery of therapeutic agents to systemic circulation *via* the transdermal route (Prausnitz & Langer, 2008). To achieve the desired drug delivery *via* transdermal route, MBC, which is produced as swollen membrane, may be applied without plasticizer for delivery of transdermal therapeutic candidates. Furthermore, the properties of BC, such as being skin moisturizing and having the highest tolerance, make it an ideal candidate for transdermal drug delivery (Almeida *et al.*, 2014). Literature has very limited data available for the applications of BC as transdermal drug delivery material. The release of drugs from the BC can be controlled by modification in the pore size of the BC membrane using physical, chemical, or radiation techniques. In this regard, studies were reported by Olyveira *et al.* (2013), where BC treated by gamma-radiation was used for *in vitro* drug release. The result showed that radiation-treated BC has pore density higher as compared to non-irradiated one, and thus resulted in the lower drug release (Olyveira *et al.*, 2013). Similarly, Stoica-Guzun *et al.* (2007) evaluated the effects of electron beam irradiations on the release of tetracycline from BMC for transdermal drug delivery. It was observed that electron beam irradiation had considerably reduced the tetracycline diffusion into the BC during the *in vitro* experiments. These results suggested that CMC can be used in the form of patches for potential transdermal drug delivery applications (Stoica-Guzun *et al.*, 2007). It was suggested that using irradiation as a technique to physically modify the BMC membrane porosity helps in achieving controlled drug release.

Similarly, BMC was used for the delivery of lidocaine hydrochloride (LHC) and Ibuprofen *via* human epidermis. The results obtained showed that BC loaded with LHC has slower drug release in comparison to conventional formulations. In case of ibuprofen, the drug release from BC was 3 times higher than conventional gel or solution prepared with PEG400 (Trovatti *et al.*, 2012). Another drug from the NSAIDS, diclofenac sodium (DS), was also studied using BMC as a supporting membrane (Silva *et al.*, 2014). In this case, glycerol was applied as plasticizer and loaded into the BMC in addition to DS salt. It was observed that drug loading into BMC was almost uniform. The drug-loaded BMC *in vitro* studies were conducted using human epidermal membranes and found to have better permeation than DS marketed patches but lower than commercially available gel formulation. This study showed that BMC can be loaded with drugs using simple techniques to give desired drug release as compared to other formulations. These features make BMC superior to conventional plant-derived cellulose (PC)-based formulations for DS and similar drugs deliveries (Ul-Islam *et al.*, 2019; Silva *et al.*, 2014). Pandey *et al.* (2013) also reported similar procedures using irradiated BC to prepare superabsorbent hydrogels of BC-PAM with the help of cross-linker and process of microwave irradiation. The hydrogels were studied for swelling at different pH and showed better swelling at pH 7 (ca. 23–25 times) for BC solution hydrogels than BC dispersion (ca. 900%). The drug release study showed that these hydrogels retained theophylline for 24 hours in buffers at pH 7.4 (Pandey *et al.*, 2013). This study revealed that such hydrogels have the potential for theophylline transdermal delivery. However, further investigations are required to further clarify the feasibility.

Huang *et al.* (2013) loaded berberine in BM and conducted *in vitro* experiments using SGF, SIF, and acidic and alkaline solutions in addition to transdermal studies. The obtained results showed that drug release rate was low in acidic environment compared to alkaline one, and fastest in neutral solution, i.e., SIF. They suggested that such type of BMC-based formulations could find potential applications in the pH-dependent swelling and drug release (Huang *et al.*, 2013). It was suggested that BMC-based hydrogels have the potential for applications in transdermal delivery of therapeutic agents to achieve better percutaneous bioavailability. In another approach, Evdokimova *et al.* (2018) worked on the preparation of a hybrid system of NFC. The system was consisting of nanoparticles prepared from combination of NFC and titania. The nanoparticles were loaded with triclosan as model drug. The composite showed controlled drug release and higher antibacterial efficacy with the model drug (Evdokimova *et al.*, 2018). Combination of CNF with other polymers such as HPMC has the ability to create composites for applications in transdermal drug delivery. In one such study, the CNF, HPMC, and Ketorolac tromethamine combination was prepared. The studies conducted showed that such combination with increase in the concentration of CNF up to 1% have the ability to control the drug release (Orasugh *et al.*, 2018). Briefly, it was observed that BC-based drug delivery systems are able to control the drug release *via* diffusion. It was also noted that in majority of cases, the drug release was temperature- and pH-dependent. Swelling at different pH showed great impact on the alteration in porosity of BMC, and thus effected the drug release (Huang *et al.*, 2013; Pandey *et al.*, 2013).

2.4.7 *Drug delivery on the basis of enantioselectivity*

Enantioselective drug delivery and avoidance of toxicity from the racemic mixtures is becoming common and important in the modern era of drug delivery and gaining importance in the pharmaceutical fields (Ullah *et al.*, 2016a). Being therapeutic agents, enantiomers and diastereomers often possesses different bioactivities (Nguyen *et al.*, 2006). In order to avoid the presence of unwanted isomers in pharmaceutical products, provide rational therapeutic control to patients and achieve the desired therapeutic effects, deracemization becomes mandatory. In this area, Bodhibukkana *et al.* (2006) applied molecularly imprinted polymers (MIPs) with BC to fabricate enantioselective *S*-propranolol-loaded composite for transdermal delivery. They prepared co-polymer of BC with methacrylic acid using ethylene glycol as a cross-linker. The prepared composite was able to upload selective enantiomers of *S*-propranolol due to specific binding sites. The release of enantioselective *S*-propranolol was studies in *in vitro* experiments using rat skin and it was observed that the matrix system consisting of BC-MIP exhibited good transport across the membrane (Bodhibukkana *et al.*, 2006). This study showed that BC-based MIPs have the potential for applications in the selective delivery of enantiomers through the transdermal route.

2.4.8 *Dental drug delivery*

Treatment of the dental caries requires extensive care, and any contamination or carelessness may cause relapse of the infection. In this regard, root canal treatment (RCT) is one of the most commonly used procedures used for the treatment of dental caries. The objective of the RCT is the decontamination of the root canal and prevention of pulp infections. The conventional RCT procedure involves PC or cellulose obtained from cotton for the drying and cleaning of root canal. To achieve the desired results in RCT, there is further need to develop materials with high absorbing capacity, biocompatibility, and ability to deliver the drug in the desired fashion. To meet the above-mentioned desired features, Yoshino *et al.* (2013) applied BC as a biomaterial for dental treatment. They observed that BC not only worked as biocompatible material but also exhibited high expanding ability, which resulted in the higher absorption of pulp during RCT as compared to conventional paper (Yoshino *et al.*, 2013). When the drug release studies were carried out, it was observed that BC released higher amount of drugs as compared to conventional cellulose. On the basis of the obtained results, it was concluded that BC possesses potential for applications in drug delivery during the RCT treatment. The combination of BNC with biocompatible polymers for mucoadhesive drug release was studied by Laurén *et al.* (2018). The film was prepared by adding mucin, pectin, and chitosan to BNC and loaded with model drug metronidazole. The film showed excellent bioadhesion, swelling ability, and non-cytotoxic nature. The drug release data showed that mucoadhesive film was a good candidate for the oral delivery of drugs to treat periodontitis (Laurén *et al.*, 2018).

2.4.9 *Macromolecular-based prodrug delivery*

Gastric disturbance is one of the main problem associated with NSAIDS (Radi and Khan, 2006; Ullah *et al.*, 2016a). Ibuprofen, (an NSAIDS) has a short half-life in additon to causing gastric irration. To achieve the desired therapeutic drug concentration in the body, frequent drug dosing is required, which may lead to increase in side effects, i.e., stomach burning due to ulceration (Wright, 2002). In order to overcome the adverse effects related to NSAIDs, the conjugates of ibuprofen with BC by esterification were reported by Shi *et al.* (2013). BC-Ibuprofen macromolecular prodrug during *in vitro* drug release studies showed that due to hydrolysis of the conjugates at basic pH, the drug was released in controlled manner. The hydrolysis of ester bond was observed to be faster in the alkaline medium and slower in the acidic environment and relatively slow at colonic pH (Shi *et al.*, 2013). The pH-based release of the drug from BC-Ibuprofen conjugates suggests a great potential for controlled drug release and has the potential to be applied to other NSAIDS with carboxylic functional groups. An example of such an NSAID is aspirin, which can be easily converted into acid-stable conjugates with BC.

All the studies discussed regarding NFC-based drug delivery are summarized in Table 3.

Table 3. An overview of NFC-based different drugs delivery systems.

Purpose	Therapeutic candidate	Approach	Outcomes	Refs.
Tissue engineering drug delivery	BMP_2	BC based protein composites	Biocompatible and showed the ability of bone formation as well fibroblast differentiation	Shi *et al.* (2012)
	GM vancomycin	BC based bone cement	Prevented fracture fragility and compression. Improved life fatigue and improved antibiotic elution	Mori *et al.* (2011)
Controlled drug delivery	SS	BC capsule shells	Controlled and immediate drug release	Ullah *et al.* (2017)
	Theophylline	BC dependent hydrogels	Controlled drug release in SGF	Amin *et al.* (2014)
	Paracetamol	BC coated tablets	Sustained drug release from the NFC film	Amin *et al.*, (2012a)
	—	BC based gelatine hydrogels	Potential for gastro-retentive drug delivery due to higher swelling ability	Păvăloiu *et al.* (2015)
	IbuNa	BC built hydrogels	pH- dependent controlled drug release	Pavaloiu *et al.* (2014a)
	Ibu	BC dependent hydrogels	pH- and electro-dependent drug release	Shi X *et al.* (2014)
	Berberine hydrochloride	BC-polyaniline hydrogel	pH-electroactive controlled drug release	Li *et al.* (2018)
	Propranolol	MIP matrix	*S*-propranolol selective transport and release	Bodhibukkana *et al.* (2006)
	IbuNa	BC built hydrogels	pH-dependent controlled drug release	Pavaloiu *et al.* (2014c)
	BSA	BCM-BSA matrices	Higher BSA loading into hydrated BCM without loss of activities	Müller *et al.* (2013)
	BSA	Cross-linker independent BC based hydrogels	BSA protected from stomach acidic environment and sustained release	Ahmed *et al.* (2014)

	BSA	BC-hydrogels	Higher drug release in alkaline environment than acidic	Amin *et al.* (2012b)
	BSA	BCM-BSA matrices	Higher BSA loading into hydrated BCM without loss of activities	Müller *et al.* (2013)
	BSA	BC hydrogels	Higher drug release in alkaline environment than acidic	Amin *et al.* (2012b)
Dental drug delivery	Trypan blue	BC based points	Comparatively higher drug release than paper	Yoshino *et al.* (2013)
Stimuli responsive drug deliver	Theophylline	Co-polymers	pH dependent drug release	Amin *et al.* (2014)
	5-flourouracil	Microspheres	Temperature dependent drug release	Zhang *et al.* (2016)
Transdermal drug delivery	Tetracycline	Drug-loaded BCM	Irradiated BCM showed slower tetracycline release than pure BC	Stoica-Guzun *et al.* (2007)
	DS	Drug loaded BCM	Slower permeation as rate as compared to commercial patches	Silva *et al.* (2014)
	Berberine	BC membrane	Faster drug release at neutral pH alkaline and acidic pH	Huang *et al.* (2013)
	Ibu, LCH	Drug-loaded BCM	Faster Ibu penetration and controlled LCH permeation in comparison to conventional formulations	Trovatti *et al.* (2012)
	Insulin	Irradiated-BCM	Irradiated BC has controlled drug release as compared to pure BCM	Olyveira *et al.* (2013)
	Theophylline	BC-hydrogels	Controlled the release of theophylline	Pandey *et al.* (2013)
Macromolecular prodrug delivery	Ibuprofen	Conjugated esters	Controlled release	Shi *et al.* (2013)
	OSTEO-1	Conjugates	Controlled release	De Carvalho *et al.* (2016)

2.4.10 *Emulsion stabilization*

Emulsions find various applications in different fields such as cosmetics, food, paint industry, and pharmaceutical industry for drug formulations and delivery. A number of emulsifying agents have been used for the stabilization of emulsions; however, the associated allergic reactions and potential carcinogenicity are still the major constraints. To overcome such discrepancies, extensive research was carried out to find non-toxic, inert, highly stable, and compatible emulsifying agents as formulations ingredients (Yang *et al.*, 2017). Nanocrystals of NBC are biodegradable, non-toxic, and highly biocompatible biomaterials with ability to be used as emulsifier. Yan *et al.* (2017) prepared NBC *via* treatment of BC membrane with sulphuric acid and oxidation with hydrogen peroxide. The characterization data revealed high chemical and thermal stabilities. The studies on applications of NBCs at different pH showed high emulsion stabilizing properties and more sensitive to variations in pH (Yan *et al.*, 2017). Other than BC, wood cellulose in the modified form has the ability for stabilization of emulsions. The modified CNC with higher phenyl groups showed more hydrophobic properties than native wood cellulose. The studies carried out for emulsion stabilization showed that modified CNC was 13 times more effective than pristine cellulose. They suggested that modified CNC has the potential for applications in personal care products, cosmetics, and pharmaceutical drug products (Gong *et al.*, 2017). The effect of combination of other polymers with CNC for emulsion stabilization has strong impact on the stability. In this regard, CNC was mixed with quinoa starch (QS) granules and studied for its effects on emulsion stability. The concentration of CNC and MFC on the stability of the emulsion was investigated in combination with limonene, eugenol, and cinnamaldehyde. The combination was studied for stability of emulsion during storage for eight weeks. It was observed that concentration of CNC and MFC in the range of 0.1–1.0% has good stability to mild centrifugation (Mikulcová *et al.*, 2016). Similarly, the increase in concentration of CNC and combination with TEMPO-oxidized cellulose and unmodified cellulose was also investigated for emulsion stability. The studies were conducted for a duration of one month, and it was observed that the emulsion was highly stable during this period. It was further noted that the creamy layer of the emulsion showed more stability to coalescence, higher temperature, and low shear than emulsion-stabilized with CNC alone (Gestranius *et al.*, 2017). Change in sources and method for

CNC preparation effects emulsion stabilizing features. In this regard, Liang & Vriesekoop (2014) reported a study where they prepare CNC from corncob cellulose *via* ammonium persulfate hydrolysis and studied it for emulsion stabilization. The experimental data showed the stability of emulsion with rise in temperature. However, lower pH and increase in salt concentration reduced the emulsion stability, which was overcome with increase in CNC concentration (Liang & Vriesekoop, 2014). The pre-treatment of CNC with a combination of polymers has effect on the stability of emulsion. The study showed that CNC treatment with MC and HEC has higher emulsion stability as compared to CNC alone. This study revealed that the above-mentioned combination of cellulose has higher stability of emulsion at temperature higher than 70°C and no coalescence was observed (Hu *et al.*, 2015a). The lower concentration of CNF has good emulsion stability as compared to conventional stabilizers such as surfactant. This was proved by preparing emulsion using CNF with model drugs ibuprofen and naproxen, which was stabilized by the hydrobhobin protein. The CNF acted to increase the viscosity and stabilized the emulsion and carried out protein encapsulation. Thus, studies showed that CNF-based emulsions are more stable as compared to conventional emulsion stabilizers (Paukkonen *et al.*, 2017b).

In addition to the source and purity of CNCs, the particle size is also a matter of interest in emulsion stability (Matos *et al.*, 2017). The effects of particle size of CNC on emulsion stability have higher impact. In this regard, Wang *et al.* (2016) prepared different size CNCs and studied their effect on emulsion stability. The CNCs were in the range of 178.2–261.8 nm. The aqueous dispersion with CNCs showed that particles having size of 1–10 μm have highly stable emulsions as compared to particles with large diameter. It was also observed that increase in the concentration of CNCs, addition of salts, and presence of casein greatly affected the emulsion stability (Wang *et al.*, 2016). The presence of surfactant in combination with CNC has an effect on the emulsion stability. In this regard Hu *et al.* (2015) reported the application of didecyldimethylammonium bromide (DAB) and CTAB in combination with CNCs. It was observed that addition of surfactant affected the emulsion stability, changed the wetting ability of CNCs, and controlled the internal phase of CNC emulsion. It was observed that the addition of surfactant to CNCs emulsion for stability has the potential for stabilizing the pharmaceutical formulations of drugs (Hu *et al.*, 2015b). In general, it was concluded that CNCs properties, particle size, surface charges, and addition of salts are the basic factors that have strong impact on the pharmaceutical emulsion stability. The nanospheres of CNF and polystyrene were prepared without the addition of any surfactant for Pickering emulsion stability. The results of SEM characterization showed that CNF-polystyrene beads have the ability to expand in emulsion and attract more water droplets, which has the potential to make the emulsion highly stabilized (Nikfarjam *et al.*, 2015).

3. Summary and Future Perspectives

Nanocellulose is an interesting material that has received immense consideration in last few decades due to its unique structural, physicochemical, mechanical, thermal, optical, and biological features. Owing to these features, it has received immense consideration for different applications in paper industry, food industry, biomedical devices and wound dressing, cosmetics, regenerative medicines, stabilizing agent in dispersion, petrochemical industry, and environmental remediation. The presence of free hydroxyl (OH) groups on its surface allows its surface to be modified by different chemical and physical methods to bestow it with unique functional properties in drug delivery applications. It offers several advantages in drug delivery applications such as non-toxicity, stability, excess availability, biodegradability, biocompatibility, ease of processing, and non-allergic nature. Different forms of cellulose including plant cellulose, bacterial nanocellulose, cellulose nanocrystals, microfibrillated cellulose, nanocrystalline cellulose (NCC), and cellulose nanowhiskers are used in various pharmaceutical applications where these play roles of filler, stabilizer, thickening agent, coating agent, in the preparation of transdermal patches, controlled drug delivery systems, delivery of proteins, and delivery of macromolecular pro-drugs.

Currently, the application cosmetic product for beautification is largely expanded. To this end, the application of therapeutic agents and anti-oxidants can potentially shift the trend towards the nanocellulose-based cosmetics. Similarly, the application of cellulose and its derivatives in controlled delivery of proteins can be further expanded to higher therapeutic value to explore their loading and release abilities and potential therapeutic applications. In addition, there is plenty of potential available for other materials to combine with cellulose to explore their potential for drug delivery applications. Currently, most studies lack *in vivo* experiments using animal models to study the possible irritation of skin to BC-based drug delivery systems and the bioavailability of drug from the drug-loaded BC with and without the addition of penetration enhancer(s). Further studies are needed to explore the nanocellulose-based drug delivery systems with body stimuli responsiveness such as fever, micro-environment of the tumor, blood glucose level, etc. The concept of pro-drug delivery could be extended to other drugs with *in vitro* experiments using cellulose degrading enzymes and be tested in *in vivo* animal and human studies. Being biocompatible, nanocellulose can be used for implantable devices that need therapy for month to years.

References

Abdalla, A.M.E., Xiao, L., Ullah, M.W., Yu, M., Ouyang, C., and Yang, G. (2018). Current challenges of cancer anti-angiogenic therapy and the promise of nanotherapeutics. *Theranostics*, 8(2): 533–548.

Abraham, E., Weber, D.E., Sharon, S., Lapidot, S., and Shoseyov, O. (2017). Multifunctional cellulosic scaffolds from modified cellulose nanocrystals. *ACS Appl. Mater. Int.*, 9(3): 2010–2015.

Adel, A.M., El-Gendy, A.A., Diab, M.A., Abou-Zeid, R.E., El-Zawawy, W.K., and Dufresne, A. (2016). Microfibrillated cellulose from agricultural residues. Part I: Papermaking application. *Ind. Crops Prod.*, 93: 161–174.

Ahmad, N., Amin, M.C.I.M., Mahali, S.M., Ismail, I., and Chuang, V.T.G. (2014). Biocompatible and mucoadhesive bacterial cellulose-g-poly (acrylic acid) hydrogels for oral protein delivery. *Mol. Pharm.*, 11: 4130–4142.

Ai, H., Jones, S.A., Lvov, Y.M. (2003). Biomedical applications of electrostatic layer-by-layer nano-assembly of polymers, enzymes, and nanoparticles. *Cell. Biochem. Biophys.*, 39(1): 23–43.

Aljohani, W., Ullah, M.W., Zhang, X., and Yang, G. (2018). Bioprinting and its applications in tissue engineering and regenerative medicine. *Int. J. Biol. Macromol.*, 107: 261–275.

Almeida, I.F., Pereira, T., Silva, N.H., Gomes, F.P., Silvestre, A.J., Freire, C.S., Sousa, Lobo, J,M., and Costa, P.C. (2014). Bacterial cellulose membranes as drug delivery systems: An *in vivo* skin compatibility study. *Eur. J. Pharm. Biopharm.*, 86: 332–336.

Amin, M.C.I.M., Abadi, A.G., Ahmad, N., Katas, H., and Jamal, J.A. (2012a). Bacterial cellulose film coating as drug delivery system: Physicochemical, thermal and drug release properties. *Sains Malays*, 41: 561–568.

Amin, M.C.I.M., Ahmad, N., Halib, N., and Ahmad, I. (2012b). Synthesis and characterisation of thermo-and pH-responsive bacterial cellulose/acrylic acid hydrogels for drug delivery. *Carbohydr. Polym.*, 88: 465–473.

Amin, M.C.I.M., Ahmad, N., Pandey, M., and Xin, C.J. (2014). Stimuli-responsive bacterial cellulose-g-poly (acrylic acid-co-acrylamide) hydrogels for oral controlled release drug delivery. *Drug. Dev. Ind. Pharm.*, 40: 1340–1349.

Aramwit, P. and Bang, N. (2014). The characteristics of bacterial nanocellulose gel releasing silk sericin for facial treatment. *BMC Biotechnol.*, 14: 104.

Badshah, M., Ullah, H., Khan, A.R., Khan, S., Park, J.K., and Khan, T. (2018). Surface modification and evaluation of bacterial cellulose for drug delivery. *Int. J. Biol. Macromol.*, 113: 526–533.

Badshah, M., Ullah, H., Khan, S.A., Park, J.K., and Khan, T. (2017). Preparation, characterisation and *in vitro* evaluation of bacterial cellulose matrices for oral drug delivery. *Cellulose*, 24(11): 5041–5052.

Barud, H.O., Barud, H.D.S., Cavicchioli, M., do Amaral, T.S., de Oliveira Junior, O.B., Santos, D.M., Petersen, A.L.D.O.A., Celes, F., Borges, V.M., and de Oliveira, P.F. (2015). Preparation and characterisation of a bacterial cellulose/silk fibroin sponge scaffold for tissue regeneration. *Carbohydr. Polym.*, 128: 41–51.

Bhandari, J., Mishra, H., Mishra, P.K., Wimmer, R., Ahmad, F.J., and Talegaonkar, S. (2017). Cellulose nanofiber aerogel as a promising biomaterial for customized oral drug delivery. *Int. J. Nanomed.*, 12: 2021.

Bideau, B., Bras, J., Saini, S., Daneault, C., and Loranger, E. (2016). Mechanical and antibacterial properties of a nanocellulose-polypyrrole multilayer composite. *Mater. Sci. Eng. C*, 69: 977–984.

Bodhibukkana, C., Srichana, T., Kaewnopparat, S., Tangthong, N., Bouking, P., Martin, G.P., and Suedee, R. (2006). Composite membrane of bacterially-derived cellulose and molecularly imprinted polymer for use as a transdermal enantioselective controlled-release system of racemic propranolol. *J. Control. Rel.*, 113: 43–56.

Bodin, A., Concaro, S., Brittberg, M., and Gatenholm, P. (2007). Bacterial cellulose as a potential meniscus implant. *J. Tissue. Eng. Regen. Med.*, 1: 406–408.

Boufi, S., González, I., Delgado-Aguilar, M., Tarrès, Q., Pèlach, M.À., and Mutjé, P. (2016). Nanofibrillated cellulose as an additive in papermaking process: A review. *Carbohydr. Polym.*, 154: 151–166.

Brown, A.J. (1886a). XIX. — The chemical action of pure cultivations of *bacterium aceti*. *J. Chem. Soc. Trans.*, 49: 172–187.

Brown, A.J. (1886b). XLIII. — On an acetic ferment which forms cellulose. *J. Chem. Soc. Trans.*, 49: 432–439.

Bruno, B.J., Miller, G.D., and Lim, C.S. (2013). Basics and recent advances in peptide and protein drug delivery. *Ther. Deliv.*, 4: 1443–1467.

Butchosa, N., Brown, C., Larsson, P.T., Berglund, L.A., Bulone, V., and Zhou, Q. (2013). Nanocomposites of bacterial cellulose nanofibers and chitin nanocrystals: Fabrication, characterisation and bactericidal activity. *Green Chem.*, 15: 3404–3413.

Campano, C., Balea, A., Blanco, A., and Negro, C. (2015). Enhancement of the fermentation process and properties of bacterial cellulose: A review. *Cellulose*, 23: 1–35.

Carlsson, D.O., Hua, K., Forsgren, J., and Mihranyan, A. (2014). Aspirin degradation in surface-charged TEMPO-oxidized mesoporous crystalline nanocellulose. *Int. J. Pharm.*, 461(1): 74–81.

Chang, C., Duan, B., Cai, J., and Zhang, L. (2010). Superabsorbent hydrogels based on cellulose for smart swelling and controllable delivery. *Eur. Polym. J.*, 46(1): 92–100.

Chawla, P.R., Bajaj, I.B., Survase, S.A., and Singhal, R.S. (2009). Microbial cellulose: Fermentative production and applications. *Food Technol. Biotech.*, 47: 107–124.

Chen, C., Chen, X., Zhang, H., Zhang, Q., Wang, L., Li, C., Dai, B., Yang, J., Liu, J., and Sun, D. (2017). Electrically-responsive core-shell hybrid microfibers for controlled drug release and cell culture. *Acta. Biomat.*, 55: 434–442.

Chen, X., Chen, Z., Zhu, J., Xu, C., Yan, W., and Yao, C. (2011). A novel H_2O_2 amperometric biosensor based on gold nanoparticles/self-doped polyaniline nanofibers. *Bioelectrochem.*, 82: 87–94.

Chen, Y.C., Ho, H.O., Liu, D.Z., Siow, W.S., and Sheu, M.T. (2015). Swelling/floating capability and drug release characterisations of gastroretentive drug delivery system based on a combination of hydroxyethyl cellulose and sodium carboxymethyl cellulose. *PloS One*, 10 (2015) e0116914.

Cheng, M., Qin, Z., Hu, S., Dong, S., Ren, Z., and Yu, H.Y. (2017). Achieving long–term sustained drug delivery for electrospun biopolyester nanofibrous membranes by introducing cellulose nanocrystals. *ACS Biomater. Sci. Eng.*, 3(8): 1666–1676.

Czaja, W., Krystynowicz, A., Bielecki, S., and Brown, R.M. (2006). Microbial cellulose — the natural power to heal wounds. *Biomaterials*, 27: 145–151.

Czaja, W., Romanovicz, D., and Brown, R.M. (2004). Structural investigations of microbial cellulose produced in stationary and agitated culture. *Cellulose*, 11: 403–411.

Czaja, W.K., Young, D.J., Kawecki, M., and Brown, R.M. (2007). The future prospects of microbial cellulose in biomedical applications. *Biomacromol.*, 8: 1–12.

De Carvalho, R.A., Veronese, G., Carvalho, A.J.F., Barbu, E., Amaral, A.C., Trovatti, E. (2016). The potential of TEMPO-oxidized nanofibrillar cellulose beads for cell delivery applications. *Cellulose*, 23 (6): 3399–3405.

de Castro, D.O., Bras, J., Gandini, A., and Belgacem, N. (2016). Surface grafting of cellulose nanocrystals with natural antimicrobial rosin mixture using a green process. *Carbohydr. Polym.*, 137: 1–8.

De Groot, A.S. and Martin, W. (2009). Reducing risk, improving outcomes: Bioengineering less immunogenic protein therapeutics. *Clin. Immunol.*, 131: 189–201.

Debotton, N. and Dahan, A. (2017). Applications of polymers as pharmaceutical excipients in solid oral dosage forms. *Med. Res. Rev.*, 37(1): 52–97.

Demircan, D. and Zhang, B. (2017). Facile synthesis of novel soluble cellulose-grafted hyperbranched polymers as potential natural antimicrobial materials. *Carbohydr. Polym.*, 157: 1913–1921.

Di, Z., Shi, Z., Ullah, M.W., Li, S., and Yang, G. (2017). A transparent wound dressing based on bacterial cellulose whisker and poly (2-hydroxyethyl methacrylate). *Int. J. Biol. Macromol.*, 105: 638–644.

Dobre, L.-M., Stoica-Guzun, A., Stroescu, M., Jipa, I., Dobre, T., Ferdeş, M., and Ciumpiliac, Ş. (2012). Modelling of sorbic acid diffusion through bacterial cellulose-based antimicrobial films. *Chem. Pap.*, 66: 144–151.

Dobre, M.L. and Stoica-Guzun, A. (2013). Antimicrobial Ag-polyvinyl alcohol-bacterial cellulose composite films. *J. Biobased. Mater. Bioenergy*, 7: 157–162.

Domingues, R.M., Silva, M., Gershovich, P., Betta, S., Babo, P., Caridade, S.G., Mano, J.F., Motta, A., Reis, R.L., Gomes, M.E. (2015). Development of injectable hyaluronic acid/cellulose nanocrystals bionanocomposite hydrogels for tissue engineering applications. *Bioconjugate Chem.*, 26: 1571–1581.

Dufresne, A. (2013). *Nanocellulose: From Nature to High Performance Tailored Materials*. Walter de Gruyter, Berlin/Boston.

Duran, N., Lemes, P.A., and Seabra, B.A. (2012). Review of cellulose nanocrystals patents: Preparation, composites and general applications. *Recent Patents Nanotechnol.*, 6(1): 16–28.

Dutta, S., Samanta, P., and Dhara, D. (2016). Temperature, pH and redox responsive cellulose based hydrogels for protein delivery. *Int. J. Biol. Macromol.*, 87: 92–100.

Ellis, B. and Smith, R. (2008). Polymers: A Property Database. CRC Press. *Boca Raton*.

Evdokimova, O.L., Svensson, F.G., Agafonov, A.V., Håkansson, S., Seisenbaeva, G.A., and Kessler, V.G. (2018). Hybrid drug delivery patches based on spherical cellulose nanocrystals and colloid titania — synthesis and antibacterial properties. *Nanomaterials*, 8(4): 228.

Farooq, U., Ullah, M.W., Yang, Q., Aziz, A., Xu, J., Zhou, L., and Wang, S. (2020). High-density phage particles immobilization in surface-modified bacterial cellulose for ultra-sensitive and selective electrochemical detection of *Staphylococcus aureus*. *Biosens. Bioelectron.*, 157: 112163.

Foong, C.Y., Hamzah, M.S.A., Razak, S.I.A., Saidin, S., and Nayan, N.H.M. (2018). Influence of Poly (lactic acid) layer on the physical and antibacterial properties of dry bacterial cellulose sheet for potential acute wound healing materials. *Fibers Polym.*, 19(2): 263–271.

Fu, L., Zhang, Y., Li, C., Wu, Z., Zhuo, Q., Huang, X., Qiu, G., Zhou, P., and Yang, G. (2012). Skin tissue repair materials from bacterial cellulose by a multilayer fermentation method. *J. Mater. Chem.*, 22: 12349–12357.

Galkina, O.L., Ivanov, V.K., Agafonov, A.V., Seisenbaeva, G.A., and Kessler, V.G. (2015). Cellulose nanofiber–titania nanocomposites as potential drug delivery systems for dermal applications. *J. Mat. Chem. B*, 3: 1688–1698.

George, J. and Sabapathi, S.N. (2015). Cellulose nanocrystals: Synthesis, functional properties, and applications. *Nanotechnol Sci. Appl.*, 8: 45.

Gestranius, M., Stenius, P., Kontturi, E., Sjöblom, J., and Tammelin, T. (2017). Phase behaviour and droplet size of oil-in-water Pickering emulsions stabilised with plant-derived nanocellulosic materials. *Coll. Surf. A: Physicochem Eng. Aspects.*, 519: 60–70.

Gong, X., Wang, Y., and Chen, L. (2017). Enhanced emulsifying properties of wood-based cellulose nanocrystals as Pickering emulsion stabilizer. *Carbohydr. Polym.*, 169: 295–303.

Haemmerle, G., Signer, M., and Mittlboeck, M. (2012). Comparison of PHMB-containing dressing and silver dressings in patients with critically colonised or locally infected wounds. *J. Wound. Care.*, 21: 13–19.

Harris, J.M. and Chess, R.B. (2003). Effect of pegylation on pharmaceuticals. *Nat. Rev. Drug Discovery*, 2: 214–221.

Hasan, N., Biak, D.R.A., and Kamarudin, S. (2012). Application of bacterial cellulose (BC) in natural facial scrub. *IJASEIT*, 2: 1–4.

Hassanpour, A., Asghari, S., and Lakouraj, M.M. (2017). Synthesis, characterisation and antibacterial evaluation of nanofibrillated cellulose grafted by a novel quinolinium silane salt. *RSC Adv.*, 7(39): 23907–23916.

Heath, B.P., Coffindaffer, T.W., Kyte III, K.E., Smith III, E.D., and McConaughy, S.D. (2012). Personal cleansing compositions comprising a bacterial cellulose network and cationic polymer. US Patent, US 8097574 B2.

Helenius, G., Bäckdahl, H., Bodin, A., Nannmark, U., Gatenholm, P., and Risberg, B. (2006). *In vivo* biocompatibility of bacterial cellulose. *J. Biomed. Mater. Res. A,* 76: 431–438.

Hou, Y., Shao, J., Fu, Q., Li, J., Sun, J., and He, Z. (2017). Spray-dried nanocrystals for a highly hydrophobic drug: Increased drug loading, enhanced redispersity, and improved oral bioavailability. *Int. J. Pharm.*, 516(1–2): 372–379.

Hu, Z., Ballinger, S., Pelton, R., and Cranston, E.D. (2015). Surfactant-enhanced cellulose nanocrystal Pickering emulsions. *J. Coll. Int. Sci.*, 439: 139–148.

Hu, Z., Patten, T., Pelton, R., and Cranston, E.D. (2015). Synergistic stabilisation of emulsions and emulsion gels with water-soluble polymers and cellulose nanocrystals. *ACS Sust. Chem. Eng.*, 3(5): 1023–1031.

Huang, L., Chen, X., Nguyen, T.X., Tang, H., Zhang, L., and Yang, G. (2013). Nano-cellulose 3D-networks as controlled-release drug carriers. *J. Mater. Chem. B*, 1: 2976–2984.

Hubbell, J.A. (1995). Biomaterials in tissue engineering. *Nature Biotechnol.*, 13: 565–576.

Jackson, J.K., Letchford, K., Wasserman, B.Z., Ye, L., Hamad, W.Y., and Burt, H.M. (2011). The use of nanocrystalline cellulose for the binding and controlled release of drugs. *Int. J. Nanomedicine*, 6: 321.

Janpetch, N., Saito, N., and Rujiravanit, R. (2016). Fabrication of bacterial cellulose-ZnO composite *via* solution plasma process for antibacterial applications. *Carbohydr. Polym.*, 148: 335–344.

Jay, J.M., Loessner, M.J., and Golden, D.A. (2008). *Modern Food Microbiology*. Springer Science & Business Media. New York.

Jeong, D., ki Kim, H., Jeong, J.P., Dindulkar, S.D., Cho, E., Yang, Y.H., and Jung, S. (2016). Cyclosophoraose/cellulose hydrogels as an efficient delivery system for galangin, a hydrophobic antibacterial drug. *Cellulose*, 23: 2609–2625.

Jipa, I.M., Stoica-Guzun, A., and Stroescu, M. (2012). Controlled release of sorbic acid from bacterial cellulose based mono and multilayer antimicrobial films. LWT-Food *Sci. Technol.*, 47: 400–406.

Kaliyaperumal, A. and Jing, S. (2009). Immunogenicity assessment of therapeutic proteins and peptides. *Curr. Pharm. Biotechnol.*, 10: 352–358.

Kaplan, E., Ince, T., Yorulmaz, E., Yener, F., Harputlu, E., and Laçin, N.T. (2014). Controlled delivery of ampicillin and gentamycin from cellulose hydrogels and their antibacterial efficiency. *J. Biomater. Tissue Eng.*, 4: 543–549.

Khalid, A., Khan, R., Ul-Islam, M., Khan, T., and Wahid, F. (2017a). Bacterial cellulose-zinc oxide nanocomposites as a novel dressing system for burn wounds. *Carbohydr. Polym.*, 164: 214–221.

Khalid, A., Ullah, H., Ul-Islam, M., Khan, R., Khan, S., Ahmad, F., and Khan, T. (2017b). Bacterial cellulose–TiO_2 nanocomposites promote healing and tissue regeneration in burn mice model. *RSC Adv.*, 7(75): 47662–47668.

Khamrai, M., Banerjee, S.L., and Kundu, P.P. (2017). Modified bacterial cellulose based self-healable polyeloctrolyte film for wound dressing application. *Carbohydr. Polym.*, 174: 580–590.

Khan, S., Ul-Islam, M., Khattak, W.A., Ullah, M.W., and Park, J.K. (2015). Bacterial cellulose-titanium dioxide nanocomposites: Nanostructural characteristics, antibacterial mechanism, and biocompatibility. *Cellulose*, 22: 565–579.

Khan, S., Ul-Islam, M., Ikram, M., Islam, S.U., Ullah, M.W., Israr, M., Jang, J.H., Yoon, S., and Park, J.K. (2018a). Preparation and structural characterization of surface modified microporous bacterial cellulose scaffolds: A potential material for skin regeneration applications *in vitro* and *in vivo*. *Int. J. Biol. Macromol.*, 117: 1200–1210.

Khan, S., Ul-Islam, M., Ullah, M.W., Israr, I., Jang, J.H., and Park, J.K. (2018b). Nano-gold assisted highly conducting and biocompatible bacterial cellulose-PEDOT:PSS films for biology-device interface applications. *Int. J. Biol. Macromol.*, 107: 865–873.

Khan, T., Park, J.K., and Kwon, J.-H. (2007). Functional biopolymers produced by biochemical technology considering applications in food engineering. *Korean J. Chem. Eng.*, 24: 816–826.

Kim, J.H., Park, S., Kim, H., Kim, H.J., Yang, Y.H., Kim, Y.H., Jung, S.K., Kan, E., and Lee, S.H. (2017). Alginate/bacterial cellulose nanocomposite beads prepared using Gluconacetobacter xylinus and their application in lipase immobilisation. *Carbohydr. Polym.*, 157: 137–145.

Kim, Y., Ullah, M.W., Ul-Islam, M., Khan, S., Jang, J.H., and Park, J.K. (2019). Self-assembly of bio-cellulose nanofibrils through intermediate phase in a cell-free enzyme system. *Biochem. Eng. J.*, 142: 135–144.

Kiprono, S.J., Ullah, M.W., and Yang, G. (2018). Surface engineering of microbial cells: Strategies and applications. *Eng. Sci.*, 1: 33–45.

Klemm, D., Heublein, B., Fink, H.P., and Bohn, A. (2005). Cellulose: Fascinating biopolymer and sustainable raw material. *Angew. Chem. Int. Edit.*, 44: 3358–3393.

Klemm, D., Schumann, D., Kramer, F., Heßler, N., Hornung, M., Schmauder, H.-P., Marsch, S. (2006). Nanocelluloses as innovative polymers in research and application. *Adv. Polym. Sci.*, 295: 49–96.

Klemm, D., Schumann, D., Udhardt, U., and Marsch, S. (2001). Bacterial synthesized cellulose — Artificial blood vessels for microsurgery. *Prog. Polym. Sci.*, 26: 1561–1603.

Kolakovic, R., Laaksonen, T., Peltonen, L., Laukkanen, A., and Hirvonen, J. (2012a). Spray-dried nanofibrillar cellulose microparticles for sustained drug release. *Int. J. Pharm.*, 430(1–2): 47–55.

Kolakovic, R., Peltonen, L., Laaksonen, T., Putkisto, K., Laukkanen, A., and Hirvonen, J. (2011). Spray-dried cellulose nanofibers as novel tablet excipient. *AAPS Pharm. Sci. Tech.*, 12(4): 1366–1373.

Kolakovic, R., Peltonen, L., Laukkanen, A., Hirvonen, J., and Laaksonen, T. (2012b). Nanofibrillar cellulose films for controlled drug delivery. *Eur. J. Pharm. Biopharm.*, 82: 308–315.

Koohi, M.K., Hejazy, M., Asadi, F., Asadian., P. (2011). Assessment of dermal exposure and histopathologic changes of different sized nano-silver in healthy adult rabbits. *J. Phys. Conf. Ser.*, 304: 012028.

Korani, M., Rezayat, S.M., and Bidgoli, S.A. (2013). Sub-chronic dermal toxicity of silver nanoparticles in guinea pig: Special emphasis to heart, bone and kidney toxicities. *Iran. J. Pharm. Res.*, 12: 511–519.

Krystynowicz, A., Czaja, W., Wiktorowska-Jezierska, A., Gonçalves-Miśkiewicz, M., Turkiewicz, M., and Bielecki, S. (2002). Factors affecting the yield and properties of bacterial cellulose. *J. Ind. Microbiol. Biotechnol.*, 29: 189–195.

Lamboni, L., Li, Y., Liu, J., and Yang, G. (2016). Silk sericin-functionalized bacterial cellulose as a potential wound-healing biomaterial. *Biomacromolecules*, 17(9): 3076–3084.

Laurén, P., Paukkonen, H., Lipiäinen, T., Dong, Y., Oksanen, T., Räikkönen, H., Ehlers, H., Laaksonen, P., Yliperttula, M., and Laaksonen, T. (2018). Pectin and mucin enhance the bioadhesion of drug loaded nanofibrillated cellulose films. *Pharm. Res.* 35(7): 145.

Lee, C.K., Hsu, K.C., Cho, J.C., Kim, Y.J., Han, S.H. (2011). Cosmetic bio-cellulose mask pack sheet and method for manufacturing same. US Patent, US 20130244977 A1.

Levinson, D.J. and Glonek, T. (2010). Microbial cellulose contact lens. US Patent, US 7832857 B2.

Lewis, L., Derakhshandeh, M., Hatzikiriakos, S.G., Hamad, W.Y., and MacLachlan, M.J. (2016). Hydrothermal gelation of aqueous cellulose nanocrystal suspensions. *Biomacromolecules*, 17(8): 2747–2754.

Li, S., Jasim, A., Zhao, W., Fu, L., Ullah, M.W., Shi, Z., and Yang, G. (2018). Fabrication of pH-electroactive bacterial cellulose/polyaniline hydrogel for the development of a controlled drug release system, *ES Mater. Manuf.*, 41–49.

Li, Y., Tian, Y., Zheng, W., Feng, Y., Huang, R., Shao, J., Tang, R., Wang, P., Jia, Y., Zhang, J., and Zheng, W. (2017). Composites of bacterial cellulose and small molecule-decorated gold nanoparticles for treating gram-negative bacteria-infected wounds. *Small*, 13(27): 1–10.

Liang, H. and Vriesekoop, F. (2014). Preparation and stabilisation of d-limonene Pickering emulsions by cellulose nanocrystals. *Carbohydr. Polym.*, 112 695–700.

Lin, N. and Dufresne, A. (2013). Supramolecular hydrogels from *in situ* host–guest inclusion between chemically modified cellulose nanocrystals and cyclodextrin. *Biomacromolecules*, 14: 871–880.

Lin, S.P., Calvar, I.L., Catchmark, J.M., Liu, J.R., Demirci, A., and Cheng, K.C. (2013). Biosynthesis, production and applications of bacterial cellulose. *Cellulose*, 20: 2191–2219.

Lin, S.P., Hsieh, S.C., Chen, K.I., Demirci, A., and Cheng, K.C. (2014). Semi-continuous bacterial cellulose production in a rotating disk bioreactor and its materials properties analysis. *Cellulose*, 21: 835–44.

Lin, S.P., Liu, C.T., Hsu, K.D., Hung, Y.T., Shih, T.Y., and Cheng, K.C. (2016). Production of bacterial cellulose with various additives in a PCS rotating disk bioreactor and its material property analysis. *Cellulose*, 3: 367–377.

Lin, Y.-C., Wey, Y.-C., Lee, M.-L., and Lin, P.-C. (2015). Cosmetic composition containing fragments of bacterial cellulose film and method for manufacturing thereof. US Patent, US 20150216784 A1.

Liu, H., Rong, L., Wang, B., Xie, R., Sui, X., Xu, H., Zhang, L., Zhong, Y., and Mao, Z. (2017). Facile fabrication of redox/pH dual stimuli responsive cellulose hydrogel. *Carbohydr. Polym.*, 176: 299–306.

Liu, Y., Sui, Y., Liu, C., Liu, C., Wu, M., Li, B., and Li, Y. (2018). A physically crosslinked polydopamine/nanocellulose hydrogel as potential versatile vehicles for drug delivery and wound healing. *Carbohydr. Polym.*, 188: 27–36.

Liyaskina, E., Revin, V., Paramonova, E., Nazarkina, M., Pestov, N., Revina, N., and Kolesnikova, S. (2017). Nanomaterials from bacterial cellulose for antimicrobial wound dressing. *Int. J. Phy. Conference Series*, 784(1): 012034. IOP Publishing.

Lu, W., Senapati, D., Wang, S., Tovmachenko, O., Singh, A.K., Yu, H., and Ray, P.C. (2010). Effect of surface coating on the toxicity of silver nanomaterials on human skin keratinocytes. *Chem. Phys. Lett.*, 487: 92–96.

Luan, J., Wu, J., Zheng, Y., Song, W., Wang, G., Guo, J., and Ding, X. (2012). Impregnation of silver sulfadiazine into bacterial cellulose for antimicrobial and biocompatible wound dressing. *Biomed. Mater.*, 7: 065006.

Luo, H., Ao, H., Li, G., Li, W., Xiong, G., Zhu, Y., and Wan, Y. (2017). Bacterial cellulose/graphene oxide nanocomposite as a novel drug delivery system. *Currt. Appl. Phys.*, 17(2): 249–254.

Luo, X., Liu, S., Zhou, J., and Zhang, L. (2009). *In situ* synthesis of Fe_3O_4/cellulose microspheres with magnetic-induced protein delivery. *J. Mater. Chem.*, (21): 3538–45.

Macuja, J.C.O., Ruedas, L.N., and España, R.C. (2015). Utilisation of cellulose from Luffa cylindrica fiber as binder in acetaminophen tablets. *Adv. Env. Chem.*, 2015: 1–8.

Mahfoudhi, N. and Boufi, S. (2017). Nanocellulose as a novel nanostructured adsorbent for environmental remediation: A review. *Cellulose*, 1–27.

Malik, N.N. (2008). Drug discovery: Past, present and future. *Drug Discovery Today*, 13: 909–912.

Manning, M.C., Chou, D.K., Murphy, B.M., Payne, R.W., and Katayama, D.S. (2010). Stability of protein pharmaceuticals: An update. *Pharm. Res.*, 27: 544–575.

Manning, M.C., Patel, K., and Borchardt, R.T. (1989). Stability of protein pharmaceuticals. *Pharmaceut Res.*, 6: 903–918.

Masruchin, N., Park, B.D., and Causin, V. (2018). Dual-responsive composite hydrogels based on TEMPO-oxidized cellulose nanofibril and poly (*N*-isopropylacrylamide) for model drug release. *Cellulose*, 25(1): 485–502.

Matos, M., Marefati, A., Bordes, R., Gutiérrez, G., and Rayner, M. (2017). Combined emulsifying capacity of polysaccharide particles of different size and shape. *Carbohydr. Polym.*, 169: 127–138.

McCarthy, R.R., Ullah, M.W., Booth, P., *et al.* (2019a). The use of bacterial polysaccharides in bioprinting. *Biotechnol. Adv.*, 37.

McCarthy, R.R., Ullah, M.W., Pei, E., and Yang, G. (2019b). Antimicrobial inks: The Anti-Infective applications of bioprinted bacterial polysaccharides. *Trends Biotechnol.*, 37: 1153–1155.

Mertaniemi, H., Escobedo-Lucea, C., Sanz-Garcia, A., Gandía, C., Mäkitie, A., Partanen, J., and Yliperttula, M. (2016). Human stem cell decorated nanocellulose threads for biomedical applications. *Biomaterials*, 82: 208–220.

Mikulcová, V., Bordes, R., and Kašpárková, V. (2016). On the preparation and antibacterial activity of emulsions stabilized with nanocellulose particles. *Food Hydrocoll.*, 61: 780–792.

Millon, L. and Wan, W. (2006). The polyvinyl alcohol–bacterial cellulose system as a new nanocomposite for biomedical applications. *J. Biomed. Mater. Res. B Appl. Biomater.*, 79: 245–253.

Mohanta, V., Madras, G., and Patil, S. (2014). Layer-by-layer assembled thin films and microcapsules of nanocrystalline cellulose for hydrophobic drug delivery. *ACS Appl. Mate. Int.*, 6(22): 20093–20101.

Mori, R., Nakai, T., Enomoto, K., Uchio, Y., and Yoshino, K. (2011). Increased antibiotic release from a bone cement containing bacterial cellulose. *Clin. Orthop. Relat. Res.*, 469: 600–606.

Müller, A., Ni, Z., Hessler, N., Wesarg, F., Müller, F.A., Kralisch, D., and Fischer, D. (2013). The biopolymer bacterial nanocellulose as drug delivery system: Investigation of drug loading and release using the model protein albumin. *J. Pharm. Sci.*, 102: 579–592.

Murphy, O. (2001). Non-polyol low-digestible carbohydrates: Food applications and functional benefits. *Br. J. Nutr.*, 85: S47–S53.

Napavichayanun, S., Yamdech, R., and Aramwit, P. (2016). The safety and efficacy of bacterial nanocellulose wound dressing incorporating sericin and polyhexamethylene biguanide: *In vitro*, *in vivo* and clinical studies. *Arch. Derm. Res.*, 308(2): 123–132.

Napavichayanun, S., Yamdech, R., and Aramwit, P. (2017). Development of bacterial cellulose incorporating silk sericin, polyhexamethylene biguanide, and glycerin with enhanced physical properties and antibacterial activities for wound dressing application. *Int. J. Polym. Mater. Polym. Biomat.*, 67(2): 61–67.

Nguyen, D.N., Ton, N.M.N., and Le, V.V.M. (2009). Optimisation of Saccharomyces cerevisiae immobilisation in bacterial cellulose by “adsorption-incubation” method. *Int. Food Res. J.*, 16: 59–64.

Nguyen, L.A., He, H., and Pham-Huy, C. (2006). Chiral drugs: An overview. *Int. J. Biomed. Sci.*, 2: 85–100.

Nikfarjam, N., Qazvini, N.T., and Deng, Y. (2015). Surfactant free Pickering emulsion polymerization of styrene in w/o/w system using cellulose nanofibrils. *Eur. Poly. J.*, 64: 179–188.

Nochos, A., Douroumis, D., and Bouropoulos, N. (2008). *In vitro* release of bovine serum albumin from alginate/HPMC hydrogel beads. *Carbohydr. Polym.*, 74(3): 451–7.

Numata, Y., Mazzarino, L., and Borsali, R. (2015). A slow-release system of bacterial cellulose gel and nanoparticles for hydrophobic active ingredients. *Int. J. Pharm.*, 486(1): 217–225.

Olyveira, G.M., Costa, L.M.M., and Basmaji, P. (2013). Physically modified bacterial cellulose as alternative routes for transdermal drug delivery. *J. Biomater. Tissue Eng.*, 3: 227–232.

Orasugh, J.T., Saha, N.R., Rana, D., Sarkar, G., Mollick, M.M.R., Chattoapadhyay, A., Mitra, B.C., Mondal, D., Ghosh, S.K., and Chattopadhyay, D. (2018). Jute cellulose nano-fibrils/hydroxypropylmethylcellulose nanocomposite: A novel material with potential for application in packaging and transdermal drug delivery system. *Ind. Crops. Prod.*, 112: 633–643.

Ougiya, H., Watanabe, K., Morinaga, Y., and Yoshinaga, F. (1997). Emulsion-stabilizing effect of bacterial cellulose. *Biosci. Biotechnol. Biochem.*, 61: 1541–1545.

Pacheco, G., Mello, C.V., Chiari-Andréo, B.G., Isaac, V.L.B., Ribeiro, S.J.L., Pecoraro, É., and Trovatti, E. (2017). Bacterial cellulose skin masks — Properties and sensory tests. *J. Cosmetic. Derm.*, 1–8.

Panday, R., Jung, A.J., Li, X., Adhikari, M., Ullah, M.W., and Yang, G. (2018). Amphiphilic core-shell nanoparticles: Synthesis, biophysical properties, and applications. *Colloid Surface B*, 172: 68–81.

Pandey, M., Amin, M.C.I.M., Ahmad, N., and Abeer, M.M. (2013). Rapid synthesis of superabsorbent smart-swelling bacterial cellulose/acrylamide-based hydrogels for drug delivery. *Int. J. Polym. Sci*., 905471.

Pandey, M., Mohamad, N., and Amin, M.C.I.M. (2014). Bacterial cellulose/acrylamide ph-sensitive smart hydrogel: Development, characterisation, and toxicity studies in ICR mice model. *Mol. Pharm.*, 11: 3596–3608.

Park, J.K., Khan, T., and Jung, J.Y. (2009). Bacterial Cellulose. In: Phillips GO, Williams PA (eds.). Handbook of Hydrocolloids. Woodhead publishing Ltd. UK, pp. 724–739.

Paukkonen, H., Kunnari, M., Laurén, P., Hakkarainen, T., Auvinen, V.V., Oksanen, T., Koivuniemi, R., Yliperttula, M., and Laaksonen, T. (2017a). Nanofibrillar cellulose hydrogels and reconstructed hydrogels as matrices for controlled drug release. *Int. J. Pharm.*, 532(1): 269–280.

Paukkonen, H., Ukkonen, A., Szilvay, G., Yliperttula, M., and Laaksonen, T. (2017b). Hydrophobin-nanofibrillated cellulose stabilized emulsions for encapsulation and release of BCS class II drugs. *Eur. J. Pharm. Sci.*, 100: 238–248.

Pavaloiu, R.-D., Stoica-Guzun, A., Stroescu, M., Jinga, S.I., and Dobre, T. (2014a). Composite films of poly(vinyl alcohol)–chitosan–bacterial cellulose for drug controlled release. *Int. J. Biol., Macromol.*, 68: 117–124.

Pavaloiu, R.-D., Stroescu, M., Parvulescu, O., and Dobre, T. (2014c). Composite hydrogels of bacterial cellulose-carboxymethyl cellulose for drug release. *Rev. Chim-Bucharest*, 65: 948–951.

Păvăloiu, R.-D., Stoica-Guzun, A., and Dobre, T. (2015). Swelling studies of composite hydrogels based on bacterial cellulose and gelatin. *UPB Sci. Bull. Ser. B*, 77: 54–62.

Peng, B.L., Dhar, N., Liu, H.L., and Tam, K.C. (2011). Chemistry and applications of nanocrystalline cellulose and its derivatives: A nanotechnology perspective. *The Canadian. J. Chem. Eng.*, 89(5): 1191–1206.

Peng, H., Ning, X., Wei, G., Wang, S., Dai, G., and Ju, A. (2018). The preparations of novel cellulose/phenylboronic acid composite intelligent bio-hydrogel and its glucose, pH-responsive behaviors. *Carbohydr. Polym.*, 195: 349–355.

Peppas, N.A., Wood, K.M., and Blanchette, J.O. (2004). Hydrogels for oral delivery of therapeutic proteins. *Expert. Opin. Biol. Ther.*, 4: 881–887.

Petersen, N. and Gatenholm, P. (2011). Bacterial cellulose-based materials and medical devices: Current state and perspectives. *Appl. Microbiol. Biotechnol.*, 91: 1277–1286.

Phisalaphong, M. and Chiaoprakobkij, N. (2012). Applications and products — nata de coco. In: Gama, M., Gatenholm, P., and Klemm, D. (eds.), Bacterial Nanocellulose: A sophisticated multifunctional material. *CRC Press, Florida*, 143–156.

Pinto, R.J., Daina, S., Sadocco, P., Pascoal Neto, C., and Trindade, T. (2013). Antibacterial activity of nanocomposites of copper and cellulose. *Biomed. Res. Int.*, 280512.

Prabhu, B.M., Ali, S.F., Murdock, R.C., Hussain, S.M., and Srivatsan, M. (2010). Copper nanoparticles exert size and concentration dependent toxicity on somatosensory neurons of rat. *Nanotoxicol.*, 2010: 150–60.

Prausnitz, M.R. and Langer, R. (2008). Transdermal drug delivery. *Nat. Biotechnol.* 26: 1261–1268.

Prausnitz, M.R., Mitragotri, S., and Langer, R. (2004). Current status and future potential of transdermal drug delivery. *Nat. Rev. Drug. Discovery*, 3: 115–124.

Qiu, K. and Netravali, A.N. (2014). A review of fabrication and applications of bacterial cellulose based nanocomposites. *Polym. Rev.*, 54: 598–626.

Qiu, Y., Qiu, L., Cui, J., and Wei, Q. (2016). Bacterial cellulose and bacterial cellulose-vaccarin membranes for wound healing. *Mater. Sci. Eng.*: C, 59: 303–309.

Radi, Z.A. and Khan, N.K. (2006). Effects of cyclooxygenase inhibition on the gastrointestinal tract. *Exp. Toxicol. Pathol.*, 58: 163–173.

Rahimian, K., Wen, Y., and Oh, J.K. (2015). Redox-responsive cellulose-based thermoresponsive grafted copolymers and *in situ* disulfide crosslinked nanogels. *Polymer*, 72: 387–394.

Ratner, B.D. and Bryant, S.J. (2004). Biomaterials: where we have been and where we are going. *Annu. Rev. Biomed. Eng.*, 6: 41–75.

Ray, P.C., Yu, H., and Fu, P.P. (2009). Toxicity and environmental risks of nanomaterials: challenges and future needs. *J. Environ. Sci. Health. A*, 27: 1–35.

Ross, P., Mayer, R., and Benziman, M. (1991). Cellulose biosynthesis and function in bacteria. *Microbiol. Rev.*, 55: 35–58.

Rouabhia, M., Asselin, J., Tazi, N., Messaddeq, Y., Levinson, D., and Zhang, Z. (2014). Production of biocompatible and antimicrobial bacterial cellulose polymers functionalized by RGDC grafting groups and gentamicin. *ACS Appl. Mater. Interfaces*, 6: 1439–1446.

Roy, S., Pal, K., Thakur, G., and Prabhakar, B. (2010). Synthesis of novel hydroxypropyl methyl cellulose acrylate — a novel superdisintegrating agent for pharmaceutical applications. *Mater. Manufac. Proc.*, 25(12): 1477–1481.

Saini, S., Belgacem, M.N., and Bras, J. (2017). Effect of variable aminoalkyl chains on chemical grafting of cellulose nanofiber and their antimicrobial activity. *Mater. Sci. Eng.*, C 75: 760–768.

Saini, S., Belgacem, N., Mendes, J., Elegir, G., and Bras, J. (2015). Contact antimicrobial surface obtained by chemical grafting of microfibrillated cellulose in aqueous solution limiting antibiotic release. *ACS App. Mater. Inter.*, 7(32): 18076–18085.

Samberg, M.E., Oldenburg, S.J., and Monteiro-Riviere, N.A. (2010). Evaluation of silver nanoparticle toxicity in skin *in vivo* and keratinocytes *in vitro*. *Environ. Health Perspect,* 118: 407–413.

Saxena, I.M. and Brown, Jr R.M. (2012). Biosynthesis of bacterial cellulose. In: Gama M, Gatenholm P, Klemm D (eds.) Bacterial nanocellulose: A sophisticated multifunctional material. Boca Raton: CRC Press 1–18.

Serafica, G., Mormino, R., Oster, G.A., Lentz, K.E., and Koehler, K.P. (2010). Microbial cellulose wound dressing for treating chronic wounds. US Patent, US 7704523 B2.

Shezad, O., Khan, S., Khan, T., and Park, J.K. (2010). Physicochemical and mechanical characterisation of bacterial cellulose produced with an excellent productivity in static conditions using a simple fed-batch cultivation strategy. *Carbohydr. Polym.*, 82: 173–180.

Shi, Q., Li, Y., Sun, J., Zhang, H., Chen, L., Chen, B., Yang, H., and Wang, Z. (2012). The osteogenesis of bacterial cellulose scaffold loaded with bone morphogenetic protein-2. *Biomaterials*, 33: 6644–6649.

Shi, X., Zheng, Y., Wang, G., Lin, Q., and Fan, J. (2014). pH-and electro-response characteristics of bacterial cellulose nanofiber/sodium alginate hybrid hydrogels for dual controlled drug delivery. *RSC Adv.*, 4: 47056–47065.

Shi, X., Zheng, Y., Zhang, W., Zhang, Z., and Peng, Y. (2013). A novel drug carrier based on functional modified nanofiber cellulose and the control release behavior. Fourth International Conference on Smart Materials and Nanotechnology in Engineering. International Society for Optics and Photonics, Gold Coast, Australia, 879304–879306.

Shi, Z., Zhang, Y., Phillips, G.O., and Yang, G. (2014). Utilisation of bacterial cellulose in food. *Food Hydrocoll.*, 35: 539–545.

Shi, Z., Gao, X., Ullah, M.W., Li, S., Wang, Q., and Yang, G. (2016). Electroconductive natural polymer-based hydrogels. *Biomaterials*, 111: 40–54.

Shoichet, M.S. (2009). Polymer scaffolds for biomaterials applications. *Macromolecules*, 43: 581–591.

Shokri, J. and Adibkia, K. (2013). Application of cellulose and cellulose derivatives in pharmaceutical industries. In: Cellulose-Medical, *Pharm. Elect. App.*, 48–66. InTech.

Shoukat, A., Wahid, F., Khan, T., *et al.* (2019). Titanium oxide-bacterial cellulose bioadsorbent for the removal of lead ions from aqueous solution. *Int. J. Biol. Macromol.*, 129: 965–971.

Siangsanoh, C., Ummartyotin, S., Sathirakul, K., Rojanapanthu, P., and Treesuppharat, W. (2018). Fabrication and characterisation of triple-responsive composite hydrogel for targeted and controlled drug delivery system. *J. Mol. Liq.*, 256: 90–99.

Silva, N.H., Rodrigues, A.F., Almeida, I.F., Costa, P.C., Rosado, C., Neto, C.P., Silvestre, A.J., and Freire, C.S. (2014). Bacterial cellulose membranes as transdermal delivery systems for diclofenac: *In vitro* dissolution and permeation studies. *Carbohydr. Polym.*, 106: 264–269.

Singla, R., Soni, S., Padwad, Y.S., Acharya, A., and Yadav, S.K. (2017b). Sustained delivery of BSA/HSA from biocompatible plant cellulose nanocrystals for *in vitro* cholesterol release from endothelial cells. *Int. J. Biol. Macromol.*, 104: 748–757.

Singla, R., Soni, S., Patial, V., Kulurkar, P.M., Kumari, A., Mahesh, S., Padwad, Y.S., and Yadav, S.K. (2017a). Cytocompatible anti-microbial dressings of S yzygium cumini cellulose nanocrystals decorated with silver nanoparticles accelerate acute and diabetic wound healing. *Sci. Rep.*, 7(1): 10457.

Siró, I. and Plackett, D. (2010). Microfibrillated cellulose and new nanocomposite materials: A review. *Cellulose*, 17: 459–494.

Solanki, A. and Thakore, S. (2015). Cellulose crosslinked pH-responsive polyurethanes for drug delivery: α-hydroxy acids as drug release modifiers. *Int. J. Biol. Macromol.*, 80: 683–691.

Stoica-Guzun, A., Stroescu, M., Tache, F., Zaharescu, T., and Grosu, E. (2007). Effect of electron beam irradiation on bacterial cellulose membranes used as transdermal drug delivery systems. *Nucl. Instrum. Methods. Phys. Res. B*, 265: 434–438.

Sulaeva, I., Henniges, U., Rosenau, T., and Potthast, A. (2015). Bacterial cellulose as a material for wound treatment: properties and modifications. A review. *Biotechnol. Adv.*, 33: 1547–71.

Svagan, A.J., Benjamins, J.W., Al-Ansari, Z., Shalom, D.B., Müllertz, A., Wågberg, L., and Löbmann, K. (2016). Solid cellulose nanofiber based foams–Towards facile design of sustained drug delivery systems. *J. Control. Rel.*, 244: 74–82.

Svensson, A., Nicklasson, E., Harrah, T., Panilaitis, B., Kaplan, D., Brittberg, M., Gatenholm P. (2005). Bacterial cellulose as a potential scaffold for tissue engineering of cartilage. *Biomaterials*, 26: 419–431.

Tam, T.T.M. and Huong, N.T. (2014). Optimisation of *Corynebacterium glutamicum* immobilisation process on bacterial cellulose carrier and its application for lysine fermentation. *IOSRJEN* 4: 33–38.

Tang, L., Persky, A.M., Hochhaus, G., and Meibohm, B. (2004). Pharmacokinetic aspects of biotechnology products. *J. Pharm. Sci.*, 93: 2184–2204.

Tiwari, G., Tiwari, R., Sriwastawa, B., Bhati, L., Pandey, S., Pandey, P., and Bannerjee, S.K. (2012). Drug delivery systems: An updated review. *Int. J. Pharm. Investig*, 2(1): 2.

Ton, N. and Le, V. (2011). Application of immobilized yeast in bacterial cellulose to the repeated batch fermentation in wine-making. *Int. Food Res. J.*, 18: 983–987.

Treesuppharat, W., Rojanapanthu, P., Siangsanoh, C., Manuspiya, H., and Ummartyotin, S. (2017). Synthesis and characterisation of bacterial cellulose and gelatin-based hydrogel composites for drug-delivery systems. *Biotechnol. Rep.*, 15: 84–91.

Trovatti, E., Freire, C.S., Pinto, P.C., Almeida, I.F., Costa, P., Silvestre, A.J., Neto, C.P., and Rosado, C. (2012). Bacterial cellulose membranes applied in topical and transdermal delivery of lidocaine hydrochloride and ibuprofen: *In vitro* diffusion studies. *Int. J. Pharm.*, 435: 83–87.

Ul-Islam, M., Khan, T., Khattak, W.A., and Park, J.K. (2013). Bacterial cellulose-MMTs nanoreinforced composite films: Novel wound dressing material with antibacterial properties. *Cellulose,* 20: 589–596.

Ul-Islam, M., Ullah, M.W., Khan, S., and Park, J.K. (2020). Production of bacterial cellulose from alternative cheap and waste resources: A step for cost reduction with positive environmental aspects. *Korean J. Chem. Eng.*, https://doi.org/10.1007/s11814-020-0524-3.

Ul-Islam, M., Ullah, M.W., Khan, S., Shah, N., Park, J.K. (2017). Strategies for cost-effective and enhanced production of bacterial cellulose. *Int. J. Biol. Macromol.*, 102: 1166–1173.

Ul-Islam, M., Khan, S., Ullah, M.W., and Park, J.K. (2019). Comparative study of plant and bacterial cellulose pellicles regenerated from dissolved states. *Int. J. Biol. Macromol.*, 137: 247–252.

Ullah, H., Badshah, M., Mäkilä, E., Salonen, J., Shahbazi, M.A., Santos, H.A., and Khan, T. (2017). Fabrication, characterisation and evaluation of bacterial cellulose-based capsule shells for oral drug delivery. *Cellulose*, 24(3): 1445–1454.

Ullah, H., Santos, H.A., and Khan, T. (2016a). Applications of bacterial cellulose in food, cosmetics and drug delivery. *Cellulose*, 23(4): 2291–2314.

Ullah, H., Wahid, F., Santos, H.A., and Khan, T. (2016b). Advances in biomedical and pharmaceutical applications of functional bacterial cellulose-based nanocomposites. *Carbohydr. Polym.*, 150: 330–352.

Ullah, M.W., Ul-Islam, M., Khan, S., Kim, Y., and Park, J.K. (2015). Innovative production of bio-cellulose using a cell-free system derived from a single cell line. *Carbohydr. Polym.*, 132: 286–294.

Ullah, M.W., Ul-Islam, M., Khan, S., Shah, N., and Park, J.K. (2017). Recent advancements in bioreactions of cellular and cell-free systems: A study of bacterial cellulose as a model. *Korean J. Chem. Eng.*, 34(6): 1591–1599.

Valo, H., Arola, S., Laaksonen, P., Torkkeli, M., Peltonen, L., Linder, M.B., and Laaksonen, T. (2013). Drug release from nanoparticles embedded in four different nanofibrillar cellulose aerogels. *Eur. J. Pharm. Sci.*, 50(1): 69–77.

Vashist, A. and Ahmad, S., (2013). Hydrogels: Smart materials for drug delivery. *Oriental J. Chem.*, 29: 861–870.

Vermonden, T., Censi, R., and Hennink, W.E. (2012). Hydrogels for protein delivery. *Chem. Rev.*, 112: 2853–2888.

Villanova, J.C.O., Ayres, E., Carvalho, S.M., Patrício, P.S., Pereira, F.V., and Oréfice, R.L. (2011). Pharmaceutical acrylic beads obtained by suspension polymerisation containing cellulose nanowhiskers as excipient for drug delivery. *Eur. J. Pharm. Sci.*, 42(4): 406–415.

Voisin, H., Bergström, L., Liu, P., and Mathew, A.P. (2017). Nanocellulose-based materials for water purification. *Nanomaterials*, 7(3): 57.

Walters, R.M., Mao, G., Gunn, E.T., and Hornby, S. (2012). Cleansing formulations that respect skin barrier integrity. *Dermatol. Res. Pract.*, 495917.

Wan, Y., Gao, C., Han, M., Liang, H., Ren, K., Wang, Y., and Luo, H. (2011). Preparation and characterisation of bacterial cellulose/heparin hybrid nanofiber for potential vascular tissue engineering scaffolds. *Polym. Adv. Tech.*, 22: 2643–2648.

Wang, L.P. and Wang, J.Y. (2014). Skin penetration of inorganic and metallic nanoparticles. *J. Shanghai Jiaotong Univ.* (*Sci.*) 19: 691–697.

Wang, T., Long, X., Cheng, Y., Liu, Z., and Yan, S. (2014). The potential toxicity of copper nanoparticles and copper sulphate on juvenile *Epinephelus coioides. Aquat. Toxicol.*, 152: 96–104.

Wang, W., Du, G., Li, C., Zhang, H., Long, Y., and Ni, Y. (2016). Preparation of cellulose nanocrystals from asparagus (Asparagus officinalis L.) and their applications to palm oil/water Pickering emulsion. *Carbohydr. Polym.* 151: 1–8.

Wang, W., Li, H.Y., Zhang, D.W., Jiang, J., Cui, Y.R., Qiu, S., Zhou, Y.L., and Zhang, X.X. (2010). Fabrication of bienzymatic glucose biosensor based on novel gold nanoparticles-bacteria cellulose nanofibers nanocomposite. *Electroanalysis*, 22: 2543–2550.

Wanling, Z., Zhe, L., Zerui, Z., Bihui, Z., Shiyan, C., Huaping, W., and Wen, Z. (2012). Preparation method for anti-virus bacteria cellulose protective. CN Patent, 102321261 A.

Wei, B., Yang, G., and Hong, F. (2011). Preparation and evaluation of a kind of bacterial cellulose dry films with antibacterial properties. *Carbohydr. Polym.*, 84: 533–538.

Wei, W., Li, J., Qi, X., Zhong, Y., Zuo, G., Pan, X., Su, T., Zhang, J., and Dong, W. (2017). Synthesis and characterisation of a multi-sensitive polysaccharide hydrogel for drug delivery. *Carbohydr. Polym.*, 177: 275–283.

Weishaupt, R., Heuberger, L., Siqueira, G., Gutt, B., Zimmermann, T., Maniura-Weber, K., Salentinig, S., and Faccio, G. (2018). Enhanced antimicrobial activity and structural transitions of a nanofibrillated cellulose-nisin biocomposite suspension. *ACS App. Mater. Inter.*, 10(23): 20170–20181.

Weishaupt, R., Siqueira, G., Schubert, M., Tingaut, P., Maniura-Weber, K., Zimmermann, T., Thöny-Meyer, L., Faccio, G., and Ihssen, J. (2015). TEMPO-oxidized nanofibrillated cellulose as a high density carrier for bioactive molecules. *Biomacromolecules*, 16(11): 3640–3650.

Wen, X., Zheng, Y., Wu, J., Yue, L., Wang, C., Luan, J., Wu, Z., and Wang, K. (2015). *In vitro* and *in vivo* investigation of bacterial cellulose dressing containing uniform silver sulfadiazine nanoparticles for burn wound healing. *Prog, Nat. Sci.*, 25: 197–203.

Wen, Y. and Oh, J.K. (2015). Intracellular delivery cellulose-based bionanogels with dual temperature/pH-response for cancer therapy. *Coll. Surf. B: Bio.*, 133: 246–253.

Wright, J.M. (2002). The double-edged sword of COX-2 selective NSAIDs. *CMAJ*, 167: 1131–1137.

Wu, S.-C. and Lia, Y.-K. (2008). Application of bacterial cellulose pellets in enzyme immobilisation. *J. Mol. Catal. B Enzym.*, 54: 103–108.

Wu, S.-C., Lia, Y.-K., and Ho, C.-Y. (2013). Glucoamylase immobilisation on bacterial cellulose using periodate oxidation method. *IJSE*, 3: 1–4.

Xin, R.H., Liang, G., and Zheng, J.F. (2017). Design and content determination of Genhuang dispersible tablet herbal formulation. *Pakistan J. Pharm. Sci.*, 30: 655–661.

Yadav, V., Paniliatis, B.J., Shi, H., Lee, K., Cebe, P., and Kaplan, D.L. (2010). Novel *in vivo*-degradable cellulose-chitin copolymer from metabolically engineered *Gluconacetobacter xylinus*. *Appl. Environ. Microbiol.*, 76: 6257–6265.

Yan, H., Chen, X., Song, H., Li, J., Feng, Y., Shi, Z., Wang, X., and Lin, Q., (2017). Synthesis of bacterial cellulose and bacterial cellulose nanocrystals for their applications in the stabilisation of olive oil Pickering emulsion. *Food Hydrocoll.*, 72: 127–135.

Yang, M. (2015). Stress and protein instability during formulation and fill/finish processes. *BioPharm Int.*, 28: 46–49.

Yang, Y., Fang, Z., Chen, X, Zhang, W., Xie, Y., Chen, Y., Liu, Z., Yuan, W. (2017). An overview of Pickering emulsions: Solid-particle materials, classification, morphology, and applications. *Front Pharmacol.*, 8: 287.

Yassin, S., Goodwin, D.J., Anderson, A., Sibik, J., Wilson, D.I., Gladden, L.F., and Zeitler, J.A. (2015). The disintegration process in microcrystalline cellulose based tablets, part 1: Influence of temperature, porosity and superdisintegrants. *J. Pharm. Sci.*, 104(10): 3440–3450.

Yoshino, A., Tabuchi, M., Uo, M., Tatsumi, H., Hideshima, K., Kondo, S., and Sekine, J. (2013). Applicability of bacterial cellulose as an alternative to paper points in endodontic treatment. *Acta. Biomater.*, 9: 6116–6122.

You, J., Cao, J., Zhao, Y., Zhang, L., Zhou, J., and Chen, Y. (2016). Improved mechanical properties and sustained release behavior of cationic cellulose nanocrystals reinforeced cationic cellulose injectable hydrogels. *Biomacromolecules*, 17(9): 2839–2848.

Zare-Akbari, Z., Farhadnejad, H., Furughi-Nia, B., Abedin, S., Yadollahi, M., and Khorsand-Ghayeni, M. (2016). PH-sensitive bionanocomposite hydrogel beads based on carboxymethyl cellulose/ZnO nanoparticle as drug carrier. *Int. J. Biol. Macromol.*, 93: 1317–1327.

Zhang, F., Wu, W., Zhang, X., Meng, X., Tong, G., and Deng, Y. (2016). Temperature-sensitive poly-NIPAm modified cellulose nanofibril cryogel microspheres for controlled drug release. *Cellulose*, 23(1): 415–425.

Zhang, L., Zhou, Y., Shi, G., Sang, X., and Ni, C. (2017). Preparations of hyperbranched polymer nano micelles and the pH/redox controlled drug release behaviors. *Mater. Sci. Eng. C*, 79: 116–122.

Zhang, T., Wang, W., Zhang, D., Zhang, X., Ma, Y., Zhou, Y., and Qi, L. (2010). Biotemplated synthesis of gold nanoparticle–bacteria cellulose nanofiber nanocomposites and their application in biosensing. *Adv. Funct. Mater.*, 20: 1152–1160.

Zhao, J., Lu, C., He, X., Zhang, X., Zhang, W., and Zhang, X. (2015). Polyethylenimine-grafted cellulose nanofibril aerogels as versatile vehicles for drug delivery. *ACS Appl. Mater. Int.*, 7(4): 2607–2615.

Zhong, C.Y. (2008). Bacterial cellulose gel face mask. CN Patent, 200610075040.8.

Zimmermann, K.A., LeBlanc, J.M., Sheets, K.T., Fox, R.W., and Gatenholm, P. (2011). Biomimetic design of a bacterial cellulose/hydroxyapatite nano-composite for bone healing applications. *Mater. Sci. Eng. C*, 31: 43–49.

Zubik, K., Singhsa, P., Wang, Y., Manuspiya, H., and Narain, R. (2017). Thermo-responsive poly (*n*-isopropylacrylamide)-cellulose nanocrystals hybrid hydrogels for wound dressing. *Polymers*, 9(4): 119.

https://doi.org/10.1142/9781786349477_0008

Chapter 8

Applications of Nanocellulose in the Food Industry

Qi Li and Shilin Liu*
College of Food Science and Technology, Huazhong Agricultural University, Wuhan, China
*slliu2013@mail.hzau.edu.cn

Abstract

Presently the entire world, especially the third-world countries, are greatly suffering from food shortages while the available food is sometimes of low quality due to poor processing and lack of storage facilities. To increase the shelf life and improve the quality, the food products need to be appropriately processed, packaged, and protected to minimize the risk of contamination and spoilage. The use of sustainable, low-cost, and lightweight biobased materials could be potential candidates used for food packaging to ensure the transportation of food to other regions and preserve it for extended duration. The idea of introducing different technologies to food science is to increase the sustainability, shelf-life, and nutritional quality of food. Currently, the development of a new generation of products with improved properties for food packaging is of utmost importance. Nanocellulose, a versatile material possessing unique structural, physico-chemical, mechanical, thermal, and biological features, has been vastly explored in food industry for various application. This chapter describes the merits and challenges of nanocellulose in food science as well as its functionalization through surface modification and *in situ* addition of reinforcement materials to meet the demand of using it as food packaging material and dietary aid, food additive, and emulsion stabilizer. It also discusses the development of active food packaging materials by using biodegradable polymers and antimicrobial materials as an innovative and attractive approach in food science to inhibit the growth of microorganism and retain the quality, freshness, and safety of food.

Keywords: Nanocellulose; food stability; food quality; active food packaging; food additive

1. Nanocellulose

Cellulose nanocrystal (CNC), cellulose nanofibril (CNF), and bacterial cellulose (BC) with differences in production modes and morphologies are the three different kinds of nanocellulose. CNC is a needle-like crystalline fibril with a length of 150–300 nm and a diameter of 5–10 nm. It is usually prepared through hydrolysis of bleached cellulose fibers with acid.[1] CNF containing both crystalline and amorphous domains is composed of nano-sized thin flexible fibril, and its production involves mechanical disintegration; in this process, the cell walls of the cellulose fiber are broken.[2] BC is made from certain types of bacteria and it contains nanosized fibrils with diameters around 20–50 nm. It is free from lignin and hemicellulose, and its water-holding capacity and degree of crystallinity are high.[3] These kinds of nanocellulose have large surface area and aspect ratio, and their crystallinity, mechanical properties, thermal stability and chemical functionality are different from those of the bulk materials. Therefore, nanocellulose is a versatile material with multifunctional properties suitable for a wide range of applications.[4–8] This chapter will be focused on the merits, benefits, and challenges of nanocellulose used in food science, as was shown in Fig. 1.

1.1 *Nanocellulose in Food Packaging*

Usually, food packages have basic functions including protection, communication, convenience, and containment; however, these functions are not sufficient as more and more people are taking care about the safety of the food-related products. To develop a new generation of products with

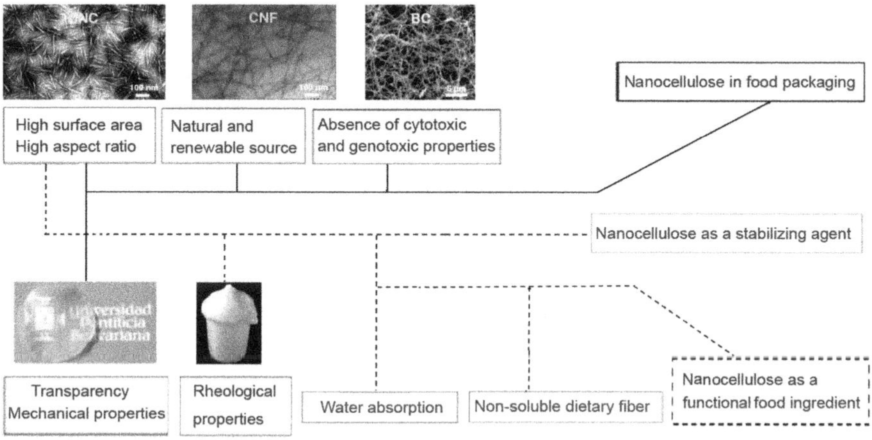

Fig. 1. Properties of nanocellulose and the applications in food science.

improved properties for food packaging technology is in great demand. The idea of the incorporation of different technologies in the food packaging field is to increase the sustainability, shelf life, and nutritional capacity of food. Bio-based materials that are light weight, low cost, sustainable have countless benefits when used in the food packaging field.[9–12] However, bio-based materials for food packaging often have poor mechanical properties.[13–15] The loading of nanocellulose as a reinforcing component in these materials can improve the properties. Nanocellulose has been applied as reinforcing phases through different pathways for the development of packaging materials, as is shown in Fig. 2, and the most studied effects of nanocellulose on polymer matrices is on the tensile properties.[16–18] This effect has often been ascribed to the merits of the strength and stiffness of the nanocellulose, and a mechanical percolation effect yielded by nanocellulose interactions through hydrogen bonding has also been developed in the research process. The percolation theory indicates that an amount of filler high enough to be dispersed within the matrix to form a continuous phase is required. It indicates that the strength and modulus are increased up to an optimum content around 5 wt% with the addition of nanocellulose, while the mechanical properties will be impaired by higher nanocellulose contents, owing to the aggregation of nanocellulose.[19] Therefore, it is challenging to achieve an appropriate percolation network in nanocellulose-filled composites for the aggregates of nanocellulose.

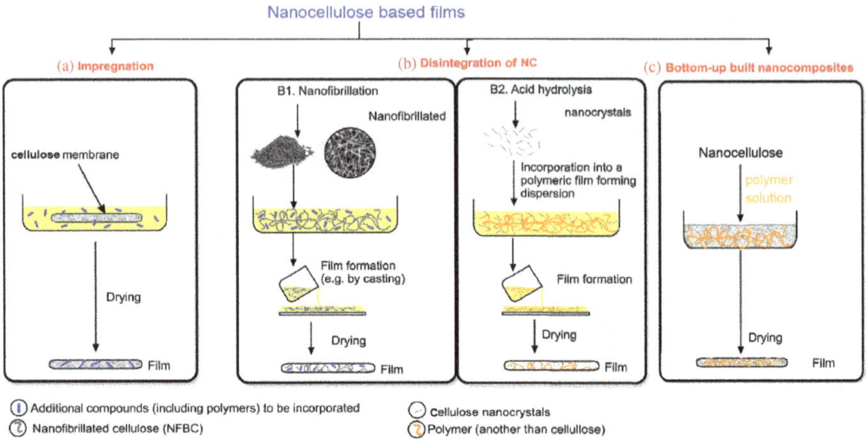

Fig. 2. Routes for use of nanocellulose for films: (a) impregnation with other film components; (b) physical and/or chemical disintegration of cellulose resources to produce nanocellulose; (c) addition to other polymer solutions to build nanocomposites.

Source: Reproduced with permission from Reference 21.

In order to from the percolation network, it is necessary to disperse nanocellulose in the polymer matrix homogeneously; in this case, the nanocellulose-polymer interfacial adhesion will be maximized. One of the possible strategies is the use of surfactants to modify the surface of the nanocellulose and to avoid its agglomeration.[20–22] This process may increase the compatibility of cellulose with a hydrophobic matrix. The use of an acid phosphate ester of ethoxylated nonylphenol as surfactant in order to enhance the dispersibility of nanocellulose and increase its compatibility with poly(lactic acid) (PLA) has been reported. Fortunati *et al.* added different contents of nanocellulose with and without surfactant modification into PLA films,[23] and they indicated that the surfactant enhanced the dispersion and interaction of nanocellulose with PLA; reductions of 34% in water vapor permeability (WVP) and 26% in the O_2 permeability (OP) were observed with a concentration of 1 wt% of surfactant. Covalent linkage of nanocellulose with the polymer matrix has also been reported.[24–26] For the bio-nanocomposite films prepared by covalently linking nanocellulose and chitosan, nanocellulose was modified with methyl adipoyl chloride first, and the then the methyl ester groups on nanocellulose surface reacted with the $-NH_2$ groups of chitosan. The tensile strength of the chitosan films was increased from 45 to 110 MPa, and the Young's modulus was increased from 1.35 to 3.75 GPa with the addition of 60% nanocellulose; furthermore, the water uptake of the nanocomposites was decreased.[27]

The surface modification of nanocellulose through acetylation,[28] esterification[29–34] and silanization/silylation[35–38] have been carried out. It was worth noting that the swelling media and the peeling effect of surface-grafted chains during the modification should be avoided, as this would otherwise result in the dissolution of modified nanocellulose in the reaction medium. Polycaprolactone[39] and DNA oligomers[40] and synthetic macromolecules, e.g., waterborne polyurethane[41] and substituted polypropylene[42] have been grafted onto the nanocellulose Grafting onto the nanocellulose is done by means of coupling agents. The modified nanocellulose had special properties ranging from excellent stability to new thermal performance. Another route called "grafting from" by using polymers grafting onto the nanocellulose was also developed; in this method, polymer monomers and initiators were used, and the polymerization process was finished on the surface of nanocellulose, and hydrophobic and hydrophilic monomers including styrene, styrene with *N, N*-dimethylaminoethyl,[43] *N,N*-di-methylaminoethyl methacrylate,[44] and caprolactone[45] have been used, and the obtained composites combine the properties of synthetic polymers with the excellent mechanical properties of nanocellulose, making nanocellulose more suitable for use in food packaging, particularly for new nanocomposites with moisture resistance. The effects of nanocellulose on water sensitivity of hydrophilic films have been characterized through different tests such as water solubility, water sorption, or water uptake. Although nanocellulose is hydrophilic, it has effective ability to increase the water resistance of alginate,[46] chitosan,[47] starch,[48] gelatin,[49] soy protein isolate,[50] and PHB.[51]

In situ modification method was also performed to increase the stability of cellulose films. It has been reported that regenerated cellulose film prepared from LiOH/urea or NaOH/urea aqueous solution had a porous structure, the cellulose nanofibrils formed a network could be used as scaffolds for the *in situ* polymerization of organic monomers, and the obtained composites had a homogeneous microstructure. Some thermally curable epoxy resins were incorporated into the cellulose matrix, and the obtained composites integrated the merits of cellulose and resins, and it could still be mechanically flexible with a resin content lower than 60 wt%.[52–53] Notably, a decrease in the water uptake and the permeability towards water vapor gas was also observed, while the transmittance of the composite films at 550 nm was kept to 85%, and the thermal and mechanical properties of the composite films were improved. Furthermore,

UV-curable polyurethane acrylate (PUA) prepolymer was also incorporated into the cellulose matrix; after being treated with UV light, the resin was cured in the cellulose matrix, and the composite films were found to have good mechanical properties, and the Young's modulus decreased with the increasing content of PUA in the composites. The basic process for this is shown in Fig. 3.[54] A decrease in the water uptake and an increase in barrier properties toward water vapor gas at different relative humidities were observed. The method was promising for the improvement of the waterproofing properties of cellulose, and it could be applied routinely for the modification of the surface properties of biomaterials.

The increase of barrier properties for food package films is especially important to reduce gas and water vapor exchanges between the food and

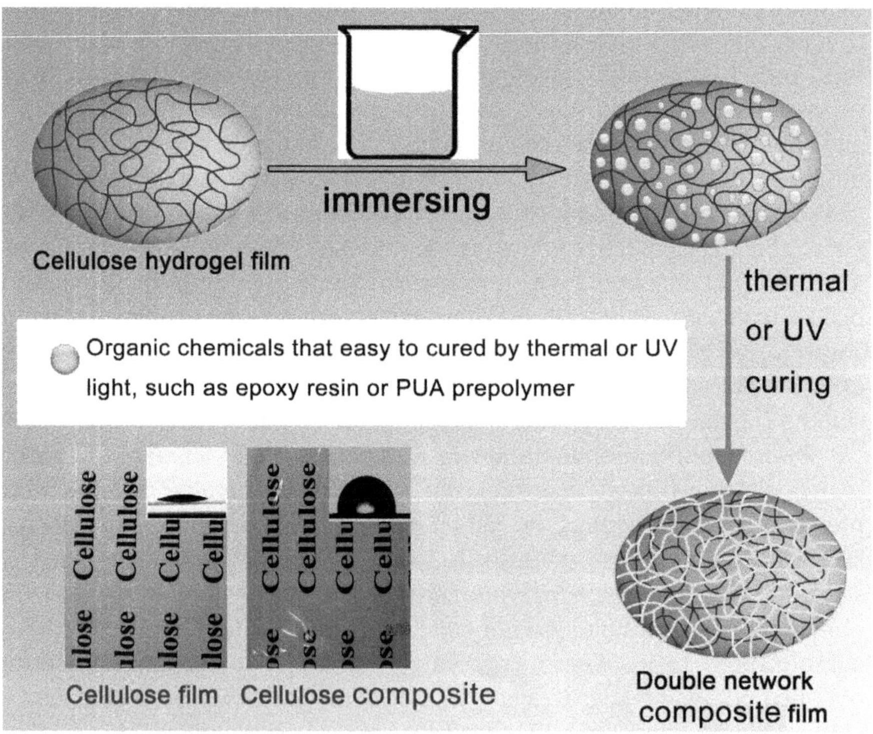

Fig. 3. Process for the hydrophobic modification of cellulose matrix through thermal or UV curing method.

Source: Reproduced with permission from Reference 54.

the surroundings, which could improve the quality of food and prolong its shelf life. Barrier-improving effects of nanocellulose have been investigated, and the mechanism was found to be related to the increased path length for diffusing molecules through the tortuosity effect. Several researchers reported that nanocellulose could reduce the WVP of alginate,[55] alginate-acerola puree,[56] chitosan,[57] gelatin,[58] hydroxypropyl methylcellulose (HPMC),[59] poly(vinyl alcohol) (PVA),[60] and poly(ε-caprolactone)(PCL).[61] Some studies reported nanocellulose could decrease OP of films remarkably. The OP of poly (3-hydroxybutyrate) (PHB) films containing 2 wt% nanocellulose could be decreased by about 65%.[62] It has been pointed out that the diffusion of molecules between two adjacent volumes separated by a thin film of solid polymer happened basically in three steps. First, the sample surface adsorbed the diffusing molecule. Second, the diffusing molecule would penetrate the film. Finally, the diffusing molecule would be desorbed from the film surface on the other side. In some case, the adsorption and desorption steps were relatively quicker than the diffusion through the film. Thus, according to the process, the OP was mainly dependent on the dissolution and diffusion rates of gas in the film.

Nanocellulose with a dense network structure formed by nanofibrils has good barrier properties. The density of nanocrystalline cellulose can be attributed to the uniform arrangement of cellulose chains; the pores within the films serve as the major path for small molecules to permeate. For nanofibrillated cellulose, the densification is related to the inherent flexibility of the material, which makes it easier to adjust its distribution. Therefore, NFC with inherent flexibility is able to form a denser film, sealing up most of the spaces between the fibrils down to the molecular scale. However, the relative humidity (RH) has an important role to play on the OP of nanocellulose films. Nanocellulose films show low oxygen permeability when compared with that of films based on plasticized starch, whey protein, and arabinoxylan at low RH (0%); however, the OP will increase exponentially at higher RH because of the swelling of the nanocellulose.[63] Aulin *et al.* also found that nanocellulose films have a low oxygen permeability of 0.086 ml·mm·m^2·day·atm, and the performance of the films was limited by their ability to attract and retain water molecules from their surroundings.[64] Ankerfors reported that the oxygen permeability values of cellulose films was in the range of 0.037–0.050 ml·mm·m^2·day·atm at 23°C and 50% relative humidity.[65] The oxygen, water vapor, and grease barrier properties of MFC films had also been evaluated,[66] and the oxygen permeability of the pure and unmodified films was below 0.6 $cm^3 \cdot \mu m \cdot m^{-2} \cdot d^{-1} \cdot kPa^{-1}$ at relative humidity below 65%. However, when the surface properties of the MFC film were modified from hydrophilic to hydrophobic with wax, the OP could be decreased to 17 $cm^3 \cdot \mu m \cdot m^{-2} \cdot d^{-1} \cdot k \cdot Pa^{-1}$ at 97.4% RH. CNC also could be coated on the surface of food packaging films. This indicated that the CNC coating could reduce the coefficient of friction while maintaining high transparency (~90%) and low haze (3–4%) values; furthermore, the modified film had an OP of about 0.003 $cm^3 \cdot \mu m \cdot m^{-2} \cdot 24h^{-1} \cdot kPa^{-1}$, and the Gelbo flex test indicated CNC-coated PET and OPA films still maintained a significantly low OP. The morphologies of coating observed after the fatigue test revealed better adhesions between CNC and PET and OPA.[67]

Coating is a powerful tool for modification of the surface properties of packaging materials. It is well known that the surface properties have a key role in controlling the permeation of gas molecules, as they can reduce the diffusion and migration rate of molecules in the film. In the packaging field, coatings may be either on the outer surface or between two substrates, forming a sandwiched structure. Normally, the thickness of the coating layer ranges from tenths of nanometers to a few micrometers. Currently, more and more research is focused on the possibility of using high-performing bio-based material as a thin functional layer to increase the sustainability of packaging materials. Nanocellulose is a good candidate for acting as a functional layer on various substrates. NFC used in spray and foam coating technologies has some advantageous, as it can be used to coat very thin and uniform layers on the paper sheet,[68] the print quality of the paper can be improved, and some new functionalities like barrier properties for oxygen, water vapor and grease/oil may be incorporated.[69] Petri dishes modified with cellulose nanofibrils showed some significant improvements in the mechanical properties, and adding cellulose nanofibrils showed a significant improvement in air permeability,[70] mechanical strength, and barrier properties of NFC, related to its highly crystalline nature and cohesive energy density.[71] Nisin as a natural antimicrobial agent was also incorporated into a cellulose coating to develop a barrier film containing the antimicrobial agent, and this was then was used for beef packaging. This showed that the coating had antibacterial ability against *L. monocytogenes*.[72] Coating technology has a great potential in producing novel packaging materials; however, the adhesion of hydrophilic cellulose onto the hydrophobic plastic substrates should be focused on surface treatments and possible primers to increase the strength adhesion and the compatibility.

Active packaging is an innovative way forward in food science and technology, as this can inhibit the growth of microbial and keep the quality, freshness, and safety of food intact. The use of biodegradable polymers may offer a good way for carrying a natural antimicrobial agent. It is well known that nanocellulose does not possess antimicrobial ability, and some methods and antimicrobial agents are needed for this added functionalization. Aminoalkyl groups, 2-benzyl-4-chlorophenol, and *L*-cysteine have been grafted onto the surface of the cellulose backbone through chemical modification.[73] Another pathway for the funcaionlization of the cellulose film with antibacterial properties was through dip-loading process. In this method, thc cellulose film was immersed into the benzalkonium chloride solution; after being dried, the obtained cellulose film had antimicrobial activity against *Staphylococcus aureus* and *Bacillus subtilis*.[74] Cellulose is a polyhydroxyl polysaccharide; it has powerful ability for the adsorption of metal ions; therefore, the *in situ* synthesis of Ag nanoparticles in the cellulose matrix was undertaken, and the Ag-cellulose composites showed antimicrobial activity against *E. coli* and *S. aureus*.[75] Ag could also be incorporated into the three-dimensional structure of cellulose materials through magnetic interaction of heterogeneous nucleation of magnetite, and the composite was found to have high antimicrobial activity against *E. coli* and *B. subtilis*. The addition of Ag/cellulose nanocrystals in the PLA matrix could also make the composites possess antibacterial property against *S. aureus* and *E. coli*.[76] CTABr surfactants have also been used for the synthesis of Ag and Au nanoparticles on CNC for the development of antibacterial materials.[77–78] ZnO-Ag/cellulose nanoparticles also showed antibacterial activity against *S. choleraesuis* and *S. aureus*.[79] Cellulose copper composites and nanoparticles had antifungal and antimicrobial potential against *S. cerevisiae* and nonpathogenic *E. coli DH5α*.[80–81] Cellulose films consisted of nanofibril cellulose that could be oxidized by Na_5IO_6, and the amino groups in chitosan could link with aldehyde group in cellulose matrix though the Schiff base reaction.[82] The composites had potential antibacterial properties against *E. coli* and *S. aureus*, and the antibacterial performance increased with the increasing the content of grafted chitosan. Figure 4 shows the antibacterial performance of the composite films used as packaging materials for sausage, and the comparison between PE and RC is also shown. The results indicate that the composites have potential application in antibacterial packaging materials.

Paper coated with CNF and/or *β*-cyclodextrin (*β*-CD) with controlled release properties was developed to deliver antibacterial chlorhexidine

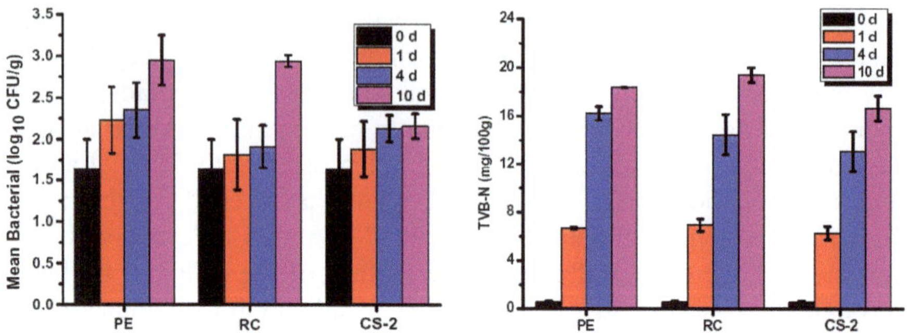

(a) (b)

Fig. 4. Total number of bacteria (a) and total volatile basic nitrogen (TVB-N) (b) of the model control group (PE wrap), control group (RC film), and test group (CS-2) in sausage during the 10-day storage period.

Source: Reproduced with permission from Reference 82.

digluconate (CHX).[83] At first, the CHX was mixed with a CNF suspension or a β-CD solution, and then it was used as coating for paper. This demonstrated that CNF had a burst effect and that β-CD could control the CHX released. The combination of β-CD and CNF presented a complementary action, and the composites presented a rapid delivery of CHX. Furthermore, this method was also used for the controlled release of carvacrol, and the β-CD-grafted paper retained antibacterial activity against *Bacillus subtilis* for 14 hours, and the CNF coating extended this activity for additional 2 hours.[84]

1.2 *Nanocellulose in Food Additive and as Dietary Aid*

Nanocellulose is a kind of insoluble dietary fiber, and it has a range of health benefits. It is well established that dietary fiber can assist in reducing the risk of diabetes, obesity, cardiovascular disease, and diverticulitis. In addition, it promotes beneficial physiological effects including laxation, blood cholesterol attenuation, and blood glucose attenuation.[85–86] Nanocellulose has been used to treat intestinal disorders; in the mentioned experiment, dried foods and drugs comprised of 15–65 wt% of nanocellulose and high content of water-soluble saccharide was used to prevent agglomeration of nanocellulose. The powdered material, after being added to foodstuff, was used in treating intestinal disorders.[87] Nanobacterial cellulose could be used as filler for the reinforcement of fragile food hydrogels, and it could decrease the stickiness of pasty foods and improve their worth. Furthermore, nanobacterial cellulose could increase the gel strength of Tofu (food made by coagulating and pressing soymilk). Moreover, nanobacterial cellulose fibers easily formed an interpenetrated network in the aqueous suspension, and the addition of bacterial cellulose to chocolate drink could prevent the precipitation of the cocoa. The stickiness of the creamy condiment could also be obviously improved by the addition of bacterial cellulose, and it could be easier to serve it quantitatively by using a spoon.

The application of nanocellulose to prepare reduced-fat formulations was also assessed. Nanocellulose could be used as a substitute of oil to produce a low-calorie salad dressing, and the color and texture of the material was similar to that of an authentic Italian dressing. Therefore, nanocellulose could be used to produce low-calorie foods to treat weight disorders.[88] Nanocellulose could also be used to reduce the energy density of many processed foods.[89] It was proposed that nanocellulose could be used to reduce the energy density of chocolate,[90] and could also be used as an additive to lower the food energy density in hamburgers,[91] toppings,[92] and dough-based products.[93] Crisp and crunchy fruit-filled cookies with water activities of 0.2 were developed by adding of 0.3–2 wt% of nanocellulose, and this could reduce the calories of fillings by 15–20 wt%.

Probiotics are used to develop a variety of foods with additional health benefits. If we want to get significant health benefit, a minimum concentration of a probiotic bacterium of about 10^6–10^7 CFU/g is required. However, the shelf life of probiotics often suffers from the inherent susceptibility of high-water activity and low pH. The effect of CNC on the enhancement of the viability of probiotics has been investigated, and it indicated that CNC acted as a filler for microencapsulation of *L. rhamnosus*.[94–95] CNC incorporation in alginate microbeads could increase the viability of probiotics by 38% compared with that of alginate microbeads without CNC, and the alginate microbeads with CNC enhanced the survivability of probiotic by 45% after storage for 6 weeks at 25°C and relative humidity of 60%, when compared with the results from that of alginate microbeads. A macro-encapsulation kit based on porous cellulose microgels also could increase the loading capacity of probiotics and afford better shelter for probiotics, as is shown in Fig. 5.[96] The merit of using porous structured cellulose was the inherently hydrophilicity, which could absorb sufficient culture medium to ensure the energy supply. However, the open structure allowed probiotic cells to be released too early in the stomach when transiting the upper gastrointestinal tract; so,

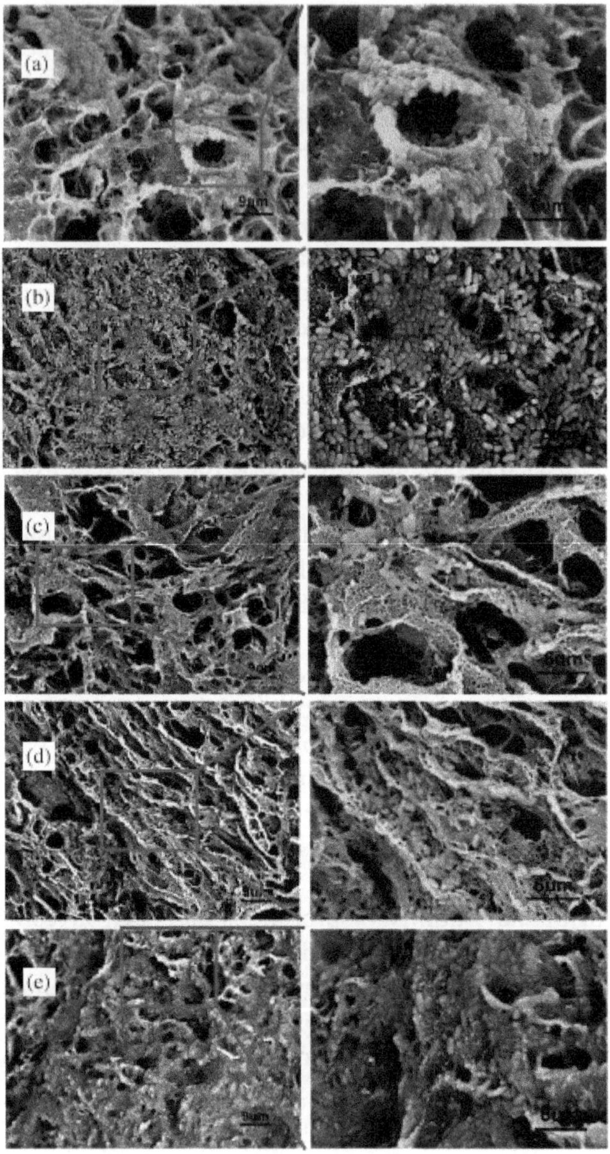

Fig. 5. SEM images and the surface morphology of microgels with *Lactobacillus plantarum*. (a), (b), (c), (d), and (e) are RCM, 0.01 RCM, 0.05 RCM, 0.1 RCM, and 0.2 RCM, respectively.

Source: Reproduced with permission from Reference 89.

the as-prepared cellulose microgels encapsulated with *L. plantarum* cells were coated with a Ca-alginate hydrogel to form a core–shell structure. The Ca-alginate shell played the part of doors that guarded the houses. When they encountered gastric acid, the door would be kept closed. But when they reached the site of action (the intestinal tracts), the doors opened and probiotic cells would be released from the cellulose microgels.[97] The core–shell gels were easy to fabricate and feasible for encapsulation, protection, and delivery of bioactive drugs and bio-nutrients, which had great potential for industrial production.

1.3 *Nanocellulose as Emulsion Stabilizer*

Nanocellulose cannot be dissolved in water and organic solvents without any treatment, which makes it a natural emulsifying and stabilizing ingredient in Pickering emulsion. Nanocellulose was used to increase the stability of a frozen dessert as a stabilizer, and the shape retention could be improved by increasing the amount of nanocellulose; when the content of nanocellulose ranged from 0.10 to 0.30 wt%, the shape retention time was extended with the other properties of the dessert being maintained.[98] Nanocellulose was also used as stabilizers and emulsifiers.[99–103] Pristine cellulosic materials with a hydrophilic nature could form O/W emulsions, and the advantages of using nanocellulose as a stabilizer have been clarified. The nanocellulose could improve its incorporation between the oil droplets, and the oil droplets were themselves incorporated in a network of nanocellulose fibers resulting in their steric stabilization, and a very firm emulsion could be developed by using 1 wt% of nanocellulose.[104] Nanocellulose with concentrations ranging from 0.05 to 0.70 wt% was also used for the formation of 10 wt% oil-in-water (O/W) emulsion.[105] Moreover, pH-responsive[106] or temperature-responsive emulsions[107] could also be obtained through grafting stimuli-responsive molecules onto cellulose surfaces. To develop a better understanding of the stabilization mechanism, the surface charge density of nanocellulose from cotton linters and bacterial cellulose were modulated to investigate the emulsifying capability, and it was found that the distribution of nanocellulose on the surface of the droplets were different, as is shown in Fig. 6. This suggested that the electrostatic interaction played a major role in the control of the interface regardless of crystalline origin. Cellulose nanocrystals with a surface charge density above 0.03 e/nm^2 were not able to efficiently stabilize at the oil/water interface, whereas a decreasing surface charge density led to stable emulsions.[108]

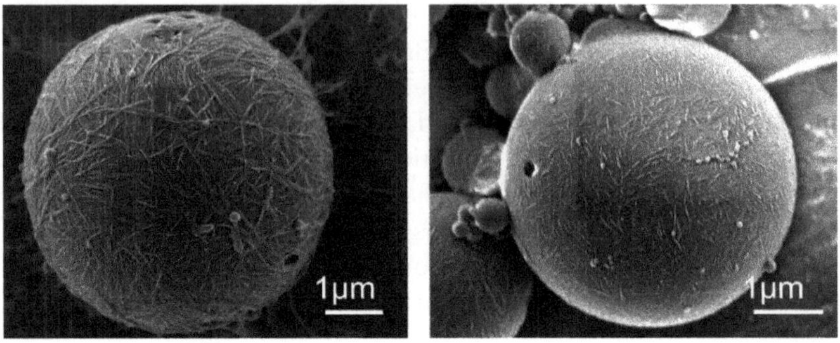

Fig. 6. SEM images of polymerized styrene/water emulsions stabilized by (a) bacterial cellulose nanocrystals (BCN) and (b) cotton cellulose nanocrystals (CCN).

Pickering emulsions stabilized by cellulose nanocrystals were also prepared by changing the physical morphology, with cellulose nanocrystals with average length from 185 to 4 mm and an aspect ratio of 13 to 16; images of the emulsions are shown in Fig. 7. The coverage ratio of the emulsion droplets was in a range from 40% to 100%.[109] It indicated that nanocellulose could be adsorbed at the oil–water interface strongly and that coalescence occurred; the capillary forces would lead to the same high bending capacity for all the cellulose nanocrystals irrespective of their cross-section and length, and the alignment of nanocellulose along the interface made the size of the droplets to be in the same range, and the amount and characteristics of the cellulose nanocrystals had a major role on the size of the droplets, which could control the coalescence process. The crystalline differences existing at the nanoscale level might have little influence on the macroscale bending capacities of all cellulose nanocrystals. However, the length of the nanocellulose had an important role in the stabilization of the formed Pickering emulsion. In Pickering emulsions, van der Waals, hydrogen bonds, and colloidal Derjaguin–Landau–Verwey–Overbeek (DLVO) interactions, as well as capillary forces, affected the particles; therefore, the wetting characteristic of the nanocrystals has an important role in the control of the interfacial stabilization. In the case of cellulose nanocrystals, the adsorption force was stronger since the two interactions were highly involved: the cooperative contact was distributed all along the crystals, and the particles trapped at the fluid–fluid interface deformed the interface, which resulted in the high deformation mediated by capillary interactions. Loudet *et al.* found that the

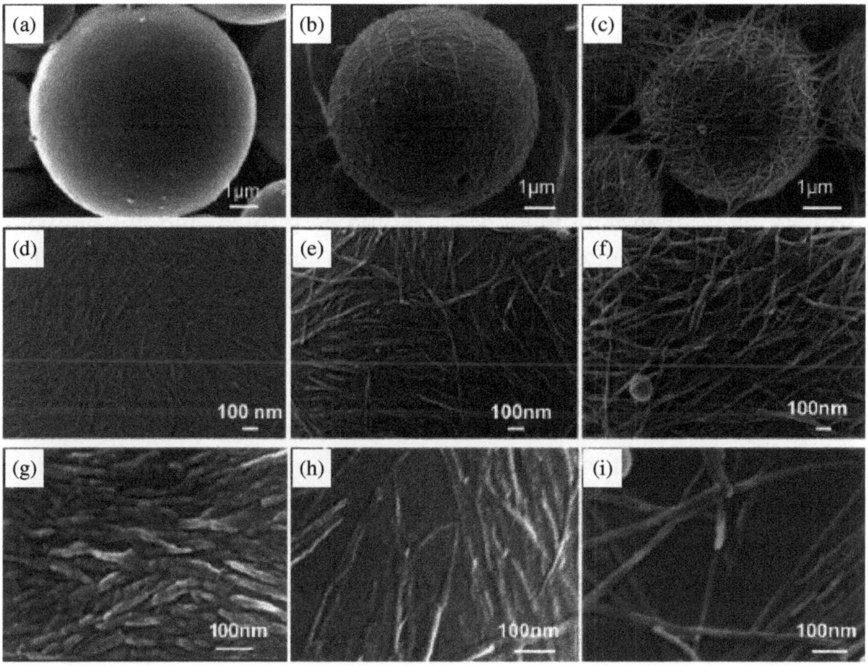

Fig. 7. SEM images of polymerized styrene–water emulsions stabilized by CCN (a, d, and g), BCN (b, e, and h), and ClaCN (c, f, and i) at two different nanocrystal concentrations, 5 gL^{-1} (d–f) and 1 gL^{-1} (g and h), revealing the coverage variation as a function of nanocrystal concentration and length.

Source: Reproduced with permission from Reference 101.

deformation of the cellulose particles could lead to an attractive strength of about 10^4 kT when cellulose particles with aspect ratios below 10 were used for the research. The strong and long-range capillary attractive forces were cooperatively distributed all along the nanocrystals, which promoted the contact between and aggregative behavior of the emulsions.[110] The maximum packing for the spherical particles used for Pickering emulsions was reached with 6 particle–particle contacts. While long particles were used, an increase in the maximum packing was induced by more particle–particle contacts; therefore, multiple interactions favored the deformation of the interface. Because shorter cellulose nanocrystals could align side by side through attractive forces, this could result in a more rigid and denser organization, and a high coverage of the droplets was obtained. However, the ability of the ClaCN nanocrystals with an aspect ratio of 160 to be used for the stabilization of droplets at fairly lower volume fractions was only 40% of coverage; it indicated that there was a critical dimension for the drop diameter controlled by the capillary strength and that the cellulose flexibility competed with the particle length for the global emulsion organization. The long cellulose nanocrystals adsorbed on neighboring droplets would simultaneously result in steric hindrance, and a stable network rather than individual drops would be formed, and this networking organization resulted in interconnected, partially covered droplets. Therefore, the use of elongated nanocrystals with higher aspect ratios made it difficult to align ClaCN nanocrystals along the surface of the droplets, and their coverage was limited.

The hydrophobic modification of cellulosic surfaces could also increase the stabilization of Pickering emulsions.[111] It has been reported that the wettability of the cellulose particles could be affected through the surface modification, which would have an impact on the stability and the emulsion size. Cellulose nanofibrils could be presented as single or dispersed fibrils or form large, network-like aggregates at the O/W interface; this network could prevent the coalescence of emulsion droplets. However, the pH or ionic strength would have an influence on the collapse of emulsions due to electrostatic screening of the negatively charged cellulose particles. For the oil-in-water emulsions stabilized solely by microcrystalline cellulose, the lipid oxidation rate could be reduced effectively, which was attributed to their ability to scavenge free radicals and form thick interfacial layers around oil droplets.[112] Protein-functionalized cellulose nanofibrils could also be assembled into tightly packed thin films at the air/water and at the oil/water interfaces, which would result in a synergistic improvement in the formation and stability of O/W emulsions.[113] The surface-modified cellulose with a fatty acid chloride, namely, lauroyl chloride, could be used for the preparation of double-layered Pickering emulsions (O/W/O).[114] This suggested that the surface accessibility and concentration of the modified nanocellulose had a crucial role in the stabilization of the double-layered emulsions, and the O/W/O double-layered emulsions demonstrated remarkable stability for over a month.

Emulsions with internal phases lower than 50% volume fractions are called low internal phase emulsions (LIPEs), while internal phases with 50–74% volume fractions are called medium internal phase emulsions (MIPEs); these two kinds emulsion have different properties and applications. High internal phase emulsions (HIPEs) are emulsion systems containing an internal phase volume fraction higher than 74.05%, in the microstructure of the emulsion, and the maximum volume ratio of mono-dispersed non-deformable spheres is packed in the most efficient manner. Recently, cellulose-based HIPEs have been obtained by using hydrophobic-modified bacterial cellulose nanofibers as the stabilizer.[115] The emulsion formation process was interesting; at first, a stable and uniform emulsion system was not formed through direct mixing of the two phases and the internal phase volume was only 50%. After 7 days, the emulsions became stable and the maximum internal phase volume could be increased up to 81%. This phenomenon was ascribed to the disentanglement of bacterial cellulose nanofiber networks into small bundles. CNCs were also used for the stabilization of O/W HIPEs through a two-step procedure.[116] In the first step, a low internal phase emulsion (10% oil) was prepared by using an ultrasonicator, and then extra oil was added by high-speed homogenization to obtain HIPEs. In the experimental process, the diameter of small oil droplets increased following a limited coalescence of the droplets, which could thus help maintain the minimum required coverage of CNCs to accommodate the added extra oil. Miao *et al.* conducted an in-depth investigation of O/W HIPEs stabilized with CNCs.[117] In that research work, the influence of the concentration of CNC and type of oil on the formation on the properties of the HIPEs was systematically investigated. They demonstrated that the maximum oil phase volume that could be stabilized by CNCs was 87% when the CNC concentration was 0.6 wt%; when the CNC concentration was increased to 1.2 wt% or higher, the oil phase volume was decreased to 83%. In addition, the oil droplets stabilized with 0.6 wt% CNC suspensions had a larger size than those stabilized with higher concentrations of CNC suspensions. It was found that the concentration of CNCs played an important role in the formation of HIPEs, and a new comprehensive CNC-stabilized HIPE formation mechanism was proposed on the basis of the experimental results, and this is shown in Fig. 8.

1.4 *Nanocellulose in Foam Stabilizer*

Long-term stability is much more difficult to achieve for aerated systems than it is for emulsions, and so the stabilization of bubbles and foams by particles has been of significant focus. The incorporation of colloidal particles as a possible way to enhance foam stability is of great technological and commercial significance; it is especially important for food systems

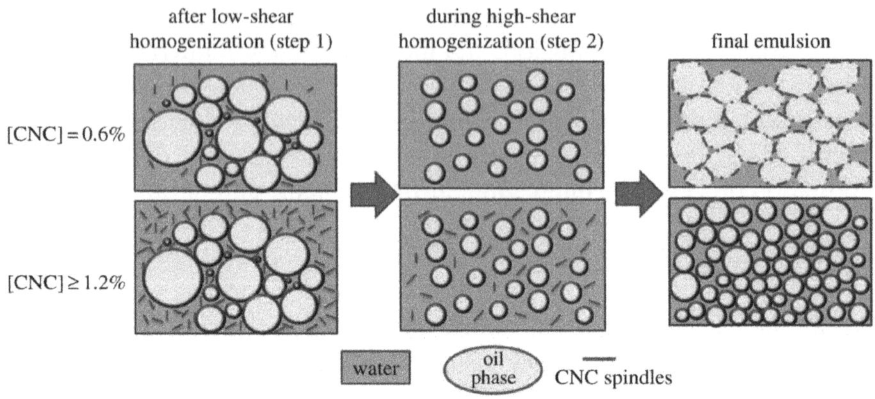

Fig. 8. Schematic of the formation mechanisms of CNC-stabilized oil/water HIPEs at different CNC concentrations in the water phase.

Source: Reproduced with permission from Reference 117.

because the presence of an aerated structure provides the essential textural characteristics of highly popular foods such as ice-cream. Furthermore, the incorporation of gas bubbles as a full or partial replacement for dispersed fat particles can play a useful role in the development of healthier products.

Foam stabilization by nanofibrillated cellulose particles from wood pulp has been investigated.[118] The hydrophobic modified cellulose particles exhibited significant surface activity at the air/water interface, and they could stabilize foams at concentrations as low as 0.5 wt%. The efficiency of foam stabilization was attributed to the high aspect ratio of the nanofibrils (~2 mm–3 nm), which facilitated the establishment of a robust gel-like network between bubbles as well as a large viscoelastic modulus at the air–water interface. The stability of the foam formed using long fibrils was substantially greater than that with some shorter cellulose nanocrystals (300 nm–4 nm) of similar chemical composition and charge density.[119] Another kind of cellulose-derived ingredient that appeared suitable for forming and stabilizing foams consisted of charged colloidal particles (~130 nm) made by anti-solvent precipitation of hydrophobic-modified ethyl cellulose.[120]

Pickering emulsions stabilized by cellulose nanocrystals were used for the preparation of foam.[121] In the experimental process, the emulsions were concentrated by centrifugation and then freeze dried to produce foam. The foams had cell sizes similar to the droplets of the starting emulsion. This indicated that the robustness of the starting emulsion drops stabilized by cellulose nanocrystals could withstand the centrifugation and freeze drying. The distribution of cellulose nanocrystals at the interface was different depending on the concentration of cellulose nanocrystals in the initial emulsions. The cell diameters of the foams produced from emulsions stabilized by cellulose nanocrystals with concentrations of 5 g/L was around 12 μm, while for the foam prepared with higher content of cellulose nanocrystals (10 g/L), the foam cells were similar in size but their walls were sheet-like structures. Functionalization of the foams was also achieved by addition of positively charged chitosan, and cell-like foam with cell sizes similar to that of the initial emulsion droplets was produced, as is shown in Fig. 9.

1.5 *Concluding Remarks and Prospects*

Nanocellulose with high surface area and crystallinity is an abundant and sustainable nanomaterial; it has high potential to maintain colloidal stability without any modification, and it also can be used for the inhibition of the growth of pathogenic bacteria after being functionalized. Nanocellulose can also be used in the development of intelligent

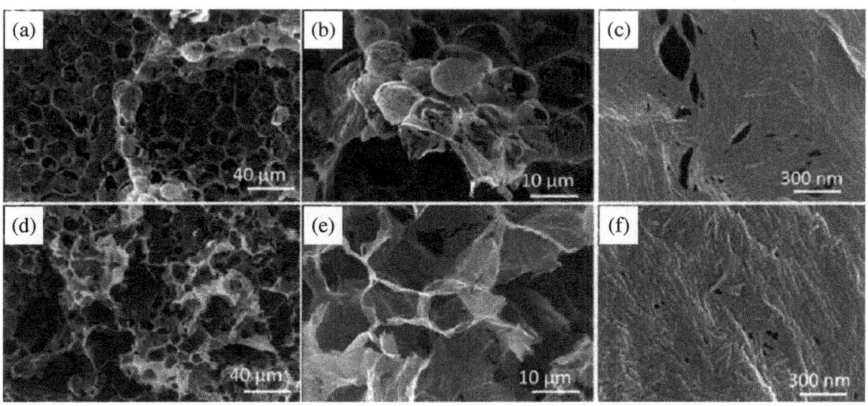

Fig. 9. SEM images of foams made from Pickering emulsions after freeze drying. The emulsions were prepared with CNCs as stabilizer at concentrations of 5 (images a–c) and 10 g/L (images d–f). Note the structuring of the CNCs in the foam lamella (c and f).

Source: Reproduced with permission from Reference 121.

packaging, which is an important feature for next-generation packaging materials. Another promising field will be the immobilization of living cells and probiotic bacteria in cellulose-based matrix, which will have highly promising applications in the functional food market when combined with the water-insoluble properties of the dietary fiber. According to the results of *in vivo* and *in vitro* tests, nanocellulose did not create cytotoxic or genotoxic concerns, and so more applications of nanocellulose will be expected in this important, yet still growing field. However, new insight into the analysis of advantages of food systems containing cellulose will help to ensure the quality, variety, and the consistent health-oriented outlook of food industry.

References

(1) Maatar, W., Boufi, S. (2017). Microporous cationic nanofibrillar cellulose aerogel as promising adsorbent of acid dyes. *Cellulose*, 24, pp. 1001–1015.
(2) Baati, R., Magnin, A., Boufi, S. (2017). High solid content production of nanofibrillar cellulose *via* continuous extrusion. *ACS Sustainable Chem. Eng.*, 5, pp. 2350–2359.
(3) Tolmachev, D., Lukasheva, N. (2012). Interactions binding mineral and organic phases in nanocomposites based on bacterial cellulose and calcium phosphates. *Langmuir*, 28, pp. 13473–13484.
(4) Habibi, Y. (2014). Key advances in the chemical modification of nanocelluloses. *Chem. Soc. Rev.*, 43, pp. 1519–1542.
(5) Isogai, A. (2013). Wood nanocelluloses: Fundamentals and applications as new bio-based nanomaterials. *J. Wood Sci.*, 59, pp. 449–459.
(6) Moon, R. J., Martini, A., Nairn, J., Simonsen, J., Youngblood, J. (2011). Cellulose nanomaterials review: Structure, properties and nanocomposites. *Chem. Soc. Rev.*, 40, pp. 3941–3994.
(7) Klemm, D., Kramer, F., Moritz, S., Lindström, T., Ankerfors, M., Gray, D., Dorris, A. (2011). Nanocelluloses: A new family of nature-based materials. *Angew. Chem., Int. Ed.*, 50, pp. 5438–5466.
(8) Sacui, I. A., Nieuwendaal, R. C., Burnett, D. J., Stranick, S. J., Jorfi, M., Weder, C., Foster, E. J., Olsson, R. T., Gilman, J. W. (2014). Comparison of the properties of cellulose nanocrystals and cellulose nanofibrils isolated from bacteria, tunicate, and wood processed using acid, enzymatic, mechanical, and oxidative methods. *ACS Appl. Mater. Interfaces*, 6, pp. 6127–6138.
(9) Bourke, P., Ziuzina, D., Boehm, D., Cullen, P. J., Keener, K. (2018). The potential of cold plasma for safe and sustainable food production. *Trends Biotechnol.*, 36, pp. 615–626.

(10) Vanderroost, M., Ragaert, P., Devlieghere, F., De Meulenaer, B. (2014). Intelligent food packaging: The next generation. *Trends Food Sci. Technol.*, 39, pp. 47–62.
(11) Mihindukulasuriya, S., Lim, L.-T. (2014). Nanotechnology development in food packaging: A review. *Trends Food Sci. Technol.*, 40, pp. 149–167.
(12) Azeredo, H. M., Rosa, M. F., Mattoso, L. H. C. (2017). Nanocellulose in bio-based food packaging applications. *Ind. Crops Prod.*, 97, pp. 664–671.
(13) Peelman, N., Ragaert, P., Verguldt, E., Devlieghere, F., De Meulenaer, B. (2016). Applicability of biobased packaging materials for long shelf-life food products. *Packag. Res.*, 1, pp. 7–20.
(14) Dudefoi, W., Villares, A., Peyron, S., Moreau, C., Ropers, M.-H., Gontard, N., Cathala, B. (2017). Nanoscience and nanotechnologies for biobased materials, packaging and food applications: New opportunities and concerns. *Innovative Food Sci. Emerging Technol.*, 46, pp. 107–121.
(15) Grumezescu, A. M. (2017). Multifunctional antimicrobial nanocomposites for food packaging applications. *Nanotech. Agri-Food Ind.*, 6, pp. 265–303.
(16) El-Wakil, N. A., Hassan, E. A., Abou-Zeid, R. E., Dufresne, A. (2015). Development of wheat gluten/nanocellulose/titanium dioxide nanocomposites for active food packaging. *Carbohydr. Polym.*, 124, pp. 337–346.
(17) Khan, A., Huq, T., Khan, R. A., Riedl, B., Lacroix, M. (2014). Nanocellulose-based composites and bioactive agents for food packaging. *Crit. Rev. Food Sci. Nutr.*, 54, pp. 163–174.
(18) Yu, H.-Y., Zhang, H., Song, M.-L., Zhou, Y., Yao, J., Ni, Q.-Q. (2017). From cellulose nanospheres, nanorods to nanofibers: Various aspect ratio induced nucleation/reinforcing effects on polylactic acid for robust-barrier food ppackaging. *ACS Appl. Mater. Interfaces*, 9, pp. 43920–43938.
(19) Cheng, D., Wen, Y., An, X., Zhu, X., Ni, Y. (2016). TEMPO-oxidized cellulose nanofibers (TOCNs) as a green reinforcement for waterborne polyurethane coating (WPU) on wood. *Carbohydr. Polym.*, 151, pp. 326–334.
(20) Gestranius, M., Stenius, P., Kontturi, E., Sjöblom, J., Tammelin, T. (2017). Phase behaviour and droplet size of oil-in-water Pickering emulsions stabilised with plant-derived nanocellulosic materials. *Colloids Surf., A*, 519, pp. 60–70.
(21) Sharma, C., Dhiman, R., Rokana, N., Panwar, H. (2017). Nanotechnology: An untapped resource for food packaging. *Front. Microbiol.*, 8, pp. 1735–1756.
(22) Mariano, M., Pilate, F., de Oliveira, F. B., Khelifa, F., Dubois, P., Raquez, J.-M., Dufresne, A. (2017). Preparation of cellulose nanocrystal-reinforced poly (lactic acid) nanocomposites through noncovalent modification with PLLA-based surfactants. *ACS Omega*, 2, pp. 2678–2688.

(23) Fortunati, E., Peltzer, M., Armentano, I., Torre, L., Jiménez, A., Kenny, J. (2012). Effects of modified cellulose nanocrystals on the barrier and migration properties of PLA nano-biocomposites. *Carbohydr. Polym.*, 90, pp. 948–956.
(24) Kramer, F., Klemm, D., Schumann, D., Heßler, N., Wesarg, F., Fried, W., Stadermann, D. (2006). Nanocellulose polymer composites as innovative pool for (Bio)material development. *Macromol. Symp.*, 244(1), pp. 136-148.
(25) Lin, N., Bruzzese, C. C., Dufresne, A. (2012). TEMPO-oxidized nanocellulose participating as crosslinking aid for alginate-based sponges. *ACS Appl. Mater. Interfaces*, 4, pp. 4948–4959.
(26) Lu, Y., Cueva, M. C., Lara-Curzio, E., Ozcan, S. (2015). Improved mechanical properties of polylactide nanocomposites-reinforced with cellulose nanofibrils through interfacial engineering *via* amine-functionalization. *Carbohydr. Polym.*, 131, pp. 208–217.
(27) Mesquita, J. P., Donnici, C. L., Teixeira, I. F., Pereira, F. V. (2012). Bio-based nanocomposites obtained through covalent linkage between chitosan and cellulose nanocrystals. *Carbohydr. Polym.*, 90, pp. 210–217.
(28) Sassi, J.-F. and Chanzy, H. (1995). Ultrastructural aspects of the acetylation of cellulose. *Cellulose*, 2, pp. 111–127.
(29) Menezes, A. J., Siqueira, G., Curvelo, A. A., Dufresne, A. (2009). Extrusion and characterization of functionalized cellulose whiskers reinforced polyethylene nanocomposites. *Polymer*, 50, pp. 4552–4563.
(30) Salam, A., Lucia, L. A., Jameel, H. (2015). Fluorine-based surface decorated cellulose nanocrystals as potential hydrophobic and oleophobic materials. *Cellulose*, 22, pp. 397–406.
(31) Rosilo, H., Kontturi, E., Seitsonen, J., Kolehmainen, E., Ikkala, O. (2013). Transition to reinforced state by percolating domains of intercalated brush-modified cellulose nanocrystals and poly(butadiene) in cross-linked composites based on thiol–ene click chemistry. *Biomacromolecules* 14, pp. 1547–1554.
(32) Chen, J., Lin, N., Huang, J., Dufresne (2015). Highly alkynyl-functionalization of cellulose nanocrystals and advanced nanocomposites thereof *via* click chemistry. *Polym. Chem.*, 6, pp. 4385–4395.
(33) Parambath Kanoth, B., Claudino, M., Johansson, M., Berglund, L. A., Zhou (2015). Biocomposites from natural rubber: Synergistic effects of functionalized cellulose nanocrystals as both reinforcing and cross-linking agents *via* free-radical thiol–ene chemistry. *ACS Appl. Mater. & Int.*, 7, pp. 16303–16310.
(34) Li, Y., Chen, H., Liu, D., Wang, W., Liu, Y., Zhou, S. (2015). pH-responsive shape memory poly (ethylene glycol)–poly (ε-caprolactone)-based polyurethane/cellulose nanocrystals nanocomposite. *ACS Appl. Mater. & Int.*, 7 12988–12999.

(35) Grunert, M., Winter, W. T. (2002). Nanocomposites of cellulose acetate butyrate reinforced with cellulose nanocrystals. *J. Polym. Environ.*, 10, pp. 27–30.
(36) Goussé, C., Chanzy, H., Excoffier, G., Soubeyrand, L., Fleury, E. (2002). Stable suspensions of partially silylated cellulose whiskers dispersed in organic solvents. *Polymer*, 43, pp. 2645–2651.
(37) Yang, J., Han, C.-R., Duan, J.-F., Ma, M.-G., Zhang, X.-M., Xu, F., Sun, R.-C., Xie, X.-M. (2012). Studies on the properties and formation mechanism of flexible nanocomposite hydrogels from cellulose nanocrystals and poly (acrylic acid). *J. of Mater. Chem.*, 22, pp. 22467–22480.
(38) Huang, J.-L., Li, C.-J., Gray, D. G. (2014). Functionalization of cellulose nanocrystal films *via* "thiol–ene" click reaction. *RSC Adv.*, 4, pp. 6965–6969.
(39) Benkaddour, A., Jradi, K., Robert, S., Daneault, C. (2013). Grafting of polycaprolactone on oxidized nanocelluloses by click chemistry. *Nanomaterials*, 3, 141–157.
(40) Razaq, A., Nyström, G., Strømme, M., Mihranyan, A., Nyholm, L. (2011). High-capacity conductive nanocellulose paper sheets for electrochemically controlled extraction of DNA oligomers. *PLoS One*, 6, p. e29243.
(41) Aranguren, M. I., Marcovich, N. E., Salgueiro, W., Somoza, A. (2013). Effect of the nano-cellulose content on the properties of reinforced polyurethanes. A study using mechanical tests and positron anihilation spectroscopy. *Polym. Test.*, 32, pp. 115–122.
(42) Yeh, S. K., Kim, K. J., Gupta, R. K. (2013). Synergistic effect of coupling agents on polypropylene-based wood–plastic composites. *J. Appl. Polym. Sci.*, 127, pp. 1047–1053.
(43) Plamper, F. A., Ruppel, M., Schmalz, A., Borisov, O., Ballauff, M., Müller, A. H. (2007). Tuning the thermoresponsive properties of weak polyelectrolytes: aqueous solutions of star-shaped and linear poly (*N,N*-dimethylaminoethyl methacrylate). *Macromolecules*, 40, pp. 8361–8366.
(44) Sui, X., Yuan, J., Zhou, M., Zhang, J., Yang, H., Yuan, W., Wei, Y., Pan, C. (2008). Synthesis of cellulose-graft-poly (*N,N*-dimethylamino-2-ethyl methacrylate) copolymers *via* homogeneous ATRP and their aggregates in aqueous media. *Biomacromolecules*, 9, pp. 2615–2620.
(45) Zoppe, J. O., Peresin, M. S., Habibi, Y., Venditti, R. A., Rojas, O. J. (2009). Reinforcing poly (ε-caprolactone) nanofibers with cellulose nanocrystals. *ACS Appl. Mater. Interfaces*, 1, pp. 1996–2004.
(46) Sirviö, J. A., Kolehmainen, A., Liimatainen, H., Niinimäki, J., Hormi, O. E. (2014). Biocomposite cellulose-alginate films: Promising packaging materials. *Food Chem.*, 151, pp. 343–351.
(47) Li, Q., Zhou, J., Zhang, L. (2009). Structure and properties of the nanocomposite films of chitosan reinforced with cellulose whiskers. *J. Polym. Sci., Part B: Polym. Phys.*, 47, pp. 1069–1077.

(48) Gaspar, M., Benkő, Z., Dogossy, G., Reczey, K., Czigany, T. (2005). Reducing water absorption in compostable starch-based plastics. *Polym. Degrad. Stab.*, 90, pp. 563–569.
(49) Yasuda, K., Gong, J. P., Katsuyama, Y., Nakayama, A., Tanabe, Y., Kondo, E., Ueno, M., Osada, Y. (2005). Biomechanical properties of high-toughness double network hydrogels. *Biomaterials*, 26, pp. 4468–4475.
(50) Tian, H., Wang, Y., Zhang, L., Quan, C., Zhang, X. (2010). Improved flexibility and water resistance of soy protein thermoplastics containing waterborne polyurethane. *Ind. Crops Prod.*, 32, p. 1320.
(51) Arrieta, M. P., Fortunati, E., Dominici, F., Rayón, E., López, J., Kenny, J. (2014). PLA-PHB/cellulose based films: Mechanical, barrier and disintegration properties. *Polym. Degrad. Stab.*, 107, pp. 139–149.
(52) Shimazaki, Y., Miyazaki, Y., Takezawa, Y., Nogi, M., Abe, K., Ifuku, S., Yano, H. (2007). Excellent thermal conductivity of transparent cellulose nanofiber/epoxy resin nanocomposites. *Biomacromolecules*, 8, pp. 2976–2978.
(53) Liu, S., Tao, D., Yu, T., Shu, H., Liu, R., Liu, X. (2013). Highly flexible, transparent cellulose composite films used in UV imprint lithography. *Cellulose*, 20, pp. 907–918.
(54) Li, W., Wu, Y., Liang, W., Li, B., Liu, S. (2014). Reduction of the water wettability of cellulose film through controlled heterogeneous modification. *ACS Appl. Mater. Interfaces*, 6, pp. 5726–5734.
(55) Huq, T., Salmieri, S., Khan, A., Khan, R. A., Le Tien, C., Riedl, B., Fraschini, C., Bouchard, J., Uribe-Calderon, J., Kamal, M. R. (2012). Nanocrystalline cellulose (NCC) reinforced alginate based biodegradable nanocomposite film. *Carbohydr. Polym.*, 90, pp. 1757–1763.
(56) Azeredo, H. M., Miranda, K. W., Rosa, M. F., Nascimento, D. M., de Moura, M. R. (2012). Edible films from alginate-acerola puree reinforced with cellulose whiskers. *LWT-Food Sci. Technol.*, 46, pp. 294–297.
(57) Khan, A., Khan, R. A., Salmieri, S., Le Tien, C., Riedl, B., Bouchard, J., Chauve, G., Tan, V., Kamal, M. R., Lacroix, M. (2012). Mechanical and barrier properties of nanocrystalline cellulose reinforced chitosan based nanocomposite films. *Carbohydr. Polym.*, 90, pp. 1601–1608.
(58) Sobral, P. D. A., Menegalli, F., Hubinger, M., Roques, M. (2001). Mechanical, water vapor barrier and thermal properties of gelatin based edible films. *Food Hydrocoll.*, 15, pp. 423–432.
(59) Bilbao-Sáinz, C., Avena-Bustillos, R. J., Wood, D. F., Williams, T. G., McHugh, T. H. (2010). Composite edible films based on hydroxypropyl methylcellulose reinforced with microcrystalline cellulose nanoparticles. *J. Agric. Food Chem.*, 58, pp. 3753–3760.
(60) Pereira, A. L. S., do Nascimento, D. M., Morais, J. P. S., Vasconcelos, N. F., Feitosa, J. P., Brígida, A. I. S., Rosa, M. D. F. (2014). Improvement of

polyvinyl alcohol properties by adding nanocrystalline cellulose isolated from banana pseudostems. *Carbohydr. Polym.*, 112, pp. 165–172.
(61) Joseph, C. S., Prashanth, K. H., Rastogi, N., Indiramma, A., Reddy, S. Y., Raghavarao, K. (2011). Optimum blend of chitosan and poly-(ε-caprolactone) for fabrication of films for food packaging applications. *Food Bioprocess Technol.*, 4, pp. 1179–1185.
(62) Dhar, P., Bhardwaj, U., Kumar, A., Katiyar, V. (2015). Poly (3-hydroxybutyrate)/cellulose nanocrystal films for food packaging applications: Barrier and migration studies. *Polym. Eng. Sci.*, 55, pp. 2388–2395.
(63) Fujisawa, S., Okita, Y., Fukuzumi, H., Saito, T., Isogai (2011). Preparation and characterization of TEMPO-oxidized cellulose nanofibril films with free carboxyl groups. *Carbohydr. Polym.*, 84, pp. 579–583.
(64) Aulin, C., Gällstedt, M., Lindström, T. (2010). Oxygen and oil barrier properties of microfibrillated cellulose films and coatings. *Cellulose*, 17, pp. 559–574.
(65) Ankerfors, M. (2012). Microfibrillated cellulose: Energy-efficient preparation techniques and key properties. Licentiate Thesis, KTH Royal Institute of Technology, Stockholm, 49.
(66) Österberg, M., Vartiainen, J., Lucenius, J., Hippi, U., Seppälä, J., Serimaa, R., Laine, J. (2013). A fast method to produce strong NFC films as a platform for barrier and functional materials. *ACS Appl. Mater. Interfaces*, 5, pp. 4640–4647.
(67) Li, F., Biagioni, P., Bollani, M., Maccagnan, A., Piergiovanni, L. (2013). Multi-functional coating of cellulose nanocrystals for flexible packaging applications. *Cellulose*, 20, pp. 2491–2504.
(68) Brodin, F. W., Gregersen, O. W., Syverud, K. (2014). Cellulose nanofibrils: Challenges and possibilities as a paper additive or coating material — A review. *Nord. Pulp Pap. Res. J.*, 29, pp. 156–166.
(69) Nygårds, S. (2011). Nanocellulose in pigment coatings: Aspects of barrier properties and printability in offset. Degree of Master Linköping University.
(70) Syverud, K., Stenius, P. (2009). Strength and barrier properties of MFC films. *Cellulose*, 16, pp. 75–85.
(71) Lindström, T., Aulin, C. (2014). Market and technical challenges and opportunities in the area of innovative new materials and composites based on nanocellulosics. *Scand. J. For. Res.*, 29, pp. 345–351.
(72) Matthews, B., Mangalasary, S., Darby, D., Cooksey, K. (2010). Effectiveness of barrier film with a cellulose coating that carries nisin blends for the inhibition of Listeria monocytogenes. *Packag. Technol. Sci.*, 23, pp. 267–273.
(73) Cassano, R., Trombino, S., Ferrarelli, T., Barone, E., Arena, V., Mancuso, C., Picci, N. (2010). Synthesis, characterization, and anti-inflammatory

activity of diclofenac-bound cotton fibers. *Biomacromolecules*, 11, pp. 1716–1720.

(74) Wei, B., Yang, G., Hong, F. (2011). Preparation and evaluation of a kind of bacterial cellulose dry films with antibacterial properties. *Carbohydr. Polym.*, 84, pp. 533–538.

(75) Pinto, R. J., Marques, P. A., Neto, C. P., Trindade, T., Daina, S., Sadocco, P. (2009). Antibacterial activity of nanocomposites of silver and bacterial or vegetable cellulosic fibers. *Acta Biomater.*, 5, pp. 2279–2289.

(76) Fortunati, E., Armentano, I., Zhou, Q., Iannoni, A., Saino, E., Visai, L., Berglund, L. A., Kenny, J. (2012). Multifunctional bionanocomposite films of poly (lactic acid), cellulose nanocrystals and silver nanoparticles. *Carbohydr. Polym.*, 87, pp. 1596–1605.

(77) Lahiji, R. R., Xu, X., Reifenberger, R., Raman, A., Rudie, A., Moon (2010). Atomic force microscopy characterization of cellulose nanocrystals. *Langmuir*, 26, pp. 4480–4488.

(78) Postek, M. T., Vladar, A., Dagata, J., Farkas, N., Ming, B., Sabo, R., Wegner, T. H. (2008). Cellulose nanocrystals the next big nano-thing? *Proc. SPIE*, 7042, PP. 70420D1–70420D11.

(79) Azizi, S., Ahmad, M. B. H., Hussein, M. Z., Ibrahim, N. A. (2013). Synthesis, antibacterial and thermal studies of cellulose nanocrystal stabilized ZnO-Ag heterostructure nanoparticles. *Molecules*, 18, pp. 6269–6280.

(80) Llorens, A., Lloret, E., Picouet, P., Fernandez, A. (2012). Study of the antifungal potential of novel cellulose/copper composites as absorbent materials for fruit juices. *Int. J. Food Microbiol.*, 158, pp. 113–119.

(81) Zhong, T., Oporto, G. S., Jaczynski, J., Jiang, C. (2015). Nanofibrillated cellulose and copper nanoparticles embedded in polyvinyl alcohol films for antimicrobial applications. *BioMed Res. Int.*, 2015, 456834.

(82) Ma, X., Lv, M., Anderson, D. P., Chang, P. R. (2017). Natural polysaccharide composites based on modified cellulose spheres and plasticized chitosan matrix. *Food Hydrocoll.*, 66, pp. 276–285.

(83) Lavoine, N., Tabary, N., Desloges, I., Martel, B., Bras, J. (2014). Controlled release of chlorhexidine digluconate using β-cyclodextrin and microfibrillated cellulose. *Colloids Surf., B*, 121, pp. 196–205.

(84) Lavoine, N., Givord, C., Tabary, N., Desloges, I., Martel, B., Bras, J. (2014). Elaboration of a new antibacterial bio-nano-material for food-packaging by synergistic action of cyclodextrin and microfibrillated cellulose. *Innovative Food Sci. Emerging Technol.*, 26, pp. 330–340.

(85) Andrade, D. R. M., Mendonça, M. H., Helm, C. V., Magalhães, W. L. E., de Muniz, G. I. B., Kestur, S. G. (2015). Assessment of nano cellulose from peach palm residue as potential food additive: Part II: Preliminary studies. *J. Food Sci. Technol.*, 52, pp. 5641–5650.

(86) Shi, Z., Zhang, Y., Phillips, G. O., Yang, G. (2014). Utilization of bacterial cellulose in food. *Food Hydrocoll.*, 35, pp. 539–545.

(87) Innami, S., Fukui, Y. Additive composition for foods or drugs. US4659388A, 1987-04-21.
(88) Cantiani, R., Knipper, M., Vaslin, S. Use of cellulose microfibrils in dry form in food formulations. US6485767B1, 2002-11-26.
(89) Robson, A. (2012). Tackling obesity: Can food processing be a solution rather than a problem? *Agro Food Industry Hi Tech*, 23, pp. 10–11.
(90) Watson, R. R., Preedy, V. R. (2013). Bioactive Foods in Chronic Disease States. Robson, A. A. (Eds). *Preventing the Epidemic of Mental Ill Health: An Overview*, Elsevier Academic Press INC.
(91) Strom, G., Ohgren, C., Ankerfors, M. (2013). Nanocellulose as an additive in foodstuff, *Innventia Report*, 403, pp. 1–25.
(92) Cantiani, R., Knipper, M., Vaslin, S. (2002). Use of cellulose microfibrils in dry form in food formulations. Patent No.: US 6,485,767 B1.
(93) Kleinschmidt, D. C., Roberts, B. A., Fuqua, D. L., Melchion, J. R. Filling-containing, dough-based products containing cellulosic fibrils and microfibrils. US4774095A, 1988-09-27.
(94) Huq, T., Vu, K. D., Riedl, B., Bouchard, J., Han, J., Lacroix, M. (2016). Development of probiotic tablet using alginate, pectin, and cellulose nanocrystals as excipients. *Cellulose*, 23, pp. 1967–1978.
(95) Huq, T., Fraschini, C., Khan, A., Riedl, B., Bouchard, J., Lacroix, M. (2017). Alginate based nanocomposite for microencapsulation of probiotic: Effect of cellulose nanocrystal (CNC) and lecithin. *Carbohydr. Polym.*, 168, pp. 61–69.
(96) Li, W., Zhu, Y., Ye, F., Li, B., Luo, X., Liu, S. (2017). Probiotics in cellulose houses: Enhanced viability and targeted delivery of Lactobacillus plantarum. *Food Hydrocoll.*, 62, pp. 66–72.
(97) Li, W., Luo, X., Song, R., Zhu, Y., Li, B., Liu, S. (2016). Porous cellulose microgel particle: A fascinating host for the encapsulation, protection, and delivery of lactobacillus plantarum. *J. Agric. Food Chem.*, 64, pp. 3430–3436.
(98) Yano, H., Nogi, M. Nanofiber sheet and method for manufacturing the same. US9012010B2, 2015-04-21.
(99) Cunha, A. G., Mougel, J.-B., Cathala, B., Berglund, L. A., Capron, I. (2014). Preparation of double Pickering emulsions stabilized by chemically tailored nanocelluloses. *Langmuir*, 30, pp. 9327–9335.
(100) Mikulcová, V., Bordes, R., Kašpárková, V. (2016). On the preparation and antibacterial activity of emulsions stabilized with nanocellulose particles. *Food Hydrocoll.*, 61, pp. 780–792.
(101) Ojala, J., Sirviö, J. A., Liimatainen, H. (2016). Nanoparticle emulsifiers based on bifunctionalized cellulose nanocrystals as marine diesel oil–water emulsion stabilizers. *Chem. Eng. J.*, 288, pp. 312–320.
(102) Wang, W., Du, G., Li, C., Zhang, H., Long, Y., Ni, Y. (2016). Preparation of cellulose nanocrystals from asparagus (*Asparagus officinalis L.*) and

their applications to palm oil/water Pickering emulsion. *Carbohydr. Polym.*, 151, pp. 1–8.
(103) Jiménez Saelices, C., Capron, I. (2018). Design of Pickering micro and nanoemulsions based on the structural characteristics of nanocelluloses. *Biomacromolecules*, 19, pp. 460–469.
(104) Winuprasith, T., Suphantharika, M. (2013). Microfibrillated cellulose from mangosteen (*Garcinia mangostana L.*) rind: Preparation, characterization, and evaluation as an emulsion stabilizer. *Food Hydrocoll.*, 32, pp. 383–394.
(105) Winuprasith, T., Suphantharika, M. (2015). Properties and stability of oil-in-water emulsions stabilized by microfibrillated cellulose from mangosteen rind. *Food Hydrocoll.*, 43, pp. 690–699.
(106) Tang, J., Lee, M. F. X., Zhang, W., Zhao, B., Berry, R. M., Tam, K. C. (2014). Dual responsive Pickering emulsion stabilized by poly [2-(dimethylamino) ethyl methacrylate] grafted cellulose nanocrystals. *Biomacromolecules*, 15, pp. 3052–3060.
(107) Zoppe, J. O., Venditti, R. A., Rojas, O. J. (2012). Pickering emulsions stabilized by cellulose nanocrystals grafted with thermo-responsive polymer brushes. *J. Colloid Interface Sci.*, 369, pp. 202–209.
(108) Kalashnikova, I., Bizot, H., Cathala, B., Capron, I. (2011). Modulation of cellulose nanocrystals amphiphilic properties to stabilize oil/water interface. *Biomacromolecules*, 13, pp. 267–275.
(109) Kalashnikova, I., Bizot, H., Bertoncini, P., Cathala, B., Capron, I. (2013). Cellulosic nanorods of various aspect ratios for oil in water Pickering emulsions. *Soft Matter.*, 9, pp. 952–959.
(110) Loudet, J.-C., Yodh, A. G., Pouligny, B. (2006). Wetting and contact lines of micrometer-sized ellipsoids. *Phys. Rev. Lett.*, 97, p. 018304.
(111) Melle, S., Lask, M., Fuller, G. G. (2005). Pickering emulsions with controllable stability. *Langmuir*, 21, pp. 2158–2162.
(112) Kargar, M., Fayazmanesh, K., Alavi, M., Spyropoulos, F., Norton, I. T. (2012). Investigation into the potential ability of Pickering emulsions (food-grade particles) to enhance the oxidative stability of oil-in-water emulsions. *J. Colloid Interface Sci.*, 366, pp. 209–215.
(113) Varjonen, S., Laaksonen, P., Paananen, A., Valo, H., Hähl, H., Laaksonen, T., Linder, M. B. (2011). Self-assembly of cellulose nanofibrils by genetically engineered fusion proteins. *Soft Matter.*, 7, pp. 2402–2411.
(114) Zhu, W., Ma, W., Li, C., Pan, J., Dai, X. (2015). Well-designed multihollow magnetic imprinted microspheres based on cellulose nanocrystals (CNCs) stabilized Pickering double emulsion polymerization for selective adsorption of bifenthrin. *Chem. Eng. J.*, 276, pp. 249–260.
(115) Lee, K.-Y., Blaker, J. J., Murakami, R., Heng, J. Y., Bismarck, A. (2014). Phase behavior of medium and high internal phase water-in-oil emulsions

stabilized solely by hydrophobized bacterial cellulose nanofibrils, *Langmuir*, 30, pp. 452–460.
(116) Capron, I., Cathala, B. (2013). Surfactant-free high internal phase emulsions stabilized by cellulose nanocrystals. *Biomacromolecules*, 14, pp. 291–296.
(117) Miao, C., Tayebi, M., Hamad, W. Y. (2018). Investigation of the formation mechanisms in high internal phase Pickering emulsions stabilized by cellulose nanocrystals. *Phil. Trans. R. Soc. A*, 376, p. 20170039.
(118) Cervin, N. T., Johansson, E., Larsson, P. A., Wågberg, L. (2016). Strong, water-durable, and wet-resilient cellulose nanofibril-stabilized foams from oven drying. *ACS Appl. Mater. & Int.*, 8, pp. 11682–11689.
(119) Cervin, N. T., Johansson, E., Benjamins, J.-W., Wågberg, L. (2015). Mechanisms behind the stabilizing action of cellulose nanofibrils in wet-stable cellulose foams. *Biomacromolecules*, 16, pp. 822–831.
(120) Jin, H., Zhou, W., Cao, J., Stoyanov, S. D., Blijdenstein, T. B., de Groot, P. W., Arnaudov, L. N., Pelan, E. G. (2012). Super stable foams stabilized by colloidal ethyl cellulose particles. *Soft. Matter.*, 8, pp. 2194–2205.
(121) Tasset, S., Cathala, B., Bizot, H., Capron, I. (2014). Versatile cellular foams derived from CNC-stabilized Pickering emulsions. *Rsc Advances*, 4, pp. 893–898.

https://doi.org/10.1142/9781786349477_0009

Chapter 9

Nanocellulose as a Green Material to Eradicate Environment and Renewable Energy Issues

Mazhar Ul-Islam[*], Muhammad Wajid Ullah[†,||], Shaukat Khan[‡], Sehrish Manan[†], Waleed Ahmad Khattak[§], Tahseen Kamal[¶], Aref Ahmad Wazwaz[*], and Ahmed Saadi Abrehem[*]

[*]*Department of Chemical Engineering, College of Engineering, Dhofar University, Salalah, Sultanate of Oman*

[†]*Department of Biomedical Engineering, College of Life Science and Technology, Huazhong University of Science and Technology, Wuhan, PR China*

[‡]*Materials Science Institute, PCFM Lab and GDHPRC Lab, School of Chemistry, Sun Yat-Sen University, Guangzhou, PR China*

[§]*Hyperdrive Lubricants Limited, Certas Energy, UK*

[¶]*Center of Excellence for Advanced Materials Research, King Abdulaziz University, Jeddah, Saudi Arabia*

[||]*wajid_kundi@hust.edu.cn*

Abstract

The human environment is at high risk of contamination by the extensive use of non-degradable resources as well as exhaustion of naturally available resources. To combat the environmental and energy issues, recent developments in nanotechnology have open gateways for the sustainable development of eco-friendly, biodegradable, and renewable polymeric materials. Nanocellulose, possessing unique features such as fibrous structure, high mechanical strength, large surface area, low visual scattering, low-cost, renewability, non-toxicity, biocompatibility, and easy availability, serves as an ideal material for diverse environmental applications. In addition, its unique three-dimensional fibril arrangement allows the impregnation of variety of nanosize materials to enable the development of nanocomposites in the form of hydrogels, aerogels, and films, papers, or membranes. Such substrates serve as templates for inorganic nanoparticles and polymers, or a combination of both. Such unique features make nanocellulose-based materials more efficient, robust, stable, reliable, and environmentally-friendly, thus enabling it to find potential applications in the development of antimicrobial filters and devices for removal of heavy metals, in water treatment and wastewaters purification, in the development of pollutant sensors, as well as in potential applications in catalysis and renewable energy. This chapter provides a comprehensive picture of the recent developments in nanocellulose-based materials to address various issues associated with environment and renewable energy.

Keywords: Nanocellulose; environmental issues; renewable energy; nanocomposites; bio-filters; biodegradation

1. Environmental Challenges

Living things including plants, animals, human beings, and microbes as well as their physical surroundings constitute the components of our environment. These components are dependent on each other, and any imbalance in the interaction between them lead to environmental hazards. Above all, human beings themselves are greatly disturbing the environmental balance to accomplish their own needs. This growing imbalance is leading to various environmental problems including pollution, depletion of natural resources, floods, soil erosion, global warming, overpopulation, waste disposal, deforestation, acid rain, etc. [1, 2]. Environmental pollution is known for causing undesirable changes in the physical, chemical, and biological characteristics of air, water, and land and is currently at an alarming situation. Broadly, deforestation, overpopulation, industrialization, and waste secretion to air and water are the major causes of such environmental hazards. The pollutants responsible for the environmental problems can be broadly divided into two main categories, i.e., degradable and non-degradable. The biodegradable pollutants are usually organic materials like domestic wastes which are easily broken into their simplest units by different microbes, and subsequently enter the respective biochemical cycles, while non-biodegradable pollutants have strong chemical bonds and cannot be broken through natural sources [3].

The chemical industry is routinely adding enormous amounts of chemical materials to the environment, mainly water, as pollutants, most of which are non-biodegradable. Further, the synthesis of these chemical extensively utilizes natural energy sources: for example, fossil fuels are used for the synthesis of various chemical products such as plastic. Although plastic synthesis from petroleum resources has brought a revolution in industry due to low price of raw materials, limited processing cost, and low price of final product [4], it is non-degradable and thus embracing the world with disposal problems. There is no exact definition of biodegradation, and it is even more difficult to observe this process. It mainly refers to the biological or chemical changes in the material nature caused by the actions of living entities such as microbes or enzymes [5]. This process is governed by different biotic and abiotic factors and is mainly comprised of three stages: biodeterioration, biofragmentation, and assimilation. During this process, the large products are broken down into small fractions that could cross the microbial cell wall or membrane. Even if some chemical products are degradable, their degradation process is very slow and thus occupies the land for a long time. Therefore, the utilization of these plastic-based materials, for packaging containers, trash bags, and daily fast food utensils, and their subsequent disposal are creating serious environmental issues. As a result of the fossil fuel crisis, alternative energy and raw materials sources such as biomass are propagated and being investigated intensively. Biomaterials including biopolymers and polymer-based composites from renewable sources are considered as alternatives for sustainable development and reduction of environmental hazards. The most important aspect of using natural bioresources is their quick and easy degradation, although some of the products might need years for complete dissimilation in the ecosystem. To get rid of plastic wastes, one important alternative is the mixing of biopolymer with polymer matrix although it can result in reduced physical and mechanical properties.

Currently, different materials obtained from renewable resources have received immense consideration owing to their low-cost, environmentally-friendly nature, easy availability, renewability, and non-toxicity [6]. Such materials can thus serve as an alternative for petroleum-based synthetic materials. In line with this, the nanocellulose-based materials have emerged as green materials in a wide range of applications for solving environmental issues. Cellulose, produced by plants [7], microbial cells [8], and cell-free systems [9–11], is considered as the most important polymer because of its easy availability, simple processing, cost-effective end products, and widespread applications [12–14]. Low price, multiple forms, widespread availability, high mechanical properties, and biodegradable nature advocate the use of cellulose in the synthesis of various composite materials for resolving environmental issues and obtaining economic benefits [12, 15, 16]. Its production resources, structure, properties, and applications are detailed in the following sections. To date, extensive efforts have been made to develop strategies for cost-effective cellulose production and to explore its widespread applications to resolve environmental and renewable energy issues [17, 18]. The nanocellulose-based materials are rapidly replacing the conventional ones and minimizing the risk to the environment and remediating environmental pollution in a more sustainable way.

2. Cellulose: Structure and Features

Being the primary component of cell wall, cellulose is the most important structural component of plants and certain bacteria, fungi, and algae. It is a hard, fibrous, and water-insoluble polymer that bestows highest plant cells with lofty mechanical properties. The chemical structure of cellulose indicates that it is produced by condensation polymerization reactions of β-1,4-linked glucopyranose units [19]. An exciting and chemically important feature of cellulose is the presence of multiple free hydroxyl (OH) groups on its surface which allow its surface modification by various methods to prepare different functional materials. Further, the hydrogen bonding capability of these OH groups have a vital role in directing the crystalline packing of cellulose. Generally, cellulose is produced as a long chain of individual molecules; however, it undergoes spinning at the synthesis site in hierarchical order, resulting in assemblies of cellulose fibers. Individual cellulose chains combine to produce elementary fibers. Their arrangement and final diameter range from 2 to 20 nm depending on different biosynthetic conditions. Naturally produced cellulose is of cellulose I form in which the chains are in parallel confirmation [20]. The proto-fibrils then further aggregate *via* coalescence to produce micro-fibrils, in which configuration there is reduced free surface energy. The fibril aggregation is mainly dependent on the Van der Waals and intra- and inter-molecular hydrogen bonding interaction owing to the presence of OH moieties in the cellulose skeleton. The orientation of cellulose molecules that aggregate to form microfibrils differs based on their production source. The aggregation of microfibrils results in closely packed cellulose chains that produce crystallites stabilized by the strong hydrogen bonding. Besides, there exist some less ordered cellulose chains that govern the amorphous part of cellulose [21, 22] (Fig. 1).

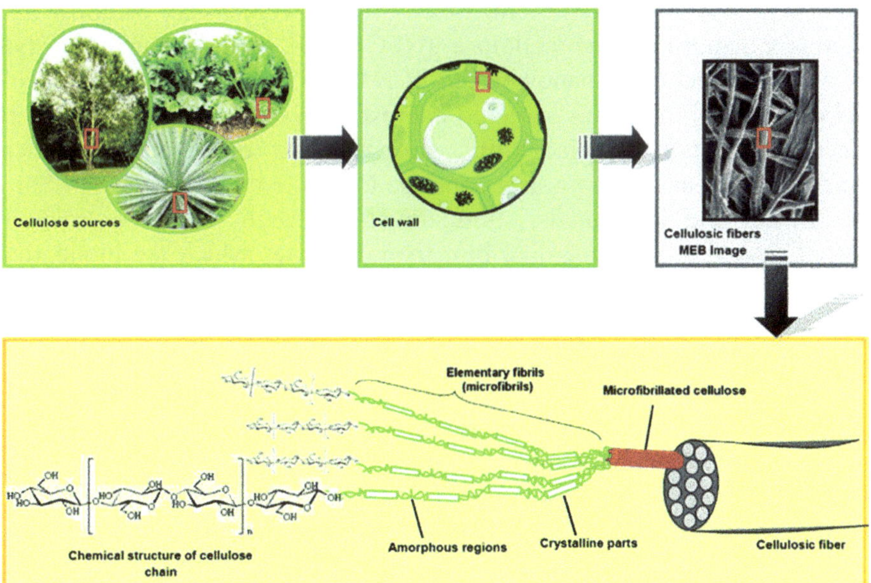

Fig. 1. Schematic illustration of synthesis of cellulose molecules from the plant-derived cellulose. The scheme shows a detailed view of the cellulose fiber structure with emphasis on the cellulose microfibrils.

Source: Figure reproduced from Ref. [29] with permission from Elsevier.

Based on molecular orientation and difference in inter- or intra-molecular interaction, cellulose can exist as different allomorphs and polymorphs. To date, six different polymorphs including cellulose I, II, III, III_I, III_{II}, IV_I, and IV_{II} have been reported [23]. Natural cellulose usually has the crystal structure of cellulose I, that is further divided into Iα (triclinic structure) and $I\beta$ (monoclinic structure) [24]. The nanofibers of cellulose further unite and assemble in the form of microfibers that, on further aggregation, generate cellulose fibers. Cellulose fibers are arranged in a web shape which ultimately produce an interconnected porous net-like structure [25–28].

3. Nanocellulose as a Green Material

Literature survey reveals that there is no standard nomenclature for nanocellulose; however, it is broadly defined as a set of particles having at least one dimension in the nanoscale and is obtained by different chemical/mechanical or only chemical processes, or through acid hydrolysis. Nanocellulose is a generic term for rod-like or ribbon-like cellulose and contains nanofibrillated cellulose (NFC) [30], cellulose nanocrystals (CNC) [31], cellulose nanowhiskers (CNW) [32], and bacterial nanocellulose (BNC) [33]. It has garnered interest from the scientific community because of its biodegradability, strength, and other characteristics. Sustainability and green issues continue to be the top priorities for many businesses and individuals, stimulating the search for non-petroleum-based structural materials like bionanocomposites which are biodegradable, highly efficient, and lightweight [34]. Our society is evolving as a technology-oriented society; therefore, environmental issues become more important for the desires of easiness, comforts, and safety. Modern technologies, such as information technology, biotechnology, and nanotechnology, have all contributed to develop our society in different fields. However, future technologies should be concerned more about the living environment and harmonize with it, and is related to sustainability in future technologies. Nanocellulose can be used as a green material, which means that is should be environment-friendly, safe, non-toxic, and absolutely from natural sources.

Nanocellulose is used in different sectors like the automobile industry for reduction of greenhouse gas emission as a result of the reduced weight of vehicles, in the building industry to reduce the greenhouse gas emission either by an increase of bio-based, recyclable, and lightweight building materials or by decreasing the energy consumption during the lightweight building materials. These advantages would effectively improve the insulation properties, information, and communication in industries where nanocellulose can be used as a green material to enhance their functionalities. In the packaging industry, there is a trend towards sustainable lightweight packaging combined with the demand for biodegradable material, particularly for food packaging application. These demands are partially driven by legislation, for example to even ban the non-biodegradable plastic materials. Nanocellulose can also be used as a green dispersant for two-dimensional energy materials [35] as well as a green and sustainable emerging material in energy applications [36].

4. Nanocellulose-Based Composites Synthesis Strategies

Cellulose is the most abundant natural polymer reflecting almost in exactable raw material that can be used for production of various environmental-friendly, economical, and biocompatible products. Besides its direct applications in numerous products, it has been largely employed in the synthesis of various nanocomposites. The nanocellulose-based composites are prepared by various strategies. Broadly, these strategies are classified into two main categories, discussed in the following.

5. Using the Disintegrated Cellulose Microfibrils/ Nanoscale Whiskers for Reinforcing Polymers

Cellulose, more specifically bio-cellulose, is known to contain highly crystalline and pure microfibers consisting of perfect crystals. These microfibers can be disintegrated and used as fillers for reinforcing various polymers [37]. There are several reports on the synthesis of various composite materials using cellulose crystals/whiskers as reinforcement components [38]. PLA nanocomposites reinforced with cellulose whiskers were synthesized *via* melt extrusion process. The suspension of nanocrystals was pumped in to the polymer-melt during the extrusion process [39]. In a similar study, PVA was used as compatibilizer to enhance the dispersion of cellulose whiskers in PLA matrix [40]. In another study,

Cao *et al.* [41] reported that flax cellulose nanocrystals reinforced waterborne polyurethane nanocomposite materials [41]. Hussain *et al.* (2014) reported around 50% enhancement in flexural properties of PLA reinforced with cellulose nanocrystals [42]. Besides plant cellulose, bacterial cellulose (BC) has also been utilized as a reinforcement partner in numerous nanocomposites. Park *et al.* (2007) prepared BC fibers-reinforced POE nanofibers where BC nanowhiskers were incorporated into POE *via* electrospinning process. They observed that fibers were well incorporated and distributed in the polymer fibers [43]. Similarly, cellulose nanowhiskers were incorporated into electrospun PVA fiber mats, which enhanced the mechanical properties of the developed composite [44, 45].

6. Cellulose and Cellulose Derivatives as Matrix Reinforced by Nanomaterials

Such cellulose and cellulose-based composites are extensively available. Owing to the web-like fibrous structure of cellulose and its hydrophilic nature, the reinforcement materials can attach to its surface as well as penetrate within the matrix. Depending upon the type of reinforcement material, the nanocomposites are synthesized by using different strategies such as *in situ*, *ex situ*, and blending methods, as comprehensively described by Shah *et al.* [26]. The choice of synthesis methods mainly depends on type and nature of the reinforcement materials. The non-toxic and small size materials can be impregnated *in situ* into the cellulose matrix; however, the toxic or very dense materials are impregnated *ex situ* into the matrix [46, 47]. Nanoparticles are the main reinforcement materials used to synthesize composites of cellulose with different materials for broad-spectrum applications. For example, nanocomposites of cellulose have been synthesized with silver [48], zinc oxide [49, 50], montmorillonite clay [51], titanium [47, 52], gold [53], and several other nanoparticles. Besides, different polymers have also been employed as reinforcement materials to strengthen the cellulose skeleton, improved its existing properties, and impart new features. Examples of polymer reinforcement includes collagen [54], silk-sericin [55], gelatin [56], alginate [57], chitosan [58], etc. Furthermore, there are several reports where composites have been synthesized by dissolving cellulose and then adding reinforcement materials, casting them in the form of sheets. This strategy

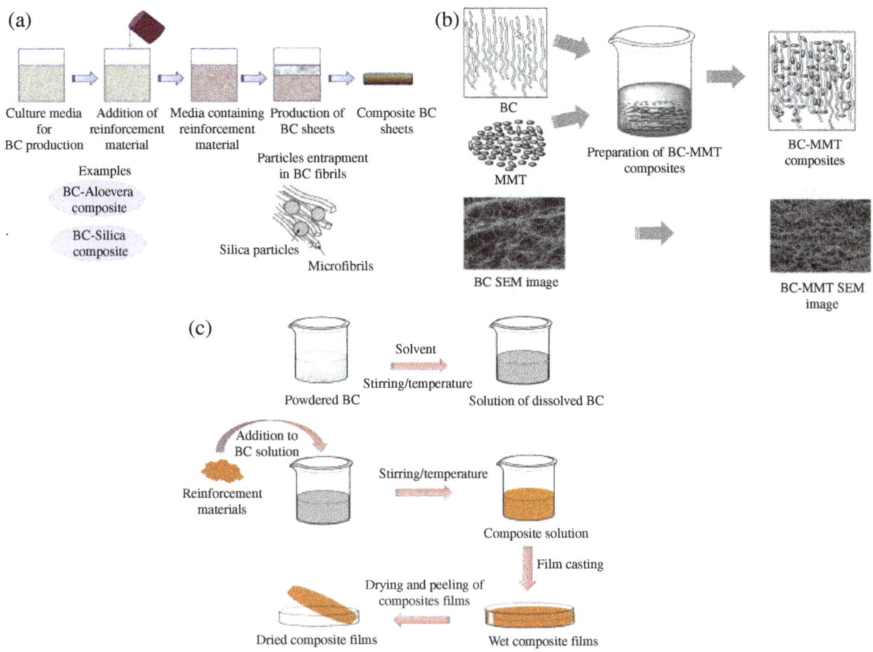

Fig. 2. (a) Schematic representation of BC composites synthesized through an *in situ* synthetic strategy. Particles entrapped between growing BC fibrils are visible in the figure, (b) schematic representation of BC composites synthesized through an *ex situ* synthetic strategy. The example illustrates the penetration of particles in the BC matrix, and (c) schematic representation of BC composites synthesized from dissolved BC solutions. The composite solution was casted to prepare BC films.

Source: Figure reproduced from Ref. [26] with permission from Elsevier.

allows the reinforcement of all kind of materials, irrespective of their size, shape, and nature, such as nano- and micro-particles as well as polymers. Figure 2 summarizes various methods used for synthesizing nanocomposites using cellulose as matrix.

7. Motivation for Development of Nanocellulose Composites to Solve Environmental Issues

Global economic and environmental hazards caused by non-degradable materials are issues of serious concern. Numerous efforts dedicated to resolve these issues argue the seriousness of the problem. Polymer composites developed from renewable resources offer a solution to this issue as well as for the sustainable progress of economically and ecologically attractive technologies. Cellulose, being a biocompatible and biodegradable polymer, is of interest to be applied instead of conventional non-renewable and non-degradable materials. It has received prime importance since two decades to serve as an alternative to the fossil fuel-based materials. Besides being an alternative for plastic materials, another important prospect of cellulose is its addition to synthetic polymers to provide them biodegradability. Afterwards, these materials can be easily degraded in the natural ecosystem, thus eliminating the issues of environmental hazards.

Although completely degradable, cellulose degradation process is too slow. This could be attributed to its highly crystalline microfibrils and their presence within the matrix, making it difficult for the hydrolyzing enzymes to be able to access and degrade it. This makes cellulose a slowly degrading polymer *via* enzymatic hydrolysis. The cellulose chains hydrolyze both in basic and acidic media *via* chemical hydrolysis reactions. Microbial degradation, especially by fungi, involves different enzymes such as endoglucanases (EGs), cellobiohydrolases (CBHs), and β-glucosidase (BGL) [59]. Different studies have evaluated the degradability of cellulose-based composites. Soil burial is one method to monitor the biodegradation process. In such a method, the mass of a material is measured before and after burial is soil and the loss of mass indicates the amount of biodegraded material and its integration into the microbial metabolism. Kalka *et al.* [60] observed the biodegradation behavior of all-cellulose composites using soil burial experiments in order to investigate the end-of-life disposal of this relatively new class of bio-based composite materials. They reported that all-cellulose composites possessed exceptionally high biodegradability and lost more than 70% mass after 70 day of soil burial [60]. In another study, Schröpfer *et al.* (2015) observed the degradation behavior of BC, vegetable cellulose, and poly(hydroxybutyrate). They buried the controlled mass samples in pots containing soil for various times intervals from one to six months and observed that despite the initial mass increase in first month, the polymers showed considerable degradation with passage of time as evidenced by loss of mass [61]. Another method of monitoring the biodegradation process is to observe the loss of mass of a material by placing it in a culture medium with a potentially known microorganism capable of biodegradation [5].

8. Challenges in Processing of Cellulose Nanocomposites

There exist several challenges in the development of cellulose-based nanocomposites which need to be addressed before these can be used for practical applications. For example, the moisture sensitivity of the nanocellulose-based materials must be alleviated for their application as packaging material and several other high-end applications. Besides, the high aspect ratio of the fibrils makes it a gel at low concentration: this property is useful in biomedical applications where it is used as a scaffold [62]; however, it is a limitation for environmental application as it reduces its capacity to remove water during the preparation of dispersion. Furthermore, the technological platform and economically efficient production of films, aerogels, and filaments are also lacking [63]. Therefore, it is imperative to identify possible solutions to overcome such obstacles.

There are some popular nanocellulose processing methods like solution casting in which the polymer is dissolved and mixed with nanocellulose followed by evaporation of solvent to obtain flat films. The polymer solution is casted in different molds to obtain composites of desired shape and thickness. Foaming is another strategy where blowing agents, gas, or pressure are used to complete the process following which fiber spinning is carried out by using different strategies such as dry spinning, melt spinning, wet spinning, or electrospinning. Finally, impregnation is carried out, in which the nanofiber network is impregnated with polymer. Despite the usefulness of these strategies, these often encounter several discrepancies. For example, it is hard to get uniform distribution and dispersion of nanocellulose in polymer. The interfacial adhesion between nanocellulose and polymer is also a major challenge. Sometimes, it is hard to completely remove the solvent without affecting the polymer interface and alignment after dissolution of nanocellulose. The mixing and extrusion process control of nanomaterials in the extruder is also quite difficult. The freeze-dried nanocellulose can cause health problem. Similarly, the impregnation of a dense nanofiber network with the polymer usually affects its properties like brittle fracture surface morphology. To some extent, these issues have been addressed through 3D printing of nanocellulose, which improves its alignment and tailored properties [64–66].

9. Environmental Applications of Nanocellulose-Based Composites

Air and water are the important components of human environment which are largely contaminated by wastewater, pollutants, microorganisms, viruses, heavy metals, etc. These contaminants need to be filtered out for safe use of such natural resources. To date, different strategies have been developed for the filtration of these contaminants including adsorption, filtration, etc.; however, these strategies are often overshadowed by several limitations such as high cost of filters, low efficiency, selectivity of filtration membrane, etc. Currently, nanotechnology has emerged as alternative to various environmental issues by developing advanced approaches and equipment. Recent developments in membrane technology have effectively addressed the issues associated with air and water filtration caused by microbial and heavy metal contamination owing to the unique advantages offered by the nanofibers constituting the membrane [67, 68]. These nanofibers minimize the pressure drop, improve the efficiency, and offer high surface adsorption of contaminants due to large surface area-to-volume ratio. The development of nanocellulose-based composite materials is one such example that is largely replacing the conventional equipment and helping develop new green materials with improved performance. This technology utilizes minimum energy and materials in addition to minimizing the harm to environment and also leads to remediation of environmental pollution in a green and sustainable way. Broadly, this technology devises solutions to the existing environmental issues, offer preventive measures to avoid future problems, and highlights the issues caused by this technology itself [16]. The following sections describe the latest developments in nanocellulose-based filters for the removal of microorganisms, viruses, and heavy metals in addition to their use in medium filtration, wastewater treatment, and renewable energy and are summarized in Table 1.

10. Antimicrobial, Virus, and Air Filters

Nanocellulose-based composites with different materials have been developed by different methods, and their filtration capabilities have been explored for the removal of unwanted microorganisms and viruses from drinking water and air [1]. In the development of such filters, nanocellulose offers high surface-to-volume ratio for the uniform impregnation of

Table 1. Environmental and energy applications of nanocellulose.

Applications	Nanocellulose-type	Reinforcement materials	Features/Uses	Refs.
Antimicrobial filtration	Poly (methacrylic acid)-functionalized CNF	Ag nanoparticles	Supramolecular structure, fluorescence marker, and antibacterial properties; application in wound-healing pads	[72]
	TEMPO-oxidized CNF-based hydrogel, aerogel, and films	Ag nanoparticles	Uninform distribution of Ag nanoparticles, mechanical properties, porosity, reusability	[70]
	CNF	TiO_2-chitosan-Ag	Layer-by-layer distribution of reinforcement materials, cable-like core–shell structure, antibacterial activity against Gram-positive and Gram negative bacteria	[69, 73, 98]
	BC	Ag nanoparticles	Antibacterial activity against Gram-positive and Gram-negative bacteria	[99]
	Thio-oxidized BC	Ag nanoparticles	Improved particle size	[71]
Virus filtration	CNF		Absolutely natural and unmodified CNF membrane, tailored pore size distribution comparable to viruses	[75]
Removal of heavy metals	Diethylaminoethyl (DEAE) cellulose	Separately coated Cu, Cd, Co, Cr, Fe, Ni, and Zn nanoparticles	Simple, reliable, and efficient	[79]
	Carboxymethyl cellulose	Clay materials	Void volume, high surface area, electrically conductive, and efficient removal of Pb from wastewater	[80]
	BC	Fe_3O_4 nanoparticles	Efficient removal of Pb, Mn, and Cr from wastewater, reusability	[81]
	BC	Chitosan	Efficient removal of Cu from wastewater, reusability	[82]

(*Continued*)

Table 1. *(Continued)*

Applications	Nanocellulose-type	Reinforcement materials	Features/Uses	Refs.
	NCC	2-mercaptobenzamide modified itaconic Acid and magnetite grafting	Efficient removal of Hg from aqueous solutions	[85]
Water purification	NCC	Poly (itaconic acid/ methacrylic acid)-grafting	Efficient removal of Co from nuclear industry wastewater samples	[86]
	CNF	—	Rod-like elongated nanofibrillated morphology, reusability, cost-effective, and efficient and environment-friendly green remediation of Cd, Ni, and Pb	[89]
	CNF	Resins and organic solvent	Hollow fiber membrane, low-cost, reverse osmosis, high productivity, and efficient	[90]
	CNF	TiO_2 nanoparticles	Good photocatalytic activity and long-term stability, efficient removal of methyl orange and paracetamol	[97]
Dye removal and oil	CNF	Chitosan	Efficient removal of ethyl orange dye from wastewater	[87]
	CNF	Fe_3O_4 nanoparticles	Efficient adsorption of Congo red dye	[88]
	CNF	Ag nanoparticles	Reusable, cost-effective, multifunctional (water purification and removal of dyes, heavy metals, and microbes)	[93]
	CNF	Methyltrimethoxysilane	Efficient, ultralight, flexible, hydrophobic, and oleophilic sponge	[94]
Sensing	CNF	—	High porosity, toughness, and mechanical properties, optical and chemically stable, application in security paper	[100, 101]
	BC	Ag nanoparticles	Natural sponges-like structure upon immersion in aqueous or ethanolic solutions, detection of thiosalicylic acid and 2,2′-dithiodipyridine	[102]

	NCC	Au nanoparticles	Cost effective, high enzyme immobilization ability (cyclodextrin glycosyl transferase and alcohol oxidase)	[103]
	BC	Au, horseradish peroxidase and heme proteins including hemoglobin and myoglobin	Detection of traces of H_2O_2, light weight, flexible, and shape stable in both aqueous acidic and alkaline solutions	[104]
	BC	Au horseradish peroxidase and glucose oxidase	Light weight, flexible, and shape stable in both aqueous acidic and alkaline solutions, detection of traces of glucose	[105]
	CNF	Au, cationic polyelectrolyte (CPAM)	SERS reproducibility, sensitivity, and high enhancement factor	[106]
Catalysis	CNF	TiO_2, Au nanoparticles	High mechanical properties, optical transparency, and efficiency	[107, 108]
	NCC	Separately coated CuO, Au, Pd, Pt, and Ag nanoparticles	Facile and green synthesis, enhanced catalytic activity, reduction of 4-nitrophenol	[109, 110]
	CNF	TiO_2 nanoparticles	Photocatalytic decomposition of organic pollutants such as methylene blue	[107] [111]
	NCC	Pd nanoparticles	Conversion of phenol to cyclohexane	[112]
	NCC	Au nanoparticles	Enhanced catalytic activity, degradation of 4-nitrophenol	[110]
	TEMPO-oxidized CNF	Bimetallic (Au and Pd)	Efficient degradation of 4-nitrophenol	[113]
Energy	MFC	Si clusters	Transparent and conductive fibers, applications in solar cell	[114]
	MFC	Graphite	Highly flexible Li-ion battery	[115]
	MFC	Si and CNT	Flexible and low cost Li-ion battery	[116]

(*Continued*)

Table 1. (*Continued*)

Applications	Nanocellulose-type	Reinforcement materials	Features/Uses	Refs.
	MFC	Graphite, SiO_2, and $LiFePO_4$	Highly flexible Li-ion battery with enhanced cycling performance	[117]
	CNF	Si and CNT	Highly flexible single paper-based Li-ion battery with high performance	[117]
	BC	CNT and ionic liquid	Supercapacitor with high flexibility, electrochemical properties, mechanical stability, and cyclability	[118]
	BC	Multi-walled carbon nanotubes (MWCNTs) and polyaniline	Supercapacitor with high flexibility, facile and low-cost, porous, high specific capacitance and excellent cycling stability, and low capacitance degradation	[119]
	CNF	MWCNTs	User-friendly rechargeable batteries	[120]
	BC	Pd nanoparticles	Membrane electrode assembly (MEA) with high thermal stability and lower gas crossover, application in fuel cell	[121]
	CNF	Indium tin oxide nanoparticles	High power conversion efficiency, applications in solar cells and photovoltaics	[114]

bactericidal materials such as silver, chitosan, and titania, to impart antibacterial properties to the filter [69]. Further, these nanofilters prevent the risk of nanoparticle leakage due to the compact fibrous structure of nanocellulose [70]. The nanocellulose-based composite filters can be developed either through impregnation of nanoparticles into the matrix or be immobilized on its surface. This is usually achieved through reduction in the presence of nanocellulose by different methods such as UV irradiation, by addition of reducing chemicals, and through *in situ* reduction by the OH groups of cellulose. Wan and Guhados developed Ag nanoparticles-coated BC nanofilter by incorporating nanoparticles into the BC matrix. During the preparation of nanofilters, the cellulose fibers were surface oxidized by adding the thio-group to BC, which reacted with silver proteinate and enhanced the Ag nanoparticles size [71]. In addition to Ag nanoparticles, the nanocellulose-based composites filters have also been developed with other Ag-based nanomaterials such as silver chloride (AgCl) nanoparticles and silver nanoclusters. Díez *et al.* (2011) developed a simple method for the fabrication of Ag nanocluster on the surface of nanocellulose. The Ag nanoclusters were immersed in aqueous AgCl suspension agitated in the dark. The synthesis was confirmed by the fluoresence activity because of the electronic transitions between quantized energy levels [72]. In another study, Hu *et al.* (2009) developed an *in situ* method for the synthesis of AgCl nanoparticles in BC matrix by soaking BC in $AgNO_3$ solution followed by the addition of NaCl [73]. All these Ag-based nanofilters showed strong antimicrobial properties against a range of microorganisms. In addition to microorganisms, nanocellulose-based filters have shown great efficiency in removal of viruses from water to a level comparable to industrial virus filters [74, 75]. Filter paper is comprised of highly crystalline nanocellulose and is routinely used filter in laboratory and industrial research. Metreveli *et al.* (2014) developed specialized filter paper with tailored pore size, which allowed effective diffusion of different sized polystyrene beads (500, 100, and 30 nm), comparable to the size of different viruses; thus, these can be potentially used for filtration of viruses [75].

The routinely used biological media for the growth of microbial and animal cells need to be filtered prior to their use in order to avoid contamination by unwanted microbial cells and other contaminants. The nanocellulose-based filters for medium filtration are generally composed of two layers: the outer and inner layers. The outer layer is comprised of melt blown filter medium while the inner layer is comprised of cellulose and provides support to the inner layer. These bilayer filters can find potential applications in the filtration of fluids and fiber-type media. The cellulose in the outer layer provides only sub-micron filtration. Efficient filtration by using such bilayer filters for filtration of particles can be achieved through careful deposition of additional fine fibers on the cellulose support medium to avoid any damage to the delicate fiber layers [76].

Compared to conventional dust capturing filters such as glass fibers, the nanofiber-based mats have recently received immense consideration for air filtration applications owing to their low cost, large surface area, simplicity, and high efficiency due to inertial impaction and interception, both alone or in combination with other filtration medium [67, 77]. The nanofibers used in the preparation of such mats are usually 800 times smaller than the conventional filtration fibers. These nanofiber-based mats effectively capture the dust particles on their surfaces and remove them by back-flushing or through mechanical means [67]. In such filters, the filter medium is comprised of nanocellulose fibers corrugated with some other materials. These filters possess greater life span and can be effectively reused while maintaining their high efficiency.

11. Removal of Heavy Metals

Microorganisms present in aquatic system accumulate a variety of heavy metals such as Cu, Pb, Cd, Cr, Co, Ni, Se, Sr, etc. to toxic levels. These heavy metals are non-biodegradable, and although some of them are essential for human health, their excess intake can cause serious risk to aquatic life as well as human health [78]. Therefore, extensive efforts are required to develop their detection methods and minimize their concentration to acceptable levels, around few tenths of parts per billion, to make it safe for drinking. For Cu detection, Marwani *et al.* (2014) developed a simple, reliable, and efficient method by modifying the nanocellulose surface with different amounts of lanthanum hydroxide that served as a selective marker [79]. Composites of clay materials with cellulose have also shown potential for removing biological, organic, and inorganic contaminants from water. Ake *et al.* (2001) developed a porous clay-based nanocellulose composite for sorption of Pb from water. The developed nanocomposite showed higher Pb sorption efficiency by providing void volume, high surface area, and high electrical conductivity [80]. In another study, Zhu *et al.* (2011) developed spherical magnetic nanoparticles-functionalized BC nanocomposites which showed high efficiency in removing Pb, Mn, and Cr from wastewaters. The nanocomposites were reusable after elution of heavy metal ions [81]. In a recent study, Urbina *et al.* (2018) developed a BC-Chitosan membrane by using the 3D network of BC as a physical support and chitosan as functional entity. The developed membrane showed high efficiency in removing Cu from wastewaters in batch operation [82]. The contaminant removal efficiency of these nanocomposites can be improved by various combinations of sorbent materials, careful designing, optimum concentration of nanoparticles, and right choice of polymeric system.

12. Water Treatment and Wastewater Purification

Natural water is largely contaminated by microorganisms, industrial wastes, human wastes, heavy metals, oils, and different chemicals [83]. To data, several commercial and non-commercial approaches, devices, and technologies have been developed to improve the water purification, wastewater treatment flow, and amount of pollutants in wastewater with the aim to improve the quality of water for humans, animals, and aquatic life [84].

Nanocellulose-based adsorbent materials can potentially remove different toxic materials such as mercury [85], cobalt [86], ethyl orange dye [87], Congo red dye [88], cadmium, nickel, and lead [89] from wastewater [84]. Lee *et al.* (2013) developed a hallow fiber membrane, comprised of cellulose-based resins, a plasticizer, and an organic solvent, for wastewater treatment. The developed low-cost membrane demonstrated reverse osmosis and offered high productivity and efficiency [90]. Electrospun nanofiber membranes (ENMs) can be effectively used for filtration application as these offer fouling tendency due to their high surface area and roughness [91]. However, these are hydrophobic in nature and need to be hydrophilized. The hydrophilicity of ENMs can be achieved by using inorganic nanoparticles owing to their highly hydrophilic nature and these can act as fillers and increase the hydrophilicity of polymer matrix such as cellulose. In one study, Anirudhan *et al.* (2015) developed a composite adsorbent based on poly(itaconic acid)-poly(methacrylic acid)-grafted nanocellulose/nanobentonite which effectively removed uranium from aqueous samples. The adsorption efficiency of the composite was directly related to the concentration of carboxylic groups on the surface of adsorbents as well as its high surface area [92]. The nanocellulose-based composites can also be used for effective removal of dyes, heavy metals, microbes, organic solvents, oils, etc., from wastewater samples. Suman *et al.* (2015) developed a nanocellulose impregnated with Ag nanoparticles for cost-effective and sustainable removal of different contaminants such as organic dyes and heavy metals from wastewater. In the developed nanocomposite, nanocellulose specifically removed dyes and heavy metals while Ag nanoparticles killed the microbes due to their antibacterial activity [93]. In another study, nanocellulose and methyltrimethoxysilane sols-based ultralight, flexible, hydrophobic, and oleophilic sponges were prepared for removal of organic solvents and oils. The developed sponges showed high efficiency by adsorbing contaminants up to 100 times their own weight [94]. Titanium dioxide doped nanocellulose also possesses oleophilic properties, low density, photocatalytic activity, and high adsorption capacity [95, 96]; therefore, TiO_2-based nanocomposites can be used for water purification if CNFs are impregnated with TiO_2 nanoparticles in aqueous environment. In one such study, a highly porous CNF/TiO_2 nanocomposite showed good photocatalytic activity and long-term stability and effectively decomposed methyl orange and paracetamol [97].

13. Pollutant Sensors

Our environment is at risk to several contaminants present in water and air. Natural water contains very low concentrations of pollutants such as pharmaceutics, cosmetics, hormones, heavy metals, pathogens, viruses, etc., which cause serious health issues. Similarly, different gases such as nitrogen oxides (NOx), sulfur oxides (SOx), carbon dioxide, and methane, produced by industries are added to our environment, which significantly contaminate air. Therefore, environmental management system requires the detection of such contaminants at very low concentrations by using detection systems, commonly known as sensors. A sensor detects the contaminants up to a certain threshold value depending upon its level of sensitivity [122]. Generally, two classes of sensors are used for detection of such contaminants: sensors in liquid applications and sensors in gas applications.

Currently, the nanofiber-based sensors have received utmost attention owing to their high efficiency and sensitivity towards traces of contaminants. However, nanocellulose fibers are non-conductive; therefore, different nanoparticles are impregnated into its matrix to impart it electrical conductivity. These sensors can be potentially used for the detection of trace contaminants in aqueous solutions. Wang *et al.* (2011) developed a gold nanoparticles-coated BC (Au/BC) nanocomposite, and horseradish peroxidase and heme proteins including hemoglobin and myoglobin were immobilized on it. The developed sensor was used for the detection of traces of H_2O_2 [104]. In a similar study, Zhang *et al.* (2014) developed Au/BC sensor and immobilized horseradish peroxidase and glucose oxidase for the detection of glucose [105]. In these systems, the biosensors use the conductance property of Au nanoparticles and biocompatible properties of BC to act as an electrode for the electrochemical detection of pollutants and chemicals [105, 123]. These BC and Au nanoparticle-based sensors are lightweight, flexible, and shape stable in both aqueous acidic and alkaline solutions [1]. Nanocellulose-based biosensors coated with Au and Ag nanoparticles have also been developed by using surface-enhanced Raman spectroscopy (SERS). Ngo *et al.* (2013) developed an Au nanoparticle-treated nanocellulose paper and adsorbed cationic polyelectrolyte (CPAM) on its surface. The CPAM pre-treated nanocellulose/Au nanocomposite contributed towards excellent SERS reproducibility, sensitivity, and high enhancement factor [106]. A honeycomb-shaped Au substrate comprised of clusters was developed where each cluster contained 50–70 Au nanoparticles and showed high Raman enhancement efficiency. The developed substrate was highly reproducible, low-cost, and simple, which further demonstrate its application for SERS quantitative diagnosis and analyte detection [124]. Marques *et al.* (2008) developed a BC/Ag nanocomposite through *in situ* reduction of Ag^+ by Na_3-citrate for the detection of thiosalicylic acid and 2,2′-dithiodipyridine at a concentration as low as 10^{-4} M [102]. Nanocellulose-based sensors have also been developed for the detection of toxic chemicals such as propranolol [125], and theophylline and 17β-estradiol [126] in drinking water and other samples. It is believed that with further advancement in nanotechnology, advanced nanocellulose-based sensors can be developed with high selectivity, sensitivity, and efficiency.

14. Catalysis

Nanocellulose-based composites have received immense consideration to catalytically degrade organic pollutants. For this purpose, composites of nanocellulose are prepared with two types of nanoparticles:

photocatalysts such as titanium dioxide nanoparticles and precious metals such as Au, Ag, and Pt, etc., where nanocellulose serves only as support to prevent agglomeration or aggregation of such nanoparticles.

The metal-coated nanocellulose composites are prepared through salt reduction by using different reducing agents such as H_2, $NaBH_4$, and Na_3-citrate, and the OH groups present in nanocellulose [1]. Schütz *et al.* (2012) developed hard and transparent nanocellulose/TiO_2 film for its application in photocatalytic decomposition of organic pollutants [107]. In another study, TiO_2 layer was deposited on nanocellulose aerogel *via* chemical vapor deposition method. The developed nanocellulose/TiO_2 nanocomposite effectively decomposed methylene blue under UV illumination [111]. In another study, nanocrystalline cellulose and Pd nanoparticle-based composite was developed, which effectively converted phenol to cyclohexane *via* catalytic hydrogenation [112]. Compared to a nanocomposite of microfibrillated cellulose (MFC) with Au nanoparticles, the composite of nanocrystalline cellulose (NCC) with Au nanoparticles showed better performance in degrading 4-nitrophenol [110]. The improved catalytic activity of NCC/Au nanocomposite could be attributed to the improved distribution of nanoparticles in the cellulose matrix. In a more advanced strategy, Azetsu *et al.* (2011) developed a bimetallic nanocomposite of Au and Pd by using TEMPO-oxidized cellulose fibers as support. Both types of nanoparticles were uniformly distributed in the cellulose matrix and catalysts were synthesized *in situ* on the surface of cellulose due to high density of carboxylate groups in the modified cellulose. Compared to Pd, the Au nanoparticles deposited on the surface of cellulose showed higher catalytic activity in the degradation of 4-nitrophenol [113]. The improved catalytic activity of the bimetallic nanocomposite could be attributed to the electronic mediated ligand effect, i.e., the improved catalytic reactions induced by the electronic interactions between the two types of metal nanoparticles relative to the single-type metal nanoparticles. The catalytic activity of metal-based nanocellulose nanocomposite can be further improved by increasing the surface coverage of metals on the nanocellulose, which can be achieved by using cetyltrimethylammonium bromide (CTAB) during the nanoparticles' synthesis process. The cationic quaternary groups in the CTAB interact with the electron-rich OH and anionic charged groups on nanocellulose surface which result in formation of nanoparticles along the fibrils [127].

15. Renewable Energy Applications

There is a growing demand for renewable energy resources due to rapid consumption of the available energy resources and increased energy demand for carrying out routine activities and running industries. To this end, nanocellulose and nanocellulose-based composite materials can be effectively utilized for various energy storage applications such as fuel cells, solar cells, supercapacitors, and storage batteries [1]. However, nanocellulose by nature is a non-conductive material; therefore, different electroactive materials have been impregnated into the nanocellulose matrix to impart it electrical properties. In such composites, nanocellulose acts as a matrix and provides flexibility, whereas the electroactive material provides electrical properties. The developed composite materials can find potential applications as flexible electrodes, sensing electrodes, flexible display, and energy storage capacitors [128].

Nanocellulose aerogel with high surface area and surface charge can be used for layer-by-layer assembly on various conductive polymers such as polyaniline and polypyrrol; conducting materials such as carbon nanotubes; and conductive nanoparticles such as TiO_2, Au, ZnO, Al_2O_3, etc., to enhance their charge capacity, flexibility, and chemical and mechanical properties for energy storage and other electronic applications [95, 129–132]. Carbon nanotubes, being highly conductive, have been extensively used in the development of nanocellulose-based electroconductive materials. In one study, Kang *et al.* (2012) developed an all-solid-state flexible supercapacitor based on BNC, CNT, and ionic liquid-based polymer gel electrolytes. The supercapacitor showed high flexibility, electrochemical properties, mechanical stability, and cyclability [118]. In another study, Li *et al.* [119] developed BC-based flexible supercapacitors by adding multiwalled carbon nanotubes (MWCNTs) and polyaniline. The developed facile and low-cost porous supercapacitor showed high specific capacitance and excellent cycling stability, and low capacitance degradation [119]. In a similar study, Lee *et al.* (2015) developed an ultrathin CNF/MWCNTs electrode for fabrication of user-friendly rechargeable batteries. The electrode was comprised of a unitized separator/electrode assembly (SEA) [120]. The nanocellulose-based materials can also be used for electrical insulator applications, thanks to the high degree of crystallinity and water holding capacity of nanocellulose, which significantly influence its dielectric properties [133]. Besides, these materials can find potential applications in printed electronics [134, 135]. Nanocellulose can also host different nano-sized anode catalysts, thus finding potential applications in the development of fuel cells. For example, Pd nanoparticles were impregnated into the BC matrix in the presence of ammonium hexachloropalladate. The resulting Pd/BC nanocomposite was used as membrane electrode assembly (MEA) in the fuel cell. In addition, the developed Pd/BC nanocomposite showed high thermal stability and lower gas crossover [121]. Further, the transparent nature, smoothness, and high mechanical properties support the development of nanocellulose-based solar cells. In one study, Hu *et al.* (2013) coated indium tin oxide (ITO) nanoparticles on nanocellulose paper using radiofrequency sputtering of In_2O_3 (90%) and SnO_2 (10%). The developed solar cells showed a higher power conversion efficiency (0.4%), indicating its potential applications in photovoltaics [114]. Nanocellulose-based materials also find potential applications in the development of Li-ion batteries (LIBs) where it serves as electrode, separator, and electrolyte. Leijonmarck *et al.* (2013) developed LIBs by using Si-coated CNT/NFC paper where CNT imparted electrical conductivity to NFC [117]. The nanocellulose-based LIBs are thin, flexible, and have high performance, thus finding potential applications in cellular phones, laptops, and other chargeable electronic devices.

16. Conclusions and Benchmark Future Perspectives

Currently, sustainability and green issues are the top priorities for different businesses, industries, as well as individuals because of the serious environmental issues caused by the petroleum-based products, in addition to their non-renewability and depletion. To this end, nanocellulose has received substantial consideration over the last couple of decades because of its biodegradability, strength, and other structural, physico-chemical, and biological features. Nanocellulose has been extensively used in the paper industry and food industry, in biomedical devices, in cosmetics, in regenerative medicines, as a stabilizing agent in dispersion, in the petrochemical industry, for energy and environmental remediation including separation and purification technologies, biosensors, and catalysis. The nanocellulose-based materials are biodegradable, highly efficient, and lightweight and so can be modified to act as filters for removal of contaminants such as pathogens, viruses, heavy metals, and toxic chemicals from culture medium, water, and air.

Currently, the large-scale applications of nanocellulose-based composites have been limited by its high production cost, environmental concerns, and limited real-time evaluation. Therefore, further efforts are required to develop strategies for low-cost production of nanocellulose from different waste materials, improve its structural, physico-chemical, and biological features, and develop of nanocellulose-based composites with novel materials for diverse environmental and other applications. These advancements will result in the development of novel materials with tailored structural features and advanced functions and applications.

Acknowledgments

This work was supported by TRC (FURAP/DU/17/003) Oman and China Postdoctoral Science Foundation (2016M602291).

References

[1] H. Wei, K. Rodriguez, S. Renneckar, and P.J. Vikesland, Environmental science and engineering applications of nanocellulose-based nanocomposites. *Environ. Sci. Nano.*, 1 (2014) 302–316.

[2] M.W. Ullah, W.A. Khattak, M. Ul-Islam, S. Khan, and J.K. Park, Encapsulated yeast cell-free system: A strategy for cost-effective and sustainable production of bio-ethanol in consecutive batches. *Biotechnol. Bioprocess Eng.*, 20 (2015) 561–575.

[3] H. Tian, Z. Tang, X. Zhuang, X. Chen, and X. Jing, Biodegradable synthetic polymers: Preparation, functionalization and biomedical application. *Prog. Polym. Sci.*, (2012). doi:10.1016/j.progpolymsci.2011.06.004.

[4] S. Mecking, Nature or petrochemistry?-biologically degradable materials. *Angew. Chem. Int. Ed.*, (2004). doi:10.1002/anie.200301655.

[5] D.R. Ruka, G.P. Simon, and K.M. Dean, Bacterial cellulose and its use in renewable composites, in *Nanocellulose Polym. Nanocomposites Fundam. Appl.*, (2014) 89–130.

[6] S. Mondal, Preparation, properties and applications of nanocellulosic materials. *Carbohydr. Polym.*, (2017). doi:10.1016/j.carbpol.2016.12.050.

[7] S. Mondal, P. Memmott, and D. Martin, Preparation and characterization of green bio-composites based on modified spinifex resin and spinifex grass fibres. *J. Compos. Mater.*, (2014). doi:10.1177/0021998313486500.

[8] M.W. Ullah, M. Ul Islam, S. Khan, N. Shah, and J.K. Park, Recent advancements in bioreactions of cellular and cell-free systems: A study of bacterial cellulose as a model. *Korean J. Chem. Eng.*, 34 (2017) 1591–1599.

[9] M.W. Ullah, M. Ul-Islam, S. Khan, Y. Kim, and J.K. Park, Innovative production of bio-cellulose using a cell-free system derived from a single cell line. *Carbohydr. Polym.*, 132 (2015) 286–294.
[10] M.W. Ullah, M. Ul-Islam, S. Khan, Y. Kim, and J.K. Park, Structural and physico-mechanical characterization of bio-cellulose produced by a cell-free system. *Carbohydr. Polym.*, 136 (2016) 908–916.
[11] Y. Kim, M.W. Ullah, M. Ul-Islam, S. Khan, J.H. Jang, and J.K. Park, Self-assembly of bio-cellulose nanofibrils through intermediate phase in a cell-free enzyme system. *Biochem. Eng. J.*, 142 (2019) 135–144.
[12] A.K. Mohanty, M. Misra, and L.T. Drzal, Sustainable bio-composites from renewable resources: Opportunities and challenges in the green materials world. *J. Polym. Environ.*, (2002). doi:10.1023/A:1021013921916.
[13] M. Ul-Islam, S. Khan, M.W. Ullah, and J.K. Park, Comparative study of plant and bacterial cellulose pellicles regenerated from dissolved states. *Int. J. Biol. Macromol.*, 137 (2019) 247–252.
[14] U. Farooq, M.W. Ullah, Q. Yang, A. Aziz, J. Xu, L. Zhou, and S. Wang, High-density phage particles immobilization in surface-modified bacterial cellulose for ultra-sensitive and selective electrochemical detection of *Staphylococcus aureus*. *Biosens. Bioelectron.*, 157 (2020) 112163.
[15] M.U. Islam, S. Khan, M.W. Ullah, and J.K. Park, 21 Recent advances in biopolymer composites for environmental issues, *Scrivener Publ. LLC*, (2017) 673–690.
[16] M. Ul-Islam, M.W. Ullah, S. Khan, T. Kamal, S. Ul-Islam, N. Shah, and J.K. Park, Recent advancement in cellulose based nanocomposite for addressing environmental challenges. *Recent Pat. Nanotechnol.*, 10 (2016). doi:10.2174/1872210510666160429144916.
[17] M.U. Islam, M.W. Ullah, S. Khan, N. Shah, and J.K. Park, Strategies for cost-effective and enhanced production of bacterial cellulose. *Int. J. Biol. Macromol.*, 102 (2017) 1166–1173.
[18] M. Ul-Islam, M.W. Ullah, S. Khan, and J.K. Park, Production of bacterial cellulose from alternative cheap and waste resources: A step for cost reduction with positive environmental aspects. *Korean J. Chem. Eng.*, 37 (2020) 925–937.
[19] J. George and S. S N, Cellulose nanocrystals: Synthesis, functional properties, and applications. *Nanotechnol. Sci. Appl.*, 45 (2015).
[20] C. Woodcock and A. Sarko, Packing analysis of carbohydrates and polysaccharides. 11. Molecular and crystal structure of native ramie cellulose. *Macromolecules*, (1980). doi:10.1021/ma60077a030.
[21] K. Mazeau and L. Heux, Molecular dynamics simulations of bulk native crystalline and amorphous structures of cellulose. *J. Phys. Chem. B.*, (2003). doi:10.1021/jp0219395.

[22] K. Spence, Y. Habibi, and A. Dufresne, Nanocellulose-based composites. in: *Cellul. Fibers Bio- Nano-Polymer Compos.*, (2011). doi:10.1007/978-3-642-17370-7.

[23] L. Brinchi, F. Cotana, E. Fortunati, and J.M. Kenny, Production of nanocrystalline cellulose from lignocellulosic biomass: Technology and applications. *Carbohydr. Polym.*, (2013). doi:10.1016/j.carbpol.2013.01.033.

[24] R.H. Atalla and D.L. VanderHart, Native cellulose: A composite of two distinct crystalline forms. *Science*, (1984). doi:10.1126/science.223.4633.283.

[25] M. Ul-Islam, T. Khan, and J.K. Park, Water holding and release properties of bacterial cellulose obtained by *in situ* and *ex situ* modification. *Carbohydr. Polym.*, 88 (2012) 596–603.

[26] N. Shah, M. Ul-Islam, W.A. Khattak, and J.K. Park, Overview of bacterial cellulose composites: A multipurpose advanced material. *Carbohydr. Polym.*, 98 (2013) 1585–1598.

[27] M. Ul-Islam, S. Khan, M.W. Ullah, and J.K. Park, Structure, chemistry and pharmaceutical applications of biodegradable polymers. *Handb. Polym. Pharm. Technol.*, 3 (2015) 517–540.

[28] M. Ul-Islam, W.A. Khattak, M. Kang, S.M. Kim, T. Khan, and J.K. Park, Effect of post-synthetic processing conditions on structural variations and applications of bacterial cellulose. *Cellulose*, 20 (2013) 253–263.

[29] N. Lavoine, I. Desloges, A. Dufresne, and J. Bras, Microfibrillated cellulose — Its barrier properties and applications in cellulosic materials: A review. *Carbohydr. Polym.*, (2012). doi:10.1016/j.carbpol.2012.05.026.

[30] M.S. Jahan, A. Saeed, Z. He, and Y. Ni, Jute as raw material for the preparation of microcrystalline cellulose. *Cellulose*, (2011). doi:10.1007/s10570-010-9481-z.

[31] X. An, Y. Wen, D. Cheng, X. Zhu, and Y. Ni, Preparation of cellulose nanocrystals through a sequential process of cellulase pretreatment and acid hydrolysis. *Cellulose*, (2016). doi:10.1007/s10570-016-0964-4.

[32] Z. Di, Z. Shi, M.W. Ullah, S. Li, and G. Yang, A transparent wound dressing based on bacterial cellulose whisker and poly(2-hydroxyethyl methacrylate). *Int. J. Biol. Macromol.*, 105 (2017) 638–644.

[33] M. Iguchi, S. Yamanaka, and A. Budhiono, Bacterial cellulose — a masterpiece of nature's arts. *J. Mater. Sci.*, 35 (2000) 261–270.

[34] J.H. Kim, B.S. Shim, H.S. Kim, Y.J. Lee, S.K. Min, D. Jang, Z. Abas, and J. Kim, Review of nanocellulose for sustainable future materials. *Int. J. Precis. Eng. Manuf. — Green Technol.*, (2015). doi:10.1007/s40684-015-0024-9.

[35] Y. Li, H. Zhu, F. Shen, J. Wan, S. Lacey, Z. Fang, H. Dai, and L. Hu, Nanocellulose as green dispersant for two-dimensional energy materials. *Nano Energy.*, (2015). doi:10.1016/j.nanoen.2015.02.015.

[36] N. Muhd Julkapli, and S. Bagheri, Nanocellulose as a green and sustainable emerging material in energy applications: A review. *Polym. Adv. Technol.*, (2017). doi:10.1002/pat.4074.
[37] V. Favier, G.R. Canova, J.Y. Cavaillé, H. Chanzy, A. Dufresne, and C. Gauthier, Nanocomposite materials from latex and cellulose whiskers. *Polym. Adv. Technol.*, (1995). doi:10.1002/pat.1995.220060514.
[38] M.A.S. Azizi Samir, F. Alloin, and A. Dufresne, Review of recent research into cellulosic whiskers, their properties and their application in nanocomposite field. *Biomacromolecules*, (2005). doi:10.1021/bm0493685.
[39] K. Oksman, A.P. Mathew, D. Bondeson, and I. Kvien, Manufacturing process of cellulose whiskers/polylactic acid nanocomposites. *Compos. Sci. Technol.*, (2006). doi:10.1016/j.compscitech.2006.03.002.
[40] D. Bondeson and K. Oksman, Polylactic acid/cellulose whisker nanocomposites modified by polyvinyl alcohol. *Compos. Part A Appl. Sci. Manuf.*, (2007). doi:10.1016/j.compositesa.2007.08.001.
[41] X. Cao, H. Dong, and C.M. Li, New nanocomposite materials reinforced with cellulose nanocrystals in waterborne polyurethane. *Biomacromolecules*, (2007). doi:10.1016/j.polymertesting.2013.04.005.
[42] K.M.Z. Hossain, R.M. Felfel, C.D. Rudd, W. Thielemans, and I. Ahmed, The effect of cellulose nanowhiskers on the flexural properties of self-reinforced polylactic acid composites. *React. Funct. Polym.*, (2014). doi:10.1016/j.reactfunctpolym.2014.09.012.
[43] W. Il Park, M. Kang, H.S. Kim, and H.J. Jin, Electrospinning of poly(ethylene oxide) with bacterial cellulose whiskers, in *Macromol. Symp.*, (2007). doi:10.1002/masy.200750347.
[44] E.S. Medeiros, L.H.C. Mattoso, E.N. Ito, K.S. Gregorski, G.H. Robertson, R.D. Offeman, D.F. Wood, W.J. Orts, and S.H. Imam, Electrospun nanofibers of poly(vinyl alcohol) reinforced with cellulose nanofibrils. *J. Biobased Mater. Bioenergy.*, (2008). doi:10.1166/jbmb.2008.411.
[45] M.S. Peresin, Y. Habibi, J.O. Zoppe, J.J. Pawlak, and O.J. Rojas, Nanofiber composites of polyvinyl alcohol and cellulose nanocrystals: Manufacture and characterization. *Biomacromolecules*, (2010). doi:10.1021/bm901254n.
[46] M. Ul-Islam, W.A. Khattak, M.W. Ullah, S. Khan, and J.K. Park, Synthesis of regenerated bacterial cellulose-zinc oxide nanocomposite films for biomedical applications. *Cellulose*, 21 (2014) 433–447.
[47] M.W. Ullah, M. Ul-Islam, S. Khan, Y. Kim, J.H. Jang, and J.K. Park, *In situ* synthesis of a bio-cellulose/titanium dioxide nanocomposite by using a cell-free system. *RSC Adv.*, 6 (2016) 22424–22435.
[48] T. Maneerung, S. Tokura, and R. Rujiravanit, Impregnation of silver nanoparticles into bacterial cellulose for antimicrobial wound dressing. *Carbohydr. Polym.*, (2008). doi:10.1016/j.carbpol.2007.07.025.
[49] M. Ul-Islam, W.A. Khattak, M.W. Ullah, S. Khan, and J.K. Park, Synthesis of regenerated bacterial cellulose-zinc oxide nanocomposite films for

biomedical applications. *Cellulose*, 21 (2014). doi:10.1007/s10570-013-0109-y.
[50] A. Khalid, R. Khan, M. Ul-Islam, T. Khan, and F. Wahid, Bacterial cellulose-zinc oxide nanocomposites as a novel dressing system for burn wounds. *Carbohydr. Polym.*, 164 (2017) 214–221.
[51] M. Ul-Islam, T. Khan, and J.K. Park, Nanoreinforced bacterial cellulose-montmorillonite composites for biomedical applications. *Carbohydr. Polym.*, 89 (2012) 1189–1197.
[52] S. Khan, M. Ul-Islam, W.A. Khattak, M.W. Ullah, and J.K. Park, Bacterial cellulose-titanium dioxide nanocomposites: Nanostructural characteristics, antibacterial mechanism, and biocompatibility. *Cellulose*, 22 (2015) 565–579.
[53] S. Khan, M. Ul-Islam, M.W. Ullah, M. Israr, J.H. Jang, and J.K. Park, Nano-gold assisted highly conducting and biocompatible bacterial cellulose-PEDOT: PSS films for biology-device interface applications. *Int. J. Biol. Macromol.*, 107 (2018) 865–873.
[54] N. Takeda, K. Tamura, R. Mineguchi, Y. Ishikawa, Y. Haraguchi, T. Shimizu, and Y. Hara, *In situ* cross-linked electrospun fiber scaffold of collagen for fabricating cell-dense muscle tissue. *J. Artif. Organs.*, (2016). doi:10.1007/s10047-015-0871-8.
[55] L. Lamboni, Y. Li, J. Liu, and G. Yang, Silk sericin-functionalized bacterial cellulose as a potential wound-healing biomaterial. *Biomacromolecules*, 17 (2016) 3076–3084.
[56] S. Khan, M. Ul-Islam, M. Ikram, S.U. Islam, M.W. Ullah, M. Israr, J.H. Jang, S. Yoon, and J.K. Park, Preparation and structural characterization of surface modified microporous bacterial cellulose scaffolds: A potential material for skin regeneration applications *in vitro* and *in vivo*. *Int. J. Biol. Macromol.*, 117 (2018) 1200–1210.
[57] S. Kirdponpattara, A. Khamkeaw, N. Sanchavanakit, P. Pavasant, and M. Phisalaphong, Structural modification and characterization of bacterial cellulose–alginate composite scaffolds for tissue engineering. *Carbohydr. Polym.*, 132 (2015) 146–155.
[58] M. Ul-Islam, N. Shah, J.H. Ha, and J.K. Park, Effect of chitosan penetration on physico-chemical and mechanical properties of bacterial cellulose. *Korean J. Chem. Eng.*, 28 (2011) 1736–1743. doi:10.1007/s11814-011-0042-4.
[59] P. Béguin and J.-P. Aubert, The biological degradation of cellulose. *FEMS Microbiol. Rev.*, (1994). doi:10.1111/j.1574-6976.1994.tb00033.x.
[60] S. Kalka, T. Huber, J. Steinberg, K. Baronian, J. Müssig, and M.P. Staiger, Biodegradability of all-cellulose composite laminates. *Compos. Part A Appl. Sci. Manuf.*, (2014). doi:10.1016/j.compositesa.2013.12.012.
[61] S.B. Schröpfer, M.K. Bottene, L. Bianchin, L.C. Robinson, V. de Lima, V.D. Jahno, H. da S. Barud, and S.J.L. Ribeiro, Biodegradation evaluation

of bacterial cellulose, vegetable cellulose and poly (3-hydroxybutyrate) in soil. *Polímeros*, (2015). doi:10.1590/0104-1428.1712.

[62] M. Ul-Islam, F. Subhan, S.U. Islam, S. Khan, N. Shah, S. Manan, M.W. Ullah, and G. Yang, Development of three-dimensional bacterial cellulose/chitosan scaffolds: Analysis of cell-scaffold interaction for potential application in the diagnosis of ovarian cancer. *Int. J. Biol. Macromol.* 137 (2019) 1050–1059.

[63] K. Oksman, Y. Aitomäki, A.P. Mathew, G. Siqueira, Q. Zhou, S. Butylina, S. Tanpichai, X. Zhou, and S. Hooshmand, Review of the recent developments in cellulose nanocomposite processing. *Compos. Part A Appl. Sci. Manuf.*, (2016). doi:10.1016/j.compositesa.2015.10.041.

[64] W. Aljohani, M.W. Ullah, X. Zhang, and G. Yang, Bioprinting and its applications in tissue engineering and regenerative medicine. *Int. J. Biol. Macromol.*, 107 (2018) 261–275.

[65] R.R. McCarthy, M.W. Ullah, E. Pei, and G. Yang, Antimicrobial inks: The anti-infective applications of bioprinted bacterial polysaccharides. *Trends Biotechnol.*, 37 (2019) 1153–1155.

[66] R.R. McCarthy, M.W. Ullah, P. Booth, E. Pei, and G. Yang, The use of bacterial polysaccharides in bioprinting. *Biotechnol. Adv.*, 37 (2019) 107448.

[67] V. Thavasi, G. Singh, and S. Ramakrishna, Electrospun nanofibers in energy and environmental applications. *Ener. Environ. Sci.*, (2008). doi:10.1039/b809074m.

[68] J.A. Howell, Future of membranes and membrane reactors in green technologies and for water reuse. *Desalination*, (2004). doi:10.1016/S0011-9164(04)00021-9.

[69] W. Xiao, J. Xu, X. Liu, Q. Hu, and J. Huang, Antibacterial hybrid materials fabricated by nanocoating of microfibril bundles of cellulose substance with titania/chitosan/silver-nanoparticle composite films. *J. Mater. Chem. B.*, (2013). doi:10.1039/c3tb20303d.

[70] H. Dong, J.F. Snyder, D.T. Tran, and J.L. Leadore, Hydrogel, aerogel and film of cellulose nanofibrils functionalized with silver nanoparticles. *Carbohydr. Polym.*, (2013). doi:10.1016/j.carbpol.2013.03.041.

[71] G.G. Wan and W, Nanosilver coated bacterial cellulose, *US Pat.* 2 (2012). doi:10.1021/n10602701.

[72] I. Díez, P. Eronen, M. Österberg, M.B. Linder, O. Ikkala, and R.H.A. Ras, Functionalization of nanofibrillated cellulose with silver nanoclusters: Fluorescence and antibacterial activity. *Macromol. Biosci.*, (2011). doi:10.1002/mabi.201100099.

[73] W. Hu, S. Chen, X. Li, S. Shi, W. Shen, X. Zhang, and H. Wang, *In situ* synthesis of silver chloride nanoparticles into bacterial cellulose membranes. *Mater. Sci. Eng. C.*, (2009). doi:10.1016/j.msec.2008.09.017.

[74] M.T. Amin, A.A. Alazba, and U. Manzoor, A review of removal of pollutants from water/wastewater using different types of nanomaterials. *Adv. Mater. Sci. Eng.*, (2014). doi:10.1155/2014/825910.

[75] G. Metreveli, L. Wågberg, E. Emmoth, S. Belák, M. Strømme, and A. Mihranyan, A size-exclusion nanocellulose filter paper for virus removal. *Adv. Healthc. Mater.*, (2014). doi:10.1002/adhm.201300641.

[76] R.J. Osendorf, Layered air filter medium having improved efficiency and pleatability, (1995) US005427597A. doi:US005485919A.

[77] M. Ul-Islam, M.W. Ullah, S. Khan, T. Kamal, S. Ul-Islam, N. Shah, and J.K. Park, Recent advancement in cellulose based nanocomposite for addressing environmental challenges. *Recent Pat. Nanotechnol.*, 10 (2016) 169–180.

[78] E.N. Zare, A. Motahari, and M. Sillanpää, Nanoadsorbents based on conducting polymer nanocomposites with main focus on polyaniline and its derivatives for removal of heavy metal ions/dyes: A review. *Environ. Res.*, (2018). doi:10.1016/j.envres.2017.12.025.

[79] H.M. Marwani, M.U. Lodhi, S.B. Khan, and A.M. Asiri, Cellulose-lanthanum hydroxide nanocomposite as a selective marker for detection of toxic copper. *Nanoscale Res. Lett.*, (2014). doi:10.1186/1556-276X-9-466.

[80] C.L. Ake, K. Mayura, H. Huebner, G.R. Bratton, and T.D. Phillips, Development of porous clay-based composites for the sorption of lead from water. *J. Toxicol. Environ. Heal. — Part A.*, (2001). doi:10.1080/152873901300343489.

[81] H. Zhu, S. Jia, T. Wan, Y. Jia, H. Yang, J. Li, L. Yan, and C. Zhong, Biosynthesis of spherical Fe3O4/bacterial cellulose nanocomposites as adsorbents for heavy metal ions. *Carbohydr. Polym.*, (2011). doi:10.1016/j.carbpol.2011.06.061.

[82] L. Urbina, O. Guaresti, J. Requies, N. Gabilondo, A. Eceiza, M.A. Corcuera, and A. Retegi, Design of reusable novel membranes based on bacterial cellulose and chitosan for the filtration of copper in wastewaters. *Carbohydr. Polym.*, (2018). doi:10.1016/j.carbpol.2018.04.007.

[83] Z. Li, B. Wang, X. Qin, Y. Wang, C. Liu, Q. Shao, N. Wang, J. Zhang, Z. Wang, C. Shen, and Z. Guo, Superhydrophobic/superoleophilic polycarbonate/carbon nanotubes porous monolith for selective oil adsorption from water. *ACS Sustain. Chem. Eng.*, (2018). doi:10.1021/acssuschemeng.8b01637.

[84] A.D. Ragab Esmail Abou-Zeid, R. Khiari, and N. El-Wakil, Current state and new trends in the use of cellulose nanomaterials for wastewater treatment. *Biomacromolecules*, (2018). doi:10.1021/acs.biomac.8b00839.

[85] T.S. Anirudhan, and F. Shainy, Effective removal of mercury(II) ions from chlor-alkali industrial wastewater using 2-mercaptobenzamide

modified itaconic acid-grafted-magnetite nanocellulose composite. *J. Colloid Interface Sci.*, (2015). doi:10.1016/j.jcis.2015.05.052.

[86] T.S. Anirudhan, J.R. Deepa, and J. Christa, Nanocellulose/nanobentonite composite anchored with multi-carboxyl functional groups as an adsorbent for the effective removal of Cobalt(II) from nuclear industry wastewater samples. *J. Colloid Interface Sci.*, (2016). doi:10.1016/j.jcis.2016.01.023.

[87] N.J. Arsad and N. Ngadi, Chitosan-grafted nanocellulose derived from empty fruit bunch for ethyl orange removal. *Appl. Mech. Mater.*, (2014). doi:10.4028/www.scientific.net/AMM.625.784.

[88] K. Srasri, M. Thongroj, P. Chaijiraaree, S. Thiangtham, H. Manuspiya, P. Pisitsak, and S. Ummartyotin, Recovery potential of cellulose fiber from newspaper waste: An approach on magnetic cellulose aerogel for dye adsorption material. *Int. J. Biol. Macromol.*, (2018). doi:10.1016/j.ijbiomac.2018.07.123.

[89] A. Kardam, K.R. Raj, S. Srivastava, and M.M. Srivastava, Nanocellulose fibers for biosorption of cadmium, nickel, and lead ions from aqueous solution. *Clean Technol. Environ. Policy.*, (2014). doi:10.1007/s10098-013-0634-2.

[90] L.S. Lee C, Kim M, Oh H, and J. Park, Preparation method of hollow fiber membrane for water treatment using cellulose-based resn, *US Pat.*, (2013) US0248441A1. doi:10.1016/j.(73).

[91] S. Homaeigohar and M. Elbahri, Nanocomposite electrospun nanofiber membranes for environmental remediation. *Materials* (*Basel*)., (2014). doi:10.3390/ma7021017.

[92] T.S. Anirudhan and J.R. Deepa, Binusreejayan, Synthesis and characterization of multi-carboxyl-functionalized nanocellulose/nanobentonite composite for the adsorption of uranium(VI) from aqueous solutions: Kinetic and equilibrium profiles. *Chem. Eng. J.*, (2015). doi:10.1016/j.cej.2015.03.007.

[93] Suman, A. Kardam, M. Gera, and V.K. Jain, A novel reusable nanocomposite for complete removal of dyes, heavy metals and microbial load from water based on nanocellulose and silver nano-embedded pebbles. *Environ. Technol.*, (United Kingdom). (2015). doi:10.1080/09593330.2014.959066.

[94] Z. Zhang, G. Sèbe, D. Rentsch, T. Zimmermann, and P. Tingaut, Ultralightweight and flexible silylated nanocellulose sponges for the selective removal of oil from water. *Chem. Mater.*, (2014). doi:10.1021/cm5004164.

[95] J.T. Korhonen, M. Kettunen, R.H.A. Ras, and O. Ikkala, Hydrophobic nanocellulose aerogels as floating, sustainable, reusable, and recyclable oil absorbents. *ACS Appl. Mater. Interfaces.*, (2011). doi:10.1021/am200475b.

[96] A. Shoukat, F. Wahid, T. Khan, M. Siddique, S. Nasreen, G. Yang, M.W. Ullah, and R. Khan, Titanium oxide-bacterial cellulose bioadsorbent

for the removal of lead ions from aqueous solution. *Int. J. Biol. Macromol.*, 129 (2019) 965–971.

[97] M. Lucchini, E. Lizundia, S. Moser, M. Niederberger, and G. Nystrom, Titania-cellulose hybrid monolith for in-flow purification of water under solar illumination. *ACS Appl. Mater. Interfaces.*, (2018). doi:10.1021/acsami.8b09735.

[98] G. Yang, J. Xie, F. Hong, Z. Cao, and X. Yang, Antimicrobial activity of silver nanoparticle impregnated bacterial cellulose membrane: Effect of fermentation carbon sources of bacterial cellulose. *Carbohydr. Polym.*, (2012). doi:10.1016/j.carbpol.2011.08.079.

[99] H.S. Barud, T. Regiani, R.F.C. Marques, W.R. Lustri, Y. Messaddeq, and S.J.L. Ribeiro, Antimicrobial bacterial cellulose-silver nanoparticles composite membranes. *J. Nanomater.*, (2011). doi:10.1155/2011/721631.

[100] M. Henriksson, L.A. Berglund, P. Isaksson, T. Lindström, and T. Nishino, Cellulose nanopaper structures of high toughness. *Biomacromolecules*, (2008). doi:10.1021/bm800038n.

[101] R.J.B. Pinto, P.A.A.P. Marques, M.A. Martins, C.P. Neto, and T. Trindade, Electrostatic assembly and growth of gold nanoparticles in cellulosic fibres. J. *Colloid Interface Sci.*, (2007). doi:10.1016/j.jcis.2007.03.043.

[102] P.A.A.P. Marques, H.I.S. Nogueira, R.J.B. Pinto, C.P. Neto, and T. Trindade, Silver-bacterial cellulosic sponges as active SERS substrates. *J. Raman Spectrosc.*, (2008). doi:10.1002/jrs.1853.

[103] K.A. Mahmoud, K.B. Male, S. Hrapovic, and J.H.T. Luong, Cellulose nanocrystal/gold nanoparticle composite as a matrix for enzyme immobilization. *ACS Appl. Mater. Interfaces.*, (2009). doi:10.1021/am900331d.

[104] W. Wang, T.J. Zhang, D.W. Zhang, H.Y. Li, Y.R. Ma, L.M. Qi, Y.L. Zhou, and X.X. Zhang, Amperometric hydrogen peroxide biosensor based on the immobilization of heme proteins on gold nanoparticles-bacteria cellulose nanofibers nanocomposite. *Talanta*, (2011). doi:10.1016/j.talanta.2010.12.015.

[105] T. Zhang, W. Wang, D. Zhang, X. Zhang, Y. Ma, Y. Zhou, and L. Qi, Biotemplated synthesis of gold nanoparticle-bacteria cellulose nanofiber nanocomposites and their application in biosensing. *Adv. Funct. Mater.*, 20 (2010) 1152–1160.

[106] Y.H. Ngo, D. Li, G.P. Simon, and G. Garnier, Effect of cationic polyacrylamides on the aggregation and SERS performance of gold nanoparticles-treated paper. *J. Colloid Interface Sci.*, (2013). doi:10.1016/j.jcis.2012.09.080.

[107] C. Schütz, J. Sort, Z. Bacsik, V. Oliynyk, E. Pellicer, A. Fall, L. Wågberg, L. Berglund, L. Bergström, and G. Salazar-Alvarez, Hard and transparent films formed by nanocellulose-TiO_2 nanoparticle hybrids. *PLoS One*, (2012). doi:10.1371/journal.pone.0045828.

[108] H. Koga, E. Tokunaga, M. Hidaka, Y. Umemura, T. Saito, A. Isogai, and T. Kitaoka, Topochemical synthesis and catalysis of metal nanoparticles exposed on crystalline cellulose nanofibers. *Chem. Commun.*, (2010). doi:10.1039/c0cc02754e.

[109] Z. Zhou, C. Lu, X. Wu, and X. Zhang, Cellulose nanocrystals as a novel support for CuO nanoparticles catalysts: Facile synthesis and their application to 4-nitrophenol reduction. *RSC Adv.*, (2013). doi:10.1039/c3ra43006e.

[110] X. Wu, C. Lu, Z. Zhou, G. Yuan, R. Xiong, and X. Zhang, Green synthesis and formation mechanism of cellulose nanocrystal-supported gold nanoparticles with enhanced catalytic performance. *Environ. Sci. Nano.*, (2014). doi:10.1039/c3en00066d.

[111] M. Kettunen, R.J. Silvennoinen, N. Houbenov, A. Nykänen, J. Ruokolainen, J. Sainio, V. Pore, M. Kemell, M. Ankerfors, T. Lindström, M. Ritala, R.H.A. Ras, and O. Ikkala, Photoswitchable superabsorbency based on nanocellulose aerogels. *Adv. Funct. Mater.*, (2011). doi:10.1002/adfm.201001431.

[112] C.M. Cirtiu, A.F. Dunlop-Brière, and A. Moores, Cellulose nanocrystallites as an efficient support for nanoparticles of palladium: Application for catalytic hydrogenation and Heck coupling under mild conditions. *Green Chem.*, (2011). doi:10.1039/c0gc00326c.

[113] A. Azetsu, H. Koga, A. Isogai, and T. Kitaoka, Synthesis and catalytic features of hybrid metal nanoparticles supported on cellulose nanofibers. *Catalysts*, (2011). doi:10.3390/catal1010083.

[114] L. Hu, G. Zheng, J. Yao, N. Liu, B. Weil, M. Eskilsson, E. Karabulut, Z. Ruan, S. Fan, J.T. Bloking, M.D. McGehee, L. Wågberg, and Y. Cui, Transparent and conductive paper from nanocellulose fibers. *Energy Environ. Sci.*, (2013). doi:10.1039/c2ee23635d.

[115] L. Jabbour, C. Gerbaldi, D. Chaussy, E. Zeno, S. Bodoardo, and D. Beneventi, Microfibrillated cellulose-graphite nanocomposites for highly flexible paper-like Li-ion battery electrodes. *J. Mater. Chem.*, (2010). doi:10.1039/c0jm01219j.

[116] L. Hu, N. Liu, M. Eskilsson, G. Zheng, J. McDonough, L. Wågberg, and Y. Cui, Silicon-conductive nanopaper for Li-ion batteries. *Nano Ener.*, (2013). doi:10.1016/j.nanoen.2012.08.008.

[117] S. Leijonmarck, A. Cornell, G. Lindbergh, and L. Wågberg, Single-paper flexible Li-ion battery cells through a paper-making process based on nano-fibrillated cellulose. *J. Mater. Chem. A.*, (2013). doi:10.1039/c3ta01532g.

[118] Y.J. Kang, S.J. Chun, S.S. Lee, B.Y. Kim, J.H. Kim, H. Chung, S.Y. Lee, and W. Kim, All-solid-state flexible supercapacitors fabricated with bacterial nanocellulose papers, carbon nanotubes, and triblock-copolymer ion gels. *ACS Nano.*, 6 (2012) 6400–6406.

[119] S. Li, D. Huang, B. Zhang, X. Xu, M. Wang, G. Yang, and Y. Shen, Flexible supercapacitors based on bacterial cellulose paper electrodes. *Adv. Energy Mater.*, 4 (2014) 1301655.
[120] S.Y. Lee, S.J. Cho, K.H. Choi, J.T. Yoo, J.H. Kim, Y.H. Lee, S.J. Chun, S.B. Park, D.H. Choi, Q. Wu, and S.Y. Lee, Hetero-nanonet rechargeable paper batteries: Toward ultrahigh energy density and origami foldability. *Adv. Funct. Mater.*, (2015). doi:10.1002/adfm.201502833.
[121] B.R. Evans, H.M. O'Neill, V.P. Malyvanh, I. Lee, and J. Woodward, Palladium-bacterial cellulose membranes for fuel cells. *Biosens. Bioelectron.*, 18 (2003) 917–923.
[122] U. Farooq, Q. Yang, M.W. Ullah, and S. Wang, Bacterial biosensing: Recent advances in phage-based bioassays and biosensors. *Biosens. Bioelectron.*, 118 (2018) 204–216.
[123] M. Park, H. Chang, D.H. Jeong, and J. Hyun, Spatial deformation of nanocellulose hydrogel enhances SERS. *Biochip J.*, (2013). doi:10.1007/s13206-013-7306-5.
[124] W. Leng and P.J. Vikesland, Nanoclustered gold honeycombs for surface-enhanced Raman scattering. *Anal. Chem.*, (2013). doi:10.1021/ac301028w.
[125] K. Yoshimatsu, L. Ye, J. Lindberg, and I.S. Chronakis, Selective molecular adsorption using electrospun nanofiber affinity membranes. *Biosens. Bioelectron.*, (2008). doi:10.1016/j.bios.2007.12.002.
[126] I.S. Chronakis, A. Jakob, B. Hagström, and L. Ye, Encapsulation and selective recognition of molecularly imprinted theophylline and 17β-estradiol nanoparticles within electrospun polymer nanofibers. *Langmuir*, (2006). doi:10.1021/la0613880.
[127] S. Padalkar, J.R. Capadona, S.J. Rowan, C. Weder, Y.H. Won, L.A. Stanciu, and R.J. Moon, Natural biopolymers: Novel templates for the synthesis of nanostructures. *Langmuir*, (2010). doi:10.1021/la904439p.
[128] Z. Shi, Y. Zhang, G.O. Phillips, and G. Yang, Utilization of bacterial cellulose in food. *Food Hydrocoll.*, 35 (2014) 539–545.
[129] M. Hamedi, E. Karabulut, A. Marais, A. Herland, G. Nyström, and L. Wågberg, Nanocellulose aerogels functionalized by rapid layer-by-layer assembly for high charge storage and beyond. *Angew. Chemie Int. Ed.*, 52 (2013) 12038–12042.
[130] M. Ul-Islam, S. Khan, M.W. Ullah, and J.K. Park, Bacterial cellulose composites: Synthetic strategies and multiple applications in bio-medical and electro-conductive fields. *Biotechnol. J.*, 10 (2015) 1847–1861.
[131] S. Khan, M. Ul-Islam, W.A. Khattak, M.W. Ullah, and J.K. Park, Bacterial cellulose-poly(3,4-ethylenedioxythiophene)-poly(styrenesulfonate) composites for optoelectronic applications. *Carbohydr. Polym.*, (2015). doi:10.1016/j.carbpol.2015.03.055.

[132] A. Jasim, M.W. Ullah, Z. Shi, X. Lin, and G. Yang, Fabrication of bacterial cellulose/polyaniline/single-walled carbon nanotubes membrane for potential application as biosensor. *Carbohydr. Polym.*, 163 (2017) 62–69.

[133] D. Le Bras, M. Strømme, and A. Mihranyan, Characterization of dielectric properties of nanocellulose from wood and algae for electrical insulator applications. *J. Phys. Chem. B.*, (2015). doi:10.1021/acs.jpcb.5b00715.

[134] A. Penttilä, J. Sievänen, K. Torvinen, K. Ojanperä, and J.A. Ketoja, Filler-nanocellulose substrate for printed electronics: Experiments and model approach to structure and conductivity. *Cellulose*, (2013). doi:10.1007/s10570-013-9883-9.

[135] K. Torvinen, J. Sievänen, T. Hjelt, and E. Hellén, Smooth and flexible filler-nanocellulose composite structure for printed electronics applications. *Cellulose*, (2012). doi:10.1007/s10570-012-9677-5.

https://doi.org/10.1142/9781786349477_0010

Chapter 10

Applications of Nanocellulose in the Paper Industry

Hye Jung Youn* and Hak Lae Lee

Department of Agriculture, Forestry and Bioresources,
Seoul National University,
Gwanak-ro 1, Gwanak-gu, Seoul 08826, Korea

**page94@snu.ac.kr*

Abstract

Nanocellulose, due to its sustainability, biodegradability, and high performance such as high tensile stiffness and low thermal expansion coefficient, can potentially replace the fossil fuel-based materials for different applications. Among them, paper and the board industry is the most promising field because of the massive product tonnage of paper. It can be used both as a dry and wet-end in the paper industry such as papermaking, paper coating, packaging, and for the production of hygienic and absorbent products. In this chapter, the first part of the chapter discusses the recent progress in the applications of nanocellulose as a dry- and wet-strength agent. The second part describes the use of nanocellulose in coating and barrier packaging. It needs to provide prospects of nanocellulose by reducing its production cost and post-processing, and explore novel application for paper-based products.

Keywords: Nanocellulose; sustainability; papermaking; dry and wet-strength agent; paper coating; surface sizing; paper industry

1. Introduction

Nanocellulose is considered one of the most promising materials that can replace fossil fuel-based materials for many applications because of its sustainability, biodegradability, and high performance such as high tensile stiffness and low thermal expansion coefficient. Nanocellulose is classified into three groups depending on preparation method: cellulose nanocrystals (CNC), cellulose nanofibrils (CNF), and bacterial cellulose (BC). Cellulose nanocrystals, which are prepared by acid hydrolysis using sulfuric acid, hydrochloric acid, or phosphoric acid, have the unique characteristics of high crystallinity, high tensile stiffness, and chiral nematic behavior.[1–3] In contrast, cellulose nanofibrils, which are prepared by mechanical shearing action using a grinder, homogenizer, or microfluidizer, have both crystalline and amorphous regions, a high aspect ratio, and good network forming ability.[4–7] Because these two types of nanocellulose each have unique properties, engineers or scientists can choose the proper type of nanocellulose that is suitable for the specific end-use purpose.

Many studies have been performed on the diverse applications of nanocellulose (e.g., reinforcing elements for biocomposite, optically transparent film, barrier film, hydrogel, aerogel, and papermaking additive).[8,9] In addition to the conventional industrial applications, many new sectors of organic LED, flexible electronics, 3D printing, and photonic films have been suggested as emerging fields for nanocellulose applications.[10] Among them, the paper and board industry is one of the most important potential markets for nanocellulose applications. According to the RISI report, the volume of the nanocellulose market for the paper and board industry is estimated to reach 20×10^6 tons.[11] It has been suggested that nanocellulose can be applied in the wet-end of papermaking, paper coating, packaging, and for the production of hygienic and absorbent products.[12–15] It is of great importance to look into the possible applications of nanocellulose in the paper and board industry. In the first part of this chapter, the recent progress in the applications of nanocellulose as a dry- and wet-strength agent will be described. In the second part, the use of nanocellulose in coating and barrier packaging will be described.

2. Application of Nanocellulose as Dry- or Wet-Strength Agent

2.1 *Wet-End Application*

A variety of organic and inorganic additives are being used in the papermaking process to produce paper or board of high quality at a low cost. Dry- and wet-strength agents, sizing agents, retention aids, drainage aids, and defoaming agents are examples of chemicals widely used in today's papermaking industry. Inorganic fillers, such as calcium carbonate, clay, and talc, are also important additives for many paper grades, particularly the printing and writing grades. The paper industry has strived to develop new functional additives to upgrade the paper quality and impart specialized functional properties to a product. In addition, improvements in process cleanliness and productivity are another critical issue that can be addressed with the use of additives. Not only high performance but also safety, environmental, and economic targets should be met when using papermaking additives. Because nanocellulose is a safe material to use and is composed of environmentally-friendly natural products, it has drawn attention as a new additive for papermaking.

Many studies have investigated the application of nanocellulose as a dry- or wet-strength agent in the wet-end. In addition, several review papers have described the performance of nanocellulose as a wet-end additive. Nanocelluloses prepared from diverse types of raw materials have been applied to bleached kraft pulp (BKP) stock, unbleached pulp stock, or recycled pulp stock. Among the different types of nanocellulose (e.g., CNF, CNC, BC), CNF, including microfibrillated cellulose (MFC), has been most widely investigated as a strengthening agent. In addition to the excellent affinity for cellulose pulp fibers, Boufi *et al.*[12] suggested that the use of CNF is motivated by a number of reasons: (1) high specific surface area, (2) semi-crystalline structure, (3) high intrinsic mechanical strength and good flexibility, (4) high potential to interact with fibers by hydrogen bonding, and (5) inherent tendency to form a strong network. In general, the effects of CNF as a wet-end additive are summarized as follows:

- Increase in mechanical strength such as tensile strength.
- Increase in density.

- Decrease in air permeability.
- Decrease in opacity.
- Decrease in drainage.

The effect of nanocellulose on the properties of paper may vary depending on the pulp type, use of other polymeric additives such as cationic polyacrylamide (C-PAM) and cationic starch, and filler content. The strengthening effects of CNF and MFC are well summarized in the review articles.[12,14]

2.1.1 *Paper properties*

MFC was investigated as an additive for papermaking before CNF because it could be produced with less mechanical energy consumption than CNF. MFC, which has a micrometer-sized width and longer length than CNF, showed good performance in paper strength tests.[16–22] Eriksen *et al.*[16] used MFC produced from kraft pulp to enhance the strength of TMP paper. MFC was produced by grinding or homogenization of bleached kraft pulp, which was added to a TMP slurry without a retention aid. The addition of 4% MFC into a TMP handsheet increased the tensile index up to 34% but decreased the Gurley air permeability, light scattering, opacity, and brightness. Taipale *et al.*[17] prepared different types of MFC using a microfluidizer or grinder and investigated their effects on drainage and paper properties when added to a softwood bleached kraft pulp suspension. Cationic starch was used as an additive for handsheet making. Figure 1

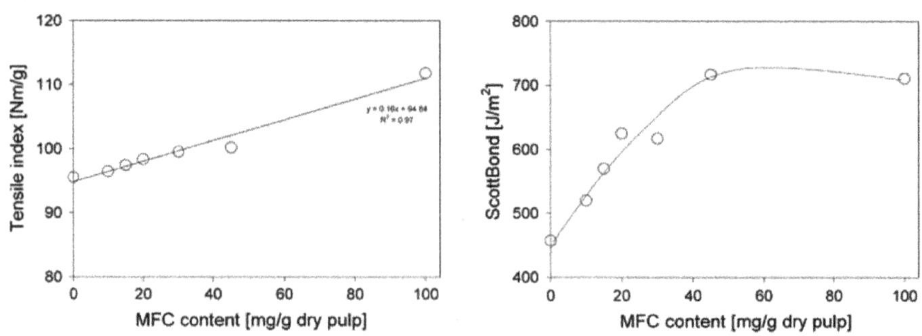

(a) (b)

Fig. 1. Tensile index (a) and Scott bond strength (b) of paper with MFC content.
Source: Reprinted by permission from Springer Nature, Cellulose, Taipale *et al.*[17]

shows the tensile index and Scott bond strength of paper versus the MFC content. The tensile strength of the paper increased linearly with MFC content, and the relative increase in tensile strength was 17% for an MFC content of 100 mg/g. The out-of-plane strength (i.e., Scott Bond strength) was linearly increased until reaching a maximum value. The increment in Scott bond strength was 57% for an MFC content of 45 mg/g. The addition of MFC together with cationic starch had a synergistic effect on strength. These results suggested that the optimum MFC content was about 30 mg/g pulp for both strength and drainage. The optimum use of MFC is dependent on the raw materials, additives, and the conditions used for papermaking.

The addition of MFC or CNF can improve the tensile strength of wood-free paper grades and wood-containing paper grades.[18,23–24] According to Manninen *et al.*[18] the role or mechanism of MFC in improving the density and mechanical strength of paper is similar to kraft fines. However, MFC showed remarkably greater improvement in mechanical strength than kraft fines.

González *et al.*[23] compared the physical and mechanical properties of sheets prepared with bleached eucalyptus pulp and CNF to those made of beaten pulp. CNF was prepared using a high-pressure homogenizer after TEMPO-oxidation and added into pulp stock at dosages of 3, 6, and 9 wt%. The pulp slurry that was reinforced with CNF showed properties similar to those of beaten pulp. The highest mechanical strength was obtained from the pulp stock with 6 to 9% CNF. However, the use of this amount of CNF was difficult to drain. When CNF was added to pulp stocks beaten to different freeness, the addition of 3% CNF to slightly beaten pulp showed comparable properties to the printing and writing paper grade. These results indicate that the addition of a small amount of CNF increased the physical and mechanical properties without the deterioration of porosity and opacity and that it also reduced the energy consumption in the beating.

The combination of enzymatic treatment (biobeating) and CNF addition to pulp slurry is an alternative to traditional beating.[25] CNF enhances the mechanical strength of enzyme-treated pulps while maintaining the opacity. With the addition of 4.5% CNF, the tensile index increased by approximately 67%, and the Gurley porosity decreased nine-fold. The drainage rate also decreased after CNF addition, which is acceptable in the papermaking industry if the CNF is added at a dosage of 1.5%.

The strengthening effect of nanocellulose in paper is synergistic if CNF is used along with a wet-strength resin. Ahola *et al.*[26] used carboxymethylated CNF together with poly(amideamine) epichlorohydrin (PAE) to improve the wet and dry strength of paper composed of bleached pine kraft long fibers. The CNF was prepared from carboxymethylated pulp using a microfluidizer. When a bi-layer of PAE and CNF was formed by the successive adsorption of PAE and CNF onto pulp fibers, both wet and dry tensile strengths were greatly improved. The significant improvement in these tensile strengths may be attributed to the use of carboxymethylated CNF, which has more carboxyl groups than untreated CNF and allows the formation of a good bi-layer with PAE. The improvement in strength was not significant when nanoaggregates of PAE and CNF were added to the pulp slurry.

Su *et al.*[22] also compared the effect of MFC with or without PAE on the properties of paper prepared from hardwood bleached kraft fibers. MFC was added up to 75%. The addition of MFC and PAE to the pulp fibers increased the dry and wet strengths of the paper by an order of magnitude. The air permeability of the paper decreased by up to four orders of magnitude in the presence of MFC. However, the refining of the fibers showed equivalent strength properties with a better formation and greater drainage benefits than the addition of MFC alone. With the use of PAE, the drainage of the stock with MFC was improved by coagulating MFC into a bigger structure, flocculating MFC with fibers, and modifying the floc properties such as density and size.

CNF can be modified by chemical pretreatment prior to homogenization. Ondaral *et al.*[27] prepared cationic CNF (CCNF) and anionic CNF (ACNF) by surface modification and investigated the effect on the tensile index of paper. With the addition of 10% CCNF, the tensile index was increased by approximately 50%. In particular, the tensile index of paper was higher when CCNF was used together with ACNF compared to the single addition of CCNF to pulp fibers.

Kajanto and Kosonen[28] produced paper reinforced with CNF using a high-speed pilot paper machine. They used mill pulp (i.e., a mixture of softwood and hardwood furnishes without filler) and mill process water to produce a sheet with a basis weight of 50–65 g/m^2, and evaluated the performance and quality of the produced paper made with 0, 1, or 2% cellulose nanofibrils. Despite the small amount of nanocellulose added, the nanocellulose addition improved the tensile strength, which could reduce the grammage by up to 8 g/m^2. Although the nanocellulose reduced the air permeability and opacity, it could be beneficial as packaging grade. In addition, it can be applied to printing and writing paper grades by compensating the decreased opacity by the use of filler.

Nanocellulose from BKP has been widely used and investigated. However, some researchers have studied the lignin-containing nanocellulose from mechanical pulps, which are referred to as lignocellulose nanofibers (LCNF) or nano-lignin-cellulose (NLC). Osong *et al.*[24] prepared LCNF from chemi-thermomechanical pulp (CTMP) and CNF from BKP. They compared the properties of the handsheet of CTMP fibers blended with LCNF with those of handsheets of BKP fibers containing CNF. Both the LCNF and CNF improved many strength properties, including tensile strength, *z*-strength, and burst strength of the paper. The chemical pulp-based CNF improved the strength better than LCNF. However, it may still be possible to further improve the performance of LCNF by controlling the degree of nanofibrillation.

The strengthening mechanism of CNF is similar to that of fibrillar fines. The improvement of strength by CNF addition can be attributed to the increased bonding area and high load-bearing capacity.[25] CNF can promote fiber–fiber bonding by bridging the adjacent fibers and increasing the bonded area because CNF has a very large specific surface area and a strong affinity for the fibers. In addition, CNF may increase the load-bearing capacity of the paper by forming different networks that are embedded among the fibers. A comparative study on the effect of CNF and cationic starch on tensile strength was performed and showed that the addition of CNF improved the tensile strength with an increasing addition rate while the use of the cationic starch reached a maximum and did not show any improvement with further addition rate (Fig. 2). These results indicate that the dry strength agent has a limit in the improvement of paper

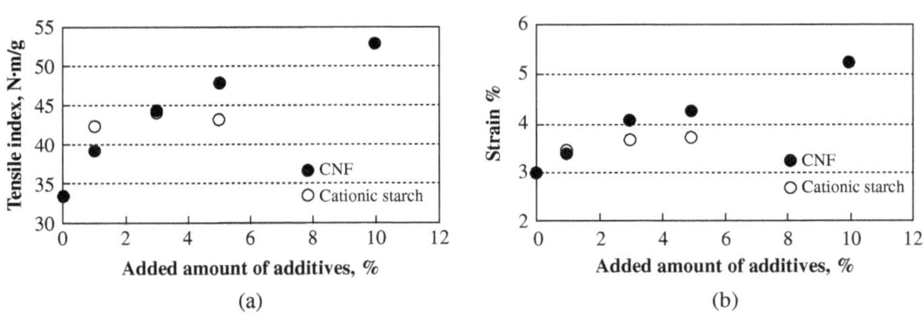

Fig. 2. Tensile index (a) and strain at break (b) of Hw-BKP sheets with CNF or cationic starch.

strength, but CNF can increase the strength consistently with dosage. However, the drainage of the pulp stock must be improved to expand the application of nanocellulose to the paper industry.

CNF and MFC markedly reduce air permeability by filling up or plugging the pores between fibers. They also increase pore tortuosity and decrease pore volume and pore connectivity.[22]

In addition, CNF and MFC affect the drying shrinkage and dimensional stability of paper. Manninen *et al.*[18] investigated the drying shrinkage and dimensional stability of wood-free paper containing different amounts of MFC and the effect of MFC with starch. The dry shrinkage and hygroexpansion were measured using image analysis. Figure 3 shows the dry shrinkage of a freely dried paper sheet. MFC increased the drying shrinkage and decreased the dimensional stability of the freely dried sheets because of its increased bonding. The impact on the dimensional stability became more severe with increasing amounts of MFC. MFC had a greater effect on the drying shrinkage and dimensional stability compared to kraft fines. However, when the cationic starch was added as a retention aid, the drying shrinkage was decreased and the dimensional stability was improved.

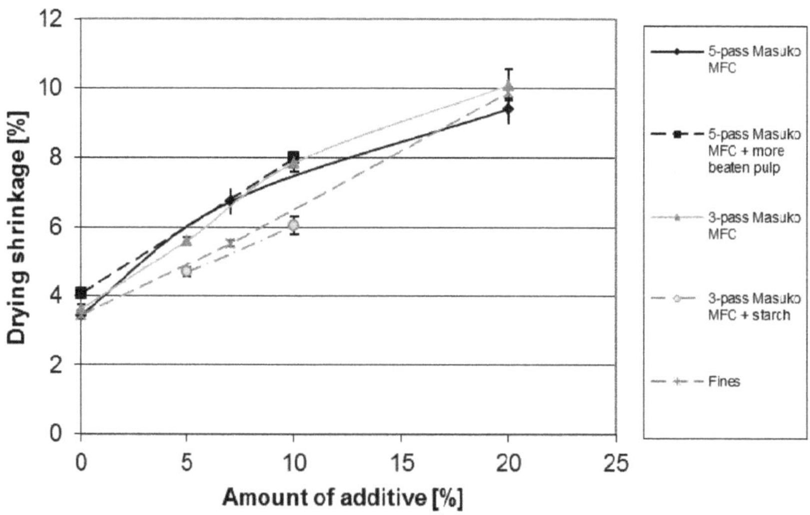

Fig. 3. Drying shrinkage of freely dried sheets as a function of additive amount. The sheets contained different amounts of 5- or 3-pass Masuko MFC or kraft fines.

Source: Reprinted by permission from De Gruyter; SPCI, *Nordic Pulp & Paper Research Journal*, M. Manninen *et al.*[18]

The addition of CNF improves the strength of filled paper, which can be used to increase the filler content of paper and lead to cost savings in production. Mörseburg and Chinga-Carrasco[19] demonstrated the potential synergistic effects of clay–CNF interactions for the improvement of the strength and optical properties of TMP-based sheets by forming appropriate layer-structured sheets. They suggested the use of CNF for high-filled papers instead of bleached softwood kraft pulp fibers. Hii *et al.*[20] also reported that the addition of MFC improved the strength and air resistance of TMP paper filled with GCC. The binding of the filler-MFC-fines aggregates to the fiber network improved the tensile strength and internal bond strength at high ash content.

Ankerfors *et al.*[21] studied the strength enhancing capabilities of MFC in highly filled papers on a pilot scale using realistic industrial conditions. A paper web was produced using the FEX pilot paper machine with a retention system composed of cationic PAM and silica. The MFC (2.5–5.0%) improved the mechanical properties of the paper. In particular, the *Z*-strength and fracture toughness of the highly filled paper were significantly increased. Further improvement in strength was achieved with the combination of MFC and cationic starch. The light scattering coefficient and opacity were not affected by the addition of MFC alone or together with cationic starch and were mainly dependent on filler content. The study suggested that the filler content could be increased from 20 to 33% without losing tensile index when 5.0% MFC and 2.0% cationic starch were added.

In recent years, a combination of filler particles and MFC has been introduced to make filled paper with high strength. A mineral company developed the microfibrillated cellulose (MFC)–mineral composite. This "FiberLean® Technologies" composite is produced by co-grinding minerals such as calcium carbonate ($CaCO_3$) and kaolin, with pulp fibers. The MFC–mineral composite improves the initial wet strength because of increased hydrogen bonding, which originates from the MFC, and provides better optical properties and low drying energy consumption, which is due to the incorporated filler particles. A total of 10–15% of the pulp fibers in the base paper could be replaced by the MFC–mineral composite. The increased strength would make it possible to reduce the refining energy or use bulkier pulp, such as CTMP.

2.1.2 *Dewatering in forming and pressing*

CNF has a high water holding capacity because of its high specific surface area and hydrophilicity. As shown in Fig. 4,[29] the dewatering time

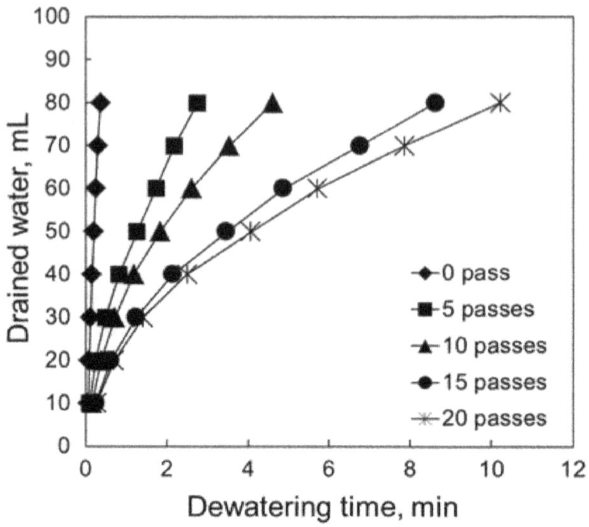

Fig. 4. Drained water volume of a 0.02% CNF suspension as a function of dewatering time for the different grinding pass numbers.

Source: Figure reproduced from Ref. 29 under the Creative Commons Attribution License.

increased and the amount of drained water decreased with an increasing number of passes through a grinder. For a pulp fiber suspension (0 passes), 20 seconds were required for the 80 mL of water to pass through the mesh wire of the pressurized dewatering equipment. In contrast, it took greater than 10 min (20 passes) to remove that volume of water from the CNF suspension, which clearly shows that the addition of CNF to pulp stock causes poor dewatering performance.

However, the poor dewatering of a CNF suspension can be improved by the addition of salt or polyelectrolytes.[29,30] Sim *et al.*[30] reported that the dewatering of a CNF suspension increased significantly at a low concentration (0.01 M) of NaCl solution. Bivalent cations (e.g., $CaCl_2$) were more effective in dewatering at the same salt concentration compared to monovalent cations (e.g., NaCl) (Fig. 5). However, there was no change or reversal in drainage above a certain salt concentration. The drainage property of the CNF suspension was greatly affected by its zeta potential.[30] Figure 6 shows the effect of the type and dosage of polyelectrolyte on the drainage of a CNF suspension. Polyethyleneimine (PEI) with a low or high molecular mass and cationic polyacrylamide (PAM) with a low or high charge density were added to a CNF suspension. The drainage time

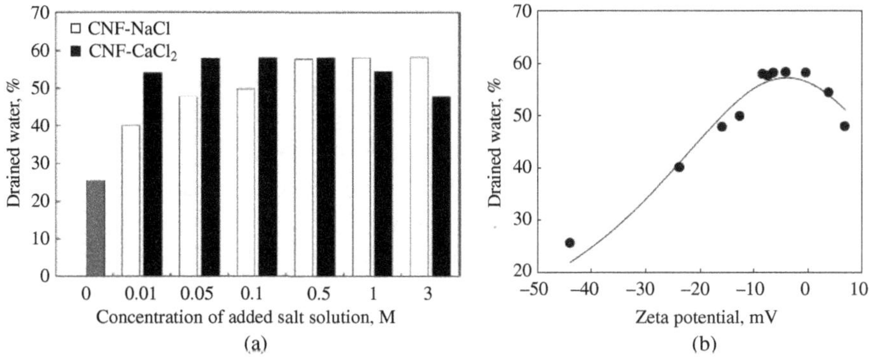

Fig. 5. Dewatering of 1.5% CNF suspensions based on the type and concentration of salt (a) and the relationship between the zeta potential and drained water (b).

Source: Reprinted by permission from Springer Nature, Cellulose, Sim *et al.*[30]

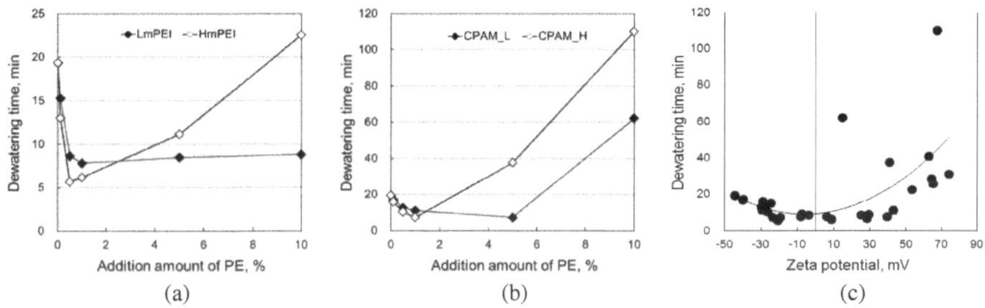

Fig. 6. Effect of polyelectrolyte addition on the drainage of 1.5% CNF suspension: (a) Addition of PEI; (b) addition of cationic PAM; (c) the relationship between the zeta potential and drainage by polyelectrolyte addition.

Source: Figure reproduced from Ref. 29 under the Creative Commons Attribution License.

was decreased by a small polyelectrolyte dosage (~1 wt%) but deteriorated again above a certain dosage. There was a clear relationship between the zeta potential of the CNF suspension and drainage.

Although drainage of a CNF suspension can be controlled to some extent with salt or polyelectrolytes, it is one of the problems associated with the use of CNF in the papermaking process. In particular, when CNF is used for wet-end application, drainage should be checked in addition to the strength properties. Taipale *et al.*[17] evaluated the drainage behavior of a pulp suspension with MFC using a dynamic drainage jar (DDJ) under different pH, salt, and polyelectrolyte conditions. In their study, a cationic starch was used to retain MFC and fines in the fiber network. In general, the addition of MFC decreased the drainage rate of the pulp suspension with a linear relationship (Fig. 7). As both MFC and pulp fibers are charged materials, the drainage behavior of the suspension was affected by pH, electrolytes, and polyelectrolytes. Figure 8 illustrates the effect of pH and salt on the drainage of a pulp slurry with an MFC content of 30 mg/g. As the pH increased, the drainage became slow due to an enhanced affinity between the cationic starch and the deprotonated carboxylic groups of the fibers, which led to more adsorption of MFC to the fibers and a decrease in dewatering rate. A similar phenomenon was observed when the salt concentration was increased.

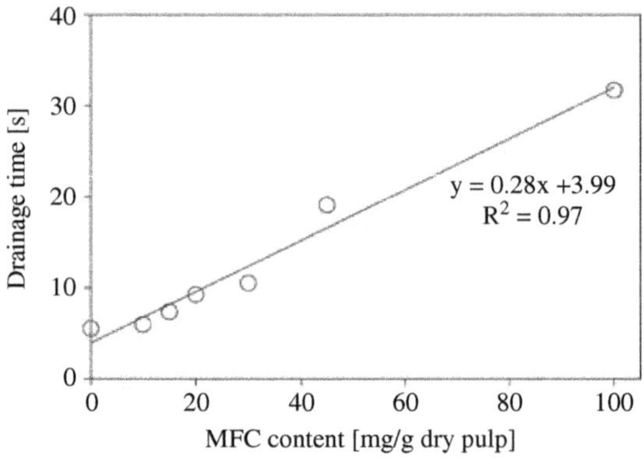

Fig. 7. The effect of MFC content on the drainage of a pulp suspension.
Source: Reprinted by permission from Springer Nature, Cellulose, Taipale *et al.*[17]

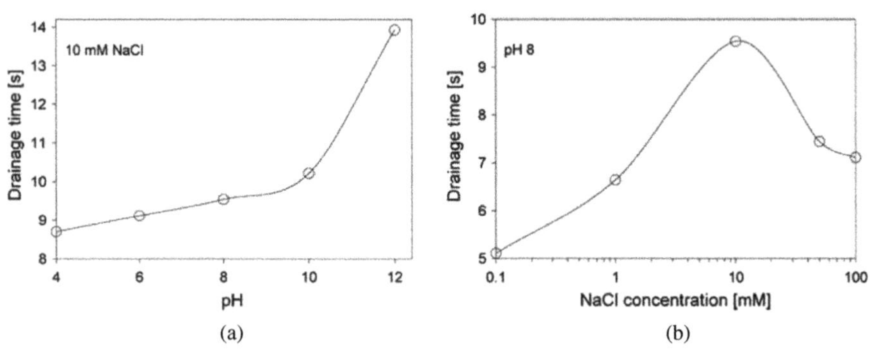

Fig. 8. The effect of pH (a) and salt concentration (b) on the drainage of a pulp suspension.
Source: Reprinted by permission from Springer Nature, Cellulose, Taipale *et al.*[17]

When the salt concentration was increased from 0.1 to 10 mM, the amount of adsorbed cationic starch on the fibers increased due to a more coiled conformation of the polymer. This enhanced cationic starch adsorption led to the formation of a swollen fiber–fibril network, which resists water flow during dewatering. This effect of the salt concentration on the drainage of the suspension was different from the results obtained by Sim *et al.*[30] for a CNF suspension. As shown in Fig. 5, the addition of salt to CNF suspension improved drainage by flocculating the fibrils. This discrepancy might be due to the type of furnish and the presence of cationic polymer. In the work of Taipale *et al.*[17] 10% of the MFC was added to the Sw-BKP suspension with a cationic starch. However, Sim *et al.*[30] investigated the CNF suspension without any cationic polymer. Therefore, the added salt behaves differently depending on the system examined. Taipale *et al.*[17] showed that the drainage of the pulp suspension with MFC was strongly affected by the molar mass and charge density of the polyelectrolyte and that the drainage rate was not improved with the addition of the polyelectrolyte. In general, polyelectrolytes with a low molecular mass and high charge density are better at improving drainage compared to those with a high molecular mass. This property is related to the thickness of the adsorption layer of the polymer on the fibers (Fig. 9). Carboxymethylated MFC showed better drainage when it was used along with cationic starch.

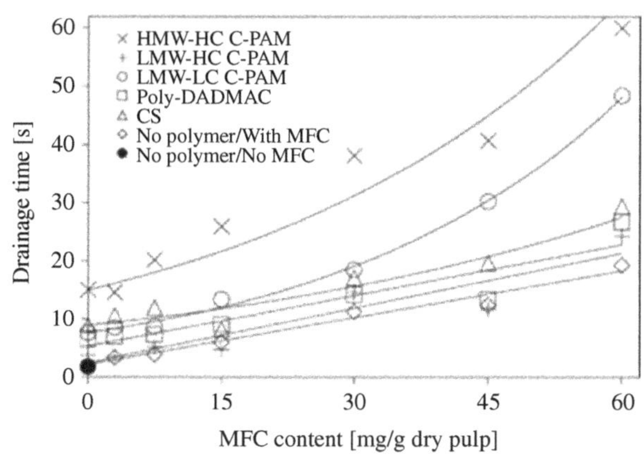

Fig. 9. The effect of MFC content and cationic polyelectrolytes on the drainage of a pulp suspension with fines.

Source: Reprinted by permission from Springer Nature, Cellulose, Taipale *et al.*[17]

Manninen *et al.*[18] evaluated the retention and drainage of wood-free pulp stock containing MFC or kraft fines with or without starch. MFC had a lower retention value compared to kraft fines, but their retention increased up to 100% with the addition of cationic starch as a retention aid. The cationic starch also improved drainage. Therefore, the proper use of a retention aid can solve both retention and drainage problems when used along with MFC in the pulp stock.

In most cases, dewatering during the web forming process deteriorates following the addition of nanocellulose. However, its impact on press dewatering seems negligible. Rantanen and Maloney[31] suggested that press dewatering is affected by the sheet grammage and the morphology of the nanofibers. The dewatering time in forming and press dewatering was investigated using a mill furnish (25% softwood bleached kraft pulp + 75% hardwood bleached kraft pulp). It was shown that the dewatering time during the forming and sheet air permeability correlated with the nanocellulose content, which was consistent with the results of other studies.[16–17,23] The restriction of water flow channels and reduction of sheet permeability by nanocellulose caused a change in dewatering and air permeability. Press dewatering decreased with the addition of nanocellulose, but the moisture ratio after pressing was affected by the sheet grammage (Fig. 10). When the sheet grammage was low (less than 60 g/m^2, in a pressure-controlled regime) and the CNF content was less than 2%, CNF had a negligible effect on press dewatering. However, at a grammage above 60 g/m^2, the moisture ratio after pressing increased with increasing CNF content. In particular, the finer CNF (i.e., TEMPO-oxidized CNF) had a greater impact on press dewatering than MFC. Rewetting of the sheet during the nip expansion and separation phases was also mainly affected by grammage.

In the case of highly filled paper, the addition of CNF or MFC does not deteriorate drainage and press dewatering. As MFC is added to a pulp slurry, the drainage rate is decreased. Because fillers, such as calcium carbonate or clay, have quite a low water holding capacity, they make dewatering easier during the papermaking process. In other words, the low drainage rate of MFC can be compensated by an increase in the ash content.[20] According to Hii *et al.*[20] a TMP paper sample that contained 5% MFC and 29% ash had similar dryness after wet pressing to that of a paper sample with 85% TMP and 15% ash.

A papermaking trial using high-speed pilot paper machine supported this effect of CNF and MFC on the retention and drainage of pulp stock.

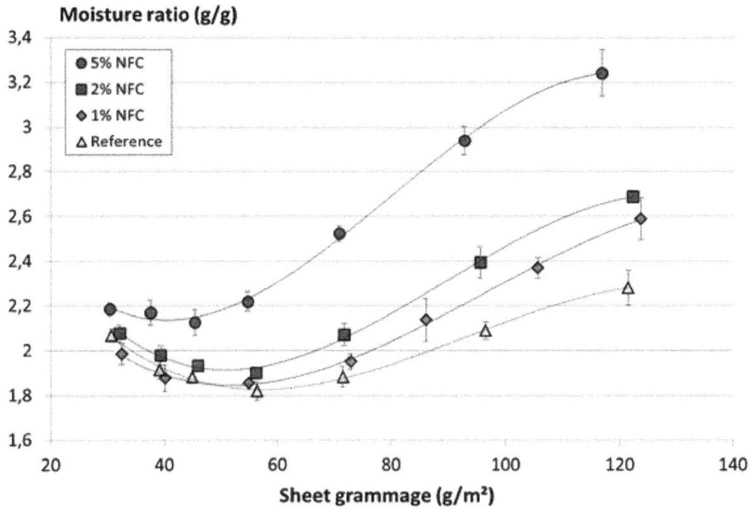

Fig. 10. The effect of CNF dosage and sheet grammage on the moisture ratio after the press.

Source: Reprinted by permission from De Gruyter; SPCI, *Nordic Pulp & Paper Research Journal*, Rantanen and Maloney.[31]

Kajanto and Kosonen[28] evaluated the process performance such as retention and dewatering on a high-speed paper machine. They prepared paper webs with basis weights of 50–65 g/m^2 with the use of 1 or 2% nanocellulose as an additive along with cationic starch or a two-component retention system. The total retention and formation were not deteriorated by the addition of nanocellulose. After the web forming section, the web dryness was reduced by 0.5–1%, which was not a significant loss. In the case of press dewatering, dryness of the paper web was increased with the addition of CNF. This improvement in press dewatering was not clearly explained, but it might be affected by the press configuration and pressure pulse. Ankerfors *et al.*[21] produced highly filled paper on a FEX pilot paper machine and found that MFC had a negative effect on dewatering in the forming process. However, MFC did not change the dryness after the press section. In addition, there was no impact on the retention at the lowest addition level of MFC (2.5%). At a higher dosage of MFC (5%), the retention decreased.

While nanocellulose can impair dewatering in forming and pressing processes, its impact is dependent on the composition and grammage of the sheet, retention aid, and other factors. When a sheet is composed of pulp with a high drainage resistance (e.g., highly beaten pulp) or the sheet grammage is higher (e.g., flow-controlled regime), press dewatering is decreased with an increase in CNF content because of the plugging of pores and a reduction in the sheet permeability by CNF. When the moisture content is sufficiently low (e.g., low grammage sheet) or the drainage resistance is low (e.g., high filled paper), a small amount of nanocellulose would not have a detrimental impact on press dewatering.

2.2 *Surface Sizing or Impregnation of Nanocellulose*

Nanocellulose can be applied to paper through surface sizing treatment or impregnation. Surface application has many advantages over wet-end application. It is a highly effective treatment with more specific action, less sensitive to the changes in wet-end operations, and allows good retention of additives and improves paper quality.[32] The surface application of nanocellulose was made for two purposes: (i) strengthening of old paper archives and (ii) reducing the linting of newsprint or mechanical pulp-based paper.

CNF has been used as a strengthening agent for old paper documents.[33] Traditionally, methylcellulose (MC) and carboxymethyl cellulose (CMC) have been used as strength agents for aged papers. The CNF suspension was treated on the surface of two kinds of aged paper samples with a brush: an accelerated aged commercial paper made of bamboo fibers and wood fibers, and an old book. Figure 11 shows the brushing process for applying CNF to the aged paper. The coat weight of CNF was controlled by the consistency of the CNF suspension. The CNF treatment improved the strength of the accelerated aged paper and old book. The tensile index and folding endurance were linearly increased with the coat weight of the CNF (Fig. 12). Figure 13 shows the surface of the CNF-treated papers. The surface of the aged paper was covered by CNF at a coat weight of 1.7 g/m^2, which also increased the bonding of the fibers by bridging the adjacent fibers. In addition, when the ink density on the printed area was measured before and after CNF treatment, similar values were obtained at a coat weight of 1.5 g/m^2 (Fig. 14). Therefore, the CNF surface treatment did not deteriorate readability of the old book. These results suggest that CNF is suitable as a strengthening agent for old paper documents.

Newsprint and mechanical pulp-based paper suffer from linting, dusting, and picking problems in the printing process because the

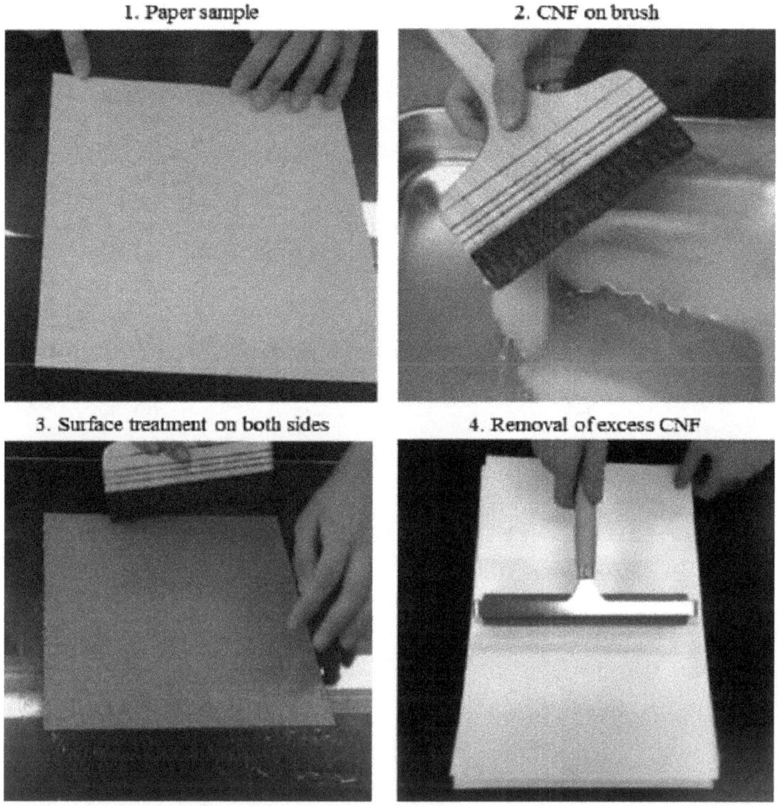

Fig. 11. Brushing process of CNF on an accelerated aged paper.

Source: Figure reproduced from Ref. 33 under the Creative Commons Attribution License.

mechanical pulp is short in length and has a low bonding ability compared to the chemical pulp. Song *et al.*[34] investigated the reduction of the linting and dusting propensity of newspaper using starch and MFC. Compared to the internal treatment with cationic starch, the surface treatment of MFC alone or together with anionic starch is more efficient in decreasing the linting tendency of TMP sheets (Fig. 15). When the mixture of MFC and anionic starch (50:50, mass ratio) was applied onto the paper surface, there was a strong synergistic effect in the reduction of linting by reinforcing the surface strength. This reduced linting propensity could increase the printing speed, which increases the printing productivity.

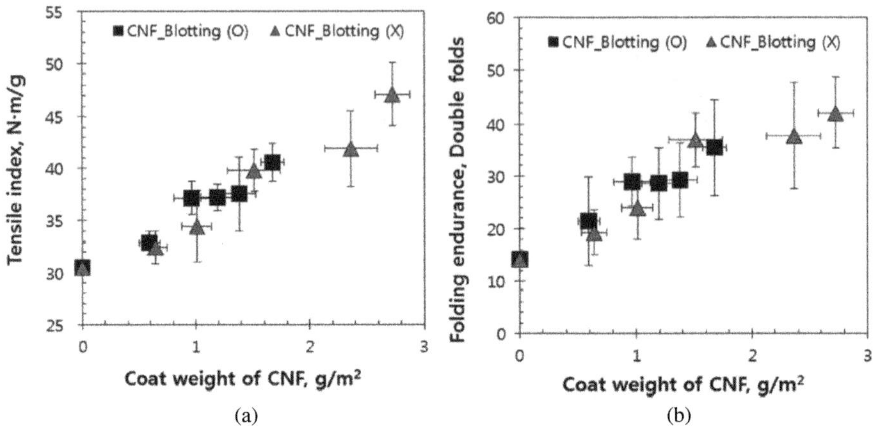

Fig. 12. Tensile index (a) and folding endurance (b) of aged paper with CNF treatment.

Source: Figure reproduced from Ref. 33 under the Creative Commons Attribution License.

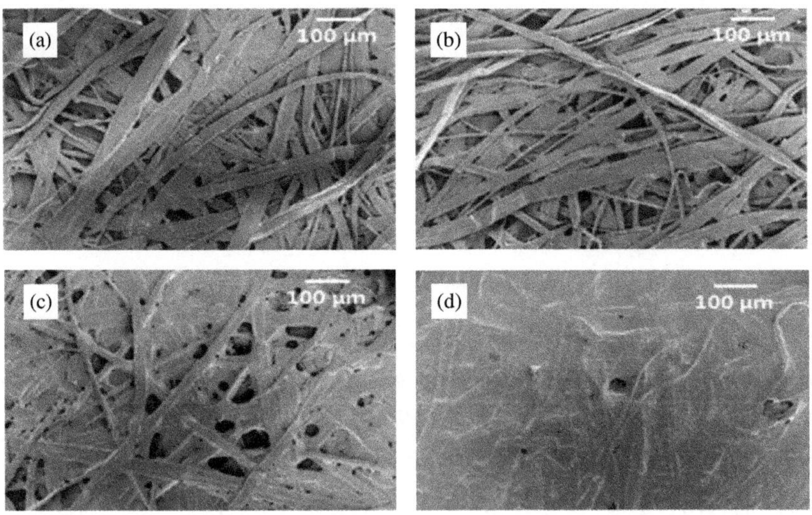

Fig. 13. SEM images on a CNF-treated surface of paper: (a) untreated bamboo paper and CNF-treated bamboo paper with a coat weight of (b) 0.6 g/m^2, (c) 1.2 g/m^2, or (d) 1.7 g/m^2.

Source: Figure reproduced from Ref. 33 under the Creative Commons Attribution License.

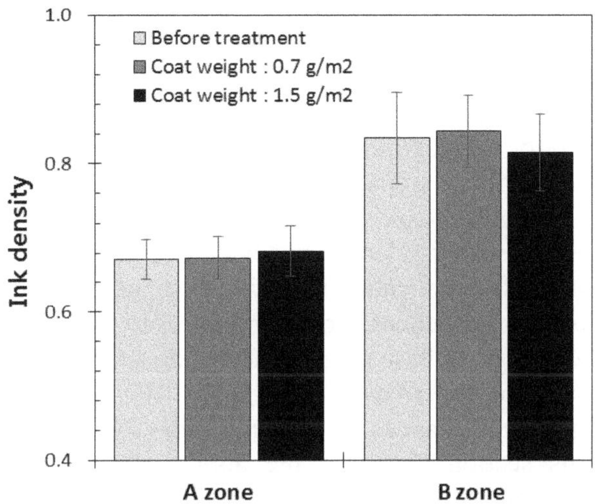

Fig. 14. Ink density of printed area before and after CNF treatment.

Source: Figure reproduced from Ref. 33 under the Creative Commons Attribution License.

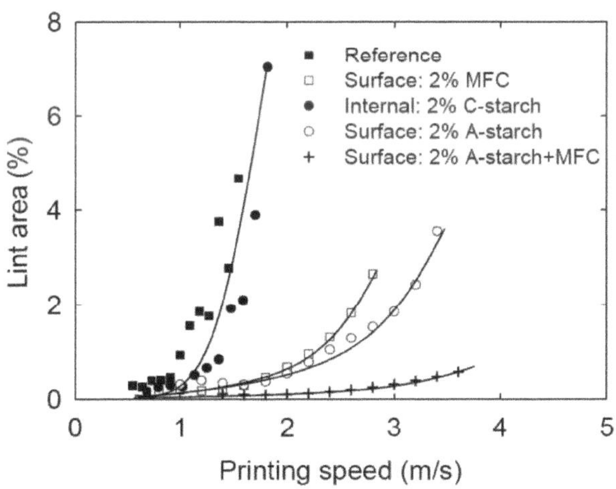

Fig. 15. Linting propensity of reference TMP sheets with different chemical treatments at the same addition level (2%).

Source: Reprinted by permission from De Gruyter; SPCI, *Nordic Pulp & Paper Research Journal*, H. Song *et al.*[34]

3. Application of Nanocellulose to Paper Coating

The function and quality of paper can be modified and improved by coating the paper with the proper materials. Pigment coating is a principal coating method for paper, which makes glossy, bright, and smoother paper with better printability. A coating of wax or other functional polymers makes the paper hydrophobic or provides new functionality. New paper-based composites, such as electronic device and decorative sheet, can be created by coating for application in a wide range of industrial fields. CNF and MFC films have excellent barrier properties to oxygen and oil, which makes these coatings on base paper a promising approach to improve the barrier properties of paper-based packaging. It is possible to make high value-added paper substances by coating with CNF or other functional materials. This section describes the application of nanocellulose in barrier, pigment, and functional coating.

3.1 *Barrier Coating*

Resistance to the transmission of gas and absorption of water or oil is the most important barrier property that protects the packaged goods and maintains their quality. In general, the barrier properties of packaging paper are obtained by coating or lamination with petroleum-based or synthetic polymers, such as polyethylene (PE), ethylene vinyl alcohol (EVA), polyvinylidene chloride (PVDC), and wax. These polymers have good barrier properties against water and oxygen. However, fossil-derived polymer-laminated packaging has the fatal weakness of a lack of biodegradability and poor recyclability. Therefore, interest in the development of environmentally-friendly, biodegradable barrier coating materials has increased. The barrier properties of various natural or biodegradable polymers have been investigated for paper packaging. Polysaccharides (e.g., starch, chitosan, and methyl cellulose), proteins, lipids, and polylactic acid (PLA) are some examples of eco-friendly barrier coatings.[35] The biomaterial CNF has drawn significant interest as an alternative to replace petroleum-based coating materials. In particular, a coating of CNF on paper or board would be a good approach to produce environmentally-friendly packaging for the next generation because paper and board are the second largest packaging materials after plastic. Because of its hydrophilic nature, CNF should be chemically modified or used with hydrophobic materials to provide a barrier to water vapor.

It is well known that CNF film provides an excellent barrier to oxygen transmission and air permeability.[36] Syverud and Stenius[37] investigated the strength and barrier properties of MFC films. MFC films (35 g/m^2) had a tensile index of 146 ± 18 Nm/g and an elongation of 8.6% ± 1.6%. The oxygen transmission rate (OTR) of the MFC film was approximately 17 ml/m^2·day, which was quite low compared to low- and high-density polyethylene (7800 ml/m^2·day and 2600 ml/m^2·day, respectively). The MFC was deposited as a surface layer atop the wet base paper using a dynamic sheet former. The strength of the base paper increased significantly with a surface layer of less than 10% MFC, and the air permeability decreased remarkably. Although the OTR value of the paper with the MFC layer was not reported, the reduced surface porosity suggested that MFC could act as a barrier coating material against oxygen transmission. Aulin *et al.*[38] also evaluated the oxygen and oil barrier properties of MFC film and MFC-coated paper. The MFC was obtained by microfluidizing carboxymethylated dissolving pulp. The OTR (0.0006 cm^3·μm/m^2·day kPa at 0% RH) of the MFC film was much lower than that of polyethylene and biopolymers, such as plasticized starch and proteins, and was in the same range as that of commercial ethylene vinyl alcohol. However, the barrier property of the MFC film was affected by the relative humidity because of the hydrophilic nature of cellulose (Fig. 16). The OTR of the MFC film increased sharply at an RH higher than 70%. The MFC-coated paper had a low air permeability and good oil barrier property even at a

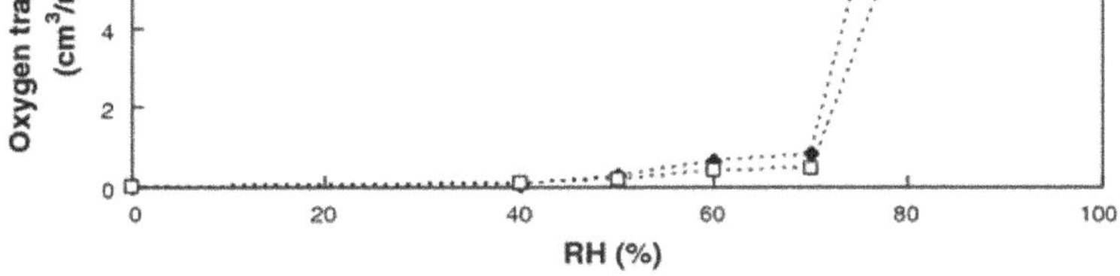

Fig. 16. Effect of relative humidity on the oxygen transmission rate (OTR) of MFC film. *Source*: Reprinted by permission from Springer Nature, Cellulose, Aulin *et al.*[38]

low coat weight of 1.8 g/m^2, which indicated the possibility of using it as packaging material with the limitation of the MFC under high RH conditions.

The poor water vapor barrier properties of the CNF coating and its sensitivity to moisture can be resolved by the incorporation of hydrophobic material. Hult *et al.*[39] suggested the use of shellac, a natural resin, as a way to enhance the barrier property of MFC-coated paper and paperboard. The paper and paperboard were coated with a single layer of an MFC-shellac blend or with a multilayer coating consisting of MFC followed by shellac as the top layer. This multilayer coating decreased the air permeability more effectively because it provided more efficient closure of nanopores on the paper surface. Even though the OTR was decreased significantly by the MFC coating, the value was not sufficient for high oxygen barrier packaging. In contrast, a good water vapor barrier property (WVTR < 10 g/m^2·day at 25°C and 50% RH) was achieved, which is applicable to the food packaging field.

The good barrier properties of CNF and MFC films or CNF-coated paper, particularly to oxygen and oil, can be explained by the change in pore structure and porosity and the intermolecular or interfibrillar bonding. Aulin *et al.*[38] explained that the excellent oxygen barrier of CNF film is derived from the dense film structure and low porosity, which increases the tortuosity and non-connectivity of pores. Because the crystallinity of wood-derived CNF is quite high (50–70%) and diffusion of gas or oil through the crystalline region is not possible, CNF contributes to the increase in the path length and decrease in permeability. CNF also forms a dense and strong network by increased inter-fibrillar bonding, which provides a more packed structure and a lower free volume for gas diffusion. However, the CNF film structure is destroyed by the sorption of water vapor at high RH.

In general, the barrier property of CNF-coated paper and paperboard can be improved by increasing the coat weight (Fig. 17). The application method (e.g., bar coating, size press, spray coating), CNF properties (e.g., morphology, electric charge), and CNF suspension properties (e.g., consistency, viscosity) determine the coat weight.

CNF can be applied onto paper by various means, including bar coating, spray coating, size pressing, foam coating, and roll-to-roll coating methods. Lavoine *et al.*[40] compared the bar coating and size press applications and found that the bar coating method provided higher coat weights than the size press method. The two-roll size press method

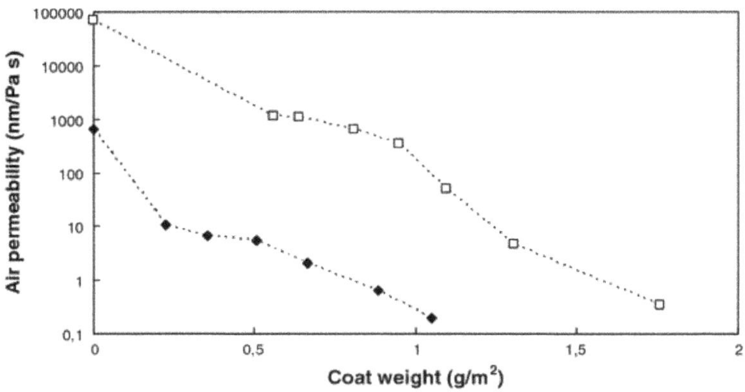

Fig. 17. Air permeability as a function of MFC coat weight for unbleached (open square) and greaseproof paper (filled diamond).

Source: Reprinted by permission from Springer Nature, Cellulose, Aulin *et al.*[38]

promoted the penetration of CNF into the paper structure. However, the high viscosity of CNF prevented an increase in the coat weight and made the CNF remain on the paper surface. According to Lavoine *et al.*[40] a coat weight of 7 g/m^2 was achieved by bar coating five times. In contrast, the coat weight remained at 3 g/m^2 following five applications of CNF using the size press method. Consequently, the size press method did not improve the paper properties as much as with bar coating. In contrast, the air barrier properties and bending stiffness were considerably improved by 90–97% and greater than 50%, respectively, after MFC bar coating with a coat weight of 7 g/m^2. The Cobb size degree, an indicator of the resistance to water absorption, was not affected by the MFC coating.

Foam coating has been introduced as a useful method for CNF coating.[41] Foams of a mixture of CNF, anionic surfactant and air (80–95%) coated evenly onto a paper surface. Despite the low coat weight (<1 g/m^2 for a single coat), a more even distribution of CNF on the paper was achievable, which changed the paper properties. Recently, the roll-to-roll coating of a CNF suspension was suggested as an eco-friendly barrier application method (Fig. 18).[42] This process enables the continuous coating of CNF at a large production capacity compared to what has previously been reported. The sheared state of CNF allows it to be easily metered with the slot die coating head, thereby controlling the final coat weight. It was shown that CNF coating improved the strength properties

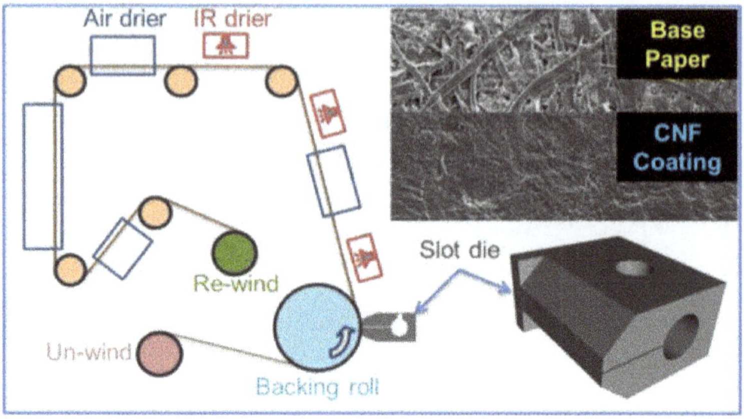

Fig. 18. Schematic diagram for applying a CNF coating to paper with the use of a narrow slot die applicator.

Source: Reprinted by permission from ACS Publications, *Industrial & Engineering Chemistry Research*, Kumar *et al.*[42]

to some extent even at very low coat weights. In particular, the barrier property of the coating against mineral oil, grease, and air increased substantially after CNF coating.

The unique and complex rheological properties of CNF are one of the hurdles to obtaining a high coat weight and uniform coating coverage of a paper surface. As the fibers are nanofibrillated by mechanical shearing, the viscosity of the CNF suspension increases rapidly (Fig. 19).[43] The storage modulus also increased by nanofibrillation of fibers. However, owing to the shear thinning characteristics of a CNF suspension, it can be used to process them into thin films and coatings. Once the CNF has been shear thinned by passing it through a narrow slot gap (Fig. 18), it can be applied as a thin film on the moving web even at low coating speeds. The viscosity of a CNF suspension can also be controlled by the use of an additive. Figure 20 shows the effect of the addition of a non-ionic surfactant on the viscosity of a CNF suspension. The low shear viscosity was markedly reduced by 65%, but the high shear viscosity remained the same.[42] The addition of a proper rheology modifier to a CNF suspension would provide a more uniform coating surface even at a low coat weight (i.e., when the low shear coating equipment is used). Carboxymethyl cellulose (CMC) improved the dispersion of CNF in the coating suspension

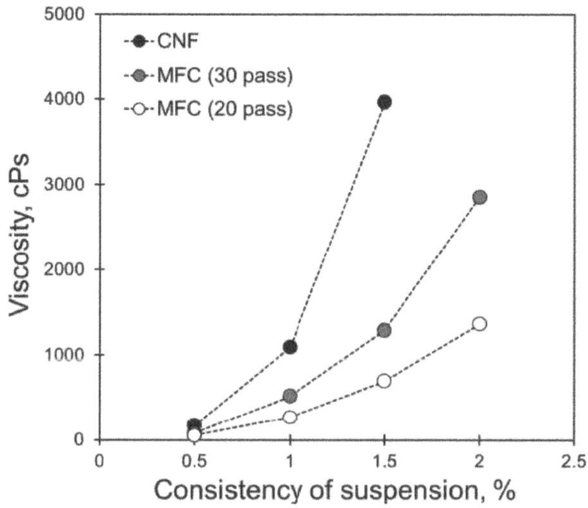

Fig. 19. Low shear viscosity of MFC and CNF suspensions.

Source: Figure reproduced from Ref. 43 under the Creative Commons Attribution License.

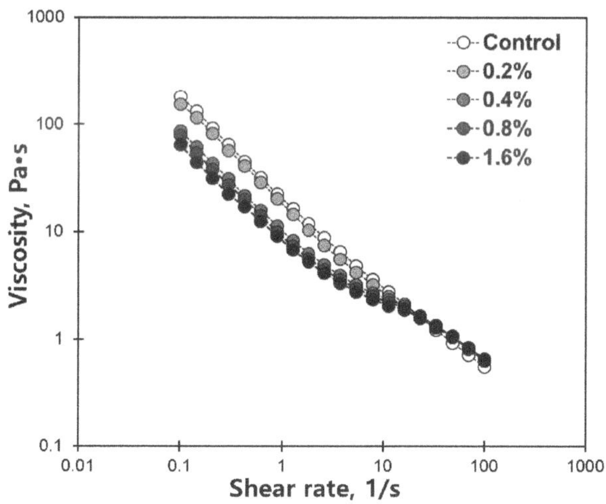

Fig. 20. Viscosity of CNF suspension containing non-ionic surfactant.

Source: Figure reproduced from Ref. 43 under the Creative Commons Attribution License.

because its highly charged anionic groups made the CNF disentangle in the suspension, which enables the coating of CNF with a higher solids content and leads to good coverage of the CNF on paperboard.[44]

The characteristics of CNF, such as morphology and ionicity, affect the coatability and the quality of coated paper. With the application of a finer CNF, barrier properties are improved. Ground CNF (average diameter, 0.036 μm) gave paperboard better barrier properties against water and oil absorption, and much lower air permeability than refined CNF (average diameter, 0.166 μm).[44] Carboxymethylation of pulp before nanofibrillation produced CNF that was narrower than that obtained from untreated pulp. These finer fibrils formed a denser, smoother, and more transparent film. Therefore, the application of carboxymethylated CNF onto paper or paper board provides more resistance against air permeability.

The properties of the base paper (e.g., grammage, porosity, surface roughness, and contact angle) influence the coat weight and quality of the coated paper substrate. Most previous studies suggested that improvements in the barrier properties to air and oxygen transmission could be obtained with CNF coating. However, Lavoine *et al.*[45] reported that an MFC coating on cardboard with a grammage of 300 g/m^2 did not affect the barrier properties but changed the bending stiffness and compressive strength, in contrast to previous reports. They explained that this difference was likely due to the high grammage of cardboard and suggested that an increase in the coat weight of MFC or multilayer structure coating should be considered to achieve good barrier properties.

3.2 *Pigment Coating Additive*

Pigment coating is applied to paper and paperboard to improve its printability and appearance. An aqueous suspension composed of pigment(s), binder(s), and other additives is prepared and applied with the use of several coating methods, including blade, rod, roll, air knife, curtain and spray coating. Among these coating techniques, blade coating is the most widely used because it allows the use of high solids coating at high speed with great efficiency. A coating suspension consisting of pigments, binders, and various additives should have good rheological properties and water retention because these factors are critical for the runnability of the coating color on the modern high-speed coater. Coating color composition also influences the printing and optical properties of the final product. Many developments and new products have been made over the past several decades to produce high-quality coated products at a low cost. Many technological achievements to develop and commercialize new diverse pigments and binders for paper coating have been made because pigments and binders are two of the most important ingredients for paper coating. The development of various types of ground or precipitated calcium carbonates with different particle shape and size distributions, and the introduction of various latex binders that provide advanced printing quality for offset, rotogravure, and other printing processes are two of the most important technological advancements.

Even though pigments and binders are the major components of the coating formulation, some additives have a critical effect on coating runnability and coated paper properties. The thickener is a minor but essential component of the coating color because it controls the rheological properties and water retention of the coating color, which are two critical properties that control coating runnability.

Recent attempts have been made to use CNF as a thickener in pigment coating because of its gel-like and shear-thinning behaviors.[46,47] Several studies have reported the rheological properties of coating colors that contain CNF.[46–51] The viscosity and viscoelastic behaviors of the coating color prepared with CNF as an additive were analyzed under various evaluation conditions to predict spreading, levelling, and dewatering of the coating color.[46,49,50] The coating color with CNF had less elasticity and slower viscosity recovery after coating application than the coating color with CMC, which was attributed to the lower flocculation between the particles in the CNF coating. Thus, the CNF coating color spread easily, which increased the gloss of the coated paper.[50]

In general, the addition of CNF increased the viscosity of the coating color. The addition of CMC, which is the most widely used thickener in paper coating, also increased coating viscosity. However, the CMC coating increased the viscosity more than the CNF coating because CMC makes the coating components flocculated via a depletion mechanism.[52–55] The addition of CNF also increased the low shear viscosity, but this increase was marginal compared to that of CMC because it does not cause any flocculation of pigments (Fig. 21).[56]

The addition of CNF causes a reduction in water retention while it increases the viscosity (Fig. 22), which indicates that CNF causes the formation of a bulkier filter cake of the coating than CMC. The higher dewatering of the CNF coating color may be due to the lower viscosity of the aqueous phase of the CNF coating color.[56]

The viscosity of the coating color with CNF shows a shear-thinning behavior.[47–48, 51] Figure 23 represents the viscosity of color with or without an additive as a function of the shear rate. The CNF colors had a lower

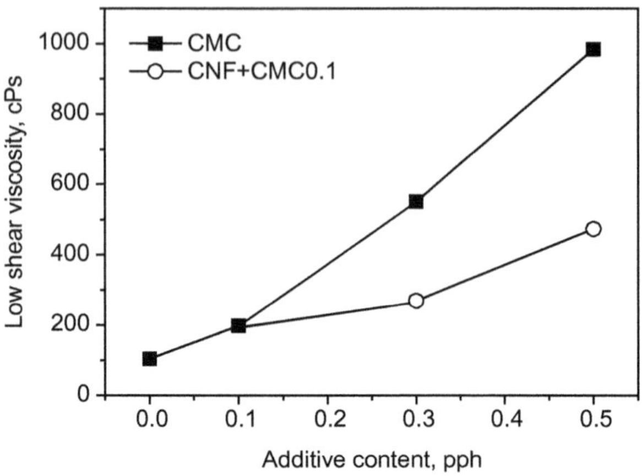

Fig. 21. Low shear viscosity of coating color with CMC and CNF (Latex content: 12 pph).

Source: Reprinted by permission from ACS Publications, *Industrial & Engineering Chemistry Research*, Oh *et al.*[56]

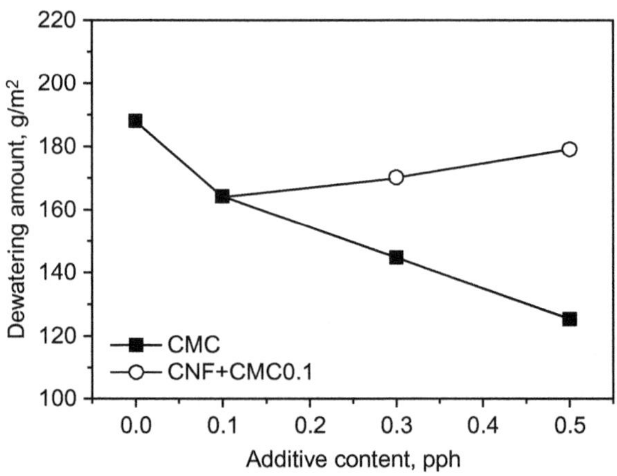

Fig. 22. Dewatering amount of coating color with CMC and CNF (Latex content: 12 pph).

Source: Reprinted by permission from ACS Publications, *Industrial & Engineering Chemistry Research*, Oh *et al.*[56]

viscosity than CMC coatings because CMC induced the flocculation of the coating component. The viscosity of the coating color containing 0.5 pph of CMC is more or less constant at low shear rates, which

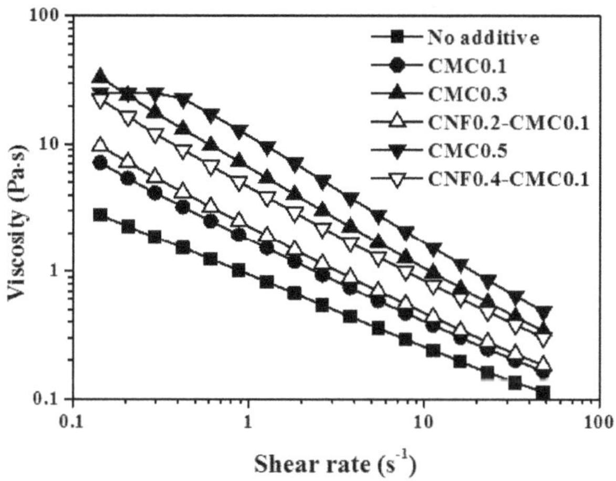

Fig. 23. Viscosity of coating color as a function of the shear rate ($\gamma = 0.1–100\ s^{-1}$).
Source: Reprinted by permission from ACS Publications, *Industrial & Engineering Chemistry Research*, Oh *et al.*[56]

indicates that this coating color had a highly structured, solid-like behavior under a shear rate < $0.3\ s^{-1}$. The increase in the viscosity of the coating color containing CNF is due to an increase of the coating solids by the water absorption and high viscosity of CNF.[56]

The coating colors containing CMC or CNF as an additive exhibited viscoelasticity. The MFC and CNF additives function as water-binding and gel-forming components rather than with the flocculating thickening action of CMC.[46]

CMC improves the dispersion of CNF in coating suspensions because highly charged anionic groups make the CNF disentangle in the suspension.[42] Amplitude sweep and frequency sweep tests, which are often used to evaluate the interaction between the coating components, showed that a linear viscoelastic region was present, and G' is larger than G'' in all coating colors (Fig. 24).[56] With an increase of CMC and CNF in the coating color, G' and G'' also gradually increased as the critical stress at which the structure was destroyed increased. These results indicate the presence of a stronger interaction between the coating components. G' and the critical stress of the CMC coating colors were higher than those of the CNF coating colors because the CMC coating is more solid-like compared to the CNF coating color.

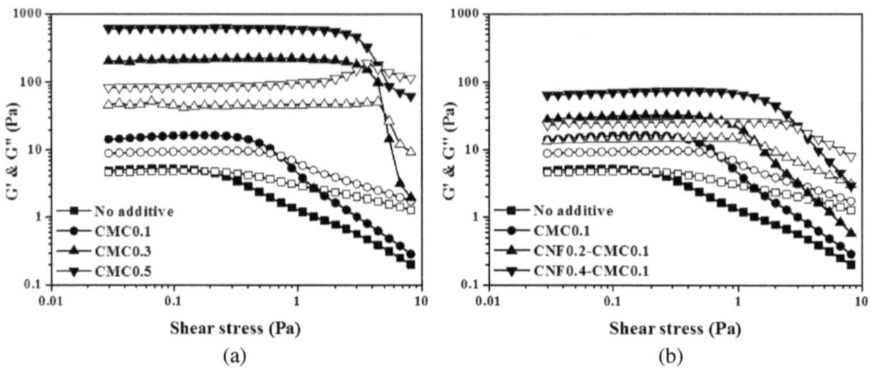

Fig. 24. Amplitude sweep of coating colors with CMC (a) and CNF (b) for a range of different stresses (0.03 to 10 Pa) at a constant angular frequency (1 Hz). Closed symbols: storage modulus (G′); open symbols: loss modulus (G″).

Source: Reprinted by permission from ACS Publications, *Industrial & Engineering Chemistry Research*, Oh *et al*.[56]

The addition of CNF to pigmented coatings increased the viscosity, generated a yield stress, and decreased the dewatering rate. CNF also increased the elasticity of the coating color and stiffness of double-coated samples by around 20%.[51] In practical coating, the viscosity increase that accompanies the addition of CNF may deteriorate coater runnability. The addition of CMC is useful for improving the coatability of CNF suspensions by improving the water retention and modifying the low shear rheology.[42]

The use of CMC increased the porosity of the coating layer due to the effect of the flocculation of CMC.[55,56] CMC and CNF did not fill the voids in the coating layer, but they expanded the coating structure, indicating that film formation and void filling were not the main effects of CMC and CNF as coating additives. CNF produced a porous coating layer with a low rate of addition (Table 1), suggesting that CNF can be used as an agent for designing porous coating structures.[56]

The final structure of the coating layer forms after the coating dries completely. During the drying process, every particle that comprised the coating structure moved until the close-packed structure was formed. When CMC was used as an additive, it prevented the rearrangement of the particles by flocculating them by depletion flocculation. The CMC allowed the formation of a loosely packed structure of coating that yielded

Table 1. Porosity of dried coating layers.

CMC (pph)	CNF (pph)	Porosity (%)
0	0	21.2
0.1	0	21.3
0.3	0	24.1
0.5	0	26.8
0.1	0.2	23.0
0.1	0.4	24.4

a porous coating layer. The addition of CNF to the pigment coating color also increases the porosity of the coating layer. The mechanism by which a porous coating structure is formed with CNF differs from that of CMC. The particles in the CNF coating form a closely packed structure because CNF does not cause depletion flocculation of pigment particles. The voluminous CNF that formed as gel-like structure color made the coating layer porous after drying.

3.3 *New Functional Coating*

The surface coating of CNF incorporated with functional additives, inorganic particles, or synthetic polymers may give paper new functional properties that can expand the applications for CNF and MFC. A few studies have been performed on the functional applications of CNF and MFC.

Arbatan *et al.*[57] investigated the use of CNF as a binder for the fabrication of superhydrophobic paper, which was prepared by a two-step dip coating method using precipitated calcium carbonate (PCC) followed by an alkyl ketene dimer (AKD). During the dip coating process of the PCC suspension, CNF was added as a binder. The function of CNF was clarified by analyzing the characteristics of the paper prepared with or without CNF. In particular, CNF played an important role in the formation of the rough PPC coating layer by retaining the PCC clusters on the paper surface, which gave the paper its superhydrophobicity.

Because MFC is used in a drug release system, a release system for food packaging was developed using MFC coating.[58] To be an effective release system for food packaging, the release rate and released amount should be controlled. MFC coating can be used to control the release of the active compounds. Lavoine *et al.*[58] investigated the release of caffeine from three paper samples: paper impregnated in caffeine solution, paper impregnated in caffeine solution and coated with MFC, and paper coated with a mixture of MFC and caffeine. MFC coatings effectively slowed down the release of caffeine, and the coating of the mixture showed the slowest release of caffeine. These results show an approach for using MFC as a component of a release system that is expected to provide many opportunities particularly in the food packaging field.

4. Concluding Remarks

Nanocellulose is a promising material to replace petroleum-based synthetic polymers. Nanocellulose is a renewable, biodegradable, and recyclable natural polymer with a high mechanical strength and specific surface area. In addition, the properties of nanocellulose can be controlled to some extent by selecting the proper preparation method (acid hydrolysis or mechanical shearing) and raw materials. Therefore, nanocellulose is expected to be applied in a wide range of industrial fields. Among them, paper and the paper-based products industry is the most promising field because of the massive product tonnage of paper. Because most paper products are inexpensive, the price of nanocellulose may be a big hurdle to its use in papermaking, which indicates that an effort should be made to produce nanocellulose at a lower cost (i.e., within the acceptable cost range for the paper industry).

In addition, more effective and efficient uses of nanocellulose should be developed to create a market for nanocellulose. The process efficiency during the application of nanocellulose, drying, and finishing must also be improved. It is already well known that the use of nanocellulose is advantageous to improve the dry- and wet-strength of paper by the wet-end application and enhance the air barrier properties and strength by the coating application. However, many quality problems remain to be resolved. Its high sensitivity to moisture and low water vapor barrier property are two examples that must be addressed to expand its use. These issues are mainly due to the hydrophilic nature of the cellulose and paper substrate. For the paper production processes, the dewatering in the forming section should be improved to use nanocellulose at the wet end. In addition, new developments of appropriate coating processes should be made to coat the nanocellulose at a high coat weight with uniform coverage.

The practical application of the newly developed nanocellulose in the conventional paper industry is a difficult challenge. However, it will open many promising new opportunities and replace fossil-based materials. It will also expand the definition of paper products and provide broad new applications that have never been explored by the paper industry.

References

[1] M.M. de Souza Lima and R. Borsali, *Macromolecular Rapid Communications* 25, (2004) 771.
[2] S. Beck-Candanedo, M. Roman, and D.G. Gray, *Biomacromolecules*, 6, (2005) 1048.
[3] S.C. Espinosa, T. Kuhnt, C. Weder, and E.J. Foster, Production and applications of cellulose nanomaterials, TAPPI Press, 17 (2013).
[4] H.P.S.A. Khalil, Y. Davoudpur, Md. N. Islam, A. Mustapha, K. Sudesh, R. Dungani, and M. Jawaid, *Carbohydr. Polym.*, 99, (2014) 649.
[5] K.L. Spence, R.A. Venditti, O.J. Rojas, Y. Habibi, and J.J. Pawlak, *Cellulose,* 18, (2011) 1097.
[6] T. Saito, M. Hirota, N. Tamura, S. Kimura, H. Fukuzumi, L. Heux, and A. Isogai, *Biomacromolecules*, 10, (2009) 1992.
[7] I. González, M. Alcalà, G. Chinga-Carrasco, F. Vilaseca, S. Boufi, and P. Mutjé, *Cellulose*, 21, (2014) 2599.
[8] M.A. Hubbe, O.J. Rojas, L.A. Lucia, and M. Sain, *Bioresources*, 3, (2008) 929.
[9] S.J. Eichhorn, A. Dufresne, M. Aranguren, N.E. Marcovich, J.R. Capadona, S.J. Rowan, C. Weder, W. Thielemans, M. Roman, S. Renneckar, W. Gindl, S. Veigel, J. Keckes, H. Ano, K. Abe, M. Nogi, A.N. Nakagaito, A. Mangalam, J. Simonsen, A.S. Benight, A. Bismarck, L.A. Berglund, and T. Peijs, *J. of Mater. Sci.*, 45, (2010) 1.
[10] J.A. Shatkin, E.M. Bilek, J. Cowie, and T.H. Wegner, *International Conference on Nanotechnology for Renewable Materials*, TAPPI (2014).
[11] J. Miller, *International Conference on Nanotechnology for Renewable Materials*, TAPPI (2014).
[12] S. Boufi, I. González, M.D. Aguilar, Q. Tarrès, and M.A. Pèlach, *Carbohydr. Polym.*, 154 (2016) 151.
[13] S.H. Osong, S. Norgren, and P. Engstrand, *Cellulose,* 23 (2016) 93.
[14] T. Lindström, A. Naderi, and A. Wiberg, *J. of Korea TAPPI*, 47(6) (2015) 5.
[15] F.W. Brodin, Ø.W. Gregersen, and K. Syverud, *Nordic Pulp and Paper Res. J.*, 29(1) (2014) 156.
[16] Ø. Eriksen, K. Syverud, and Ø. Gregersen, *Nordic Pulp and Paper Res. J.*, 23(3) (2008) 299.

[17] T. Taipale, M. Österberg, A. Nykänen, J. Ruokolainen, and J. Laine, *Cellulose*, 17 (2010) 1005.
[18] M. Manninen, I. Kajanto, J. Happonen, and J. Paltakari, *Nordic Pulp and Paper Res. J.*, 26(3) (2011) 297.
[19] K. Mörseburg and G. Chinga-Carrasco, *Cellulose*, 16, (2009) 795.
[20] C. Hii, Ø.W. Gregersen, G. Chinga-Carrasco, and Ø. Eriksen, *Nordic Pulp and Paper Res. J.*, 27(2) (2012) 388.
[21] M. Ankerfors, T. Lindström, and D. Söderberg, *Nordic Pulp and Paper Res. J.*, 29(3) (2014) 476.
[22] J. Su, W.K.J. Mosse, S. Sharman, W.J. Batchelor, and G. Garnier, *Cellulose*, 20, (2013) 1925.
[23] I. González, S. Boufi, M.A. Pèlach, M. Alcalà, F. Vilaseca, and P. Mutjé, *Bioresources*, 7(4) (2012) 5167.
[24] S.H. Osong, S. Norgren, and P. Engstrand, *Nordic Pulp and Paper Res. J.*, 29(2) (2014) 309.
[25] I. González, F. Vilaseca, M. Alcalák, M.F. PèLach, S. Boufi, and P. Mutjé, *Cellulose*, 20, (2013) 1425.
[26] S. Ahola, M. Österberg, and J. Laine, *Cellulose*, 15, (2008) 303.
[27] S. Ondaral, G. Hocaoğlu, and M.E. Ergün, *Cellulose Chem. and Technol.*, 49(7–8) (2015) 617.
[28] I. Kajanto and M. Kosonen, *J-FOR*, 2(6) (2012) 42.
[29] J. Ryu, K. Sim, and H.J. Youn, *J. of Korea TAPPI*, 46(6) (2014) 78.
[30] K. Sim, J. Lee, H. Lee, and H.J. Youn, *Cellulose*, 22, (2015) 3689.
[31] J. Rantanen and T.C. Maloney, *Nordic Pulp and Paper Res. J.*, 28(4) (2013) 582.
[32] G.A. Smook, *Handbook for Pulp & Paper Technologists*, TAPPI Press (2002) 284.
[33] J. Lee, K. Sim, K. Sim, H.J.Youn, *J. of Korea TAPPI*, 48(6) (2016) 123.
[34] H. Song, M. Ankerfors, M. Hoc, and T. Lindström, *Nordic Pulp and Paper Res. J.*, 25(4) (2010) 519.
[35] V.K. Rastogi and P. Samyn, *Coatings*, 5 (2015) 887.
[36] N. Lavoine, I. Desloges, A. Dufresne, and J. Bras, *Carbohydr. Polym.*, 90, (2012) 735.
[37] K. Syverud and P. Stenius, *Cellulose*, 16 (2009) 75.
[38] C. Aulin, M. Gällstedt, and T. Lindström, *Cellulose*, 17 (2010) 559.
[39] E.-L. Hult, M. Iotti, and M. Lenes, *Cellulose*, 17 (2010) 575.
[40] N. Lavoine, I. Desloges, B. Khelifi, and J. Bras, *J. of Mater. Sci.*, 49 (2014) 2879.
[41] K. Kinnunen, T. Hjelt, E. Kenttä, and U. Forsström, *2013 Paper Conf. 213* (2013).
[42] V. Kumar, A. Elfving, H. Koivula, D. Bousfield, and M. Toivakka, *Industrial & Eng. Chem. Res.*, 55 (2016) 3603.

[43] H. Park, K. Sim, J. Lee, and H.J. Youn, *J. of Korea TAPPI*, 49(3) (2017) 118.
[44] S.M.M. Mousavi, E. Afra, M. Tajvidi, D.W. Bousfield, and M. Dehghani-Firouzabadi, *Cellulose*, 24 (2017) 3001.
[45] N. Lavoine, J. Bras, and I. Desloges, *J. of Appl. Polym. Sci.*, 131 (2014) 40106.
[46] K. Dimic-Misic, P.A.C. Gane, and J. Paltakari, *Ind. Eng. & Chem. Res.*, 52 (2013) 16066.
[47] K. Dimic-Misic, T. Salo, J. Paltakari, and P. Gane, *Nordic Pulp & Paper Res. J.*, 29 (2014) 253.
[48] K. Dimic-Misic, C. Ridgway, T. Maloney, J. Paltakari, and P. Gane, *Transport in Porous Media*, 103(2) (2014) 155.
[49] T. Salo, K. Dimic-Misic, P. Gane, and J. Paltakari, *Nordic Pulp & Paper Res. J.*, 30 (2015) 165.
[50] K. Dimic-Misic, M. Hummel, J. Paltakari, H. Sixta, T. Maloney, and P. Gane, *J. of Collo. and Int. Sci.*, 446 (2015) 31.
[51] B. Nazari and D. Bousfield, *Nordic Pulp & Paper Res. J.*, 31 (2016) 511.
[52] G. Fadat, G. Engström, and M. Rigdahl, *Rheologica Acta*, 27 (1988) 289.
[53] P. McGenity, P. Gane, J. Husband, and M. Engley, *TAPPI Coating Conference* 133 (1992).
[54] A. Wallström, and L. Järnström, *Nordic Rheology Conference* 215 (2004).
[55] E.H. Choi, C.H. Kim, H.J. Youn, and H.L. Lee, *Bioresources*, 10, (2015) 7188.
[56] K. Oh, J.-H. Lee, W. Im, A.R. Abhari, and H.L. Lee, *Ind. & Eng. Chem. Res.*, 56, (2017) 9569.
[57] T. Arbatan, L. Zhang, X.-Y. Fang, and W. Shen, *Chemical Eng. J.*, 210 (2012) 74.
[58] N. Lavoine, I. Desloges, and J. Bras, *Carbohydr. Polym.*, 103 (2014) 528.

https://doi.org/10.1142/9781786349477_0011

Chapter 11

Application of Nanocellulose in Specialty Paper

Hao Ling[*], Xiaohui Wang[*,‡], and Runcang Sun[†]

[*]*State Key Laboratory of Pulp and Paper Engineering, South China University of Technology, China*

[†]*Centre for Lignocellulose Science and Technology and Liaoning Key Laboratory Pulp and Paper Engineering, Dalian Polytechnic University, Dalian, China*

[‡]*fewangxh@scut.edu.cn*

Abstract

A paper with special properties that can be used for specialized applications is termed as "specialty paper". The basic features of specialty paper include the specific performance, specialized application-based features, low quantity demand, small-scale production, low-cost input equipment, narrow application, and high value-added and technical threshold; it should demonstrate uniform structure and good appearance. It has been developed rapidly in the domestic paper market, which is putting a great impact on the domestic paper, design, and printing industry. The specialty paper could be classified according to performance and applications in different fields. This chapter provides a comprehensive overview of the concept, properties, and classification of specialty paper. It discusses the potential of nanocellulose, the smallest physical structural unit of cellulose possessing nanometer network structure and excellent properties, to be replaced with wood pulp in making the specialty paper. Nanocellulose can add more fillers and reduce the overall production cost of specialty paper, which can further impart stability and strength and improve the printing quality of specialty paper. It further summarizes various applications of nanocellulose-based specialty paper in developing energy storage and electronic devices as well as in nanomedicine and water cleaning.

Keywords: Nanocellulose; specialty paper; performance; characteristics; applications

1. Specialty Paper

1.1 *Definition of Specialty Paper*

In recent years, specialty paper has developed rapidly in the domestic paper market, making a great impact on the domestic paper, design, and printing industry. As a proper noun, specialty paper first appeared in the book *Industrial and Specialty Papers*, which was edited by R. H. Mosher in 1969.[1] According to the book, the paper that has special properties and is suitable for special use is named as "specialty paper". It is generally believed that specialty paper originates from the carbonless copy paper developed by Green of the National Cash Register Co. (NCR), in the United States in 1945. As shown in the schematic illustration of carbonless copy paper in Fig. 1, the back of the first piece of paper is coated with microcapsules, and the front of the second sheet is painted with a colorant. While writing, the pressure of the pen tip breaks the microcapsules, then the dye precursor in the microcapsules comes into contact with the color reagent, thus achieving the rewriting effect. With the wide application of carbonless copy paper, it is regarded as the model of specialty paper industrialization in the history of the development of specialty paper. Shortly afterward, other specialty papers continued to be introduced, expanding people's horizons and changing people's traditional views of paper.

1.2 *Properties of Specialty Paper*

There are obvious differences between specialty paper and plain paper. First, the material ratio of specialty paper is diversified. Plain paper is mostly made of plant fiber, such as wood, grass, etc. The raw materials of specialty paper, in addition to plant fibers, can also be matched with

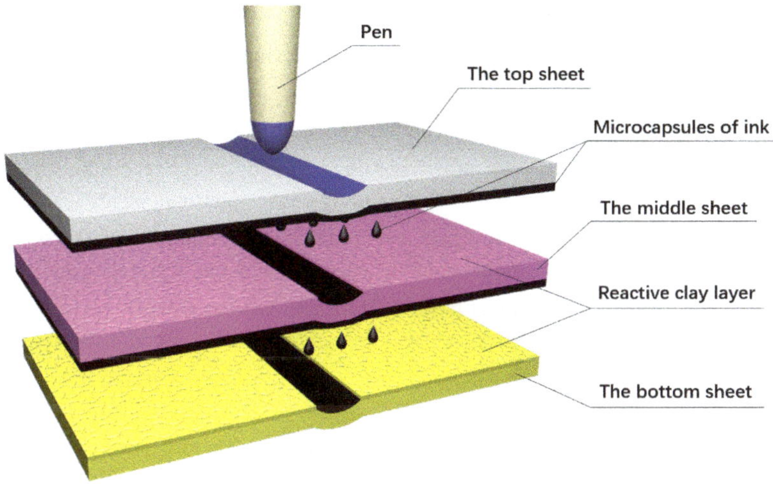

Fig. 1. Schematic illustration of carbonless copy paper.

non-plant fibers, for instance, synthetic fibers, carbon fibers, mineral fibers, animal fibers, and metal fibers, to compensate for some weaknesses of the original natural fibers and to help improve the quality of the paper, achieving the goal of optimizing combination. Second, specialty paper can be processed to increase the number of layers in addition to the fiber interleaving layer. Third, the performance of specialty paper is changed to specialization. Specialty paper can be processed through different methods and the addition of different chemicals (auxiliaries) to make the finished paper have some outstanding properties.

The basic features of specialty paper can be summarized as follows[2]: (1) It has a specific performance for special purposes; (2) Quantity demanded is generally low; (3) It can be produced with small-scale, low-cost input equipment; (4) The application of special paper is relatively narrow, and the specific variety has a specific purpose; (5) It has high added value and high technical threshold. In addition, considering the application of specialty paper, it is generally required to have a relatively uniform structure and good appearance.

1.3 *Classification of Specialty Paper*

At present, there are two ways to classify specialty paper. According to purpose, specialty paper can be divided into 9 kinds: specialty paper for

Table 1. Classification of specialty paper by purpose.

Category	Examples
Specialty paper for printing	Art paper, jet printing paper, cheque paper, silent paper thermo-sensitive paper, luminescent paper
Specialty paper for information	Electrostatic recording paper, discharge recording paper, electrolytic recording paper, electronic paper
Specialty paper for packaging	Anti-rust paper, preservative paper, anti-fungicide paper, antibacterial paper, soft compound wrapping paper
Specialty paper for industrial use	Building paper board, flame retarding paper, glass fiber paper, ceramics paper, micanite paper, activated carbon paper
Specialty paper for agriculture	Fruit cultivation bag paper, germinating paper, insecticide paper, vegetable paper
Specialty paper for daily use	Rag paper, deodorization paper, clothing paper, facial tissue, vacuum bag paper, lampshade paper
Specialty paper for medical use	Cold prevention paper, hemostatic paper, thermometry paper, water-soluble paper, surgical instrument wrapping paper
Specialty paper for military use	Blasting fuse paper, dynamite paper, nitrocellulose paper, anti-eavesdrop paper, water transmitting paper
Other specialty papers	Calcium plastic paper, urea paper

printing, specialty paper for information, specialty paper for packaging, specialty paper for industrial use, specialty paper for agriculture, specialty paper for daily use, specialty paper for medical use, specialty paper for military use, and other specialty paper.

According to performance, specialty paper can be divided into 10 kinds: media specialty paper, seal specialty paper, electromagnetic specialty paper, insulated specialty paper, medical specialty paper, filter specialty paper, cleaning specialty paper, decorative specialty paper, costume specialty paper, and edible specialty paper. Tables 1 and 2 show specific examples.[2]

2. Nanocellulose and Its Applied Advantages in Specialty Paper

Nanocellulose is the smallest physical structure unit of cellulose with a natural nanometer network structure and excellent properties, and it is expected to be used in diverse fields.[3] It is a rigid rod cellulose with a

Table 2. Classification of specialty paper by performance.

Category	Examples
Media specialty paper	Thermo-sensitive paper, synthetic paper, electronic paper
Seal specialty paper	Fruit cultivation bag paper, fisheries waterproof paper, calcium plastic paper
Electromagnetic specialty paper	Conductive paper, shielding paper, graphite paper, magnetic paper
Insulated specialty paper	Sheathing paper, heat-resistant paper, sound-absorptive paper
Medical specialty paper	Hemostatic paper, sterilization paper, thermometry paper
Filter specialty paper	Glass fiber paper, activated carbon paper, nano-filter paper
Cleaning specialty paper	Anti-fungicide paper, air-laid paper, deodorization paper, rat preservative paper
Decorative specialty paper	Wallpaper, ceramics paper, furniture paper, windows paper
Costume specialty paper	Clothing paper, polyester paper
Edible specialty paper	Vegetable paper, oblatum

diameter of about 10 nm and a length in the range of tens to several hundreds of nanometers. Nanocellulose not only retains the advantages of natural cellulose, such as renewability, degradability, high strength, and high elastic modulus, but also has high specific surface area similar to inorganic nanoparticles, strong surface activity, and other special effects due to its nanometer size.[4]

As a new kind of biomaterials, nanocellulose has become the frontier and hotspot of cellulose research due to its unique nano-sized structure, mechanical properties, and optical properties. Compared with traditional cellulose, nanocellulose has many excellent properties, for instance, high crystallinity, high purity, high Young's modulus, high strength, high hydrophilicity, hyperfine structure, and high transparency. Furthermore, the impressive mechanical properties, reinforcing capabilities, abundance, low density, light weight, and biodegradability of nanocellulose make it an ideal candidate for applications in the fields of specialty paper, building, automobile, food, cosmetics, electronics, medicine, paint, and aviation.[5]

In terms of the application in specialty paper, the addition of nanocellulose in specialty paper making process can replace wood pulp because nanocellulose has superior structural strength. Using nanocellulose can add more fillers, and the production cost can be reduced by about 3%. Compared with traditional cellulose, the carbon footprint of nanocellulose can be reduced by 15% or more. The paper produced with nanocellulose is not only cheap but also has excellent properties, for instance, less space, high printing quality, and high strength.

Owing to the high specific surface area and abundant surface hydroxyl, nanocellulose can be tightly combined with the pulp fiber when added to the pulp, thus improving the adhesion between the pulp fibers. As a consequence, nanocellulose can be used as a reinforcing agent, retention aid, drainage aid, and so forth in the pulping and papermaking process, which leads to a good development prospect. In addition, nanocellulose can act as a binder or carrier for other functional materials, providing more opportunity for the development of novel specialty paper or other paper-based materials.

3. Application of Nanocellulose in Specialty Paper

3.1 *Nanocellulose-Based Specialty Paper for Energy Storage*

In order to meet the growing energy needs of modern society and reduce dependence on fossil fuels, energy storage devices play a vital role. As an abundant, renewable, and environmentally-friendly material, nanocellulose has great applied value in the field of energy storage devices.

3.1.1 *Nanocellulose-based Li-ion battery separator*

As a new clean energy source, Li-ion battery has been used since the 1990s; it has the advantages[6] of light, high working voltage, high conductivity, low self-discharge rate, low pollution, no memory effect, etc. In the development of major technical fields, such as new energy and environmental protection, Li-ion battery plays an important role.

Separator is considered to be the core component of Li-ion battery. Its important function is not only to maintain the electrical isolation between cathode and anode but also to act as the electrolyte energy storage layer to ensure ion transport. The separator produced by papermaking process has high porosity, excellent thermal property, good chemical stability, and cost competitiveness, which has attracted widespread attention.[7]

Zhang *et al.*[8] reported a novel nanofibrillated cellulose (NFC) membrane with excellent wettability, mechanical strength, thermal property, and electrochemical performance, which was prepared by wet papermaking process and used as a separator to assemble Li-ion battery. Compared with the conventional PP membrane, NFC membrane has higher porosity, better wettability, higher melting point, and better cycling performance. Based on the excellent physical and thermal properties of the NFC membrane, it is a promising choice to prepare the separator for high-temperature and high-safety Li-ion batteries by using NFC.

3.1.2 *Nanocellulose-based paper fuel cell*

Fuel cells are recognized as the preferred clean and efficient power generation technology in the 21st century. Fuel cells are electrochemical energy conversion devices. As long as the reactants are provided and the products are removed, the chemical energy (such as hydrogen) of the fuel can be directly converted into electric energy.[9]

Nanocellulose paper could be a suitable ionomer membrane for fuel cells due to the low cost, good gas barrier properties, mechanical toughness, and acidic oxygen functional groups. Thomas Bayer *et al.*[10] reported a nanocellulose membrane for fuel cells that retains good proton conductivity at 120°C. The hydrogen barrier properties are much better than those of conventional ionomer membranes. Fuel cells with nanocellulose membranes are successfully operated at 80°C. Besides, they compared two different membranes made of nanofibrillated cellulose (NFC) and nanocrystalline cellulose (NCC) (Fig. 2). Because of the higher conductivity, lower cell resistance, and lower hydrogen gas permeability, fuel cells incorporating NCC displayed better performance than NFC. This work shows that nanocellulose paper is promising as an environmentally-friendly and biodegradable ionomer membrane in fuel cells.

3.2 *Nanocellulose-Based Specialty Paper for Electronic Devices*

Electronic devices play an important role in our daily life. People use electronic products everywhere. Meanwhile, with the dramatic growth of the global economy, people change electronic products more and more frequently. However, most electronic products are not biodegradable and

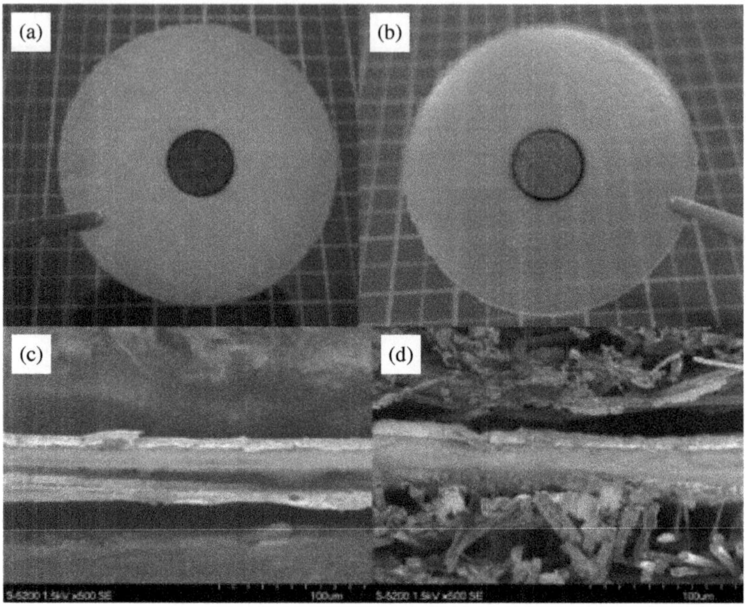

Fig. 2. Nanocellulose fuel cells. Photographs of membrane electrode assemblies prepared from (a) an NFC membrane and (b) NCC membrane. Post operation cross-sectional SEM images of (c) NFC and (d) NCC nanocellulose fuel cells.[10]

Source: Copyright 2016, American Chemical Society.

recyclable, which leads to severe ecological problems. Therefore, high-performance, biodegradable materials are of great significance in the use of electronic products, which can minimize dependence on oil and reduce environmental pollution.

Nanocellulose comes from the most abundant biopolymers on the earth with excellent biodegradability and biocompatibility. In addition to environmental protection, nanocellulose substrates have smaller bending radius, lighter mass, and higher mechanical strength than glass; compared to plastics, nanocellulose substrates have a smaller coefficient of thermal expansion, which are more suitable for roll to roll process.

3.2.1 *Nanocellulose-based transparent and conductive paper*

Recently, paper has been employed for wide applications in electronics and power areas. Thin film transistors, solar cells, and batteries have all been created using paper as a substrate.[11,12] In general, the porous structure of paper not only simplifies the manufacturing process towards low-cost techniques, but also provides additional structural advantages to manipulate electrons, ions, and photons, thereby improving device performance.[12,13]

For example, since the size of the nanocellulose is much less than the wavelength of visible light, nanocellulose paper is highly transparent with a large light scattering in the forward direction.[14] The transparent property of nanocellulose makes it very useful for electronic devices.

The nanocellulose paper can be coated with a wide variety of conductive materials, such as carbon nanotubes, silver nanowires, and tin-doped indium oxide (ITO), to produce transparent conductive paper.[14] Masaya Nogi *et al.*[15] reported a novel nanopaper made of cellulose nanofiber. This optically transparent paper exhibits high Young's modulus, high strength, ultralow coefficient of thermal expansion, and high foldability. This study shows the potential of nanocellulose paper to become an extremely important material to support future generations of electronic devices.

3.2.2 *Electroluminescent nanocellulose paper*

Electroluminescence (EL) is an optical and electrical phenomenon in which a material emits light in response to the passage of an electric current or to a strong electric field. At present, the main research direction of electroluminescence is the application of organic materials. Recently, electroluminescent paper made of nanocellulose has been studied due to its high flexibility, high porosity, excellent thermal property, and good chemical stability.

NM Park *et al.*[16] reported surface light-emitting transparent papers operated by alternating current (AC) voltages. They prepared these electroluminescent nanocellulose papers by uniformly coating of Ag NW electrodes on both sides of the nanocellulose paper and using luminescent phosphors dispersed in the nanocellulose paper (Fig. 3). These papers show good electroluminescent properties. It is the first time to apply nanocellulose paper as a light-emitting active component, which provides the foundation for the realization of commercially viable, flexible, and transparent light-emitting papers.

3.2.3 *Nanocellulose substrates for transistors*

Transistors are functional semiconductor devices that can act as an amplifier, voltage stabilizer, demodulator, rectifying signal modulator, etc. It is a basic component of various electronic devices. Nanocellulose paper is flexible and light and can be used as substrates for transistors.

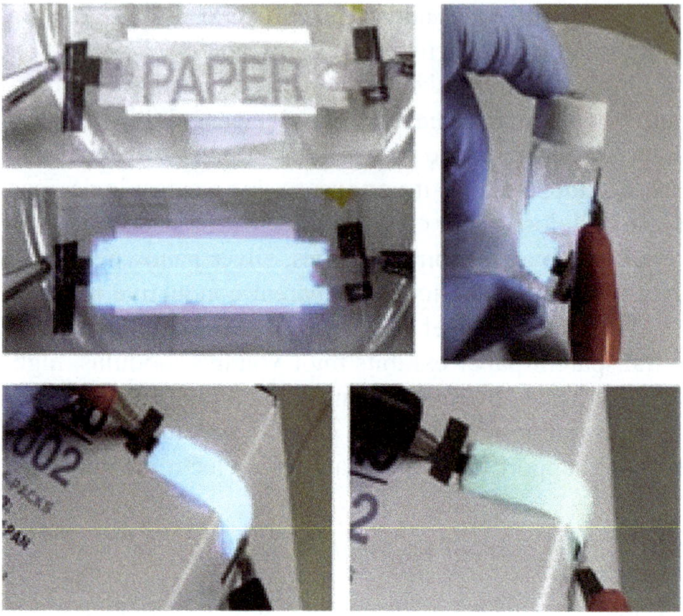

Fig. 3. Image of the nanocellulose paper emitting a blue light on both sides.[16]
Source: Copyright 2017, Elsevier.

Huang *et al.*[17] reported a transparent thin film field-effect transistor that uses transparent NFC paper as a substrate, CNT as a conductive layer, Al_2O_3 as electrolyte, and NTCDI-F15 as *n*-type semiconductor materials. This transistor exhibits good optoelectronic properties, high transmittance, and good flexibility.

Zhang *et al.* successfully fabricated MoS_2 phototransistor using highly transparent NFC as a substrate[18] (as shown in Fig. 4). Under the applied voltage, the cationic electrolyte accumulates on the surface of MoS_2 to form a double electric layer, which effectively shields charge impurity and reduces the electron scattering, thus greatly improving the ion mobility, so that the transistor shows excellent optical responsivity.

It can be seen that transparent nanocellulose paper has a golden prospect in the new generation of green electronic products.

3.2.4 *Nanocellulose-based antenna*

Antenna is a kind of converter that transforms guided and electromagnetic waves into one another. It is widely used in various communication

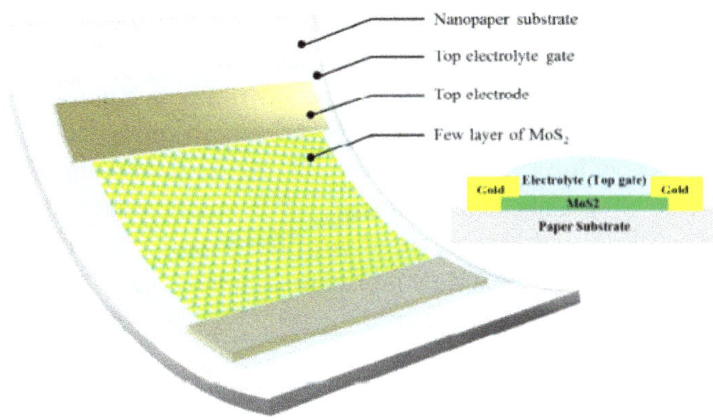

Fig. 4. Three-dimensional schematic and cross-sectional view of a fabricated flexible nanocellulose paper phototransistor.[18]

Source: Copyright 2016, Royal Society of Chemistry.

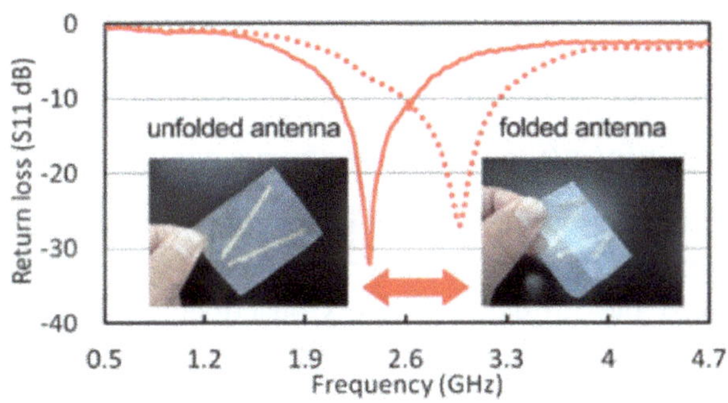

Fig. 5. Return loss of folded nanocellulose paper antennas.[19]

Source: Copyright 2013, Royal Society of Chemistry.

technologies, including television sets, mobile phones, GPS, and radio. Paper substrate can be folded into a three-dimensional structure, which has great advantages in antenna fabrication. Plain paper can be used in small and portable electronic products due to its excellent folding property. However, its high surface roughness leads to low equipment sensitivity. Compared with plain paper, nanocellulose paper has a very smooth surface, which makes it more suitable in antenna fabrication.

Nogi *et al.*[19] used silver nanowire ink to print "V" pattern on nanocellulose paper in the form of silk-screen printing. They fabricated a folded antenna successfully by using this nanocellulose paper. Figure 5 shows the relationship between the return loss and the frequency of the nanocellulose paper antenna. It can be seen that the position of the resonant peak can be achieved by folding the substrate. When the folded paper antenna unfolds and returns to its original length, the peak value returns to its original position. At the same time, the antenna echo loss is lower (less than 26 dB). This nanocellulose paper antenna exhibits good foldable performance and retains good sensitivity under folding conditions. Consequently, it can be used to fabricate antennas that receive and transmit multi-frequency signals by folding.

3.3 *Nanocellulose-Based Specialty Paper for Medicine*

Because of its special structure and excellent properties, nanocellulose has been applied to medical research in recent years. Nanocellulose paper possesses excellent mechanical properties, good barrier properties, natural nanometer network structure, as well as non-toxicity, hydrophilicity, and antibacterial activity, which makes it widely used in the area of virus removal, hemodialysis, antibiosis, etc.

3.3.1 *Nanocellulose filter paper for virus removal*

Influenza virus is a troublesome problem bothering people. Besides, the frequent mutations of viruses undermine the development of efficient vaccines against the influenza virus. Consequently, in order to prevent the spreading of viruses, thus reducing the risk caused by pandemics, it is essential to develop efficient, robust, and affordable virus removal filters.

In terms of filtration applications, cellulose gains wide attention because it is inert, non-toxic, hydrophilic, resistant to pH between 2 and 11, cost-efficient, and disposable.[20] Recently, with the progress of nanotechnology, scientists began to develop nanocellulose-based virus removal filters. Giorgi Metreveli *et al.*[20] presented a non-woven, micrometer-thick filter paper, consisting of pristine highly crystalline cellulose nanofibers and featuring a tailored pore size distribution particularly suitable for virus removal. This is the first time that a 100% natural, unmodified nano-fibrous polymer-based membrane was demonstrated to be capable of removing virus particles solely based on the size-exclusion principle, with $\log_{10}$ reduction value (LRV) ≥6.3, thereby matching the performance of industrial synthetic polymer virus removal filters. Asper *et al.*[21] reported a size-exclusion filter paper composed of 100% naturally derived cellulose, which is useful for the removal of endogenous rodent retroviruses and retrovirus-like particles during the production of recombinant proteins.

3.3.2 *Nanocellulose membranes for hemodialysis*

Hemodialysis (HD), peritoneal dialysis, and kidney transplantation are the only treatment options available for patients suffering from end-stage renal disease (ESRD). Hemodialysis substitutes only a small fraction of normal kidney function and cannot mimic the metabolic, endocrine, and immune functions of the tubule system.[22] However, the increasing prevalence of kidney failure and the shortage of organ donors further stress the need for innovative approaches for developing hemodialysis membranes and dialysis technology that resembles normal renal function.

Composites of nanocellulose and the conductive polymer polypyrrole (PPy) are presented as candidates for a new generation of hemodialysis membranes. Mihranyan *et al.*[23] reported highly porous composite membranes of *Cladophora* cellulose nanofibers coated with a thin (~50 nm) layer of PPy. The hemocompatibility of these novel membranes and the feasibility of effectively removing small uremic toxins by potential-controlled ion exchange were investigated by Ferraz *et al.*[24]. It is found that the composites are comparable with polymer polysulfone in terms of platelet adhesion and thrombin generation. In addition, with respect to complement activation, the composites are more biocompatible than commercially available membranes, showing that they are promising hemodialysis materials.

3.3.3 *Nanofibrillated cellulose-based antibacterial paper*

Different pathogenic microorganisms can cause diseases and represent a serious threat to public health.[25] Many infectious diseases occur by contamination with spoiled food or in hospital environments. Therefore, it is necessary to develop a novel antibacterial packaging material for health care and food packaging that can inhibit microbial contamination.

Metal oxide nanoparticles (NP) have attracted more and more attention as antimicrobial agents. Compared with organic antimicrobial agents, these materials have a longer life and more stable chemical properties under extreme conditions (e.g., high temperature or pressure). Over the past ten years, ZnO NP has emerged as an effective antimicrobial agent under solar light, thus attracting much attention.[26]

In the production of nanocomposites containing inorganic nanoparticles, the use of nanofibrillated cellulose (NFC) has attracted particular interest. NFC fiber has remarkable properties such as high aspect ratio, tensile strength, and modulus, which makes it a good candidate to reinforce nanocomposites. Martins *et al.*[27] reported the preparation of antibacterial NFC/ZnO paper by the electrostatic assembly of ZnO NP onto NFC using polyelectrolytes as well as their use in paper coating formulations to produce papers with antibacterial properties. It was found that in both light and dark conditions, NFC/ZnO nanocomposite antibacterial paper exhibited bacteriostatic and/or bactericidal activity against Gram-positive (*Staphylococcus aureus* and *Bacillus cereus*) and Gram-negative (*Klebsiella pneumoniae*) bacteria (see Fig. 6).

3.4 *Nanocellulose-Based Specialty Paper for Water Cleaning*

In recent years, with the acceleration of global industrialization, water pollution is becoming more and more serious. At present, membrane technologies such as ultrafiltration and nanofiltration are widely used in wastewater treatment, which can help in the decentralization of industrial and domestic water to aid in the removal of various contaminants.[28] Conventional membranes used for filtration are commonly produced from polymers derived from fossil resources, which require huge quantities of solvents, chemicals, and energy.[29] To produce a simpler, cleaner, and faster process for renewable ultrafiltration and nanofiltration membranes, nanocellulose has been studied as a novel membrane material.

Fig. 6. Antibacterial paper based on composite coatings of nanofibrillated cellulose and ZnO.[27]

Source: Copyright 2012, Elsevier.

3.4.1 *Nanocellulose composite ultrafiltration membrane*

Ultrafiltration (UF) is a separation process in which solvent and small molecular solutes are permeated through the membrane, as are macromolecular substances such as colloid and protein, and these are intercepted and concentrated by using membrane solids and surface chemical properties. In the development and application of ultrafiltration membrane technology, apart from membrane separation and permeability energy, the mechanical strength of the membrane is one of the most significant problems encountered. Polysulfone, polyethersulfone, polyvinylidene fluoride, polyacrylonitrile, etc., are commonly used ultrafiltration materials. However, they lack mechanical toughness, suffer from poor hydrophilicity, easily pollute the environment, and so on, which makes them need to be improved.

Compared with traditional ultrafiltration materials, nanocellulose paper has the advantages of low thermal expansion coefficients, high optical transparency, and good gas barrier properties,[30] which makes it suitable to act as an ultrafiltration membrane. Andreas Mautner *et al.*[31] used nanocellulose papers as tight aqueous ultrafiltration membranes. These membranes exhibit transmembrane permeance in the range of commercially available tight UF membranes with molecular weight cut-offs ranging from 6 to 25 kDa. Besides, by using different types of nanocellulose and fibril diameters, it can be possible to control the pore size and, accordingly, the permeance and molecular weight cut-off of membranes.

3.4.2 *Nanocellulose membrane for organic solvent nanofiltration*

Organic solvent nanofiltration (OSN) has gained wide attention since its emergence in the past decade. Organic solvent nanofiltration is a new type of membrane separation technique driven by pressure, which has the advantages of being green, of high efficiency, and so on.[32] The organic solvent nanofiltration membrane is the core of the nanofiltration process and directly determines the separation efficiency of the system.[33] In order to adapt to the mixed solvent system, the nanofiltration membrane should have high pass quantity, high selectivity, and high generality, and maintain high chemical stability and mechanical stability.[34]

Recently, nanofibrillated cellulose (NFC) has found widespread interest in organic solvent nanofiltration due to its excellent mechanical and chemical properties. Mautner *et al.*[35] reported solvent-stable nanofiltration (NF) membranes entirely made from nanocellulose. These membranes avoid the use of organic solvents that are commonly necessary for the production of conventional OSN polymer membranes by using a papermaking process utilizing an aqueous suspension of nanocellulose.[36] It turned out that the NFC membranes exhibited great barrier properties. In conclusion, it is feasible to prepare, in a process as simple as making paper, solvent-stable OSN membranes from renewable resources such as nanocellulose.

References

[1] Mosher R H, Davis D S. *Industrial and specialty papers*. Chemical Pub. Co., 1968.

[2] Chen, G. The developmental situation of specialty paper and its demand to chemicals in China. *China Pulp & Paper Industry*, 2008.

[3] Hoeng F, Denneulin A, Bras J. Use of nanocellulose in printed electronics: A review. *Nanoscale*, 2016, 8(27): 13131–13154.

[4] Sabo R, Yermakov A, Law C T, *et al*. Nanocellulose-enabled electronics, energy harvesting devices, smart materials and sensors: A review. *J. of Rene. Mater.*, 2016, 4(5): 297–312.

[5] Lima M M D S, Borsali R. Rodlike cellulose microcrystals: Structure, properties, and applications. *Macromol. Rapid Commun.*, 2010, 25(7): 771–787.

[6] Liu F, Qin L J, Liu Y X. Latest research progress of silicon-carbon composite anode material for lithium ion battery. *Moder. Chem. Ind.*, 2017.

[7] Liu XY. Study on polyolefin fibre battery separator paper. *China Pulp & Paper Industry*, 2007.

[8] Zhang H, Wang X, Liang Y. Preparation and characterization of a Lithium-ion battery separator from cellulose nanofibers. *Heliyon*, 2015, 1(2): e00032.

[9] Winter M, Brodd R J. What are batteries, fuel cells, and supercapacitors? *Chem. Rev.*, 2004, 35(50): 4245.

[10] Bayer T, Cunning B V, Selyanchyn R, *et al*. High temperature proton conduction in nanocellulose membranes: Paper fuel cells. *Chem. of Mater.*, 2016, 28(13).

[11] Mc B, Ja R, Rr L, *et al*. Direct monolithic integration of organic photovoltaic circuits on unmodified paper. *Adv. Mater.*, 2011, 23(31): 3500–3505.

[12] Eder F, Klauk H, Halik M, *et al*. Organic electronics on paper. *Appl. Phys. Lett.*, 2004, 84(14): 2673–2675.

[13] L H, Jw C, Y Y, *et al*. Highly conductive paper for energy-storage devices. *Proceedings of the National Academy of Sciences of the United States of America*, 2009, 106(51): 21490–21494.

[14] Hu L, Zheng G, Yao J, *et al.* Transparent and conductive paper from nanocellulose fibers. *Ener. & Env. Sci.*, 2013, 6(2): 513–518.
[15] Nogi M, Iwamoto S, Nakagaito A N, *et al.* Optically transparent nanofiber paper. *Adv. Mater.*, 2009, 21(16): 1595–1598.
[16] Park N M, Koo J B, Oh J Y, *et al.* Electroluminescent nanocellulose paper. *Mater. Lett.*, 2017, 196: 12–15.
[17] Huang J, Zhu H, Chen Y, *et al.* Highly transparent and flexible nanopaper transistors. *Acs Nano*, 2013, 7(3): 2106–2113.
[18] Zhang Q, Bao W, Gong A, *et al.* A highly sensitive, highly transparent, gel-gated MoS_2 phototransistor on biodegradable nanopaper. *Nanoscale*, 2016, 8(29): 14237–14242.
[19] Nogi M, Komoda N, Otsuka K, *et al.* Foldable nanopaper antennas for organic electronics. *Nanoscale* 2013, 5(10): 4395–4399.
[20] Metreveli G, Wagberg L, Emmoth E, *et al.* A size-exclusion nanocellulose filter paper for virus removal. *Adv. Health. Mater.*, 2014, 3(10): 1546–1550, 1524.
[21] Asper M, Hanrieder T, Quellmalz A, *et al.* Removal of xenotropic murine leukemia virus by nanocellulose based filter paper. *Biol. J. of the Int. Asso. of Biol. Stand.*, 2015, 43(6): 452–456.
[22] Zweigart C, Neubauer M, Storr M, *et al.* *Progress in the development of membranes for kidney-replacement therapy*, 2010.
[23] Mihranyan A, Nyholm L, Bennett A E, *et al.* A novel high specific surface area conducting paper material composed of polypyrrole and Cladophora cellulose. *J. of Phy. Chem. B.*, 2008, 112(39): 12249–12255.
[24] Ferraz N, Carlsson D O, Hong J, *et al.* Haemocompatibility and ion exchange capability of nanocellulose polypyrrole membranes intended for blood purification. *J. of the Royal Society Int.*, 2012, 9(73): 1943–1955.
[25] Kaittanis C, Santra S, Perez J M. Emerging nanotechnology-based strategies for the identification of microbial pathogenesis. *Adv. Drug Deli. Rev.*, 2010, 62(4): 408–423.
[26] Jones N, Ray B, Ranjit K *et al.* Antibacterial activity of ZnO nanoparticle suspensions on a broad spectrum of microorganisms. *Fems Microbiol. Lett.*, 2010, 279(1): 71–76.
[27] Martins N C T, Freire C S R, Neto C P, *et al.* Antibacterial paper based on composite coatings of nanofibrillated cellulose and ZnO[J]. *Colloids and Surfaces A: Physicochemical and Engineering Aspects*, 2013, 417: 111–119.
[28] Shannon M A, Bohn P W, Elimelech M, *et al.* Science and technology for water purification in the coming decades. *Nature*, 2008, 452(7185): 301–310.
[29] Lee K P, Arnot T C, Mattia D. A review of reverse osmosis membrane materials for desalination — Development to date and future potential. *J. of Memb. Sci.*, 2011, 370(1): 1–22.

[30] Chun S J, Lee S Y, Doh G H, *et al*. Preparation of ultrastrength nanopapers using cellulose nanofibrils. *J. of Ind. & Eng. Chem.*, 2011, 17(3): 521–526.
[31] Mautner A, Lee K Y, Tammelin T, *et al*. Cellulose nanopapers as tight aqueous ultra-filtration membranes. *Reac. & Func. Polym.*, 2015, 86: 209–214.
[32] Vandezande P, Gevers L E, Vankelecom I F. Solvent resistant nanofiltration: Separating on a molecular level. *Chem. Soc. Rev.*, 2008, 37(2): 365–405.
[33] Cheng X Q, Zhang Y L, Wang Z X, *et al*. Recent advances in polymeric solvent-resistant nanofiltration membranes. *Adv. in Polym. Technol.*, 2015, 33(S1).
[34] Dutczak S M, Cuperus F P, Wessling M, *et al*. New crosslinking method of polyamide-imide membranes for potential application in harsh polar aprotic solvents. *Separation & Purification Technol.*, 2013, 102(1): 142–146.
[35] Mautner A, Lee K Y, Lahtinen P, *et al*. Nanopapers for organic solvent nanofiltration. *Chem. Commun.*, 2014, 50(43): 5778–5781.
[36] Soroko I, Bhole Y, Livingston A G. Environmentally friendly route for the preparation of solvent resistant polyimide nanofiltration membranes. *Green Chem.*, 2011, 13(1): 162–168.

https://doi.org/10.1142/9781786349477_0012

Chapter 12

Nanocellulose-Based Composites: Optoelectronic Applications

Mingwei Zhu[*], Chao Jia[†], Zhiqiang Fang[‡], and Liangbing Hu[§,¶]

[*]*National Laboratory of Solid State Microstructures & College of Engineering and Applied Sciences, Nanjing University, Nanjing, China*

[†]*State Key Laboratory of New Ceramics and Fine Processing, School of Materials Science and Engineering, Tsinghua University, Beijing, China*

[‡]*State Key Laboratory of Pulp and Paper Engineering, South China University of Technology, Guangzhou, Guangdong, China*

[§]*Department of Materials Science and Engineering, University of Maryland, College Park, MD, USA*

[¶]*binghu@umd.edu*

Abstract

The cellulose-based materials are transparent in nature as these allow the light to pass through them; however, if it is present in the form of fibers whose size is comparable to the wavelength of light, it can greatly scatter the light. This chapter discusses the light measurement of nano-cellulose, and its optical behavior associated with dielectric materials.

The optical properties of nanocellulose endow it a large space for optoelectronic design and applications, including in flexible electronics, solar cells, displays, and touch screens. The chapter further discusses the applications of nanocellulose-based materials in photoreduction, plasmonic, and ecological building and energy-saving systems. It also provides prospects for establishing advanced approaches for developing nanocellulose-based novel materials with varying dimensions and for novel applications.

Keywords: Nanocellulose; optical properties; optoelectronic design; applications

1. Introduction

In recent years, novel nanocellulose-based materials are emerging rapidly. Many novel phenomena and effects are discovered and applied to a variety of new fields, transparency, for example. Cellulose materials themselves have no light absorption at visible wavelengths, which endows nanocellulose-based composites the ability to be used for optoelectronic designs and applications.

Basically, the interaction between light and cellulose-based composites is the fundamental basis of various optical functions and devices. These interactions belong to the category of classical geometrical optics, which can be simply described by the traditional optics. Because almost all cellulose-based composites are dielectric materials, herein, we only discuss the optical behaviors associated with dielectric materials. In principle, as shown in Fig. 1, when light is incident on the nanocellulose-based composite, reflection, absorption, refraction, and transmittance can occur, if there is no fluorescence.

(1) Reflection
Reflection usually occurs on the material's surface or interface when the incident light encounters a sharp change in refractive index. For a smooth surface or interface, the reflection is equal but opposite in angle between the light ray and the surface, and this is called specular (mirror-like) reflection. For a coarse surface, reflected light comes off the surface at any other angle, and this is termed as diffuse reflection. The normal reflection for a smooth surface can be described by the following equation:

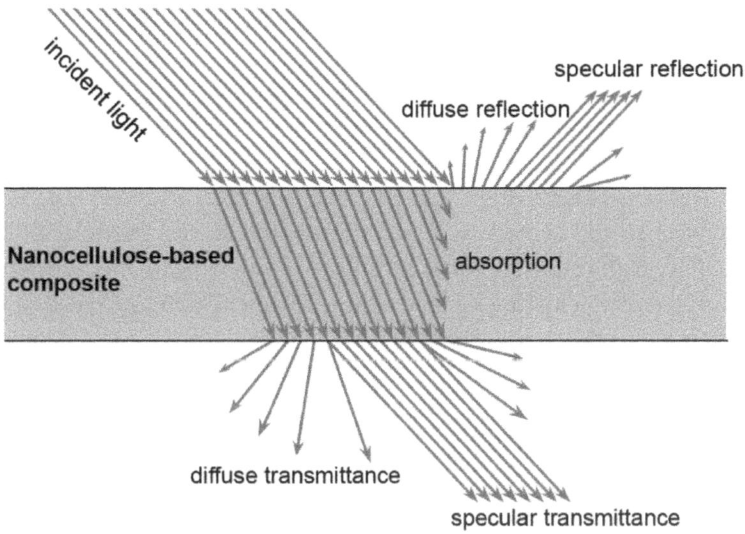

Fig. 1. The interaction of light with nanocellulose-based composite.

$$R = (n-1)^2/(n+1)^2 \times 100\%$$

where R is the reflectivity and n is the refractive index.

For example, the refractive index of most cellulose materials at optical wavelength is around 1.5. Accordingly, the normal reflection is about 4% on the surface of most cellulose materials. For a smooth, thin cellulose film without absorption, such reflection occurs two times, and a total of up to 8% of the light energy is lost by reflection.

(2) Absorbance
Light energy can be captured by some component in host material for electron transition to a higher energy state. The energy may also be lost in the form of heat, resulting in mechanical waves or acoustic energy. For plants, such as wood, the colorful lignin absorbs optical light intensively; hence, it is usually necessary to remove the lignin component before further processing of wood or other plants into transparent cellulose composites.

(3) Scattering
Light is usually scattered by particles or bubbles existing in nanocellulose-based composites. The haze in transparent materials is often

caused by light scattering. The following three types of scattering occur:

(a) When light wavelength is much larger than the particle or bubble size, light is scattered slightly and the light momentum has changed slightly but there is no energy loss, and this is called Rayleigh scattering.
(b) When light wavelength is comparable to the particle or bubble size, light momentum is strongly changed with little energy loss, and this is called Mie scattering. Mie scattering is affected by the shape of particles or bubbles.
(c) When light wavelength is much smaller than the particle or bubble size, the usual geometric scattering occurs, caused by reflection, diffraction, etc.

Therefore, the adjustment of the scattering is a very useful way to control the light momentum and light energy. This is used to achieve both high haze and high transparency in nanocellulose-based composites such as transparent paper and transparent wood composite.

(4) Transmittance
After the reflection, absorption, and scattering losses, the light that has passed through material is called transmitted light. Many applications require the maximum amount of light to be transmitted, which means as little energy loss as possible. However, for the same transmitted light energy, the spatial distribution of the transmitted light intensity can vary greatly. Here, an important parameter "haze" is introduced to characterize the extent to which the transmitted light deviates from the optical axis. The haze also reflects the amount of momentum change of the transmitted light.

Haze is defined as the amount of light that is diverted from the normal at an angle more than 2.5°. In practice, haze is measured in an indirect way. As shown in Fig. 2, the photo-detector only collects light whose direction is less than 2.5° from the optical axis. The resulting spectrum is called specular transmittance (Fig. 2a). If an integrated sphere is used to collect all scattered light at all angles, the measured light transmittance is called diffusive transmittance (Fig. 2b). The difference between the diffusive and specular transmittance gives the total forward scattering at an angle of more than 2.5° from the optical axis.

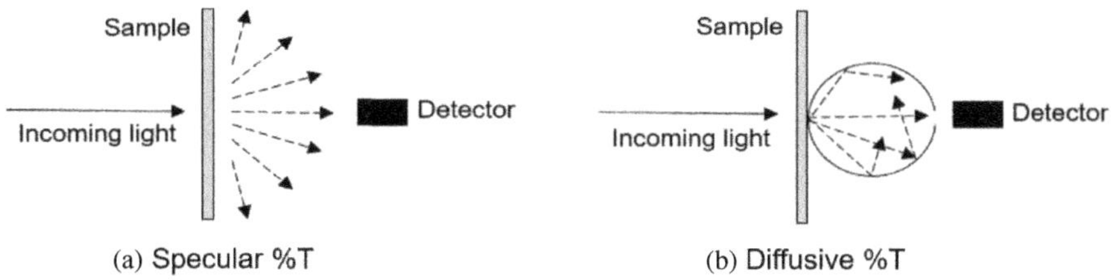

Fig. 2. The definition of specular transmittance (a) and total transmittance (diffusive transmittance) (b) [9].

2. Light Tuning by Nanocellulose-Based Composites

Cellulose materials in nature are transparent, so light can pass through them. However, if they are in the form of fibers and the size is comparable to the wavelength of light, the light will be strongly scattered.[1] For this reason, plain paper is usually not transparent. Recently, an unusual nanocellulose-based transparent paper has emerged and attracted a lot of attention.[2–6] Endowing cellulose-based paper with novel optical properties (e.g., with a high transparency, a high haze, or both) may greatly extend its unconventional applications, such as in optics and photonics, fields in which cellulose was rarely involved in in the past.

2.1 *Transparency and Haze Properties*

At the early stage, people spent a significant amount of effort to make paper as clear as possible.[7] The basic idea for achieving a high transparency is to reduce the light scattering of the unit fibers in the paper. When the fiber diameter is decreased to nano-size, its light scattering ability will be greatly reduced. The light scattering is negligible for cellulose nanofibers (CNFs) of about several nanometers in diameter, so light cannot "see" them, which is the basis for the transparency of the paper. To achieve this, the key technology is to extract the CNFs from the plants, then turn them into a transparent paper with little light scattering. These methods can be categorized as "Bottom-Up" methods for the fabrication of transparent paper. In 2009, Nogi *et al.*[7] for the first time reported an optically transparent paper prepared from CNFs with a diameter of about 15 nm.[7] The prepared transparent paper demonstrates

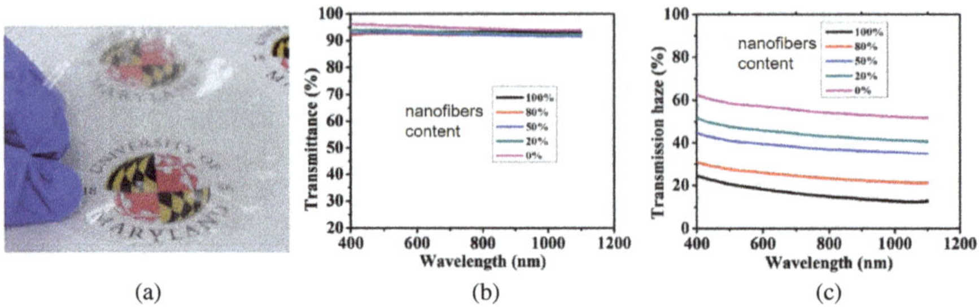

Fig. 3. Optical photos of highly transparent papers (a). By adjusting the ratio of nanofibers in paper, their optical properties can be efficiently tuned (b–c).[2]

good optical, mechanical, and thermal properties, including a high light transmittance of 71.6% at the wavelength of 600 nm, a high modulus of 13 GPa, a high mechanical strength of 223 MPa, and a low thermal expansion coefficient of 8.5 ppm K^{-1}. Fang *et al.*[2] studied the optical properties of transparent paper in detail and realized the modulation of light transmittance by adjusting the ratio of nanofibers to microfibers in paper[2] (Fig. 3).

Zhu *et al.*[8] proposed a different idea for fabricating transparent paper directly from wood *via* a "Top-Down" approach.[8] It is based on the fact that an optically uniform material with little absorption is transparent. Using delignified wood slice (i.e., the lignin component in wood has been completely or nearly completely removed) as raw material, transparent paper can be directly fabricated by removing the scattering factors (e.g., air-filled pores) by mechanical compression. The pressure can rearrange the positions of the nanofibers in wood slice, thereby eliminating the light scattering voids or spaces, and finally forming an optically uniform material. This provides another way to prepare transparent paper *via* the "Top-Down" method (Fig. 4). This method eliminates the expensive and time-consuming process of nanofiber extraction, washing, drying, etc. Hence, the "Top-Down" method greatly simplifies the preparation process of the transparent paper, and substantially reduces the manufacturing cost of the transparent paper.

For transparent nanocellulose-based composites, controlling haze while ensuring high transparency above 90% is attractive in various emerging applications. That is, the light energy can pass through the composites as much as possible; however, the light momentum could be changed and tailored.

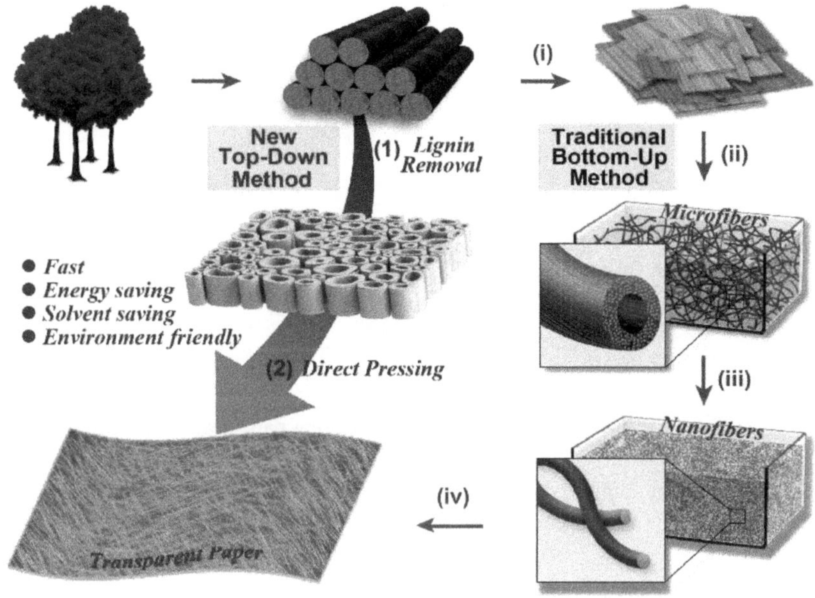

Fig. 4. Comparison of transparent paper fabrication methods: The new "Top-Down" method *vs.* the traditional "Bottom-Up" method.[8] The top-down method includes two steps. Step (1) involves the removal of lignin from the wood while preserving the cellulose framework. Step (2) includes mechanically pressing the anisotropic wood, and the anisotropic wood turns into the isotropic, transparent paper. The traditional bottom-up process includes four main steps: (i) is a delignification process; (ii) is a chemical process to obtain microsized cellulose fibers; (iii) is a mechanical process to reduce the fibers to nanometer-scale diameter; and (iv) is the process to filter and press the nanofibers, producing transparent paper.

Zhu *et al.*[9] prepared paper materials using cellulose fibers with different diameters and studied the effects of fiber diameter on the optical properties of the paper materials. The results showed that the light transmittance and scattering are related to the fiber diameter and packing density, and the transmittance gradually increases with the decrease of fiber diameter (Fig. 5a–d). Here, haze is caused by the light scattering of fibers with larger diameter or voids (Fig. 5e). Large-diameter fibers or large voids have stronger light scattering ability, but the light is back-scattered too much, resulting in large amount of energy loss of transmitted light. As the fiber diameter decreases, the back-scattering of the light decreases rapidly, but its front-scattering ability decreases slowly. Further reducing the fiber

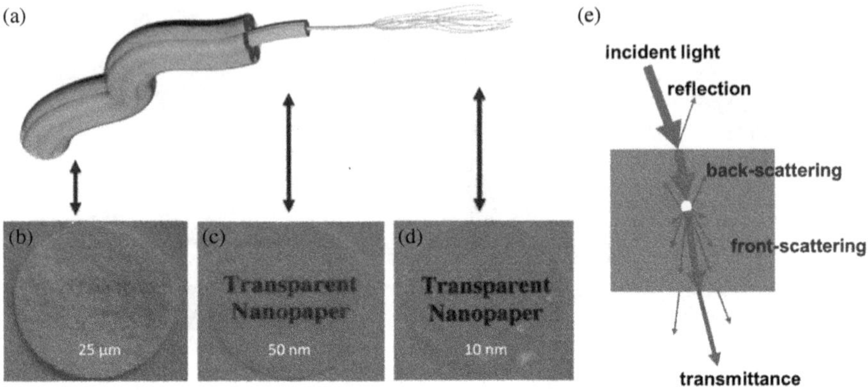

Fig. 5. Tuning of transmittance and haze in transparent paper. (a) Schematic to show the hierarchical structure of cellulose fibers. (b–d) Paper materials composed of cellulose fibers with different diameters. All the paper materials have the same thickness of 40 μm.[9] (e) Schematic of the light scattering behavior in the transparent paper.

diameter to ensure little back-scattering will have little effect on the transmittance, but it has a great influence on haze. Therefore, in this case, the haze can be tuned by the fiber diameter, without sacrificing the high transmittance.

Recently, another cellulose-based material, wood, was made transparent.[10,11] Natural wood is not transparent because of the presence of a light absorption component (e.g., lignin) and light scattering microstructures. Wood is mainly composed of lignin, hemicellulose, and cellulose. Lignin is the light absorbing component in wood, whereas hemicellulose and cellulose are optically transparent. Hence, to achieve transparent wood, lignin (or its chromophoric groups) need to be removed firstly. Various established methods have been developed to delignify wood. However, the wood structure will collapse without lignin's binding, since this acts as a binder to fix cellulose fibers in wood. Zhu *et al.*[10] solved this problem by developing an "*in situ*" lignin removal process (Fig. 6). After the removal of lignin, wood retained its natural microstructures and became "white wood" because of its strong light scattering. Next, a material that matches the refractive index of the cellulose fibers is filled into the voids of the microstructure to decrease the scattering of light. Many polymers with a similar refractive index as cellulose material in wood ($n \sim 1.53$) can be selected as the refractive index-matching materials, including polystyrene (PS) and polymethyl

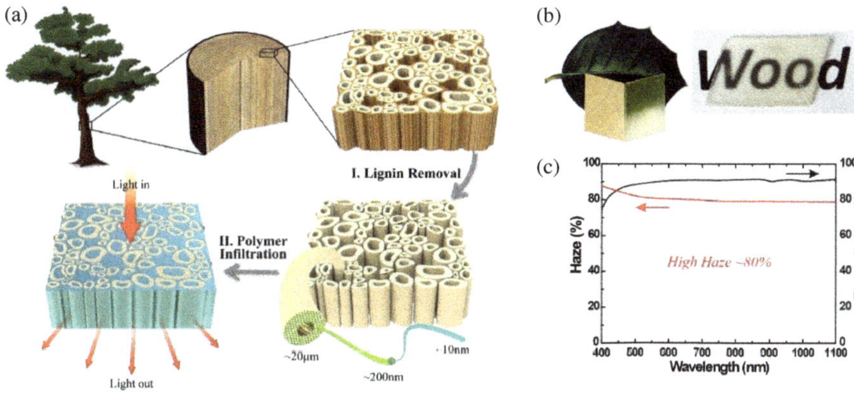

Fig. 6. (a) Fabrication processes of transparent wood composite and its (b) visual appearance and (c) optical properties (transmittance and haze).[10]

methacrylate (PMMA). The resulting transparent wood composite exhibits both high transparency (about 90%) and high haze (about 80%) (Fig. 6b–c).

Haze in nanocellulose-based composite is a kind of bulk scattering where the incident light is continuously scattered inside the nanocellulose-based composites. This is very different from the way frosted glass scatters light, as this relies on surface scattering. Frosted glass results in serious energy loss; the more severe the scattering of light, the more serious the loss of light energy. In the case of large scattering angles, even more than half of the incident light energy is lost. By contrast, the energy loss in the highly transparent and haze wood composite is low owing to the "light-guiding effect" inside the composite. This results in a much higher light scattering efficiency than that of commonly used light scattering materials such as frosted glass.

2.2 *Anisotropic Optical Properties*

Some anisotropic cellulose-based materials exhibit anisotropic optical properties.[12] For example, the lumen in transparent wood composite is aligned along the direction of tree growth, forming an anisotropic microstructure. They have an important influence on the optical properties of the transparent wood composite, as shown in Fig. 7. When the incident light direction is along the lumen axis, it is scattered in

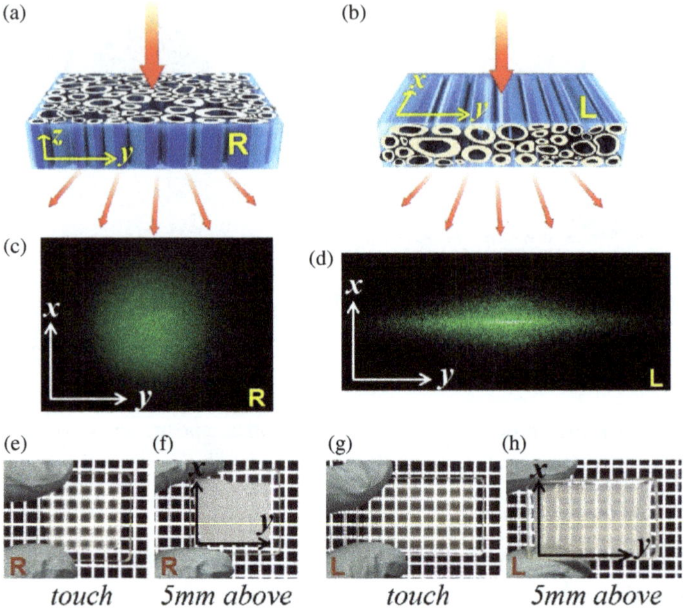

Fig. 7. Schematic representation of (a) isotropic and (b) anisotropic transparent wood composites and their corresponding optical properties (c–d).[12] (e) The isotropic transparent wood composite in contact with the grids and (f) 5 mm above the grid. (g) The anisotropic transparent wood composite in contact with the grids and (h) 5 mm above the grid.

an isotropic way (Fig. 7a, c). However, when incident light is perpendicular to the lumen axis, it is scattered in an anisotropic way (Fig. 7b, d). Light is scattered strongly along the y-axis direction, while scatter is much weaker along the x-axis direction. That is to say, the light momentum changes more along the y-axis than the x-axis. As a result, the haze is much larger along y-axis than x-axis. Such an anisotropic optical property is clearly indicated by Fig. 7e–h. When the transparent wood composite comes into contact with the grid in Fig. 7e, there is no difference in the patterns seen through the two samples, and the grids along both x-axis and y-axis direction are clearly observed. However, if it is placed 5 mm above the grid, the gird is totally blurred, as seen in Fig. 7f; however, the grid along x-axis can be still clearly observed in Fig. 7h. Such anisotropic optical properties are due to the anisotropic microstructures, especially in fiber-based materials. A similar anisotropic property is also observed in anisotropic paper with aligned cellulose fibers.[13,14]

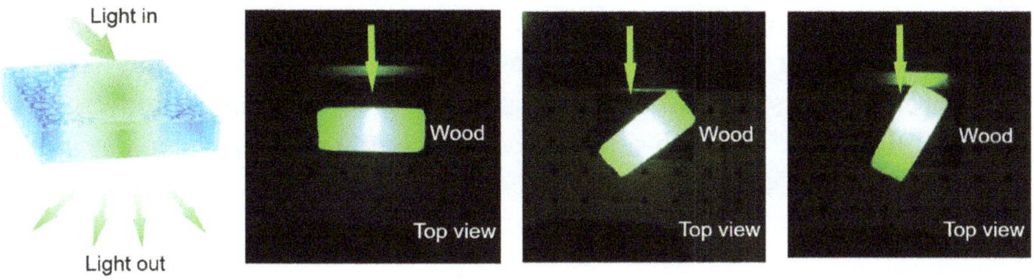

(a) (b) (c) (d)

Fig. 8. (a) Schematic showing the light scattering behavior of transparent wood composite under laser illumination. The light-guiding effect of transparent wood composite under laser illumination at different light incident angles (b) 90°, (c) 45°, and (d) 30°.[12]

2.3 *Light-Guiding Effect*

The light-guiding effect is found in transparent wood composite.[12] Light is preferably passed through along the lumen direction, so the lumen can be considered to "guide light" as shown in Fig. 8. When a laser beam is illuminated along the lumen direction of the transparent wood composite, the light path along the same direction can be clearly seen from the side. This indicates that the reflection/scattering are mainly at the interface between the cellulose and the polymer. Apparently, light is continuously reflected/scattered by the wood composite along its whole pathway. More importantly, the main light energy flux is always along the lumen direction, which seems to be independent of the incident angle, even the if incident angle reaches to 60° (Fig. 8d).

The light "guiding effect" originates from the unique microstructures of the wood raw materials. The diameter of wood fibers is far larger than the optical wavelength, and the light transport behavior in them can be considered as complex geometrical optics. For the transparent wood composite, the one-time reflection that occurs on their interface is very small because the refractive index mismatch between the polymer and wood is minimal. The light guiding behavior is the collective effect after many rounds of light reflection in the wood composite structure.

3. Optoelectronic Applications of Nanocellulose-Based Composites

In recent years, using transparent paper as a flexible substrate to construct electronic devices has attracted great attention from academia and the industry because of its excellent biodegradability, optical transparency, and mechanical properties. Flexible electronic devices developed based on transparent paper include organic solar cells,[15–17] organic light-emitting diodes (OLEDs),[18–20] touch screens,[19,21] transistors,[22–24] supercapacitors,[25,26] antennas,[27,28] and so on.

3.1 *Flexible Electronics*

Antennas are an important part of satellites, computers, and radio frequency identification devices for transmitting and receiving signals.[6] Cellulose nanopaper is suitable for use as a substrate for printing foldable antenna due to its advantages of foldability, printability, light weight, and low surface roughness. Nogi *et al.*[28] fabricated nanopaper with smooth surface and high foldability from CNFs, and then obtained a foldable antenna by mask printing silver nanowires on the nanopaper (Fig. 9a). The nanopaper-based antenna exhibits high sensitivity due to its smooth surface and high foldability. In addition, the nanopaper-based antenna is able to transmit and receive multiple frequency band signals by folding and unfolding to adjust its length without using any additional components such as condensers and coils (Fig. 9b). Therefore, the cellulose nanopaper-based antenna with printed silver nanowires can be used in foldable electronic devices. Zhu *et al.*[27] also prepared a nanopaper-based antenna by gravure printing the radio frequency identification components on the transparent nanopaper substrate had a stable shape (Fig. 9c–d). The results show that the nanopaper-based antenna demonstrates excellent radio frequency response.

In recent years, using cellulose nanopaper as a substrate to prepare transparent flexible film supercapacitors has attracted more and more attention because of its good light transmittance, low thermal expansion coefficient, high mechanical strength, and excellent flexibility. Gao *et al.*[26] developed a CNF-RGO hybridized nanopaper using CNFs and reduced graphene oxide (RGO) as the raw materials (Fig. 10a), and then prepared transparent flexible film supercapacitor (T-SC) with hybridized nanopaper as an electrode material and charge collector (Fig. 10b). The flexible device exhibits excellent electrochemical performance (about 1.73 mF cm^{-2}), and a high light transmittance of ~56% at 550 nm. Wang *et al.*[25] prepared hybrid multilayer films using transparent CNFs nanopaper as the substrate through a simple layer-by-layer

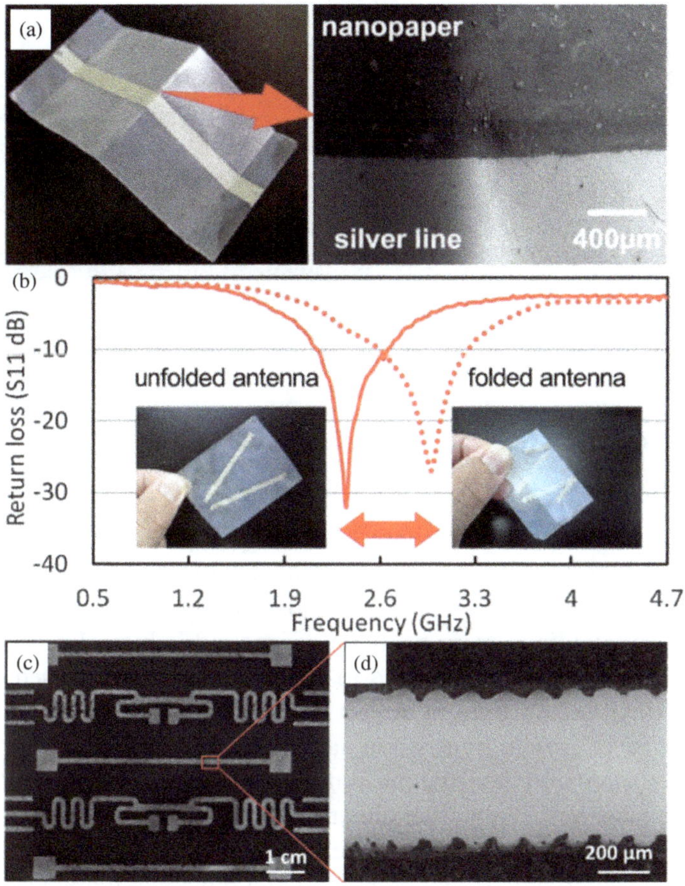

Fig. 9. (a) Digital image of the foldable nanopaper-based antenna and SEM image at the folding place. (b) Return losses of the nanopaper-based antenna (solid line) and folded nanopaper-based antenna (dotted line).[28] (c–d) Microstructure of an antenna that is gravure printed on nanopaper.[27]

self-assembly process, and these were used as electrode materials to construct transparent flexible film supercapacitors.[25] The test results show that the prepared supercapacitors possess excellent electrochemical properties (the areal capacitance is up to 5.86 mF cm^{-2} at a current density of 0.0043 mA cm^{-2}), good cyclic stability, and bending stability.

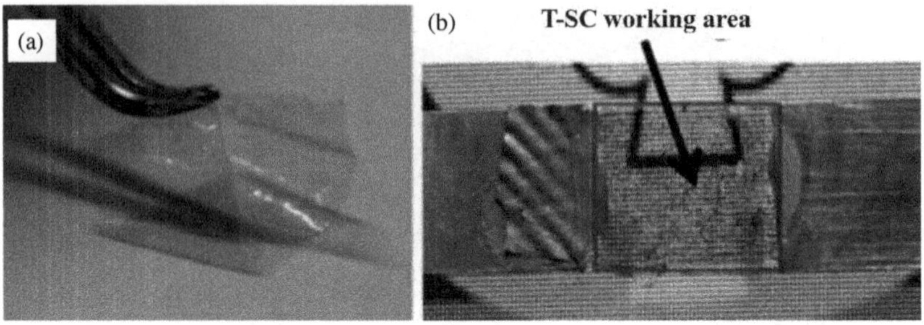

Fig. 10. Transparent paper-based film supercapacitors. (a) Digital image of a CNF-RGO hybrid nanopaper. (b) Digital image of a film supercapacitor made from CNF-RGO hybrid nanopaper.[26]

3.2 *Solar Cells*

Solar energy is a potential renewable energy source that we can utilize. Nowadays, 80% of photovoltaic solar panels are made of crystalline silicon because it is durable and has good stability. However, the cost of extracting pure silicon from a silicon compound is high. In recent years, many efforts have been dedicated to using nanocellulose-based composites as a flexible substrate to collect solar energy. Hu *et al.*[15] developed organic solar cells with transparent nanopaper as the substrate for the first time. They prepared transparent conductive nanopaper by depositing a layer of transparent conductive material on the nanopaper (Fig. 11a), and then constructed organic solar cells on the conductive nanopaper (Fig. 11b). The organic solar cell exhibited a power conversion efficiency of 0.21% (Fig. 11c). Subsequently, Zhou *et al.*[16] prepared organic solar cells on optically transparent cellulose nanocrystal (CNC) substrates (Fig. 11d–e), and the solar cells displayed good rectification performance in the dark with a high power conversion efficiency of 2.7% (Fig. 11f). In addition, they also demonstrated that these solar cells can be easily separated at room temperature and their main components can be reused, opening the door to truly recyclable solar cell technology.

Recently, transparent wood composite has also been used in the solar cell as the surface light management layer to improve its efficiency.[10] As shown in Fig. 12, it allows up to 90% of light energy to pass through, almost comparable to glass (about 92%). However, the light momentum in transparent wood composite is greatly changed, leading to a high haze of about 80%. The light path is greatly increased with a transparent wood

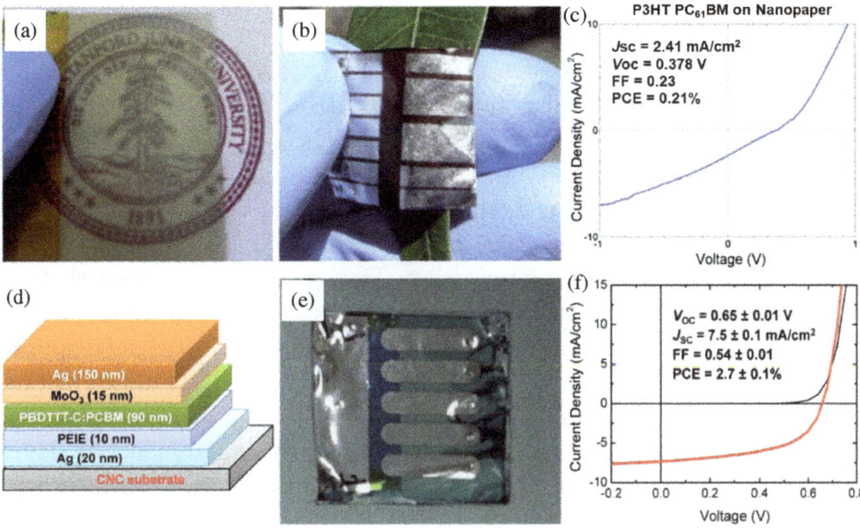

Fig. 11. Transparent nanopaper-based solar cells. (a) Conductive nanopaper deposited with indium tin oxide, which has a light transmittance of 65% at 550 nm. (b) A printed organic solar cell on the transparent conductive nanopaper. (c) I–V curve of the organic solar cell shown in (b).[15] (d) Solar cell structure based on the CNC substrates. (e) Digital image of a prepared solar cell as shown in (d). (f) J–V curves of a solar cell on the CNC substrates in the dark (black line) and under illumination (red line).[16]

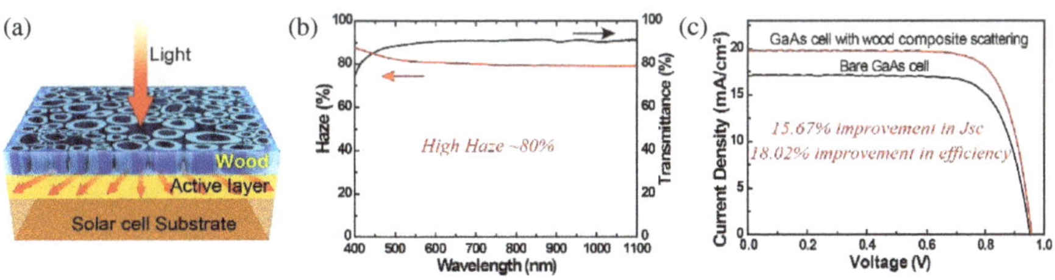

Fig. 12. The haze of transparent wood and its application on GaAs solar cell. (a) The transmittance and haze of the wood used in the experiment; (b) schematic of light distribution incident on a solar cell; (c) current density *vs*. voltage characteristics for both bare GaAs cell (black) and the GaAs cell with the light management wood composite coating (red).[10]

composite placed on the surface of the GaAs solar cell. Combined with the less light reflection loss caused by index match between the wood composite and the GaAs, the efficiency is substantially enhanced by about 18.02% than bare GaAs solar cell.

3.3 *Displays*

Organic light-emitting diodes (OLEDs) have shown great application prospect in the flexible display field due to their small thickness, light weight, and high efficiency. Currently, OLEDs are mainly built on glass or plastic films. Commonly used plastics include polyethylene naphthalate (PEN) and polyethylene terephthalate (PET). However, the high rigidity and density of glass and the high thermal expansion coefficient of plastic limit their practical application. Transparent paper prepared from nanocellulose has nanoscale surface, and its optical properties can be adjusted,[9] therefore, OLED devices made of nanocellulose can be applied in different fields. Zhu *et al.*[18] successfully developed cellulose nanopaper-based OLED devices using transparent cellulose nanopaper as substrate, carbon nanotubes (CNTs) film as transparent conductive layer, 10 nm molybdenum oxide (MoO_3) and 30 nm of poly (3,4-ethylenedioxy thiophene):poly(styrenesulfonate) (PEDOT:PSS) as hole injection layer, green polyfluorene as light emission layer, and 20 nm calcium (Ca) as electron injection layer (Figs. 13a–b). They demonstrated that the flexible nanopaper-based OLED devices can be bent into a bow shape with a

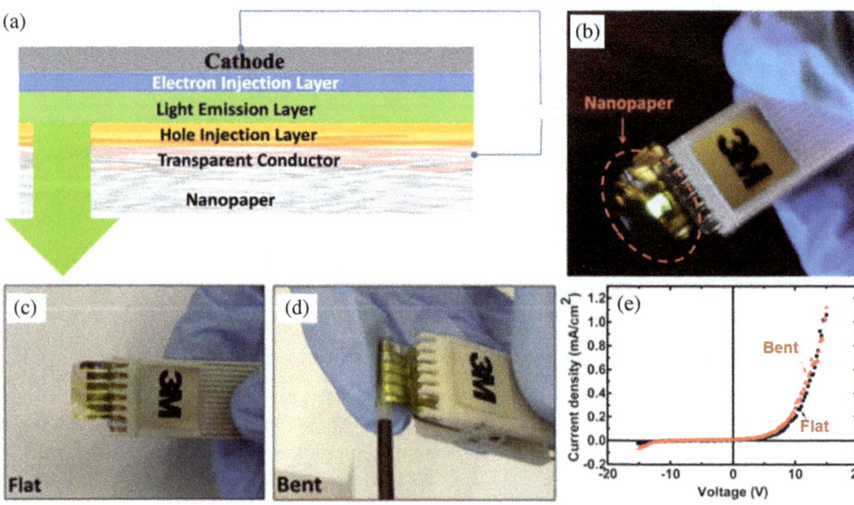

Fig. 13. Transparent nanopaper-based flexible OLED. (a) Schematic diagram of the nanopaper-based OLED. (b) Digital image of an OLED in operation. Digital images of the (c) flat and (d) bent nanopaper-based OLED. (e) J–V curves of the nanopaper-based flexible OLED in the different states.[18]

radius of 1.5 mm and that there was no significant change between the J–V curves before and after bending the devices (Figs. 13c–e).

Transistors are an important kind of semiconductor electronics mainly used to amplify and convert electronic signals. Nanopaper prepared from nanocellulose possesses excellent optical properties and low surface roughness, and has been used as substrate for preparing transistors. Huang *et al.*[23] first obtained nanopaper using nanocellulose as the raw material, and then prepared organic field-effect transistor using nanopaper as the substrate (Fig. 14a). No surface coating is required since the nanopaper has a smooth surface. The prepared devices exhibit high total light transmittance of 83.5% and excellent mechanical flexibility (Fig. 14b). In addition, the nanopaper-based transistors also show excellent electrical properties. When the nanopaper-based transistors are bent and folded, their electron mobility is reduced by only about 10% (Fig. 14c). Zhang *et al.* prepared MoS_2 phototransistor with flexible transparent nanopaper as the substrate for the first time (Fig. 14d). The nanopaper-based transistor exhibits an optical transmittance of 82%, good flexibility, and excellent photoresponsivity of ~1.5 kA W^{-1} (Fig. 14e).

3.4 *Touch Screens*

Transparent papers with high haze can effectively eliminate the glare phenomenon caused by sunlight reflection, showing great application prospect in the touch screen field. Fang *et al.*[21] designed and assembled a four-wire analogue resistance paper-based touch screen with excellent anti-glare effect by taking all-cellulose hybrid nanopaper as the substrate for the first time (Fig. 15a). The word written on the touch screen can be displayed on the computer (Fig. 15b), indicating that the transparent conductive hybrid nanopaper can potentially replace PET to prepare transparent electrodes for electronic devices. When the PET-based touch screen and paper-based touch screen are exposed to light, the PET-based touch screen displays glare effect, and the patterns below the screen become blurred, while the transparent hybrid nanopaper-based touch screen exhibits excellent anti-glare characteristics (Fig. 15c–d). Zhu *et al.*[19] prepared a high-performance capacitive paper-based multitouch screen with ultra-clear paper as the substrate and thin graphene layer as the conductive electrode (Fig. 15e). The paper-based touch screen exhibits excellent flexibility and high clarity (Fig. 15f). The graphene electrodes and metal contacts are patterned by a laser (Fig. 15g). In addition, the prepared

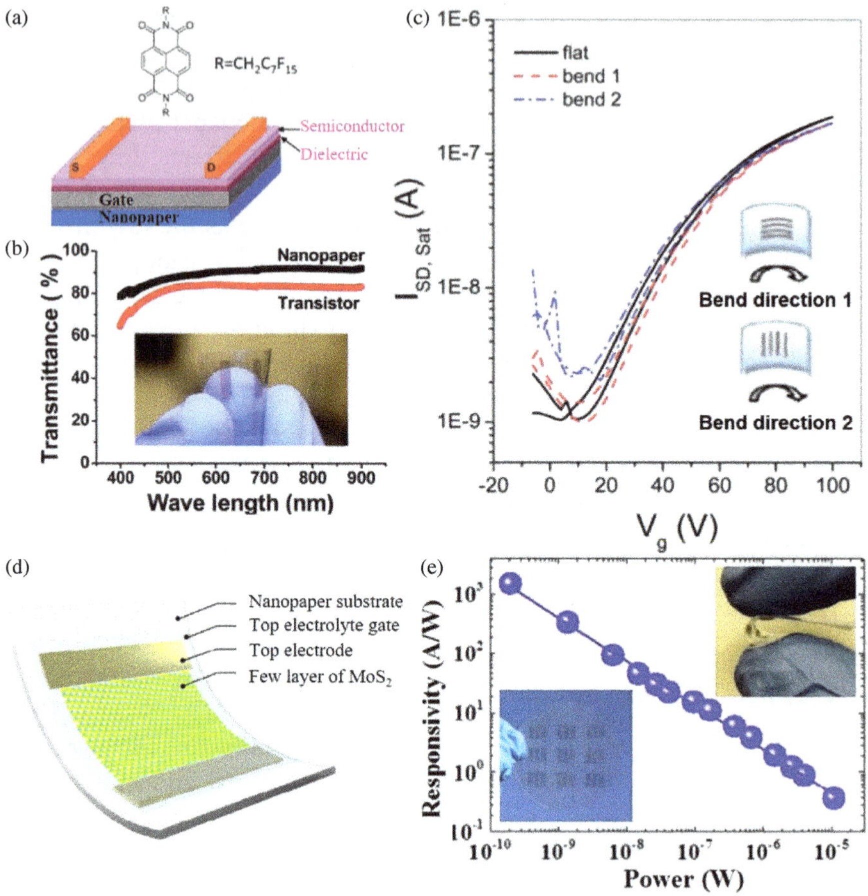

Fig. 14. (a) Schematic diagram of the nanopaper-based organic field-effect transistor; (b) optical transmittance of the nanopaper and nanopaper-based transistors. Inset is the digital image of the transparent flexible nanopaper-based transistor; (c) transfer characteristics of the nanopaper-based transistors before and after bending;[23] (d) schematic to show the structure of the MoS_2 phototransistor prepared on the transparent flexible nanopaper; and (e) photoresponsivity of the MoS_2 phototransistor as a function of illumination intensity. Insets are the digital images of the MoS_2 phototransistor to show the high light transmittance and flexibility.[24]

paper-based touch screen demonstrates excellent linearity and sensing performance that are comparable to the high-quality commercial touch screen products.

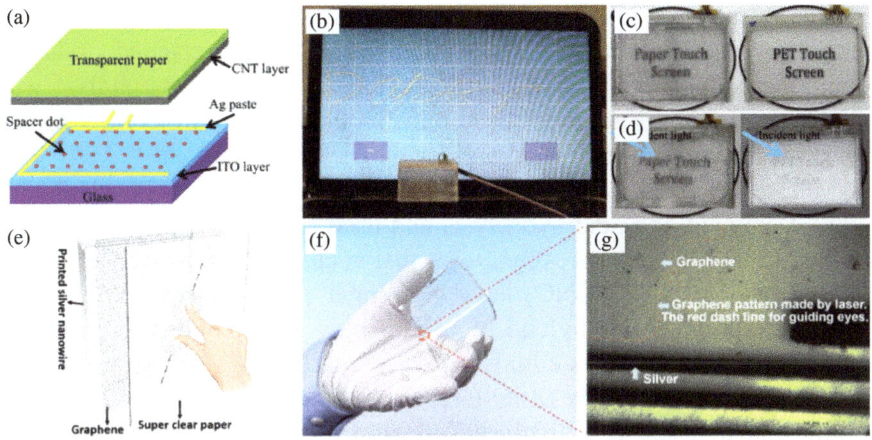

Fig. 15. (a) Schematic of a transparent four-wire touch screen based on nanopaper. (b) Digital image of an assembled nanopaper-based touch screen. (c) Digital images of a paper-based touch screen (left) and a PET-based touch screen (right). (d) Digital images of the two types of touch screens exposed to sunlight. The paper-based touch screen has no glare effect (left), while the PET-based touch screen shows strong glare (right).[21] (e) Schematic of the paper-based multipoint touch screen. (f) Digital image of the highly flexible paper-based touch screen. (g) The magnified image of the paper-based touch screen to show the laser-patterned graphene electrodes and metal contacts.[19]

3.5 *Photoreduction*

Nanocellulose has been widely used in the field of photocatalysis because of its unique properties as a supporting material. Nanocellulose, as a catalyst carrier, not only prevents the aggregation of nanoparticles but also helps to form narrow wells with clear boundaries.[29,30] Nanocellulose-based composites can be used as a photocatalyst for a new pollutant, methyl orange (MO). CdS supported on bacterial CNFs can photocatalyze about 82% of MO after 90 minutes of irradiation.[31] Similarly, the use of bacterial cellulose (BNC)/TiO_2 composite nanofibers as a photocatalyst to degrade MO under ultraviolet light is more efficient than the commercial photocatalyst P25, and nitrogen-doped catalyst can enhance its photocatalytic activity.[32]

An *in situ* synthesis method in which TiO_2 nanoparticles are directly added to a BNC carrier is employed in the biosynthesis process.[33] During the formation of the BNC network, the distribution of NPs in the mixed material is affected by the anisotropic structure of the BNC.

The photocatalytic activity of BNC-TiO_2 composites under UV irradiation was measured by methanol conversion (MC). In the repeated principal component analysis (PCA) test, the efficiency of the photocatalyst was not changed. Therefore, it can be applied in the field of porous filter media for drinking water purification and air purification.

Interestingly, nanocellulose was prepared from oil palm shells by alkali treatment and acid hydrolysis and used as a carrier to prepare zinc oxide (ZnO) nanostructures by *in situ* fusion casting. Zinc oxide/cellulose nanocomposites (ZnO/CNC) have a stronger photocatalytic activity for the degradation of methylene blue than pure ZnO nanostructures.[34] In addition, ZnO-GO/nanocellulose was prepared by adding graphene-oxidized nano-zinc oxide (ZnO-GO) to the nanocellulose matrix. It was found that this can effectively adsorb and photo-degrade the antibiotic ciprofloxacin (CF). The feasibility of the photocatalyst in the real-life situation was thus demonstrated.[35]

In another study, the straw cellulose (SC)/CeO_2 composites were obtained by hydrothermal method, and Cr(VI) was selected as the target pollutant to test the photocatalytic reduction activity of SC/CeO_2 nano-composites.[36] It was confirmed that SC/CeO_2 nanocomposites have good visible light-induced photocatalytic activity for the reduction of Cr(VI) in solution.

3.6 *Plasmonic Applications*

CNF, a unique sustainable material, has proven to be a platform for nanoparticle synthesis. In addition to its high specific surface area, the excellent colloidal stability of the negatively charged CNF makes it suitable for surface-mediated nanoparticle synthesis.[37,38] Reports on the use of nanocellulose in plasmonic materials mainly focus on the modification of cellulose with nanoparticles or nanorods.

A novel Surface-Enhanced Raman Scattering (SERS) substrate platform has been reported for microchemical and biological detection previously. The platform is based on a common filter paper that allows the nanorods to adhere evenly and stably to the microfibers and nanofibers. Due to the high transparency of CNF, low light scattering, low fluorescence background, and low Raman background of cellulose, excellent sensing performance can be obtained in low-load density plasma nanocrystals.[39–41] The experimental study found that this simple and efficient method can uniformly adsorb the anisotropic plasma nanostructure on the paper substrate and enable a SERS enhancement factor of ~5×10^6, which shows better sample collection efficiency. The paper substrate has a high sensitivity (0.5 nM trans-1,2-bis(4-pyridyl) ethene (BPE)) and excellent reproducibility (15% relative standard deviation (RSD).[39]

A separate chiral plasmonic film is formed by entropy-driven self-assembly of gold nanoparticles (GNPs) and cellulose nanocrystals (CNCs). The CNCs-GNPs composite film not only retains the photon order of the CNCs matrix but also preserves the plasmon resonance of GNPs, resulting in significant plasma-induced chiral activity and strong resonance plasma photon coupling. Switchable optical activity can be obtained by changing the incidence angle of the light or by utilizing the response characteristics of the matrix of the CNCs. The angle-dependent plasmonic resonance in chiral nematic hybrid membranes was observed for the first time, which is quite different from the GNPs embedded in three-dimensional (3D) photonic crystals, making mass production of functional optical metamaterials possible.[42]

Then, a plasma-active 3D biofilm (Plasmonic Biofoam) is created by combining plasmonic nanostructures (AuNRs) with a highly porous biomaterial aerogel. The plasmonic biofoam is a versatile optically active platform that provides higher sensing, photothermal and load efficiency than traditional two-dimensional (2D) plasma bioelectromagnetic fields. The plasmonic biofoam can find numerous applications including: (i) ultrasensitive chemical detection using surface-enhanced Raman scattering; (ii) highly efficient energy harvesting and steam generation through plasmonic photothermal heating; and (iii) optical control of enzymatic activity by triggered release of biomolecules encapsulated within the aerogel.[43,44]

In a recent study, Hu *et al.*[15] prepared a highly translucent light-diffusing film composed of multilayer nanocellulose fibers and wood pulp microfibers based on the TEMPO/NaClO/NaBr system.[45] The film exhibits a combination of high transmittance of up to 90% and tunable diffuse transmittance of up to 78% in the visible and near infrared spectra. The detailed properties of the film indicate high light transmittance and haze due to the high density of the film and the special microstructured surface.

Interestingly, the Hu group designed a new type of plasmonic material more recently, which is made by uniformly decorating fine metal nanoparticles into the 3D mesoporous matrix of natural wood (plasmonic wood).[46] The plasmonic wood exhibits high light absorption ability (≈99%) over a broad wavelength ranging from 200 to 2500 nm due to the plasmonic effect of metal nanoparticles and the waveguide effect of microchannels in the wood matrix. Due to the capillary effect, 3D mesoporous wood with many low curvature microchannels and nanochannels can effectively transport water upward from the bottom of the device. Therefore, the 3D aligned porous structure can achieve 85% high solar energy conversion efficiency at ten-sun illumination (10 kW m^{-2}). The plasmonic wood also showed excellent stability for the production of solar steam, without any degradation after 144 hours of evaluation. Its high conversion efficiency and excellent cycle stability demonstrate the potential of the newly developed plasmonic wood for solar desalination.

It is well known that localized surface plasmon resonance exhibited by noble metal nanoparticles can be tuned by changing the size of the particles and the distance between the particles. Studies have found that exposure to corrosive vapors (ammonia) greatly reduces the density of silver nanoparticles (AgNPs) embedded in bacterial cellulose, resulting in a lager distance between the remaining nanoparticles and a decrease in the UV–Vis absorbance associated with the plasmonic properties of AgNPs. The size distribution of AgNPs embedded in BC is reduced during the release of volatile compounds during food spoilage. Moreover, this plasmonic nanopaper exhibits a change from amber to light amber when exposed to corrosive vapors and changes from amber to gray or taupe when fish or meat is exposed to spoilage. These phenomena are proposed as flexible, transparent, permeable, and stable disposable nanofilms for simple visual inspection of volatile compounds. This opens the way for innovative approaches to gas sensing and smart packaging.[47]

3.7 *Ecological Building and Energy-Saving Systems*

A bright and comfortable environment is the inherent needs of human living. To achieve this goal, energy is required to adjust the brightness and temperature of the room. However, to maintain a sustainable future, energy saving is required now all over the world. As promoted by the U.S. Department of Energy (DOE), energy consumption of buildings is required to reduce 20% by 2020 and 50% as the long-term goal.[48] This is based on the fact that energy used for lighting and temperature adjustments cost more than 50% of the total energy consumption in residential and commercial buildings.[49]

Glass is the mostly used material for windows and doors in building. About one-third of the energy used for heating or cooling building is lost through windows because of the relatively high intrinsic thermal conductivity of glass.[50] Hu *et al.*[15] developed a new transparent wood-based composite, which exhibits which exhibits new light and thermal properties, as shown in Fig. 16.[51] Both transparent wood composite and glass show high total light transparency, which means that they all let most of the light energy passed through. However, the transparent wood composite shows a different light guiding effect. Although glass is very transparent, the light passes directly through it and results in a strong contrast between the interior and the dark, which causes discomfort to the eyes. Transparent wood composite does not have this problem. Because of its unique microstructures, light is "guided" through the wood composite and distributes uniformly. When used as window materials in the roof, the indoor light distribution will be both uniform and bright. Also, even though the position of the sun changes, the direction of the light transmitted from transparent wood composite does not change much. This will provide a very comfortable indoor light environment compared with the ordinarily used glass. Moreover, it has a good thermal barrier effect with a thermal conductivity of around 0.32 W m^{-1} K^{-1} along the wood growth direction and 0.15 W m^{-1} K^{-1} in the cross plane, much lower than that of glass (≈1 W m^{-1} K^{-1}). As a result, in summer, heat can be efficiently

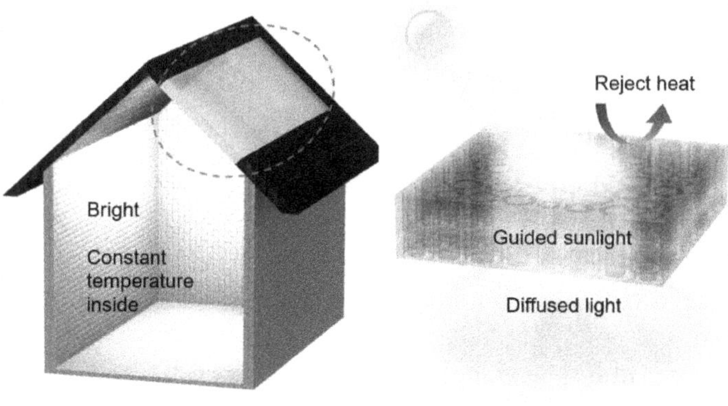

Fig. 16. Schematic of transparent wood composite as an energy-efficient building material and its effect of guiding light and heat blocking.[51]

blocked, and the light used for illumination can enter as much as possible. In winter, the heat that is indoors can also be prevented from escaping by the transparent wood composite, thus avoiding heat loss.

More recently, Hu *et al.*[52] developed a cooling wood structural material by combining delignification with densification. Most lignin and hemicellulose were removed from the wood cell wall, resulting in a mesoporous "white wood". Further pressing makes the "white wood" highly dense with intertwined cell walls connected with each other, contributing a high mechanical strength of 404.3 MPa. The cooling wood is able to backscatter solar radiation and emit strongly in mid-infrared wavelengths, resulting in continuous sub-ambient cooling during both day and night.

The transparent wood and cooling wood composites provide a new kind of energy-saving building materials with fascinating optical, mechanical, and thermal properties. This may find a range of potential applications in next-generation energy-efficient buildings and energy-saving systems.

4. Concluding Remarks and Prospects

Novel nanocellulose-based composites are emerging for both researches and applications in recent years. This is mainly based on the numerous advantages of nanocellulose materials and their structures. The research and application of cellulose-based materials has just begun. Despite the tremendous progress achieved worldwide in a wide range of application fields in the past few decades, challenges remain for their wide commercialization that can truly benefit our daily life. The material cost of nanocellulose is still relatively high, which hinders its scalable applications. Water stability is another challenge that needs continuous efforts from both academic and industrial communities. Fire safety is critical for most applications (e.g., energy-efficient buildings), which, however, is not fully satisfied in nanocellulose-based composites. Future efforts to enhance their fire retardance *via* coating, impregnation, or thermal modifications are desired.

In theory, many new cellulose-based materials and optoelectronic devices are ready to be discovered. New materials and properties may be found in different dimensions (0, 1, 2, 3) of cellulose-based composites. The correlations with other new functional materials will also lead to new optoelectronic applications. Till now, nanocellulose-based composites are mostly used as substrates in optoelectronic applications. The use of nanocellulose-based composites and their structures to achieve new functions will open up new fields of applications. For example, artificial microstructures can be introduced to realize structure-related optical, photonic, and electronic functions and applications. The "Top-Down" approach for fabricating nanocellulose-based composites directly from natural wood exhibits great potential and has attracted intensive attention in recent years. The anisotropic structure of natural wood can be largely inherited by the obtained nanocellulose-based composites; simultaneously, the mechanical, optical, and thermal properties can be manipulated *via* tuning their macro-, micro-, nano-, and molecular structures, leading to greatly enhanced or even completely new properties and functions.[53,54] Given the abundant, renewable, and sustainable resource and considerable efforts being dedicated to bridging the gap between state-of-the-art materials, devices, and systems and future low-cost and scalable applications, nanocellulose-based composites will be of benefit in our daily life in the years to come.

References

(1) Toivonen, M. S., Onelli, O. D., Jacucci, G., Lovikka, V., Rojas, O. J., Ikkala, O., Vignolini, S. Anomalous-diffusion-assisted brightness in white cellulose nanofibril membranes. *Adv. Mater.* **2018**, 30, e1704050.

(2) Fang, Z., Zhu, H., Bao, W., Preston, C., Liu, Z., Dai, J., Li, Y., Hu, L. Highly transparent paper with tunable haze for green electronics. *Energy Environ. Sci.* **2014**, 7, 3313–3319.

(3) Olsson, R. T., Samir, M. A. S. A., Salazar-Alvarez, G., Belova, L., Strom, V., Berglund, L. A., Ikkala, O., Nogues, J., Gedde, U. W. Making flexible magnetic aerogels and stiff magnetic nanopaper using cellulose nanofibrils as templates. *Nat. Nanotech.* **2010**, 5, 584–588.

(4) Jung, Y. H., Chang, T. H., Zhang, H., Yao, C., Zheng, Q., Yang, V. W., Mi, H., Kim, M., Cho, S. J., Park, D. W., Jiang, H., Lee, J., Qiu, Y., Zhou, W., Cai, Z., Gong, S., Ma, Z. High-performance green flexible electronics based on biodegradable cellulose nanofibril paper. *Nat. Commun.* **2015**, 6, 7170.

(5) Gaspar, D., Fernandes, S. N., de Oliveira, A. G., Fernandes, J. G., Grey, P., Pontes, R. V., Pereira, L., Martins, R., Godinho, M. H., Fortunato, E. Nanocrystalline cellulose applied simultaneously as the gate dielectric and the substrate in flexible field effect transistors. *Nanotechnology* **2014**, 25, 094008.

(6) Zhu, H., Fang, Z., Preston, C., Li, Y., Hu, L. Transparent paper: Fabrications, properties, and device applications. *Energy Environ. Sci.* **2014**, 7, 269–287.
(7) Nogi, M., Iwamoto, S., Nakagaito, A. N., Yano, H. Optically transparent nanofiber paper. *Adv. Mater.* **2009**, 21, 1595–1598.
(8) Zhu, M., Jia, C., Wang, Y., Fang, Z., Dai, J., Xu, L., Huang, D., Wu, J., Li, Y., Song, J., Yao, Y., Hitz, E., Wang, Y., Hu, L. Isotropic paper directly from anisotropic wood: Top-down green transparent substrate toward biodegradable electronics. *ACS Appl. Mater. Interfaces* **2018**, 10, 28566–28571.
(9) Zhu, H., Parvinian, S., Preston, C., Vaaland, O., Ruan, Z., Hu, L. Transparent nanopaper with tailored optical properties. *Nanoscale* **2013**, 5, 3787–3792.
(10) Zhu, M., Li, T., Davis, C. S., Yao, Y., Dai, J., Wang, Y., AlQatari, F., Gilman, J. W., Hu, L. Transparent and haze wood composites for highly efficient broadband light management in solar cells. *Nano Energy* **2016**, 26, 332–339.
(11) Li, Y., Fu, Q., Yu, S., Yan, M., Berglund, L. Optically transparent wood from a nanoporous cellulosic template: combining functional and structural performance. *Biomacromolecules* **2016**, 17, 1358–1364.
(12) Zhu, M., Song, J., Li, T., Gong, A., Wang, Y., Dai, J., Yao, Y., Luo, W., Henderson, D., Hu, L. Highly Anisotropic, Highly transparent wood composites. *Adv. Mater.* **2016**, 28, 5181–5187.
(13) Zhu, M., Wang, Y., Zhu, S., Xu, L., Jia, C., Dai, J., Song, J., Yao, Y., Wang, Y., Li, Y., Henderson, D., Luo, W., Li, H., Minus, M. L., Li, T., Hu, L. Anisotropic, transparent films with aligned cellulose nanofibers. *Adv. Mater.* **2017**, 29, 1606284.
(14) Jia, C., Li, T., Chen, C., Dai, J., Kierzewski, I. M., Song, J., Li, Y., Yang, C., Wang, C., Hu, L. Scalable, anisotropic transparent paper directly from wood for light management in solar cells. *Nano Energy* **2017**, 36, 366–373.
(15) Hu, L., Zheng, G., Yao, J., Liu, N., Weil, B., Eskilsson, M., Karabulut, E., Ruan, Z., Fan, S., Bloking, J. T., McGehee, M. D., Wagberg, L., Cui, Y. Transparent and conductive paper from nanocellulose fibers. *Energy Environ. Sci.* **2013**, 6, 513–518.
(16) Zhou, Y., Fuentes-Hernandez, C., Khan, T. M., Liu, J. C., Hsu, J., Shim, J. W., Dindar, A., Youngblood, J. P., Moon, R. J., Kippelen, B. Recyclable organic solar cells on cellulose nanocrystal substrates. *Sci. Rep.* **2013**, 3, 1536.
(17) Fang, Z., Zhu, H., Yuan, Y., Ha, D., Zhu, S., Preston, C., Chen, Q., Li, Y., Han, X., Lee, S., Chen, G., Li, T., Munday, J., Huang, J., Hu, L. Novel nanostructured paper with ultrahigh transparency and ultrahigh haze for solar cells. *Nano Lett.* **2014**, 14, 765–773.
(18) Zhu, H., Xiao, Z., Liu, D., Li, Y., Weadock, N. J., Fang, Z., Huang, J., Hu, L. Biodegradable transparent substrates for flexible organic-light-emitting diodes. *Energy Environ. Sci.* **2013**, 6, 2105–2111.

(19) Zhu, H., Fang, Z., Wang, Z., Dai, J., Yao, Y., Shen, F., Preston, C., Wu, W., Peng, P., Jang, N., Yu, Q., Yu, Z., Hu, L. Extreme light management in mesoporous wood cellulose paper for optoelectronics. *ACS Nano* **2016**, 10, 1369–1377.
(20) Yagyu, H., Saito, T., Isogai, A., Koga, H., Nogi, M. Chemical modification of cellulose nanofibers for the production of highly thermal resistant and optically transparent nanopaper for paper devices. *ACS Appl. Mater. Interfaces* **2015**, 7, 22012–22017.
(21) Fang, Z., Zhu, H., Preston, C., Han, X., Li, Y., Lee, S., Chai, X., Chen, G., Hu, L. Highly transparent and writable wood all-cellulose hybrid nanostructured paper. *J. Mater. Chem. C* **2013**, 1, 6191–6197.
(22) Fujisaki, Y., Koga, H., Nakajima, Y., Nakata, M., Tsuji, H., Yamamoto, T., Kurita, T., Nogi, M., Shimidzu, N. Transparent nanopaper-based flexible organic thin-film transistor array. *Adv. Funct. Mater.* **2014**, 24, 1657–1663.
(23) Huang, J., Zhu, H., Chen, Y., Preston, C., Rohrbach, K., Cumings, J., Hu, L. Highly transparent and flexible nanopaper transistors. *ACS Nano* **2013**, 7, 2106–2113.
(24) Zhang, Q., Bao, W., Gong, A., Gong, T., Ma, D., Wan, J., Dai, J., Munday, J. N., He, J. H., Hu, L., Zhang, D. A highly sensitive, highly transparent, gel-gated Mos2 phototransistor on biodegradable nanopaper. *Nanoscale* **2016**, 8, 14237–14242.
(25) Wang, X., Gao, K., Shao, Z., Peng, X., Wu, X., Wang, F. Layer-by-Layer Assembled hybrid multilayer thin film electrodes based on transparent cellulose nanofibers paper for flexible supercapacitors applications. *J. Power Sources* **2014**, 249, 148–155.
(26) Gao, K., Shao, Z., Wu, X., Wang, X., Zhang, Y., Wang, W., Wang, F. Paper-based transparent flexible thin film supercapacitors. *Nanoscale* **2013**, 5, 5307–5311.
(27) Zhu, H., Narakathu, B. B., Fang, Z., Aijazi, A. T., Joyce, M., Atashbar, M., Hu, L. A Gravure printed antenna on shape-stable transparent nanopaper. *Nanoscale* **2014**, 6, 9110–9115.
(28) Nogi, M., Komoda, N., Otsuka, K., Suganuma, K. Foldable nanopaper antennas for origami electronics. *Nanoscale* **2013**, 5, 4395–4399.
(29) Kaushik, M., Moores, A. Review: Nanocelluloses as versatile supports for metal nanoparticles and their applications in catalysis. *Green Chem.* **2016**, 18, 622–637.
(30) Wei, H., Rodriguez, K., Renneckar, S., Vikesland, P. J. Environmental science and engineering applications of nanocellulose-based nanocomposites. *Environ. Sci.: Nano* **2014**, 1, 302–316.
(31) Yang, J., Yu, J., Fan, J., Sun, D., Tang, W., Yang, X. Biotemplated preparation of CdS nanoparticles/bacterial cellulose hybrid nanofibers for photocatalysis application. *J. Hazard. Mater.* **2011**, 189, 377–383.

(32) Sun, D., Yang, J., Wang, X. Bacterial cellulose/TiO_2 hybrid nanofibers prepared by the surface hydrolysis method with molecular precision. *Nanoscale* **2010**, 2, 287–292.
(33) Wesarg, F., Schlott, F., Grabow, J., Kurland, H.-D., Hessler, N., Kralisch, D., Mueller, F. A. *In situ* synthesis of photocatalytically active hybrids consisting of bacterial nanocellulose and anatase nanoparticles. *Langmuir* **2012**, 28, 13518–13525.
(34) Lefatshe, K., Muiva, C. M., Kebaabetswe, L. P. Extraction of nanocellulose and *in-situ* casting of Zno/cellulose nanocomposite with enhanced photocatalytic and antibacterial activity. *Carbohydr. Polym.* **2017**, 164, 301–308.
(35) Anirudhan, T. S., Deepa, J. R. Nano-zinc oxide incorporated graphene oxide/nanocellulose composite for the adsorption and photo catalytic degradation of ciprofloxacin hydrochloride from aqueous solutions. *J. Colloid Interface Sci.* **2017**, 490, 343–356.
(36) Zhang, Y., Zhou, Z., Yuan, K., Wen, F., Tan, J., Hu, C., Wang, H. Fabrication of a modified straw cellulose and Cerium oxide nanocomposite and its visible-light photocatalytic reduction activity. *J. Environ. Chem. Eng.* **2017**, 5, 3734–3740.
(37) Kontturi, E., Laaksonen, P., Linder, M. B., Nonappa, Groschel, A. H., Rojas, O. J., Ikkala, O. Advanced materials through assembly of nanocelluloses. *Adv. Mater.* **2018**, 30, e1703779.
(38) Majoinen, J., Hassinen, J., Haataja, J. S., Rekola, H. T., Kontturi, E., Kostiainen, M. A., Ras, R. H. A., Torma, P., Ikkala, O. Chiral plasmonics using twisting along cellulose nanocrystals as a template for gold nanoparticles. *Adv. Mater.* **2016**, 28, 5262–5267.
(39) Lee, C. H., Hankus, M. E., Tian, L., Pellegrino, P. M., Singamaneni, S. Highly sensitive surface enhanced raman scattering substrates based on filter paper loaded with plasmonic nanostructures. *Anal. Chem.* **2011**, 83, 8953–8958.
(40) Lee, C. H., Tian, L., Singamaneni, S. Paper-based SERS swab for rapid trace detection on real-world surfaces. *ACS Appl. Mater. Interfaces* **2010**, 2, 3429–3435.
(41) Zhang, S., Xiong, R., Mahmoud, M. A., Quigley, E. N., Chang, H., El-Sayed, M., Tsukruk, V. V. Dual-excitation nanocellulose plasmonic membranes for molecular and cellular SERS detection. *ACS Appl. Mater. Interfaces* **2018**, 10, 18380–18389.
(42) Chu, G., Wang, X., Yin, H., Shi, Y., Jiang, H., Chen, T., Gao, J., Qu, D., Xu, Y., Ding, D. Free-standing optically switchable chiral plasmonic photonic crystal based on self-assembled cellulose nanorods and gold nanoparticles. *ACS Appl. Mater. Interfaces* **2015**, 7, 21797–21806.

(43) Tian, L., Luan, J., Liu, K.-K., Jiang, Q., Tadepalli, S., Gupta, M. K., Naik, R. R., Singamaneni, S. Plasmonic biofoam: A versatile optically active material. *Nano Lett.* **2016**, 16, 609–616.
(44) Tian, L., Jiang, Q., Liu, K.-K., Luan, J., Naik, R. R., Singamaneni, S. Bacterial nanocellulose-based flexible surface enhanced Raman scattering substrate. *Adv. Mater. Interfaces* **2016**, 3, 1600214.
(45) Wu, W., Tassi, N. G., Zhu, H., Fang, Z., Hu, L. Nanocellulose-based translucent diffuser for optoelectronic device applications with dramatic improvement of light coupling. *ACS Appl. Mater. Interfaces* **2015**, 7, 26860–26864.
(46) Zhu, M., Li, Y., Chen, F., Zhu, X., Dai, J., Li, Y., Yang, Z., Yan, X., Song, J., Wang, Y., Hitz, E., Luo, W., Lu, M., Yang, B., Hu, L. Plasmonic wood for high-efficiency solar steam generation. *Adv. Energy Mater.* **2018**, 8, 1701028.
(47) Heli, B., Morales-Narvaez, E., Golmohammadi, H., Ajji, A., Merkoci, A. Modulation of population density and size of silver nanoparticles embedded in bacterial cellulose *via* ammonia exposure: visual detection of volatile compounds in a piece of plasmonic nanopaper. *Nanoscale* **2016**, 8, 7984–7991.
(48) Buildings | Department of Energy, http://energy.gov/eere/efficiency/buildings, Accessed: April, 2016.
(49) Buildings Energy Data Book, http://buildingsdatabook.eren.doe.gov/, Accessed: April, 2016.
(50) High R Value Windows, http://www.gsa.gov/portal/content/180891, Accessed: April, 2016.
(51) Li, T., Zhu, M., Yang, Z., Song, J., Dai, J., Yao, Y., Luo, W., Pastel, G., Yang, B., Hu, L. Wood composite as an energy efficient building material: Guided sunlight transmittance and effective thermal insulation. *Adv. Energy Mater.* **2016**, 6, 1601122.
(52) Li, T., Zhai, Y., He, S., Gan, W., Wei, Z., Heidarinejad, M., Dalgo, D., Mi, R., Zhao, X., Song, J., Dai, J., Chen, C., Aili, A., Vellore, A., Martini, A., Yang, R., Srebric, J., Yin, X., Hu, L. A radiative cooling structural material. *Science* **2019**, 364, 760–763.
(53) Song, J., Chen, C., Zhu, S., Zhu, M., Dai, J., Ray, U., Li, Y., Kuang, Y., Li, Y., Quispe, N., Yao, Y., Gong, A., Leiste, U. H., Bruck, H. A., Zhu, J. Y., Vellore, A., Li, H., Minus, M. L., Jia, Z., Martini, A., *et al.* Processing bulk natural wood into a high-performance structural material. *Nature* **2018**, 554, 224–228.
(54) Li, T., Song, J., Zhao, X., Yang, Z., Pastel, G., Xu, S., Jia, C., Dai, J., Chen, C., Gong, A., Jiang, F., Yao, Y., Fan, T., Yang, B., Wagberg, L., Yang, R., Hu, L. Anisotropic, lightweight, strong, and super thermally insulating nanowood with naturally aligned nanocellulose. *Sci. Adv.* **2018**, 4, eaar3724.

https://doi.org/10.1142/9781786349477_0013

Chapter 13

Nanocellulose-Based Materials for Energy Conversion

Jinyu Wang[*], Tao Wu[*], Huan Chen[*], Siling Chen[*], Wenchao Yu[*], Yanan Yin[*], Zhuqun Shi[†,‡], and Quanling Yang[*,§]

[*]*School of Materials Science and Engineering, Wuhan University of Technology, Wuhan, China*

[†]*School of Chemistry, Chemical Engineering and Life Sciences, Wuhan University of Technology, Wuhan, China*

[‡]*zqshi2016@whut.edu.cn*

[§]*yangql@whut.edu.cn*

Abstract

The increasing environmental pollution and serious shortage of energy and resources have become big problems in human society in the recent years. Exploration of novel renewable and biodegradable materials based on biomass and construction of biomass-based energy-related applications have received more and more attention. Cellulose is the most abundant biomass on earth. Especially, nanocellulose possesses excellent mechanical, thermal, and optical properties. In this chapter, we review the recent advances in nanocellulose-based energy conversion materials, including piezoelectric materials, loudspeakers, antenna, phototransistors, organic light-emitting diodes (OLEDs), and touch screen.

Keywords: Nanocellulose; energy conversion; piezoelectric materials; loudspeaker; antenna; phototransistors; organic light-emitting diodes; touch screen

1. Introduction

With the rapid development of science and technology, solving the problem of lack of resources and environmental degradation has become a top priority.[1] Scholars from all over the world are working hard to develop new materials to solve the current situation and make full use of renewable energy such as solar energy and wind energy to effectively solve the problem of energy conversion and storage. At the same time, it will alleviate environmental degradation and achieve high efficiency, energy saving, environmental protection, and sustainable development. Converting energy into different forms of energy that humans need greatly alleviates the energy crisis.[2] Energy materials have become a hot spot in the research of new materials, and its development is inseparable from energy conversion materials.

The development and implementation of green electronics incorporating bio-based materials is the key to providing a more sustainable future. Bio-based materials open a new window for the establishment of disposable materials owing to their nature of easy availability, biodegradability, and renewability. Among them, cellulose is intensively investigated by researchers and cellulose-based functional devices have been reported. Transparent nanopapers have attracted considerable attention in recent years for their great application potential as substrate in touch screens, organic light-emitting diodes (OLEDs), transistors, solar cells.[3–5] Compared with regular cellulose, nanocellulose fibers have much smaller fiber diameter, even as small as 3 nm, enabling nanopaper substrate to show much better properties such as smoother surface, better optical transparency, and mechanical properties, thus endowing the resultant devices with better comprehensive performances and contributing to the construction of next-generation flexible devices.[6]

In general, nanocellulose cannot be directly employed in many fields. Thus, additional modification or other components are required to meet the requirement of application. For example, cellulose is highly hydrophilic due to the large amount of hydroxyl groups, which adsorb and hold water efficiently; therefore, a cross-linking agent is required to improve their shape stability in water. In addition, excellent conductivity is another parameter that is required by substrates of touch screens and organic light-emitting diodes; thus, highly conductive components such as carbon nanotube (CNT) and graphene were usually used to improve the conductivity of cellulose-based substrates.

This chapter aims to introduce the recent progress in nanocellulose-based energy conversion materials. We also analyze specific properties and performance of reported devices.

1.1 *Piezoelectric Materials*

The rapid development of the economy has driven the rapid rise of the electronics field, and the advent of the intelligent era has promoted the demand for advanced smart materials. The rapid rise of smart materials symbolizes the development of human wisdom and is conducive to the full development of automation. As a kind of special intelligent material, piezoelectric materials could sense the changes of surrounding environment and realize the mutual conversion of mechanical energy and electrical energy. They have been widely used in many fields, such as acoustic equipment, transducers, sensors, and infrared detectors. Piezoelectric materials are characterized by the piezoelectric effect. When a mechanical force is applied to a crystal in a certain direction, the crystal would deform and thereby the center of gravity of the positive and negative charges would separate; that is to say, the electric moment changes leading to the charging phenomenon of the crystal surface. When the crystal is compressed, a quantity of charges proportional to the magnitude of the pressure are generated at both ends of the material; when the external force is removed, the crystal returns to the uncharged state; when the direction of the external force changes, the polarity of the charge also follows. The amount of the charge generated by the force of the crystal is proportional to the magnitude of the external force.

The cellulose exhibits multiple –OH groups on the glucose from one chain to form hydrogen bonding with oxygen atoms with the same or a neighbor chains and to hold the chains together side-by-side, responsible for the spontaneous electric dipole formations in cellulose microfibers inside the crystal lattice.[7] Thus, the piezoelectric effect is produced by the displacement or reorientation of the dipoles in crystal lattice under external stress. The piezoelectricity of wood, whose main component is cellulose, has been investigated for decades.

Maiti *et al.*[8] made a piezoelectric nanogenerator based on onion skin.[8] The onion skin consists of α-cellulose, carbonyl, carboxyl, and amino

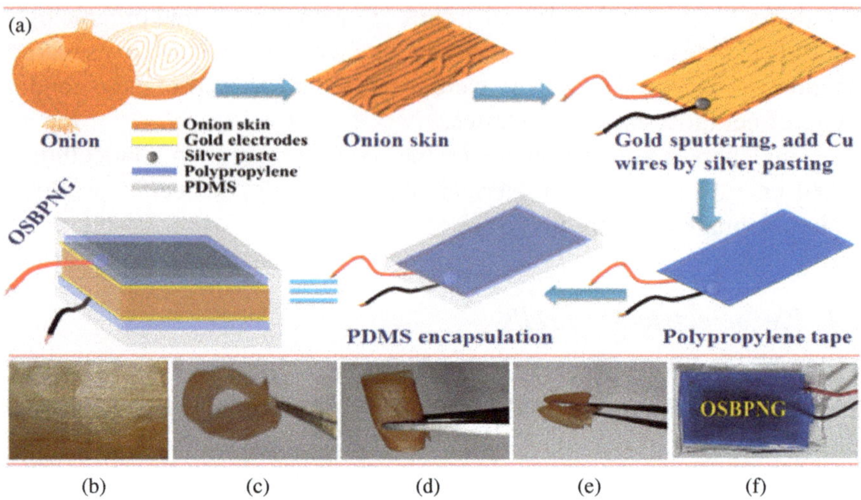

Fig. 1. (a) Schematic of fabrication of OSBPNG with cross-section view. (b) Photograph of onion skin. The demonstration of flexibility of OS during (c) bending, (d) rolling, and (e) twisting, respectively. (f) Photograph of OSBPNG.[8]

groups, as well as *N*-containing bio-ligands and quercetin as plant pigment (flavonoid). Fig. 1a gives a schematic of detailed fabrication and cross-sectional view of onion skin bio-piezoelectric nanogenerator (OSBPNG) device. The OSBPNG was fabricated with no further purificatory onion skin. The onion skin was then cut into rectangles (30 mm × 20 mm) and placed in an air oven at 40°C to remove moisture. Both sides of the onion skin were sputtered by gold (Au) coating, and then connected by two copper wires on the upper and lower surfaces. The entire device was wrapped with polypropylene tape to avoid any triboelectric effect from single electrode. Finally, OSBPNG was encapsulated by polydimethylsiloxane (PDMS) to protect the device from external environmental hazards.

The output voltage and current of OSBPNG were about 18 V and 166 nA, respectively, through the biomechanical force acting vertically on OSBPNG. Thus, high instantaneous power density (≈1.7 $\mu W/cm^2$) with excellent energy conversion efficiency (≈61.7%) of the self-powered highly durable OSBPNG lit up to 30 green LEDs.

The addition of piezoelectric fillers increases the piezoelectricity of the material. Zinc oxide (ZnO) is a well-known material for piezoelectric applications. Gullapalli *et al.*[9] fabricated a flexible piezoelectric nanocomposite material for strain sensing with ZnO nanostructures embedded in a stable matrix of paper (cellulose fibers) applied for self-powered sensors.[9] The ZnO paper was fabricated with sol-gel synthesis method. Sol A was prepared by dissolving zinc acetate dihydrate solution in ethanol, and sol B was prepared by dissolving NaOH. The paper was soaked with sol A dropwise, dried in an oven, then soaked with sol B, and dried again. Several repetitions of the above procedure with occasional washing with distilled water to remove any by-products resulted in uniform coating of ZnO crystals over individual fibers of the cellulose paper. The composite sensor can be molded as sensor devices, and they show characteristics of good sensitivity and linearity at different frequencies.

The ZnO/cellulose hybrid paper can also be applied to flexible piezoelectric nanogenerators (PENGs). Zhang *et al.*[10] designed and fabricated flexible V-ZnO/bacterial cellulose (BC) hybrid paper-based PENGs.[10] According to the process shown in the Fig. 2, zinc oxide is formed on a

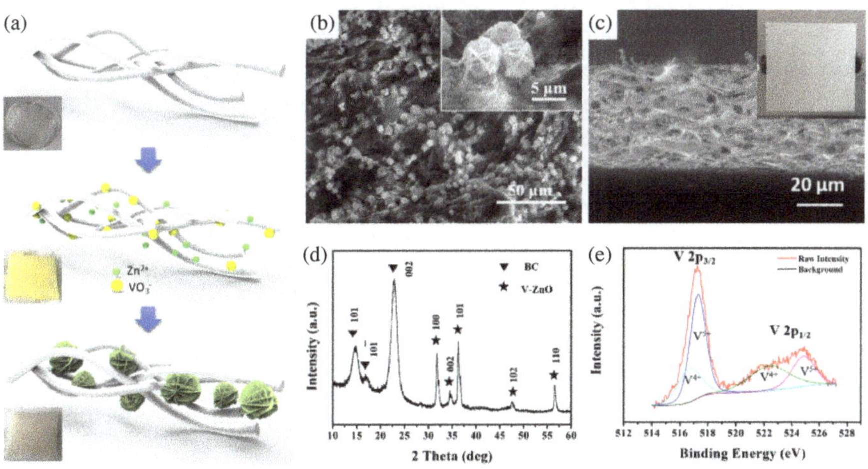

Fig. 2. (a) Fabrication process of the V-ZnO/BC hybrid paper. The insets show the optical image of the BC membrane at different stages. (b) Inside morphology of the V-ZnO/BC hybrid paper. The inset shows the magnified structure of V-ZnO microflowers. (c) Cross-sectional view of the V-ZnO/BC hybrid paper. The inset shows the optical image of the hybrid paper. (d) XRD pattern of the V-ZnO/BC hybrid membrane from which ZnO and BC can be indexed. (e) XPS spectrum of the V-ZnO confirming the presence of V^{5+} and V^{4+} in the hybrid paper.[10]

single cellulose fiber. Purified BC membranes were cut into rectangular pieces and immersed in nutrient solution containing zinc acetate and V_2O_5 for 24 hours to let the Zn^{2+} and V_2O_5 penetrate into the network matrix. Then ZnO/bacterial cellulose (BC) hybrid paper was prepared by hydrothermal reaction. Finally, V-ZnO/BC hybrid paper was spin-coated with a thin layer of PDMS on both sides to provide a smooth surface. Au electrodes were deposited on both sides followed by attaching two stripes of conductive tape on the electrodes. The device was poled with the electric field of 100 kV/cm at room temperature for 12 hours. The output voltage and current density can reach 1.5 V and 80 nA/cm^2, respectively. This possesses superior mechanical robustness and stability. The output performance demonstrates no obvious decline after 10000 bending/releasing cycles.

$BaTiO_3$ is one of the most attractive material owing to its lead-free property as well as high piezoelectricity. However, as a major problem for practical application, these bulk piezoelectric ceramics are intrinsically very brittle and thus hardly compatible with the irregular mechanical deformations. Zhang *et al.*[11] fabricated a piezoelectric paper based on $BaTiO_3$ (BTO) nanoparticles and bacterial cellulose (BC) and made a piezoelectric nanogenerator with the piezoelectric paper.[11] The preparation method was a very simple and environmentally-friendly mixing method, and then vacuum filtration was used to form a film. As shown in Fig. 3, the BTO/BC piezoelectric paper-based nanogenerator (NG) shows outstanding output performance with open-circuit voltage of 14 V and short-circuit current density of 190 nA cm^{-2}. The BTO/BC piezoelectric paper can also drive a liquid crystal display screen when it is in bending condition. The flexible BTO/BC piezoelectric paper-based NG holds great promises for achieving wearable or implantable energy harvesters and self-powered electronic devices.

Research on porous piezoelectric materials has been attracting people's attention. Porous piezoelectric material is a porous polymer film with distributed charge on the inner surface of the cavity. The porous piezoelectric electret exhibits piezoelectricity macroscopically, and its piezoelectric coefficient is comparable to that of conventional piezoelectric ceramics. Zheng *et al.*[12] fabricate high-performance flexible piezoelectric nanogenerators (NGs) using porous cellulose nanofibril (CNFs)/polydimethylsiloxane (PDMS) aerogel film.[12] As shown in Fig. 4, the porous CNF/PDMS aerogel film was prepared by coating a layer of PDMS on the porous surface of a compressed CNF aerogel film produced

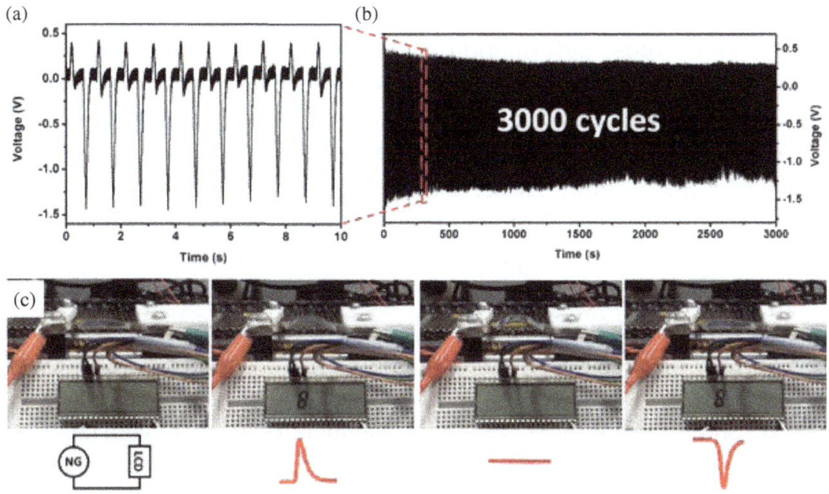

Fig. 3. (a) Output voltage of the BTO/BC piezoelectric paper based NG under the bending frequency of 1 Hz. (b) Cyclic bending test for 3000 cycles of bending/releasing motions. (c) Four states when a commercial LCD screen is driven by the NG under bending conditions spectra (measured in attenuated mode) of native cellulose.[11]

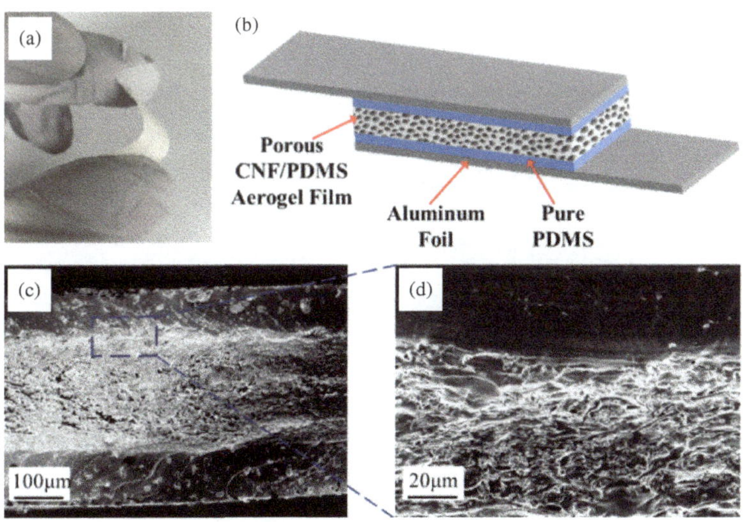

Fig. 4. (a) Photograph of the flexible porous CNF/PDMS aerogel film. (b) A schematic diagram of the flexible porous CNF/PDMS aerogel film-based NG device with a sandwich structure. Cross-sectional SEM images of a porous CNF/PDMS film-based NG at (c) a low magnification and (d) a high magnification.[12]

via an environmentally-friendly freeze-drying process. Under periodic external mechanical deformation by an oscillator, the resulting flexible porous CNF/PDMS aerogel film-based NGs exhibited very stable and high output piezoelectric signals, namely, an open-circuit voltage (V_{oc}) of 60.2 V, a short-circuit current (I_{sc}) of 10.1 μA, and a corresponding power density of 6.3 mW/cm^3. The electric power generated by these NGs was able to directly turn on 19 blue light-emitting diodes (LEDs) and charge a capacitor up to 3.7 V. The porous CNF/PDMS film-based NG may offer many promising applications in flexible electronics and biocompatible self-powered systems.

Rajala *et al.*[13] used self-standing films of native cellulose nanofibrils (CNFs) to fabricate a sensing device.[13] Cellulose fibers obtained from birch wood by sulfate digestion and subsequent bleaching were suspended in water and then processed through a grinding process. It formed a highly viscous aqueous dispersion. This was followed by pressure filtering (15–30 minutes) followed by pressing and drying in a hot-press at 100°C for 2 hours to fabricate a self-standing CNF film. The CNF and PVDF sensor assembly is described in Fig. 5. The piezoelectric sensitivity values are presented as mean sensitivity standard deviations for each sensor side. The average sensitivity value for the four CNF sensors was (5.7 ± 1.2)

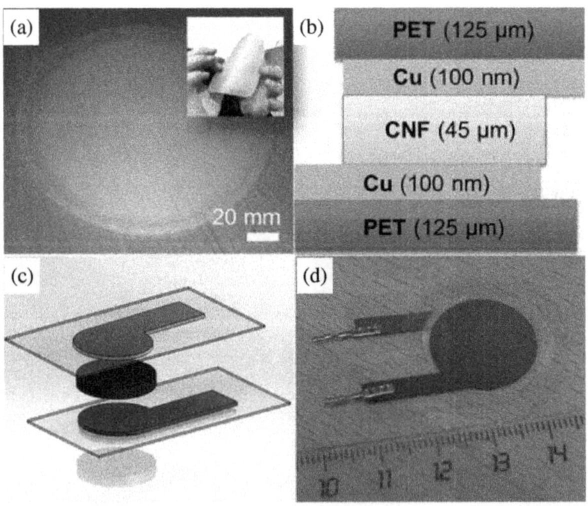

Fig. 5. (a) Photographs of a fabricated self-standing CNF film and its bending robustness (an insert). (b, c) Schematic side view and (d) a photograph of assembled sensor.[13]

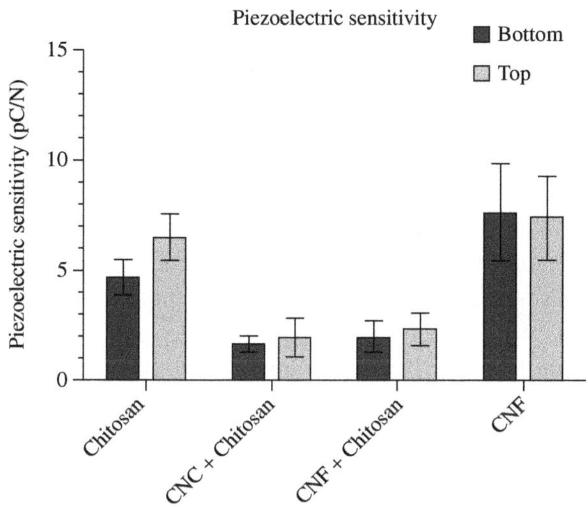

Fig. 6. The mean piezoelectric sensitivities (n = 20) and their standard deviations measured from both sides of the solvent casted films.[14]

pC/N, which was lower than the sensitivity of the PVDF sensor (27.5 ± 2.6) pC/N.

Hänninena *et al.*[14] fabricated free-standing biodegradable piezoelectric films from chitosan, nanocellulose, and their blends. The piezoelectric sensitivity values of that blend films consisting of chitosan and nanocellulose was ~2 pC/N, which was lower than the plain films of chitosan (~4–6 pC/N) and CNF (~7–8 pC/N) (Fig. 6).[14] The piezoelectric chitosan and nanocellulose films could provide biodegradable alternatives for commercial PVDF films. Even though the sensitivities of the bio-based films are somewhat lower, they could be suitable for low-cost large-area sensor applications, such as floor sensors.

1.2 *Loudspeaker*

Loudspeaker is an energy conversion device that could convert an electronic audio signal into an acoustic signal. Though the weakest devices in audio equipment, loudspeaker is of great importance to achieving superb acoustics. Audio electrical energy produces sound through electromagnetic, piezoelectric, or electrostatic effects that cause the basin or diaphragm to vibrate and resonate with the surrounding air. Membrane is the most important component of loudspeaker; requirements for materials to be used for loudspeaker membranes are high modulus, high internal loss, low density, and high sound propagation velocity. Low density plays an important role in achieving good high-frequency response, and high modulus is indispensable for effective low-frequency response. Cellulose is considered to be an extremely ideal material for loudspeaker membrane due to the fact that it can be processed into a lightweight membrane with a relatively high strength through vacuum filtration and other process.

Bacterial cellulose (BC) films have been studied as membranes of loudspeakers and are used in a few commercial products. In the study by Nishi *et al.*[15] the BC sheets showed a modulus of 30 GPa and the sound propagation velocity was estimated to be about 5000 m/s; the damping was as high as 0.04, comparable to that of cone paper, the conventional membrane material. The applicability of bacterial cellulose to diaphragms of electroacoustic transducers is discussed. Ciechanska *et al.* prepared different films of modified BC with modulus above 4 GPa and sound propagation velocity above 2000 m/s; the materials were tested as loudspeaker membranes and were reported to have acoustic properties comparable to titanium diaphragms, although the specific "metallic" sound was absent.

Traditional loudspeakers have a voice coil bonded to a geometrically shaped acoustic membrane, which is suspended in the magnetic field of a bulky permanent magnet. Membranes with the combination of magnetic and mechanical functionality may make the external magnet unnecessary, thus contributing to the construction of a novel super-thin loudspeaker prototype. Galland *et al.*[16] attached ferrite to cellulose nanofibrils by *in situ* precipitation of metal precursors, and then the membrane was prepared by vacuum filtration of the hybrid nanofibril suspensions. An ultrathin prototype loudspeaker was made, using the 60 wt% ferrite-loaded membrane mounted in a poly(methyl-methacrylate) baffle plate and with a thin copper coil (75 mm inner diameter) leveled in line with the membrane. The acoustic performance in terms of output sound pressure was of good quality, with reasonably flat frequency response in the audible range from 20 Hz to 15 kHz. It is worth mentioning that the concept established here shows great potential in small loudspeakers, hearing aids, and other noise reduction technologies.

1.3 *Antenna*

Antenna is a kind of convertor that converts the guided waves into the electromagnetic waves, or vice versa. As an essential component for radio equipment to transmit or receive information signals, antenna is widely used in electronic devices such as smart phones and laptop computers.

With the development of electronic devices toward miniaturization, electrical components are required to be lighter and thinner to meet the requirement of portability. Flexible antennas have already been fabricated using liquid metals enclosed in elastomeric plastic substrates, but all of them are not foldable. Paper is the only candidate substrate for foldable antennas because of its high foldability. A lot of research groups have already reported on paper antennas, but none of these antennas are foldable, and high surface roughness reduces the sensitivity. By reducing the diameter of microsized pulp fibers to the scale of nanosize, the surface of the obtained nanopaper would become smooth. Nogim *et al.*[17] prepared pulp paper and nanopaper by filtration of their water suspensions, then V-shaped antenna patterns with length of 30 mm were mask printed with silver nanowires inks on the surface. Compared with the pulp paper, nanopaper antennas possessed much smoother surface and exhibited higher sensitivity that was equivalent to that of a plastic antenna. In addition, the suitability for foldable antenna devices was assessed. The resonance peaks of the antennas can be adjusted by varying the length of the antenna lines. In traditional antenna devices, this length could not be changed after the fabrication of the antenna patterns, and the resonance peaks were tuned by using additional components such as condensers and coils. When the resonance peaks are tuned by folding the antennas, no additional components are necessary, and the resulting electronic devices will thus be lighter and smaller. Because the silver nanowire lines on the plastic substrates were broken during the folding process, the resonance peaks did not return to their original frequency bands, even when the substrates were recovered to their original shapes. However, the nanopaper antennas can be controlled to transmit and receive signals by folding and unfolding, and they transmitted and received multiple frequency band signals without using any other components (Fig. 7).

Conventional electronic manufacturing processes consume too much time and energy, thus increasing the cost of the fabrication of electronic devices. Printed electronics have garnered much attention since they offer a lower cost alternative to conventional fabrication processes while

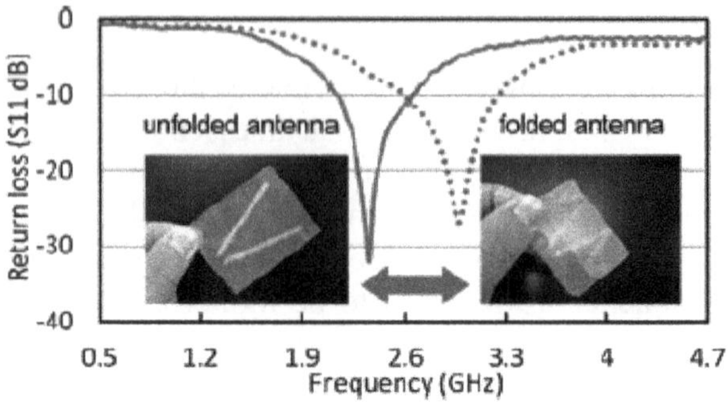

Fig. 7. Return losses of nanopaper antennas (solid line) and folded antennas (dotted line).[17]

enabling large-scale manufacturing of electronic devices. Flexible and biodegradable cellulose-based paper with excellent printability is a suitable candidate for printed substrates. However, since cellulose is hydrophilic, their poor shape stability in solution seriously restricts their application in solution-based printing processes, including gravure, screen printing, and inkjet printing. In order to overcome this obstacle, Zhu *et al.*[18] proposed that (TEMPO)-oxidized cellulose nanofibrils (TOCN) nanopaper cross-linked by glutaraldehyde could show improved shape stability in water, because glutaraldehyde can block the exposure of hydroxyl groups to the solvent. Then, a device consisting of RF components for RFIDs were fabricated by printing silver nanoparticle ink onto the stable, transparent nanopaper *via* gravure printing. During the production process, the shape of the nanocellulose paper was very stable and no wrinkles occurred. This work presents an important advancement towards the practical use of a transparent nanopaper substrate for printable and flexible electronics.

With the emergence of wearable electronics such as wearable computers, foldable electronics, and healthcare sensors, foldable and flexible antenna is urgently required for future devices, which would enable the devices to be compact in size and become portable. The development of highly foldable antenna components with high sensitivity is a major issue for future electronics.

1.4 *Phototransistor*

Transistors are fundamental building blocks of various electronic devices and have attracted significant attention in recent years due to their excellent semiconductor multifunction. Solar energy, as a clean renewable energy, has attracted considerable attention for their extremely abundant supply. The energy of the sun illuminated on the earth for only about 40 minutes is enough for global human energy consumption for one year. Therefore, it is of great significance to develop and utilize this green energy.

Phototransistor is essentially a light-sensitive Field Effect Transistor (FET) that transduces incoming photo energy to electrical current. Conventional electronics are made with a glass or plastic substrate. However, when devices are subjected to bending deformation, the appearance of cracks in brittle glass substrates would dramatically reduce conductivity, leading to severely degraded performance. Meanwhile, low processing temperature and environmental pressure limit the wide appliance of plastic substrates. Therefore, cost-efficient, renewable, and flexible paper is an excellent candidate to replace glass and plastic as a substrate for flexible electronics, especially cellulose-based nanopaper with excellent optical and mechanical properties.

In recent years, phototransistors based on two dimensional (2D) materials have become increasingly popular as a semiconducting channel for high-performance field effect transistors, especially typical semiconducting tungsten disulfide (WS_2) and molybdenum disulfide (MoS_2), because of their intriguing electrical and optical properties.[19–22] Many groups have reported on MoS_2-based phototransistors fabricated on Si/SiO_2 substrates, which were neither flexible nor environmentally friendly. Accordingly, it is of great importance to search for bio-based substrates to replace them. Zhang *et al.*[20] reported on the first MoS_2-based phototransistor gated through gel-electrolyte, highly transparent, and flexible cellulose nanopaper as both supporting substrate and passivation layer to sandwich the MoS_2 channel and the gel-electrolyte in between (Fig. 8). Excellent photo responsivity of ~1.5 kA/W under 10 nW illumination power was achieved, much higher than typical backgated devices reported in the previous studies. Besides excellent photoresponsivity, optical transmittance of 82% over a wide range of wavelengths and excellent stability under sustained voltage bias and illumination were achieved at the same time.

Solution-based printing is an ideal process due to its ultrahigh fabrication speed, which would largely promote the development of printable electronics. Substrates play an important role in the printing process for printable devices and greatly impact the mechanical performance and

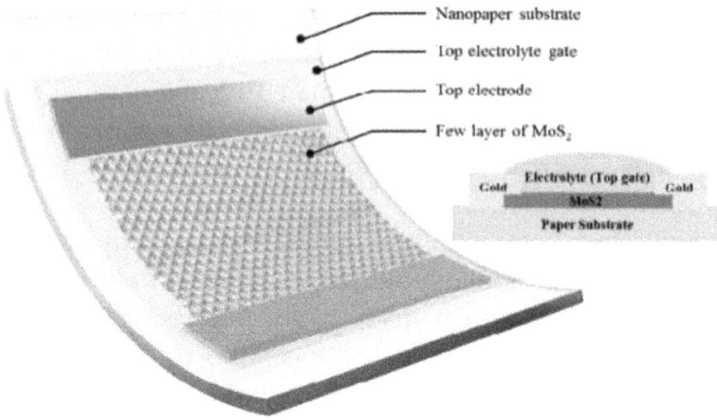

Fig. 8. Three-dimensional schematic and cross-sectional view of a fabricated flexible nanocellulose paper phototransistor.[20]

sometimes the optical and electrical properties of devices. Excellent hygroscopicity and porous nature make cellulose paper have better ink adsorption property compared with conventional plastic and glass substrates. Huang *et al.*[23] prepared highly transparent and surface-smooth CNF nanopaper by vacuum filtration of the CNF solution, and then flexible OFETs were successfully fabricated on the designed nanopaper.[23] The nanopaper OFETs exhibited good transistor electrical characteristics and optical transmittance up to 83.5%. It is believed that the fibrous structure of nanopaper can effectively release bending strain, thus providing devices with good flexibility. Only a 10% decrease in mobility was observed when the nanopaper transistors were bent and folded. These excellent optical, mechanical, and electrical properties suggest the great potential of cellulose-based nanopaper as a substrate of new-generation of green electronics. Although CNF nanopaper substrate can achieve high transparency, high optical haze still limits their application, which is caused by high diffuse transmittance and small direct transmittance. Xu *et al.*[24] developed a facile methodology to synthesize highly transparent, mechanically robust, hybrid cellulose nanopaper with a wide range of optical properties, consisting of long CNFs and short CNCs.[24] Optical properties of the nanopaper can be tailored by varying the ratio between CNFs and CNCs. The concentration of CNCs affected the surface roughness, porosity, and CNF entanglement of the nanopaper, resulting in varied optical properties.

With the development of integrating and intelligent wearable electronic devices in recent years, developing flexible electronics has become a trend. However, there can be tremendous challenges with integrating devices on nanopaper consisting of cellulose fibers with large diameter, conventional glass items, and plastic substrates due to its shape stability during processing. Flexible nanopaper based on nanosized cellulose fibers in a promising candidate for the fabrication of next-generation flexible organic field-effect transistors (OFETs).

1.5 *Organic Light-Emitting Diodes* (*OLEDs*)

Organic light-emitting diodes (OLEDs) are widely used in flexible lighting and display technology, and they have been intensively researched since they were first reported by Tang and Vanslyke in 1987,[25] due to their attractive advantages such as low electrical consumption, self-luminance, wide viewing angle, and fast time response. As mentioned above, phototransistors could convert incoming photo energy to electrical current; the biggest difference between phototransistors and light-emitting diodes lies in that the latter supplied electrical energy into photo energy.

Traditional rigid glass or flexible plastics have been widely used as substrates in OLEDs. However, the brittleness of glass substrates makes them unsuitable for flexible applications, which is strongly required by many OLEDs. Most plastics such as polyethylene terephthalate (PET), polycarbonate (PC), and polyimide (PI) have large coefficient of thermal expansion (CTE), which would damage the functional layer under high processing and operating temperature; thus, the ordinary thermal stability of plastic substrates cannot be ignored. In addition, synthetic plastics are derived from petroleum, and it takes decades, even hundreds of years, for them to decompose; therefore, they cannot satisfy the requirement of renewability of sustainable development. These disadvantages of existing substrates made it an increasingly attractive option to produce the substrates from bioresources. Compared with glass and plastic substrates, unique features of papers, such as excellent thermal stability, cost efficiency, low-weight, flexible, recyclability, and large-scale fabrication, enable them to be more attractive candidates to use as substrates of OLEDs, and they hold promise to replace conventional glass and plastic substrates. Among them, cellulose-based nanopaper received special interest due to its mechanical and optical properties.

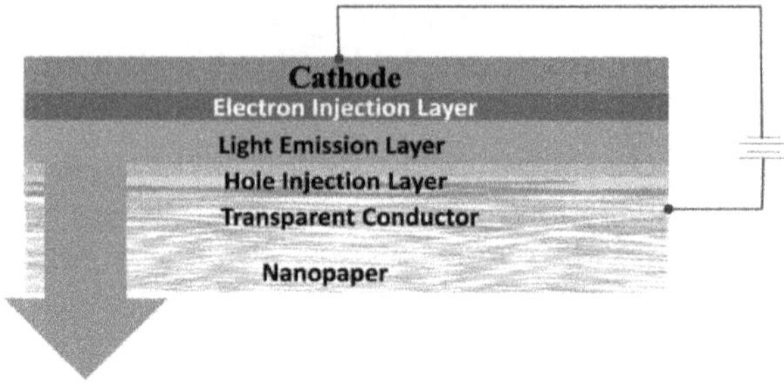

Fig. 9. Schematic drawing of a nanopaper OLED.[26]

Organic light-emitting diodes (OLEDs) fabricated on recyclable and biodegradable substrates are a step towards the realization of a sustainable OLED technology. Several studies have been reported on the construction and fabrication of organic light-emitting diodes devices with cellulose-based materials as substrates. For example, Zhu *et al.*[26] successfully prepared highly flexible OLED device on biodegradable TOCN nanopaper substrate.[26] As shown in the schematic drawing of prepared nanopaper-OLED device (Fig. 9), three functional layers, including a light emitting layer of green polyfluorene, electron injection layer, and hole injection layer, were set between the anode (carbon nanotube) and cathode (aluminum). It was demonstrated that the fibrous structure of nanopaper could release the stress and avoid cracking when devices suffer from bending, leading to stable electrical and optical performances, whereas this is difficult to achieve on plastic substrate flexible electronics. At the same time, high optical transmittance and low surface roughness required by OLED devices were obtained.

Conventional OLEDs are often bottom-emitting, with the generated light emitting through the semi-transparent anode and transparent substrate, which would inevitably lead to undesirable waveguide losses and light-trapping. Top-emitting devices with inverted layer sequence may overcome this obstacle and seem to be preferable in this case. Najafabadi *et al.*[27] reported on efficient inverted top-emitting green electro-phosphorescent OLEDs on free-standing transparent cellulose nanocrystal (CNC) substrates[27]. A maximum luminance of 74591 cd/m^2 with a current efficacy of 53.7 cd/A at a luminance of 100 cd/m^2 and 41.7 cd/A at 1000 cd/m^2 were achieved, comparable to the average performance achieved on glass substrates. It is worth mentioning that CNC substrates can dissolve in water; thus, constituent organic semiconducting materials and metal layers of abandoned OLEDs can be filtered out and recycled, which would relieve environmental stress.

Therefore, highly flexible organic light-emitting diodes based on the biodegradable substrates are paving the way for next-generation green and flexible electronics.

1.6 *Touch Screen*

With the development and popularization of computer technology, there is a promising human–computer interaction technology that has emerged since the early 1990s, namely touch screen technology. Without fussy operation of keyboard and mouse, instructions could be issued by just clicking a little on the icon or text on the display screen with this touch screen technology, making human–computer interaction more straightforward and convenient. It is worth mentioning that this intriguing technology has been widely used in many fields such as mobile phones, tablet computers, public information inquiries, and car display screens, and they have broad market prospects in the near future.

Silicon dioxide has long been the traditional material used for touch screen due to the extremely low absorption and super high transmittance. When SiO_2 is forged into crystal or glass, it is rigid at room temperature, which impedes the formation of new compelling flexible display devices. In the coming decades, flexible devices will be thoroughly developed with the integration of advanced electrical and optical functions. Plastic has been studied as a replacement for SiO_2 in flexible optical devices over the past decades, but plastic has intrinsic problems such as low thermal stability, poor accessibility for functionalization, and harm to environment. Touch screens prepared from nanocellulose substrates composites will receive more and more attention due to their advantages of flexibility, degradability, and renewability.

Generally, the four-wire analogue resistive touch screen is composed of two layers: a PET film and a glass substrate, both coated with a transparent conductor. Fang *et al.*[28] designed and assembled a four-wire analogue resistive paper touch screen with transparent and conductive two-layer paper as the top transparent electrode, one layer is pure CNF film, which can provide nano-level surface roughness, and the other layer is a mixture of ordinary wood fiber and CNF as the supporting layer. Touch screen was prepared by replacing the top indium tin oxide (ITO)-coated PET film with CNT-coated high transparent paper. A word "paper" was written on the prepared touch screen, and this was successfully displayed on the computer *via* a USB connection, which implies the PET film usually used in touch screen may be replaced by transparent and conductive hybrid cellulose paper to fabricate transparent electrodes for electronic devices. Because the relatively rough surface on the top of the double-layer paper can scatter incident light to different surfaces, the haze of the mixed paper and the anti-reflective performance of the touch screen are better than that of the PET-based touch screen, which is more suitable for outdoor electronic products and shows its great potential in the field of touch screen.

In addition, the optical clarity is critical in displays, and this requires a haze value (indication of optical scattering) less than 1%; however, the nanopaper demonstrated with cellulose nanofibers exhibits a haze value of around 10%, which is too high for high-definition displays. Zhu *et al.*[29] prepared a super-clear nanocellulose paper with a similar clarity as high-quality display glass and plastics, light transmittance greater than 90%, and haze less than 1%, which is required by touch screen. And the paper surface was covered with single-layer graphene by heat transfer method to endow it with conductivity. A commercially competitive capacitive-based multi-point touch screen was fabricated with clear paper as a replacement for plastic substrates; the sensitivity of the touch screen was comparable to that of traditional commercial touch screen. Clear paper is mesoporous with simultaneous nanometer-scale smoothness, the smooth surface of the clear paper leads to an excellent performance of the graphene-on-paper, with a sheet resistance of 445 Ω/sq and transmittance of 89% in the visible range.

Recently, in order to overcome the flammability of nanocellulose, the abundantly available plant fiber and natural clay were used as raw materials and the hydrophilic and lipophilic properties of nanocellulose made it suitable for use as a dispersing agent for dispersing and stripping natural clay, which overcame the defects of flocculation of monolayer clay.[30] Then they obtained a stable and uniformly dispersed monolayer clay dispersion, which overcame the problem of low transmittance of clay composite films. Thus, transparent and flexible nanocellulose/clay nanocomposite substrates with excellent self-extinguishing behavior were successfully prepared.

2. Conclusions

Against the backdrop of shortage of resources and environment pollution, the construction of materials based on biomass resources is of great significance for sustainable development. Nanocellulose-based materials hold great potential in the field of energy conversion due to their unique structure and properties of microporous structure, high transparency, high strength, and low density. Although great progress has made in electronics, incorporating nanocellulose-based materials, there are still many challenges in the fabrication and practical utilization of devices. The emergence of intelligent devices sets higher requirements for materials, and the performances of nanocellulose-based energy conversion devices still need to be enhanced in the future.

Acknowledgments

This work was supported by the National Natural Science Foundation of China (No. 51703177, 21704079), the Fundamental Research Funds for the Central Universities (WUT: 2018III009, 2018IVB022, 2018IVB041), and Key Laboratory of Processing and Quality Evaluation Technology of Green Plastics of China National Light Industry Council, Beijing Technology and Business University (No. PQETGP2019007).

References

[1] Gardner, G.T. and Stern, P.C. (1996). *Environmental Problems and Human Behavior*. Boston, MA: Allyn & Bacon.
[2] Cook, S.W. and Berrenberg, J.L. (1981). Approaches to encouraging conservation behavior: A review and conceptual framework. *J. Soc. Issues*, 37(2): 73–107.
[3] TobjÖRk, D. and Ouml, R. (2011). Paper electronics. *Adv. Mater.*, 23(17): 1935–1961.
[4] Nogi, M., Iwamoto, S., Nakagaito, A., *et al.* (2009). Optically transparent nanofiber paper. *Adv. Mater.*, 21(16): 1595–1598.
[5] Okahisa, Y., Yoshida, A., Miyaguchi, S., *et al.* (2009). Optically transparent wood–cellulose nanocomposite as a base substrate for flexible organic light-emitting diode displays. *Compo. Sci. and Technol.*, 69(11–12): 1958–1961.
[6] Moon, R., Martini, A., Nairn, J., *et al.* (2011). Cellulose nanomaterials review: Structure, properties and nanocomposites. *Chem. Soci. Rev.*, 40(7): 3941–3994.

[7] Alam, M.M. and Mandal, D. (2016). Native cellulose microfiber-based hybrid piezoelectric generator for mechanical energy harvesting utility. *ACS Appl. Mater. Interfaces*, 8: 1555–1558.

[8] Maiti, S., Karan, S.K., Lee, J., *et al.* (2017). Bio-waste onion skin as an innovative nature-driven piezoelectric material with high energy conversion efficiency. *Nano Ener.*, 42: 282–293.

[9] Gullapalli, H., Vemuru, V.S.M., Kumar, A., *et al.* (2010). Flexible piezoelectric ZnO-paper nanocomposite strain sensor. *Small*, 6(15): 1641–1646.

[10] Zhang, G.J., Liao, Q.L., Ma, M.Y., *et al.* (2018). Uniformly assembled vanadium doped ZnO microflowers/bacterial cellulose hybrid paper for flexible piezoelectric nanogenerators and selfpowered sensors. *Nano Ener.*, 52: 501–509.

[11] Zhang, G.J., Liao, Q., Zhang, Z., *et al.* (2016). Novel piezoelectric paper-based flexible nanogenerators composed of $BaTiO_3$ nanoparticles and bacterial cellulose. *Adv. Sci.*, 3: 1500257.

[12] Zheng, Q., Zhang, H., Mi, H., *et al.* (2016). High-performance flexible piezoelectric nanogenerators consisting of porous cellulose nanofibril (CNF)/poly(dimethylsiloxane) (PDMS) aerogel films. *Nano Ener.*, 26: 504–512.

[13] Rajala, S., Siponkoski, T., Sarlin, E., *et al.* (2016). Cellulose nanofibril film as a piezoelectric sensor material. *ACS Appl. Mater. Int.*, 8(24): 15607–15614.

[14] Hänninena, A., Sarlinb, E., Lyyra, I., *et al.* (2018). Nanocellulose and chitosan based films as low cost, green piezoelectric materials. *Carbohydr. Polym.*, 202: 418–424.

[15] Nishi, Y. and Uryu, M. (1990). The structure and mechanical properties of sheets prepared from bacterial cellulose. *J. Mater. Sci.*, 25: 2997–3001.

[16] Galland, S., Andersson, R.L., Salajkova, M., *et al.* (2013). Cellulose nanofibers decorated with magnetic nanoparticles-synthesis, structure and use in magnetized high toughness membranes for a prototype loudspeaker. *J. Mater. Sci. Technol.,* 1(47): 7963–7972.

[17] Nogi, M., Komoda, N., Otsuka, K., *et al.* (2013). Foldable nanopaper antennas for origami electronics. *Nanoscale*, 5(10): 4395–4399.

[18] Zhu, H., Narakathu, B., Fang, Z., *et al.* (2014). A gravure printed antenna on shape-stable transparent nanopaper. *Nanoscale*, 6(15): 9110–9115.

[19] Xu, M., Wu, J., Feng, Q., *et al.* (2014). High responsivity and gate tunable graphene-MoS_2 hybrid phototransistor. *Small*, 10(11): 2300–2306.

[20] Zhang, Q., Bao, W.Z., Gong, A. *et al.* (2016). A highly sensitive, highly transparent, gel-gated MoS_2 phototransistor on biodegradable nanopaper. *Nanoscale*, 8(29): 14237–42.

[21] Gong, F., Luo, W., Wang, J. *et al.* (2016). High-sensitivity floating-gate phototransistors based on WS2 and MoS_2. *Adv. Funct. Mater.*, 26: 6084–6090.

[22] Choi, W., Cho, M., Konar, A. *et al.* (2012). High-detectivity multilayer MoS_2 phototransistors with spectral response from ultraviolet to infrared. *Adv. Mater.*, 24(43): 5832–5836.
[23] Huang, J., Zhu, H.L., Chen, Y.C. *et al.* (2013). Highly transparent and flexible nanopaper transistor. *ACS Nano.*, 7(3): 2106–2113.
[24] Xu, X.Z., Zhou, J., Jiang, L. *et al.* (2016). Highly transparent, low-haze, hybrid cellulose nanopaper as electrodes for flexible electronics. *Nanoscale*, 8(24): 12294–306.
[25] Tang, C. and VanSlyke, S. (1987). Organic electroluminescent diodes. *Appl. Phys. Lett.*, 51: 913–915.
[26] Zhu, H.L., Xiao, Z.G., Liu, D.T. *et al.* (2013). Biodegradable transparent substrates for flexible organic light-emitting diodes. *Ener. Environ. Sci.*, 6(7): 2105–2111.
[27] Najafabadi, E., Zhou, Y.H., Knauer, K.A., *et al.* (2014). Efficient organic light-emitting diodes fabricated on cellulose nanocrystal substrates. *Appl. Phys. Lett.*, 105(6): 063305.
[28] Fang, Z., Zhu, H., Preston, C., *et al.* (2013). Highly transparent and writable wood all-cellulose hybrid nanostructured paper. *J. Mater. Chem. C.*, 1 (39): 6191–6197.
[29] Zhu, H., Fang, Z., Wang, Z. *et al.* (2016). Extreme light management in mesoporous wood cellulose paper for optoelectronics. *ACS Nano.*, 10(1): 1369–1377.
[30] Ming, S., Chen, G., Yan, J., *et al.* (2018). Preparation of nanofibrillated cellulose dispersed monolayer clay nanoplatelet dispersion with high stability and its application in transparent flexible films. *Acta Mater. Compo. Sinica*, 35(9): 2573–2578.

https://doi.org/10.1142/9781786349477_0014

Chapter 14

Nanocellulose-Based Flexible Energy Storage Devices

Yasir Beeran Pottathara[*,¶], Hanuma Reddy Tiyyagura[*], Hideya Kawasaki[†], Zakiah Ahmad[‡], Sabu Thomas[§], and Vanja Kokol[*]

[*]*Faculty of Mechanical Engineering, University of Maribor, Maribor, Slovenia*

[†]*Faculty of Chemistry, Materials and Bioengineering, Kansai University, Suita-shi, Osaka, Japan*

[‡]*Institute for Infrastructure Engineering and Sustainable Management, Universiti Teknologi Mara, Shah Alam, Selangor, Malaysia*

[§]*International and Inter University Centre for Nanoscience and Nanotechnology, Mahatma Gandhi University, Kottayam, Kerala, India*

[¶]*ptyasirbeeran@gmail.com*

Abstract

The demand for bio-based flexible energy devices has increased rapidly in recent decades due to their unique advantages such as flexibility, light weight nature, shape diversity, etc. Nanocellulose has been established as a promising potential candidate for high-performance flexible energy storage device fabrication due to its exceptional physical, mechanical, and optical properties. In this chapter, we discuss about the recent advances in nanocellulose-based flexible energy storage devices for a wide range of applications. Various chemical and physical modification strategies on nanocellulose have also been discussed. The unique properties of nanocellulose enable multiscale structuring of functional composites, which can be tailored to develop new concepts of energy storage devices. This chapter is a comprehensive effort focusing the recent advances on nanocellulose-based flexible energy devices, including supercapacitors, sensors, batteries, and solar cells.

Keywords: Nanocellulose; flexible energy storage; Li-ion battery; supercapacitor; solar cell

1. Introduction

The manipulation of bio-based nanostructured materials offers numerous opportunities that takes the field of nanotechnology research to an advanced level. Cellulose is one of the most abundant organic materials on earth, and it can be obtained from plants. The engineering of cellulose fibers at the nanoscale level opens the door for many promising applications for advanced energy storage systems and devices. Cellulose-based functional energy storage devices are widely reported due to their flexible, cost-effective, and light weight nature in addition to their environmentally friendly nature. Researchers are trying to develop new, versatile, and flexible energy storage materials for multifunctional applications.

Cellulose is obtained from regular wood fibers with a fiber diameter of 20–30 μm. Cellulose-based papers have been successfully employed for the fabrication of numerous flexible energy storage devices due to their outstanding properties such as high degree of porosity, renewability, mechanical strength, and biodegradability. However, the performance of cellulose paper-based energy storage devices is restricted due to the higher thickness of paper, usually greater than 100 μm. In contrast to regular cellulose fibers, nanocellulose fibers offer a diameter of 5–30 nm, which is supposed to act as a more efficient substrate or binder material for energy storage devices due to its nanoscale nature. The advantages of physico-chemical strength [1], light weight, and low cost make nanocellulose a promising material for flexible energy storage device fabrication [2–12]. Nanocellulose was isolated from cellulose by the removal of the defects associated with the hierarchical structure [13]. Generally, the term nanocellulose used to refer to cellulose nanofibrils (CNF), cellulose nanocrystals (CNC), and bacterial nanocellulose (BNC). The typical separation approaches for the isolation of nanocellulose are acid hydrolysis, mechanical treatment, and enzymatic hydrolysis [13, 14]. The nanofibrils are always isolated by the combined effect of mechanical treatment and acid hydrolysis, whereas nanocrystals are generally processed by sulfuric acid hydrolysis [15].

Better mechanical strength and optical transparencies were reported by nanocellulose structures compared to regular cellulose. Usually, nanocellulose cannot be directly employed for the energy storage device fabrication because of their non-electrically conductive nature. So, researchers focused their attention on nanocellulose-based flexible energy storage devices by incorporating cellulose with conducting polymers, metallic particles, conductive carbon materials, and functional and hybrid materials for Li-ion batteries [9, 16–20], supercapacitors [7, 10, 21–29], solar cells [30–33], etc. These composite materials with high-conductivity offer the advantages to nanocellulose and aid in developing high-quality energy storage devices (see Fig. 1).

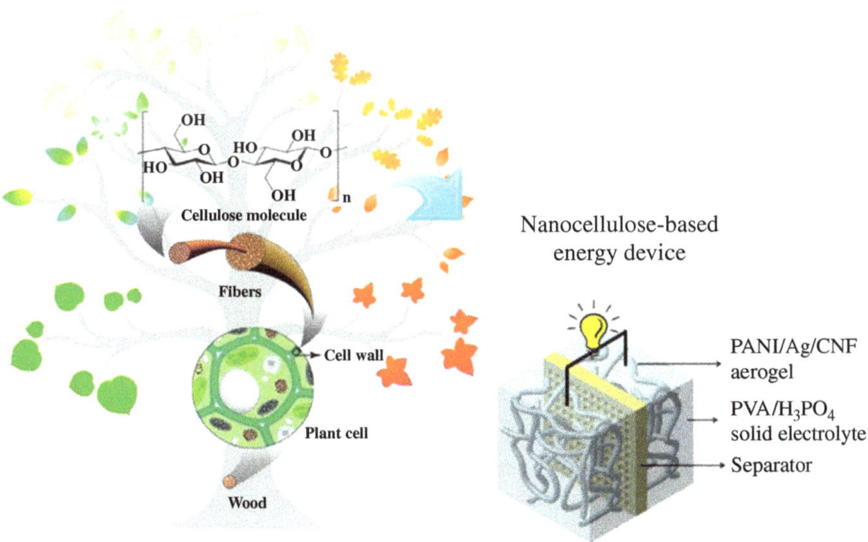

Fig. 1. Schematic illustration of the utilization of nanocellulose for energy storage devices.

Source: Reproduced with permission from Ref [34]. Copyright (2017) Elsevier.

In the subsequent sections, we review the state of the art of nanocellulose-incorporated composites or structures for conducting polymers, conducting carbon materials, metallic particles, and functional hybrids for flexible energy storage devices. We also summarize the properties and performance of nanocellulose-based supercapacitors, solar cells, and Li-ion batteries with a flexible nature. The final section focuses on the challenges and perspectives of nanocellulose-based flexible energy storage devices.

2. Nanocellulose-Based Conducting Composites

To make nanocellulose suitable for energy storage applications, it is mandatory to fabricate conducting composites or hybrids with various conducting materials such conductive polymers, conductive carbon materials, metallic particles, etc. These composite materials offer the properties of both components. Mainly, these conductive composites or hybrids are fabricated by either coating a conductive layer on the surface of nanocellulose substrates or mixing the conductive materials with nanocellulose. *In situ* polymerization and blending are the major mixing strategies for making conductive nanocellulose composites. Nanocellulose composites with conducting polymers are usually fabricated by *in situ* polymerization, whereas conductive carbon materials and metallic particles use a combination of nanocelluloses obtained by the blending process. The following section discusses the basics of conducting materials and their advantages as well as disadvantages for incorporating nanocellulose in flexible energy storage devices.

2.1 *Nanocellulose with Conducting Polymers*

Electronically conducting polymers, polymers with highly *p*-conjugated polymeric chains, such as polyaniline (PANI), polypyrrole (PPy), polyacetylene (PA), poly(phenylenevinylene) (PPV), and polythiophene (PTP), are particularly incorporated with nanocellulose materials to make them conducting [3, 25, 35–37]. The major drawback of these conductive polymers are their poor mechanical integrity, low solubility, and difficulties in the manufacturing processes. The interaction between the functional groups in conductive polymers and hydroxyl and carboxylate moieties on the nanocellulose surface provide good affinity for the composite formation. This affinity between nanocellulose and the conducting polymer offers stability, flexibility, and mechanical strength in the ultimate composite. *In situ* polymerization is generally applicable for fabricating conductive composites of nanocellulose with conducting polymers where nanocellulose is considered as a matrix. Deposition or coating techniques such as sputtering or vapor deposition are not suitable for making conductive nanocellulose composites due to the thermal degradation of the conductive polymer at evaporation temperature. Also, it is hard to find a good solvent for most of the conductive polymers to suit the option of other simple types of coating techniques.

2.2 *Nanocellulose with Metallic Particles*

Metallic particles have higher intrinsic conductivity, which makes them promising for making composites through coating or bonding with nanocellulose for energy storage applications [10, 23, 26, 27]. Metallic particles such as gold (Au), silver (Ag), copper (Cu), and tin (Sn) were widely reported for use with nanocellulose for high-conductivity applications. Apart from coating and bonding, other fabrication techniques were reported to incorporate metallic particles into nanocellulose, such as solvent exchange [38], doping [39], co-precipitation method [40], and pressured extrusion paper-making technique [41].

2.3 *Nanocellulose with Conductive Carbon Materials*

Because of the higher conductivity of carbon materials such as graphene, carbon nanotubes (CNT), carbon nanofiber, carbon black, and so on, many nanocellulose-combined carbon composites were reported for energy storage applications [6, 7, 11, 42]. Out of these materials, graphene offers high electrical and thermal conductivity, ultrahigh specific surface area, and excellent mechanical properties [43]. CNT also has the reported highest electrical and thermal conductivity and mechanical strength [44]. The fabrication of conductive nanocellulose composites is achieved by either coating or blending with carbon materials. In the coating process, the non-conductive nanocellulose paper is used as a substrate and its surface is coated by conductive carbon materials to make it conductive. Since the coating technique is a simpler process than making composites with conducting polymers, it is impossible to increase the conductivity by increasing the thickness of coating due to the weak carbon–carbon particle bonding in the coating layer. Instead of coating, blending with carbon materials can accommodate more conductive particles on the surface of nanocellulose.

2.4 *Nanocellulose with Functional Hybrids*

The functional hybrid-combined nanocellulose-based conductive composites were developed with specific properties for a wide range of energy storage applications. Generally, functional hybrid materials are composed of metallic particles along with conducting polymers or active carbon or other foreign materials in order to introduce the desired properties. Commonly employed hybrid structures are comprised of conductive polymer and inorganic nanoparticles combined in a nanocellulose matrix. Hybrid systems with two conductive polymers and two conductive carbons along with nanocellulose have also been developed [3, 10, 26, 28, 45, 46]. Numerous fabrication routes such as *in situ* polymerization, coating, doping, and blending were employed for the preparation of hybrid structures with nanocellulose.

3. Nanocellulose for Flexible Energy Storage Devices

Nanocellulose-based flexible energy storage devices have many potential applications in portable electronic devices, displays, integrated circuits, wearable systems, medical devices, etc. Flexible storage devices such as electrodes, batteries, supercapacitors, and solar cells are available with outstanding charge storage performances, capable of use as an advanced storage system. In order to introduce electrical conductivity in cellulose-based systems, conducting materials such as conductive carbon material, metal particles, and conductive polymers can be integrated. This section focuses on the recent development on flexible energy storage devices including flexible electrode materials, flexible Li-ion batteries, flexible supercapacitors, and flexible solar cells. These types of nanocellulose-based flexible energy storage devices with their technological innovations and challenges will be detailed in this section.

3.1 *Nanocellulose-Based Flexible Li-ion Batteries*

The significant features of lithium, such as its light weight and electropositive nature, usually contribute to the increased use of Li-ion batteries in the field of portable electronics [47]. In Li-ion batteries, lithium ions move from the anode (negative electrode) to the cathode (positive electrode) during charging and discharging [48]. This section highlights the recent advances in nanocellulose-based electrodes (cathode and anode), electrolytes, and separators for Li-ion batteries constructed with various conducting materials such as conducting polymers, conducting carbon materials, metal particles etc., applicable in the development of flexible energy storage devices.

The suitable designing of electrode materials facilitates ion and electron transport, and thus improves the electrode performance of Li-ion batteries. Wang *et al.* [16] reported flexible freestanding paper anodes by using carbon nanotubes, silicon nanoparticles, and cladophora nanocellulose with specific capacities of up to 800 mA hg^{-1} and good cycling performances for Li-ion batteries [16]. A lamellar graphene–nanocellulose–silicon hybrid film was fabricated by Zhou *et al.* [32] as a free-standing anode material for flexible Li-ion batteries [42]. In this hybrid system, conductive graphene matrix act as a current collector and hydroxylated silicon nanoparticles, which are entrenched within the graphene interlayer and nanocellulose, act as an adhesive to cross-link graphene and silicon nanoparticles. The hybrid anode film displays a high reversible capacity of 1251 mA hg^{-1} at 100 mA g^{-1} after 100 cycles and greater rate capability. This layered lamellar hybrid structure offers the opportunity of flexible batteries. Another layered hybrid film was reported by Cao *et al.* [18] by using lithium titanate–carbon nanotube–cellulose nanofiber as an electrode for Li-ion battery, and this shows an electrical conductivity of up to 15.0 S cm^{-1} and good cycling performance [19]. Here, cellulose nanofiber serves as building block as well as binder and the CNT–CNF layer serves as the current collector. Lu *et al* introduced TEMPO-oxidized cellulose nanofibrils as a binder material for preparing flexible electrodes for Li-ion batteries [49] (see Fig. 2).

To be used as a separator in Li-ion batteries, a material must display strong mechanical strength, good thermal stability, better porosity, good wettability, and electrochemical stability and proper thickness. Due to good thermal and mechanical properties, wettability, and high stability, regular cellulose-based materials were reported as separators in Li-ion batteries [50–52]. Though the large pore size and increased thickness of cellulose papers limits their application as a separator, nanocellulose was introduced in order to overcome these difficulties. Kim *et al.* [17] developed a smart battery separator membrane with terpyridine-functionalized

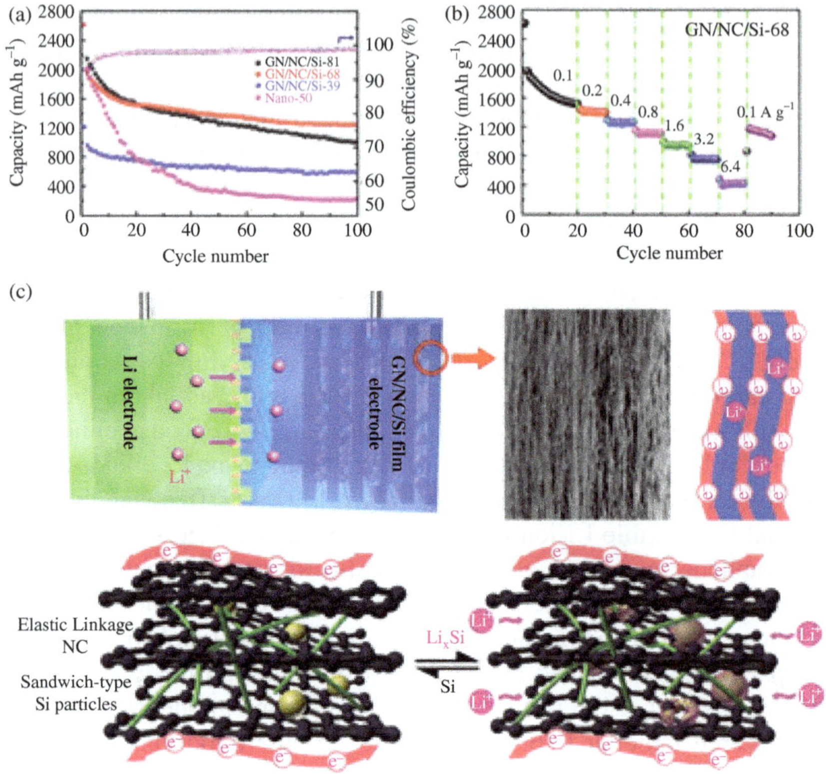

Fig. 2. (a) Cycling performances of the free-standing graphene–nanocellulose–silicon hybrid anode film anode with different Si content from cycle 1 to 100 at a current density of 100 mA g^{-1}; (b) the rate capabilities at current densities from 0.1 to 6.4 A g^{-1} and (c) schematic illustration the mechanism of lithiation/delithiation in free-standing film.

Source: Reprinted with permission from Ref [42]. Copyright (2018) American Chemical Society.

cellulose nanofibril nanoporous thin mat as the top layer and an electrospun polyvinylpyrrolidone (PVP)/polyacrylonitrile (PAN) macroporous thick mat as the support layer [17]. This c-mat separator membrane, shown in Fig. 3, provides a remarkable enhancement in the high-temperature (60°C) cycling performance (capacity retention ~80% after 100 cycles). Zhang *et al.* [51] have synthesized a thermally resistant composite separator membrane by a wet-laid process using northern bleached softwood kraft fibers, polysulfonamide (PSA) fibers, and nanofibrillated cellulose for Li-ion battery applications [53]. Here, nanofibrillated cellulose served as a pore size mediator in the composite separator

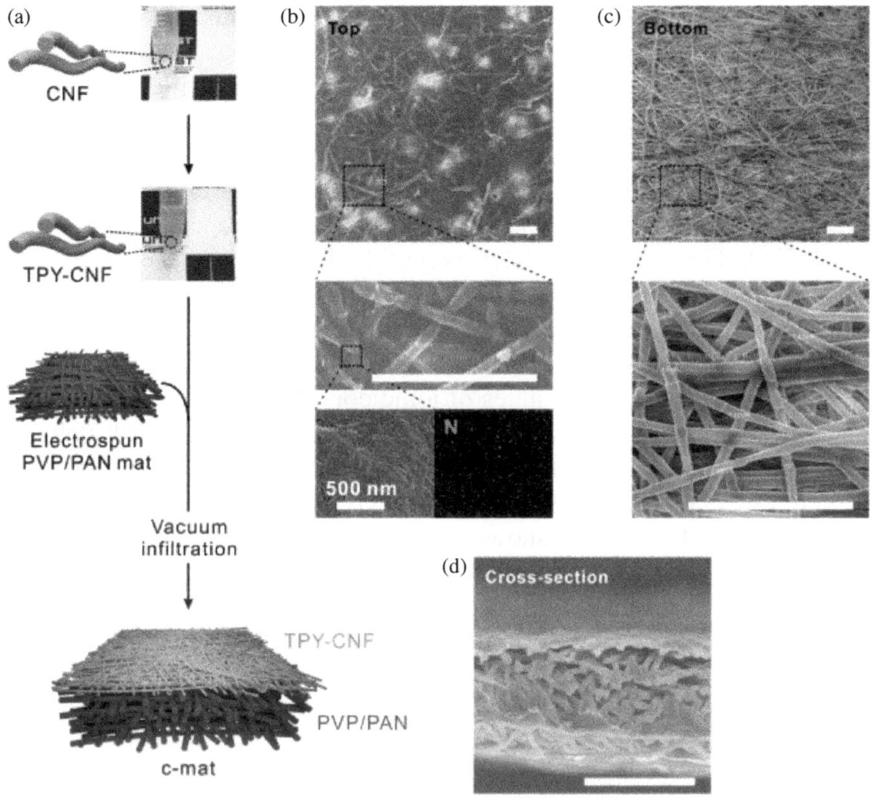

Fig. 3. (a) Schematic illustration of the fabrication procedure of the c-mat separator. SEM images of the (b) top, (c) bottom, and (d) cross-section of the c-mat separator and an elemental mapping image of nitrogen.

Source: Reprinted with permission from Ref [17]. Copyright (2016) American Chemical Society.

membranes. Pan *et al.* [54] established a tri-layer separator design to overcome the poor cycling stability and safety concerns of Li metal-based batteries [54]. For this purpose, a flexible and thermally stable cellulose nanofiber layer is directly laminated on each side of a plasma-treated polyethylene (PE) separator. The mesoporous cellulose nanofiber layer stabilizes the Li metal anodes by creating a uniform Li^+ flux toward the electrode through its homogenous nanochannels, leading to improved cycling stability.

For Li-ion batteries, an electrolyte is a gel or liquid which contains precise conductive ions that allows for the ionization or flow of electrical charge between the electrodes. In order to achieve optimum performance, the electrolyte must possess better ionic conductivity as well as mechanical strength. Nanocellulose has been reported as a nanofiller in the electrolyte polymer matrix to enhance the stability and transport properties. Two-phase electrolytes composed of ionic conductive poly(ethylene glycol) (PEG) matrix having stiff nanofibrillated cellulose (CNF) paper as reinforcement to provide mechanical integrity were reported by Willgert *et al.* [55]. The reinforcing CNF is modified in order to create covalent bonds between the phases, which is particularly beneficial when swelling the composite with a liquid electrolyte to enhance the ionic conductivity. After swelling of the composite polymer electrolyte, leading to the formation of a gelled structure, values of ionic conductivity at 5×10^{-5} S cm^{-1} and an elastic modulus around 400 MPa at 25°C were obtained. Polymer electrolyte systems reinforced with nanocellulose-laden composite were reported by Nair *et al.* [20] by using a thermally induced polymerization process [20]. This system, shown in Fig. 4, was also tested as an electrolyte separator in a Li–Sulfur battery, which demonstrated excellent ionic conductivity, thermal stability, and a stable interface. There were many reports by Samir *et al.* [57] on the enhancement of ionic conductivity of CNF-POE-based polymer electrolytes [56–58]. The CNF-POE polymer electrolyte system reduces the thickness of the liquid electrolyte and thus

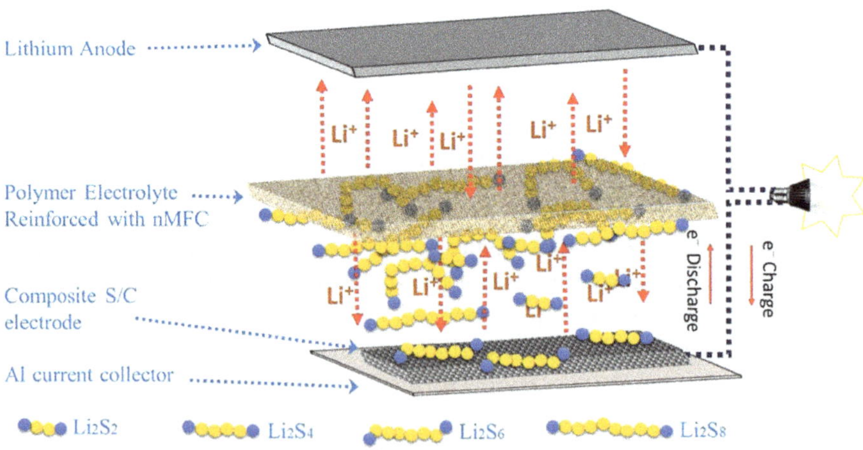

Fig. 4. Schematic illustration of the polymer electrolyte system reinforced with nanocellulose-laden for Li–Sulfur batteries.

Source: Reproduced with permission from Ref [20]. Copyright (2017) Elsevier.

decreases the internal resistance of the battery without changing the conductivity level. This promotes the increase in the performance of conductivity and mechanical strength. As a concluding remark, nanocellulose-based systems in Li-ion batteries offer the superior performance with a low cost and environmentally benign nature. However, many challenges are still faced in the large-scale fabrication of nanocellulose-based Li-ion energy storage devices.

3.2 *Nanocellulose-Based Flexible Supercapacitors*

Supercapacitors attracted much attention during the last few decades due to their unique advances such as high energy density, high power density, fast charging/discharging rate, and safe operation [59]. There are two types of supercapacitors according to the charging/discharging mechanisms, termed as electrochemical double-layer capacitors (EDLCs) and pseudocapacitors [60]. In EDLCs, energy is stored by ions adsorbing at the interfaces of electrode, and thus its capacitive performance is dominated by available surface area and conductivity of electrode [61]. But in pseudocapacitors, energy is stored by reversible redox reactions between functional groups at the electrode surface and the electrolyte, and the redox reaction rate and conductivity of electrode determine its supercapacitive behavior. Hybrid supercapacitors consist of both double-layer capacitive and pseudocapacitive charging/discharging mechanisms, and these are demonstrated to have high energy density, high power density, and good cycling stability. This section highlights the recent advances on the nanocellulose-based flexible supercapacitors containing various conducting materials such as conducting polymers, conducting carbon materials, metal particles, etc. applicable in the development of flexible energy storage devices.

Nanocellulose-based porous materials such as film and aerogels with robust mechanical properties were reported as substrates for flexible supercapacitors [11, 26, 45, 46]. These flexible electrode materials utilize conductive carbon materials, conducting polymers, metal nanoparticles, functional hybrids etc. with nanocellulosic materials. Cellulose nanofibril (CNF)–reduced graphene oxide (RGO)–carbon nanotube (CNT) hybrid aerogels, shown in Fig. 5, were established by Zheng *et al.* [29] as a flexible electrode for all-solid-state supercapacitors [45]. A poly (vinyl alcohol) (PVA)–H_2SO_4 gel has been employed as the electrolyte for this purpose. Due to porous structure of the CNF/RGO/CNT aerogel

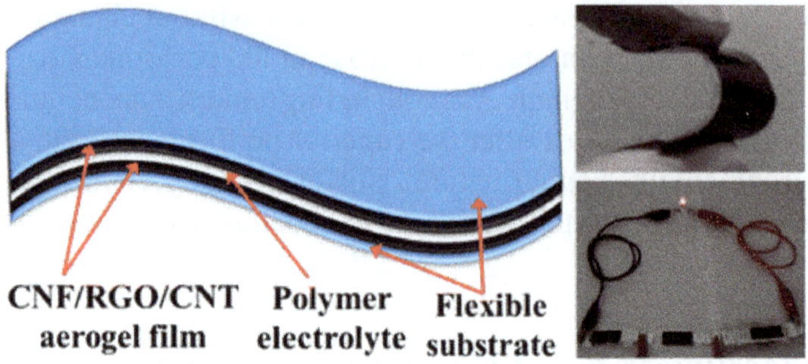

Fig. 5. Schematic illustration of the all-solid-state supercapacitor where the polymer-gel electrolyte serves as the electrolyte and the separator. The first inset image shows the flexibility of the aerogel film electrode and the second image shows a series connection of three flexible devices used to light a red LED.

Source: Reprinted with permission from Ref [45]. Copyright (2015) American Chemical Society.

electrodes and the excellent electrolyte absorption properties of the CNFs present in the aerogel electrodes, the resulting flexible supercapacitors exhibited a high specific capacitance of 252 F g^{-1} at a discharge current density of 0.5 A g^{-1} and a remarkable cycle stability, with more than 99.5% of the capacitance retained after 1000 charge–discharge cycles at a current density of 1 A g^{-1}. The reported areal capacitance of 216 mF cm^{-2}, areal power density of 9.5 mW cm^{-2}, and energy density of 28.4 μWh cm^{-2} demonstrate the outstanding performance of aerogel electrodes. The same group recently reported a flexible free-standing high porosity aerogel film electrode with cellulose nanofibril (CNF)–reduced graphene oxide (RGO)–molybdenum oxynitride (MoOxNy) aerogel film electrode synthesized by freeze-drying a CNF/GO/MoO_3 aqueous dispersion followed by subsequent *in situ* hydrazine reduction [26]. The free-standing CNF/RGO/MoOxNy aerogel film electrode delivered an outstanding specific capacitance of 680 F g^{-1} in an aqueous electrolyte and 518 F g^{-1} in an ionic liquid electrolyte at a current density of 1.0 A g^{-1}. This outstanding performance is attributed to the partial reduction and nitrogen doping of the MoO_3 nanobelts and the highly open and continuous porous structure of the hybrid aerogel film. Besides, CNF/RGO/MoOxNy-based hybrid electrodes were fabricated with either a hydrogel or ionogel as the electrolyte. The reported energy density of 114 Wh kg^{-1} and the corresponding volumetric energy density of 18.8 Wh L^{-1} obtained by the solid-state supercapacitor devices fabricated using an ionogel electrolyte are among the highest values achieved for any type of solid-state supercapacitor. Composites prepared from cellulose nanofibrils (CNF) and reduced graphene oxide (RGO) as film-like xerogels and porous cryogels were reported to be related to their dielectric performance and electrochemical storage abilities [11]. A high specific capacitance of ~460 F/g and electrochemical resistance of 12–15 Ω cm^2 was observed for well-aligned and longitudinal-oriented micro-porous cryogel, indicating its suitability as electrodes for electrochemical storage devices.

The hydroxyl groups on the nanocellulose surface contribute not only the sheet strength and flexibility of EDLC-based SCs but also are the fictional groups that can form either covalent or hydrogen bonds with conducting polymers, such as polyaniline (PANI), polypyrrole (PPY), etc., which are generally used as pseudocapacitor active materials. Cellulose nanofibers (CNF)–polyaniline (PANI) based flexible electrodes were reported by Zheng *et al.* [22] as an active material for high-performance supercapacitors [22]. The specific capacitance of 421.5 F g^{-1} is accomplished for CNF–PANI electrodes with 20 wt% CNF loading at a current density of 1 A g^{-1}. This could be due to the presence of nanocellulose, which can tailor the morphology and doping of grown PANI worm-like nanorods during *in situ* polymerization. Furthermore, good rate capability and energy/power density balance are exhibited by the hybrid electrodes. Wang *et al.* [24] demonstrated a robust and compact freestanding electrode material based on nanocellulose coupled polypyrrole@graphene oxide paper *via in situ* polymerization [25]. The electrode exhibits stable cycling over 16,000 cycles at 5 A g^{-1} as well as the largest specific volumetric capacitance of 198 F cm^{-3} so far reported for flexible polymer-based electrodes. The surface-modified nanocellulose fibers (NCFs) can be employed as substrates to synthesize supercapacitor electrodes with a gravimetric capacitance of 127 F g^{-1} and a volumetric capacitance of 122 F cm^{-3} at high current densities of 300 mA $cm^{-2} \approx 33$ A g^{-1}, as reported by Wang *et al.* [24] by using polypyrrole (PPy) [24]. The growth of manganese dioxide (MnO_2) on cellulose nanofiber-coated graphite papers *via* chemical reactions, shown in Fig. 6, reported by Qi *et al.* [23] resulted in the inferior volumetric-specific capacitance [23]. The symmetric supercapacitors displayed an extremely high volumetric energy density of 10.6 mWh/cm^3 at a power density of 0.11 W/cm^3, which is superior to that of MnO_2-based symmetric and even asymmetric devices in previous studies.

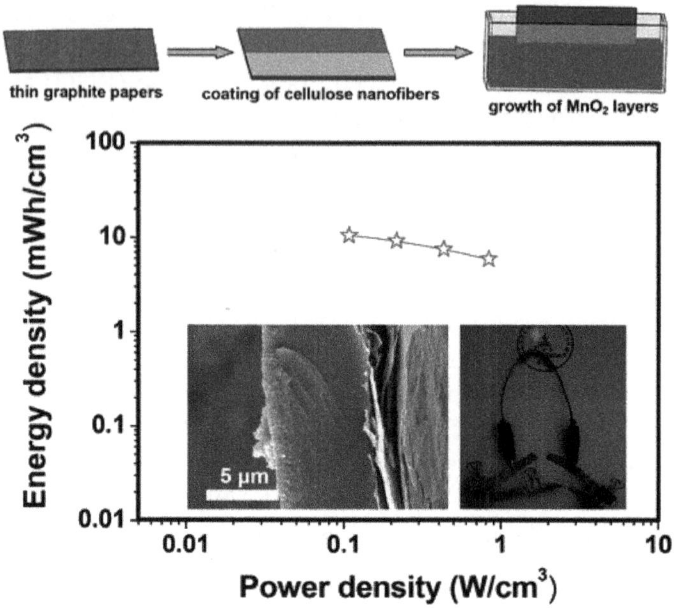

Fig. 6. Schematic design showing that cellulose nanofibers significantly promoted growth of MnO_2 on thin graphite papers, contributing to high volumetric energy density of the symmetric supercapacitors.

Source: Reprinted with permission from Ref [23]. Copyright (2018) American Chemical Society.

3.3 *Nanocellulose-Based Flexible Solar Cells*

Solar energy is potentially the largest source of renewable energy at our disposal, but significant advances are required to make photovoltaic technologies economically viable and, from a life-cycle perspective, environmentally friendly, and consequently scalable. Cellulose-based materials were reported for solar cells application due to their low cost, flexibility, and porous nature [62]. However, the regular cellulose papers are not transparent due to the larger fiber diameter, so the exploitation of nanocellulose is an excellent option for the fabrication of flexible solar cells [30–33]. This section highlights the recent advances in the nanocellulose-based flexible solar cell materials based on various conducting materials applicable in the development of flexible energy storage devices.

In 2013, Zhou *et al.* [32] demonstrated efficient polymer solar cells fabricated on optically transparent cellulose nanocrystal (CNC) substrates

[32]. The fabricated solar cell displayed good rectification in the dark and reached a power conversion efficiency of 2.7%. In addition, they demonstrated recyclable solar cell technology by using low-energy processes at room temperature. The synthesis of two types of nanocellulose substrates of nanofibrillated cellulose and nanocrystalline cellulose and their application in organic photovoltaics were demonstrated by Costa *et al.* [31]. Miettunen *et al.* [33] introduced a method for depositing electrolyte in dye solar cells such as a nanocellulose hydrogel membrane screen-printed on the counter electrode and further freeze-dried to form a highly porous nanocellulose aerogel, which acts as an absorbing sponge for the liquid electrolyte [33]. When the nanoporous dye-sensitized TiO_2 photoelectrode film is pressed against the wetted aerogel, it becomes filled with the electrolyte. The electrolyte flows inside the TiO_2 film only about ten micrometers (i.e., the TiO_2 film thickness). In addition, in this method, there is no need for electrolyte filling holes, which simplifies significantly the sealing of the cells and eliminates one common pathway for leakage. The accelerated aging tests at 1 Sun 40°C for 1000 hours showed that the nanocellulose cells were as stable as the conventional DSCs.

Cheng *et al.* [30] recently synthesized O-(2,3-Dihydroxypropyl) cellulose (DHPC) by homogeneous etherification of cellulose in 7 wt% NaOH/12 wt% urea aqueous solution [30]. The DHPC exhibited a high level of transparency, good adhesiveness, and outstanding ductility but poor mechanical properties. Thus, to construct tough nanocomposite papers, stiff tunicate cellulose nanocrystals (TCNCs) were introduced into the DHPC, as shown in Fig. 7. This study was also predicted the reinforcement of nanocomposite papers by a percolating model. The nanocomposite of TCNCs and DHPC offers high transparency, smooth surface, as well as satisfactory mechanical properties due to their excellent interfacial compatibility. The nanocomposite paper was then directly coated by tin-doped indium oxide and showed a power conversion efficiency of 4.98% for the inverted polymer solar cells. This work is important as it enabled developing clean energy by using sustainable materials derived from renewable resources. Generally, solar cells require high optical transparency and high optical haze to increase the light scattering and therefore the absorption in the active materials. Fang *et al.* [63] introduced a transparent paper that displays ultrahigh optical transparency (~96%) and ultrahigh haze (~60%) [63]. This work was accomplished by first TEMPO/NaBr/NaClO oxidizing primary wood fibers. They have reported an improved solar energy conversion efficiency of 5.88% as compared to

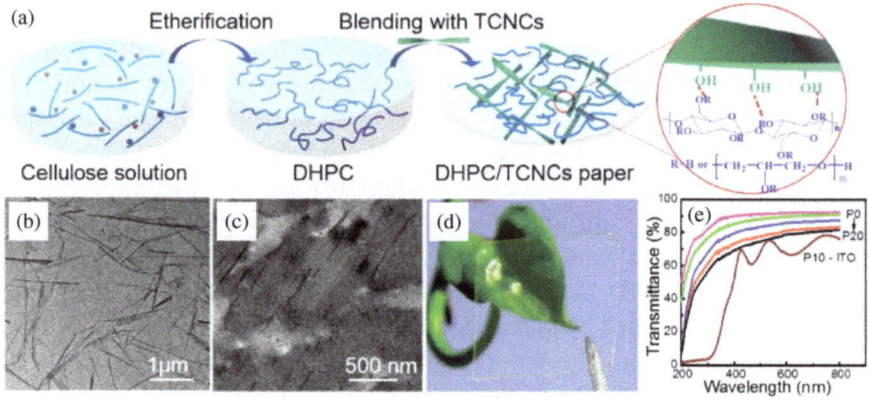

Fig. 7. (a) Schematic illustration of the fabrication of cellulose-based (TCNCs and DHPC) nanocomposite papers; (b) TEM image of TCNCs; (c) TEM image; (d) photograph of nanocomposite paper, and (e) optical transmittance under UV–Vis light of neat DHPC and nanocomposite papers.

Source: Reprinted with permission from Ref [30]. Copyright (2018) American Chemical Society.

regular CNF nanopaper. Bacterial cellulose nanofibers were introduced by modifying ethylene vinyl acetate copolymer (EVA)-based solar cell encapsulating films and reported by Yuwawech *et al.* [64]. They have reported the enhancement of thermal, mechanical, and barrier properties of EVA owing to the reinforcement of cellulose nanofibers.

4. Summary and Perspectives

The necessity of flexible energy storage devices with environmentally benign properties promotes the development of nanocellulose-based structures due to their unique properties of transparency, high strength, low density, and high aspect ratio. In order to make nanocellulose-based energy storage devices fully worthwhile, further improvements in the performance, reliability, cost reduction, and large-scale manufacturing must be done. The uniform dispersion of nano-sized fillers *via* low-cost green methods for manufacturing nanocellulose-based composites still remains a challenge. Besides, the performances of nanocellulose-based energy storage devices such as supercapacitors, Li-ion batteries, and solar cells still need to be enhanced in the future. The energy application of nanocellulose-based materials is still limited in academic research.

The large-scale production of nanocellulose structures is still relatively complicated and energy-cost. Therefore, efforts for the development of low-cost scalable production of nanocellulose is attracting more research, specifically flexible energy storage devices.

Acknowledgments

Zakiah Ahmad would like to acknowledge the financial support by the TNB Research Sdn Bhd. Malaysia (Grant No: 100-IRMI P. 37/7/ 20160419007) and Universiti Teknologi Mara, Selangor Malaysia.

References

[1] D. Tobjörk and R. Österbacka, *Adv. Mater*. (2011).
[2] Y.B. Pottathara, S. Thomas, N. Kalarikkal, T. Griesser, Y. Grohens, V. Bobnar, M. Finšgar, V. Kokol, and R. Kargl, *New J. Chem*. (2019).
[3] M. Lay, M.À. Pèlach, N. Pellicer, J.A. Tarrés, K.N. Bun, and F. Vilaseca, *Carbohydr. Polym*. (2017).
[4] M. Hasan, D.A. Gopakumar, V. Arumughan, Y.B. Pottathara, K.S. Sisanth, D. Pasquini, M. Bračič, B. Seantier, A. Nzihou, S. Thomas, S. Rizal, and H.P.S. Abdul Khalil, *Polymers* (*Basel*) 11 (2019).
[5] D.A. Gopakumar, A.R. Pai, Y.B. Pottathara, D. Pasquini, L. Carlos De Morais, M. Luke, N. Kalarikkal, Y. Grohens, and S. Thomas, *ACS Appl. Mater. Int.* (2018).
[6] Y. Beeran P. T., V. Bobnar, S. Gorgieva, Y. Grohens, M. Finšgar, S. Thomas, and V. Kokol, *RSC Adv*. 6 (2016) 49138–49149.
[7] Q. Jiang, C. Kacica, T. Soundappan, K. Liu, S. Tadepalli, P. Biswas, and S. Singamaneni, *J. Mater. Chem. A* (2017).
[8] Q. Yang, C. Zhang, Z. Shi, J. Wang, C. Xiong, T. Saito, and A. Isogai, *ACS Appl. Nano Mater*. (2018).
[9] H. Lu, V. Guccini, H. Kim, G. Salazar-Alvarez, G. Lindbergh, and A. Cornell, *ACS Appl. Mater. Int.* (2017).
[10] K.K. Liu, Q. Jiang, C. Kacica, H.G. Derami, P. Biswas, and S. Singamaneni, *RSC Adv*. (2018).
[11] Y.B. Pottathara, V. Bobnar, M. Finšgar, Y. Grohens, S. Thomas, and V. Kokol, *Polymer (Guildf)* (2018).
[12] O. Koshy, Y.B. Pottathara, S. Thomas, B. Petovar, and M. Finsgar, *Curr. Anal. Chem*. (2017).
[13] R.J. Moon, A. Martini, J. Nairn, J. Simonsen, and J. Youngblood, *Chem. Soc. Rev*. 40 (2011) 3941–3994.

[14] I. Siró and D. Plackett, *Cellulose*. (2010).
[15] L. Chen, J.Y. Zhu, C. Baez, P. Kitin, and T. Elder, *Green Chem*. (2016).
[16] Z. Wang, C. Xu, P. Tammela, J. Huo, M. Strømme, K. Edström, T. Gustafsson, and L. Nyholm, *J. Mater. Chem. A* (2015).
[17] J.H. Kim, M. Gu, D.H. Lee, J.H. Kim, Y.S. Oh, S.H. Min, B.S. Kim, and S.Y. Lee, *Nano Lett*. (2016).
[18] S. Cao, X. Feng, Y. Song, H. Liu, M. Miao, J. Fang, and L. Shi, *ACS Appl. Mater. Int.* (2016).
[19] S. Cao, X. Feng, Y. Song, X. Xue, H. Liu, M. Miao, J. Fang, and L. Shi, *ACS Appl. Mater. Int.* (2015).
[20] J.R. Nair, F. Bella, N. Angulakshmi, A.M. Stephan, and C. Gerbaldi, *Ener. Storage Mater*. (2016).
[21] Z. Li, J. Liu, K. Jiang, and T. Thundat, *Nano Ener*. 25 (2016) 161–169.
[22] W. Zheng, R. Lv, B. Na, H. Liu, T. Jin, and D. Yuan, *J. Mater. Chem. A* (2017).
[23] W. Qi, R. Lv, B. Na, H. Liu, Y. He, and N. Yu, *ACS Sustain. Chem. Eng*. (2018).
[24] Z. Wang, D.O. Carlsson, P. Tammela, K. Hua, P. Zhang, L. Nyholm, and M. Strømme, *ACS Nano* (2015).
[25] Z. Wang, P. Tammela, M. Strømme, and L. Nyholm, *Nanoscale* (2015).
[26] Q. Zheng, A. Kvit, Z. Cai, Z. Ma, and S. Gong, *J. Mater. Chem. A* (2017).
[27] M. Park, D. Lee, S. Shin, H.J. Kim, and J. Hyun, *Carbohydr. Polym*. (2016).
[28] Y. Liu, J. Zhou, J. Tang, and W. Tang, *Chem. Mater*. (2015).
[29] Q. Zheng, Z. Cai, Z. Ma, and S. Gong, *ACS Appl. Mater. Int.* (2015).
[30] Q. Cheng, D. Ye, W. Yang, S. Zhang, H. Chen, C. Chang, and L. Zhang, *ACS Sustain. Chem. Eng*. (2018).
[31] S.V. Costa, P. Pingel, S. Janietz, and A.F. Nogueira, *J. Appl. Polym. Sci.* (2016).
[32] Y. Zhou, C. Fuentes-Hernandez, T.M. Khan, J.C. Liu, J. Hsu, J.W. Shim, A. Dindar, J.P. Youngblood, R.J. Moon, and B. Kippelen, *Sci. Rep*. (2013).
[33] K. Miettunen, J. Vapaavuori, A. Tiihonen, A. Poskela, P. Lahtinen, J. Halme, and P. Lund, *Nano Ener*. (2014).
[34] X. Du, Z. Zhang, W. Liu, and Y. Deng, *Nano Ener*. (2017).
[35] J. Tian, D. Peng, X. Wu, W. Li, H. Deng, and S. Liu, *Carbohydr. Polym*. (2017).
[36] A. Jasim, M.W. Ullah, Z. Shi, X. Lin, and G. Yang, *Carbohydr. Polym*. 163 (2017) 62–69.
[37] U. Farooq, M.W. Ullah, Q. Yang, A. Aziz, J. Xu, L. Zhou, and S. Wang, *Biosens. Bioelectron*. 157 (2020) 112163.
[38] J. Gutierrez, S.C.M. Fernandes, I. Mondragon, and A. Tercjak, *Chem. Sus. Chem.* (2012).
[39] Z.-Y. Wu, B.-C. Hu, P. Wu, H.-W. Liang, Z.-L. Yu, Y. Lin, Y.-R. Zheng, Z. Li, and S.-H. Yu, *NPG Asia Mater*. (2016).
[40] K. Liu, J. Nasrallah, L. Chen, L. Huang, and Y. Ni, *Carbohydr. Polym*. (2015).
[41] Y. Song, Y. Jiang, L. Shi, S. Cao, X. Feng, and M. Miao, *J. Fang. Nanosc.* (2015).

[42] X. Zhou, Y. Liu, C. Du, Y. Ren, X. Li, P. Zuo, G. Yin, Y. Ma, X. Cheng, and Y. Gao, *ACS Appl. Mater. Int.* (2018).
[43] Y. Wu, Z. Wang, X. Liu, X. Shen, Q. Zheng, Q. Xue, and J.K. Kim, *ACS Appl. Mater. Int.* (2017).
[44] A. Xie, Y. Wang, P. Jiang, S. Li, and X. Huang, *Compos. Sci. Technol.* 154 (2018) 154–164.
[45] Q. Zheng, Z. Cai, Z. Ma, and S. Gong, *ACS Appl. Mater. Int.* 7 (2015) 3263–71.
[46] D. Zhao, Q. Zhang, W. Chen, X. Yi, S. Liu, Q. Wang, Y. Liu, J. Li, X. Li, and H. Yu, *ACS Appl. Mater. Int.* (2017).
[47] L. Jabbour, R. Bongiovanni, D. Chaussy, C. Gerbaldi, and D. Beneventi, *Cellulose* (2013).
[48] T.H. Nguyen, A. Fraiwan, and S. Choi, *Biosens. Bioelectron.* (2014).
[49] H. Lu, J. Hagberg, G. Lindbergh, and A. Cornell, *Nano Energy* (2017).
[50] Q. Xu, Q. Kong, Z. Liu, X. Wang, R. Liu, J. Zhang, L. Yue, Y. Duan, and G. Cui, *ACS Sustain. Chem. Eng.* (2014).
[51] J. Zhang, Z. Liu, Q. Kong, C. Zhang, S. Pang, L. Yue, X. Wang, J. Yao, and G. Cui, *ACS Appl. Mater. Int.* (2013).
[52] Z. Ren, Y. Liu, K. Sun, X. Zhou, and N. Zhang, *Electrochim. Acta* (2009).
[53] H. Zhang, J. Liu, M. Guan, Z. Shang, Y. Sun, Z. Lu, H. Li, X. An, and H. Liu, *ACS Sustain. Chem. Eng.* (2018).
[54] R. Pan, X. Xu, R. Sun, Z. Wang, J. Lindh, K. Edström, M. Strømme, and L. Nyholm, *Small* (2018).
[55] M. Willgert, S. Leijonmarck, G. Lindbergh, E. Malmström, and M. Johansson, *J. Mater. Chem. A* (2014).
[56] M.A.S.A. Samir, F. Alloin, J.Y. Sanchez, and A. Dufresne, *Macromolecules*, 37 (2004) 4839–4844.
[57] M.A.S.A. Samir, F. Allein, W. Gorecki, J.Y. Sanchez, and A. Dufresne, *J. Phys. Chem. B* (2004).
[58] M.A.S. Azizi Samir, F. Alloin, J.Y. Sanchez, and A. Dufresne, *Polymer (Guildf).* (2004).
[59] M. Winter and R.J. Brodd, *Chem. Rev.* (2004).
[60] D. Yu, Q. Qian, L. Wei, W. Jiang, K. Goh, J. Wei, J. Zhang, and Y. Chen, *Chem. Soc. Rev.* (2015).
[61] P. Yang and W. Mai, *Nano Energy*, 8 (2014) 274–290.
[62] L. Hu, G. Zheng, J. Yao, N. Liu, B. Weil, M. Eskilsson, E. Karabulut, Z. Ruan, S. Fan, J.T. Bloking, M.D. McGehee, L. Wågberg, and Y. Cui, *Ener. Environ. Sci.* (2013).
[63] Z. Fang, H. Zhu, Y. Yuan, D. Ha, S. Zhu, C. Preston, Q. Chen, Y. Li, X. Han, S. Lee, G. Chen, T. Li, J. Munday, J. Huang, and L. Hu, *Nano Lett.* (2014).
[64] K. Yuwawech, J. Wootthikanokkhan, and S. Tanpichai, *Polym. Test.* (2015).

https://doi.org/10.1142/9781786349477_0015

Chapter 15

Functional Nanocellulose Aerogels for Varying Technological Applications

Deepu A. Gopakumar[*], Daniel Pasquini[†], Abdul Khalil H.P.S.[*], and Sabu Thomas[‡, §]

[*]*School of Industrial Technology, Universiti Sains Malaysia, Gelugor, Penang, Malaysia*

[†]*Chemistry Institute, Federal University of Uberlândia-UFU, Uberlândia, Brazil*

[‡]*International and Inter University Centre for Nanoscience and Nanotechnology, Mahatma Gandhi University, Kottayam, India*

[§]*sabuthomas@mgu.ac.in*

Abstract

Nanocellulose is a renewable and biocompatible nanomaterial with a high strength, low density, and tunable surface chemistry. Since a lot of hydroxyl groups are present on the surface of the nanocellulose, this gives them various surface chemistries that can be tailor-made for specific application. This unique property makes them a promising candidate for various applications. Nanocellulose aerogels have gained considerable interest in the scientific community globally due to their renewability, good mechanical properties, low density, high porosity, natural biodegradability, and environmental friendliness. These fascinating properties makes them an attractive material in environmental and engineering applications. This chapter reviews recent attempts to apply nanocellulose aerogels as efficient materials for thermal insulation, flame retardancy, EMI shielding, oil absorption etc.

Keywords: Nanocellulose; aerogels; thermal insulation; flame retardation; electromagnetic shielding; oil absorption

1. Introduction

Modern technologies such as information technology, biotechnology, and nanotechnology have contributed to the development of our society in various different fields. However, future technologies should take into account our ecosystem and thereby focus on the cost-effective, eco-friendly, sustainable materials.[1] Development of new materials is a cornerstone of sustainable future technologies. Robust and ultra-lightweight sustainable materials could reduce the weight of automobiles and aircrafts, which might substantially enhance the performance and reduce energy consumption as well as air pollution.[2] To meet the technological requirement, there is every need to develop cost-effective, ecofriendly, stronger, and lighter materials. To date, petroleum-based polymeric materials have been extensively developed and used for various technological applications. These materials are competitive from the economic point of view. However, the economic aspects, life cycle assessment, and eco-friendly nature must be taken in account for the betterment of our society. In this context, the current, commercial petroleum-based polymeric materials are imposing a great burden on our society, especially in terms of sustainability and non-biodegradability. At the end of the life cycle of the materials, products should be able to be reused, recycled, or disposed. However, most of the petroleum-based polymeric materials take too long to be composted. In this context, the materials which are derived from nature have gained considerable interest due to their ecofriendly and sustainable nature.[3]

As the most important skeletal component in plants, cellulose is an almost inexhaustible polymeric raw material with fascinating structure and properties. Formed by the repeated connection of *D*-glucose building blocks, the highly functionalized, linear stiff-chain homopolymer is characterized by its hydrophilicity, chirality, biodegradability, broad chemical modifying capacity, and formation of versatile semi-crystalline fiber morphologies.[4] Cellulose is the most common organic polymer, representing about 1.5 trillion tons of the total annual biomass production, and is considered as an almost inexhaustible source of raw material for the increasing demand for environmentally friendly and biocompatible products. Nanocellulose can be classified into three types: (1) Cellulose nanofibers (CNFs)/Microfibrillated cellulose (MFCs), (2) Cellulose nanocrystals (CNCs)/Cellulose nanowhiskers (CNWHs), and (3) Bacterial Nanocellulose (BNC)[5] All these nanocellulose have been isolated and extracted from native cellulose fiber suspensions with or without pretreatment by mechanical disintegration. Recently, the novel nanocellulose aerogels have gained considerable interest in the scientific community globally due to their renewability, good mechanical properties, low density, high porosity, natural biodegradability, and environmental friendliness [6]. These fascinating properties make it possible to use these in varying technological applications. This chapter reviews recent advancement of nanocellulose aerogels and discusses about its possibility for sustainable future materials. These materials can be applied for environmentally friendly engineering composites that surpass current technological limits of composite materials.

2. Comparison Between Cellulose Nanocrystals (CNCs) and Cellulose Nanofibers (CNFs)

Two main types of nanocelluloses are (i) cellulose nanocrystals and (ii) cellulose microfibrils or nanofibers. The major difference between the CNCs and CNFs lies in their amorphous portions and dimensions. The CNCs have nano dimensions in both length and diameter, whereas CNFs have length in the micro dimension and diameter in the nano dimensions, as shown in Fig. 1. CNFs can be manufactured by mild acid hydrolysis coupled with steam explosion, whereas CNCs need a strong acid like sulfuric acid that destroys all the amorphous portion (dis-ordered region) and leads to the nanocrystal structure as shown in Fig. 1. Both CNCs and CNFs can be isolated from the plant cell walls. The abundant availability of plant biomass and the superior mechanical properties of nanocellulose, including CNCs and CNFs, makes it a desirable reinforcing material for polymer nanocomposites.[7]

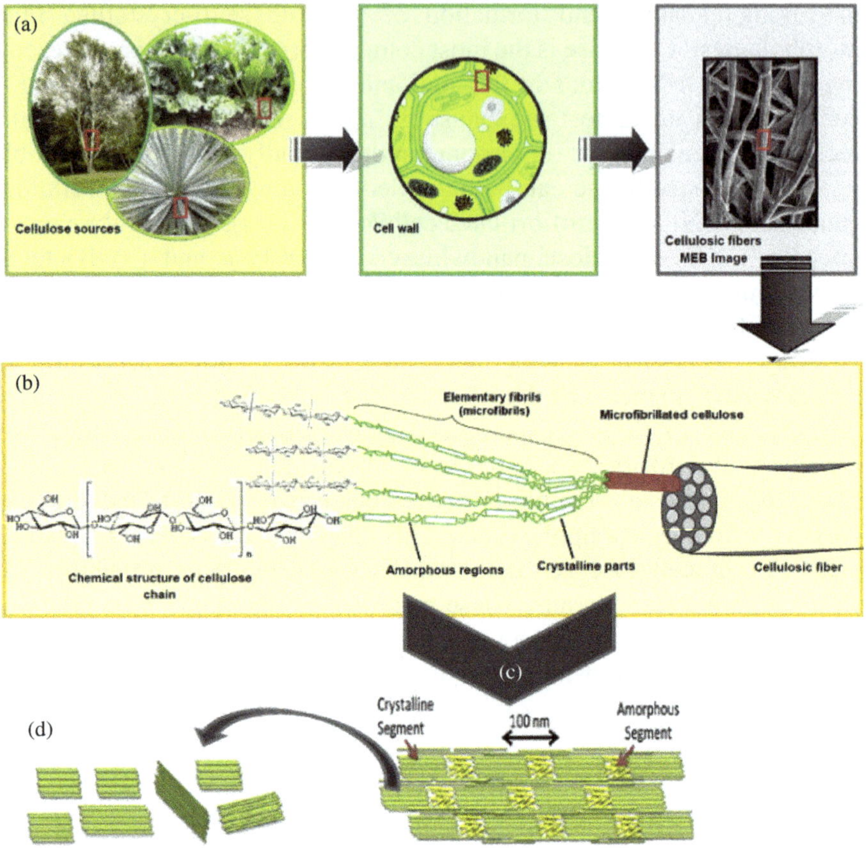

Fig. 1. Schematics of cellulose fiber structure with emphasis from the (a) cellulosic plant sources and (b) cellulose fiber to the (c) cellulose nanofibers (CNFs) bundles consisted of crystalline and amorphous regions and to the (d) cellulose nanocrystals (CNCs).[8]

Source: Reproduced with permission from Elsevier Ltd.

3. Extraction of Cellulose Nanofibers and Cellulose Nanocrystals

3.1 *Extraction of Cellulose Nanofibers (CNFs)*

Many investigations have been done on the isolation and characterization of cellulose nanofibers from numerous sources. Cellulose nanofibers can be extracted from the cell walls *via* simple mechanical shearing or by chemo-mechanical routes. Renewable resources are considered as a potential and advantageous source of CNF. The use of such natural materials is becoming more completely understood and appreciated due to their high strength, biocompatibility, etc. The extraction of the cellulose nanofibers from the plant cell wall can be accomplished by a variety of means.

3.2 *Cryocrushing Treatment*

Cryocrushing is a method for producing nanofibers where the fibers are frozen using liquid nitrogen, and then high shear forces are applied. This process exerts high energy on the frozen fibers, sufficient to break the microfibrils off the fiber walls and form individual fibrils of very low diameter. Cryocrushing combined with a high-pressure fibrillation process was used by Wang and Sain (2007) for the isolation of nanofibers with diameters in the range 50–100 nm from soybean stock.[9] Alemdar *et al.* (2008) reported the extraction of cellulose nanofibers from wheat straw and soy hulls by mechanical treatment involving cryocrushing. They investigated the chemical composition, morphology, and physical and thermal properties of the nanofibers to study their usability in biocomposite applications. They showed that the produced wheat straw nanofibers' diameters were in the range of 10–80 nm with lengths of a few thousand nanometers.[10] Chakraborty *et al.* (2005) also reported a novel technique to produce cellulose microfibrils through mechanical methods. The technique involved a combination of severe shearing in a refiner, followed by high-impact crushing under liquid nitrogen.[11]

3.3 *Grinding Procedure*

Grinding process comprises of passing the cellulose fibers through a gap between two specially modified grooved discs, one of which is static and other is revolving at about 1,500 rpm. During this process, it has been shown that the cell wall structure is broken down by the high shearing forces, resulting in nanosized individualized fibers. The distance between the discs, the morphology of the disc channels, and the number of passages through the grinder play important roles in the extent of the fibrillation of cellulose fibers. Cellulose nanofibers with a uniform width of 15 nm from wood *via* grinding in an undried state were reported by Abe *et al.* (2007). They obtained the nanocellulose fiber with a diameter of

15 nm. This report clearly demonstrated that grinding is an efficient and simple method for the production of cellulose nanofibers from biomass.[12] When homogenized cellulosic pulp was subjected to grinder treatment by Iwamoto *et al.* (2005) the fibril bundles were further fibrillated, and 10 repetitions of the grinder treatment resulted in uniform nanofibers 50–100 nm wide.[13]

3.4 *Electrospinning Technique*

Electrospinning is a quite simple and cost-effective method for the fabrication of nanofibers. Here, the solution is extruded and electrospun under the action of a high electric field. Once the voltage is high enough, a charged stream of matter is ejected following a rather complicated loop. During this process, the solvent evaporates, leaving behind randomly oriented nanofibers accumulated on the collector. Cellulose nanofibers have been produced by using electrospinning technique by dissolving cellulose fibers in solvents, such as ethylene diamine, with a salt selected from the group consisting of potassium thiocyanate, potassium iodide. Only few fundamental studies are currently dedicated to this method. Ramakrishna *et al.* (2005) prepared cellulose acetate (CA) nanofibers with fiber diameter ranging from 200 nm to 1 μm by electrospinning technique, and the solvent employed was acetone/DMF/trifluoroethylene (3:1:1) mixture.[14]

3.5 *Enzymatic Pretreatments*

This process enables the manufacture of cellulose nanofibers with significantly reduced consumption. In nature, cellulose is not degraded by a single enzyme but a set of cellulases. These can be classified as A- and B-type cellulases, which are able to attack highly crystalline cellulose, and C- and D-type cellulases, which generally require some disordering of the structure in order to degrade cellulose. Henriksson *et al.* (2007) and Paakko *et al.* (2007) reported that endoglucanase pre-treatment facilitates disintegration of cellulosic wood fiber pulp into cellulose nanofibers.[15,16] Another study was done by Tibolla *et al.* (2014) and they isolated cellulose nanofibers from banana peel bran using chemical treatment (involves alkaline treatment, bleaching, and acid hydrolysis) and enzymatic treatment (alkaline treatment and hydrolysis with xylanase). Nanofibers produced from chemical treatment and enzymatic treatment had an average diameter of 10.9 and 7.6 nm and a length of 454.9 and 2889.7 nm, respectively. They showed that the nanoparticles isolated by both the treatments had potential application as reinforcing elements in composites.[17]

3.6 *Ultrasonic Technique*

This process comprises of a combination of chemical pretreatment and high-intensity ultrasonication. The ultrasonic impact can gradually disintegrate the micron-sized cellulose fibers into nanofibers. Before ultrasonication, the plant fibers are purified to prepare the cellulose fibers; this is done by mild acid hydrolysis followed by alkali treatment and bleaching treatment. This process removes the non-cellulosic materials like lignin, hemicelluloses, etc. After chemical pretreatment, the purified cellulose fibers are soaked in distilled water. About 110 ml of solution containing chemical-purified cellulose fibers are then placed in an ultrasonic generator of 20–25 KHz frequency. The ultrasonication is conducted for 30 minutes to isolate the nanofibers. Li *et al.* (2014) prepared nanocellulose fibers by pretreating cellulose in NaOH/urea/thiourea solution and then defibrillating the fibers by ultrasonication. They achieved a high yield of 85.4%. The obtained nanocellulose fibers were about 30 nm in diameter with cellulose II crystal structure. They possessed high thermal stability with an onset of thermal degradation at 270°C and a maximum degradation temperature of 370°C.[18] Chen *et al.* (2014) reported that the cellulose nanofibers were individualized from poplar wood by explosive chemical pretreatment and high-intensity ultrasonication. When the output power of ultrasonic treatment used for the chemical-purified cellulose fibers was greater than 1000 W, cellulose nanofibers of 5–20 nm in width and several microns in length were obtained.[19]

3.7 *Steam Explosion Coupled with Mild Acid Hydrolysis*

This process comprises of a short vapor phase cooking at temperatures in the range of 180–210°C, followed by explosive decompression and sudden release of pressure. In this process, the cellulosic biomass is pressurized for a short period of time in an autoclave with steam, and then explosively discharged to atmospheric pressure, which results in a sudden disintegration of the starting material into a fibrous dispersed solid. This process results in a substantial breakdown of lignocellulosic structure, hydrolysis of the hemicellulose fraction, depolymerization of the lignin components, and defibrillation. Deepa *et al.* (2014) reported the extraction of the cellulose nanofibers from the banana plant by the steam explosion process in an autoclave.[20] Chirayil *et al.* (2014) employed the steam explosion process for extracting cellulose nanofibers from isora fiber in an autoclave. This technique involves alkaline treatment, bleaching, acidic steam treatment, and homogenization. The results showed that the prepared cellulose nanofibers have a nanofibrillar network-like structure with high crystallinity and good thermal stability.[21] Cherian *et al.* (2010) also reported that the steam explosion process was employed for the successful extraction of cellulose nanofibrils from pineapple leaf fibers for the first time. Steam-coupled acid treatment on the pineapple leaf fibers was found to be effective in the depolymerization and defibrillation of the fiber to produce nanofibrils of these fibers.[22] Recently, Fortunati *et al.* (2016) successfully extracted cellulose nanofibers from sunflower stalks *via* steam explosion coupled with mild oxalic acid hydrolysis, and the schematic of procedure is shown in Fig. 2 [23].

3.8 *Homogenization via High Pressure*

This is the method extensively used for the large-scale production of cellulose nanofibers. This method has been widely used by the dairy and

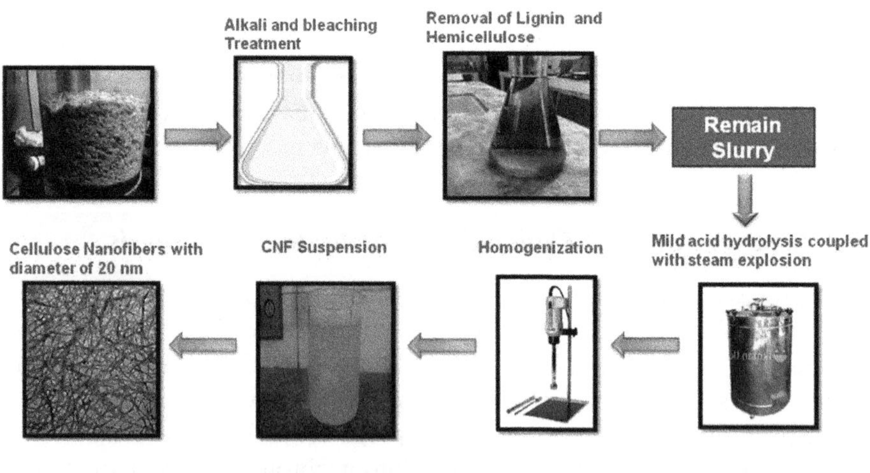

Fig. 2. Scheme of the extraction procedure of cellulose nanofibrils from sunflower stalks *via* steam explosion coupled with mild acid hydrolysis.[23]

Source: Reproduced with permission from Elsevier Ltd.

food industry mainly to stabilize food emulsion. This process comprises of the pushing of the cellulose fibers at high force to be fed through a spring-loaded valve assembly. As this valve opens and closes at a very fast rate, the fibers are exposed to a large pressure drop with shearing and impacting forces. This combination of forces will lead to a high degree of microfibrillation of cellulose fibers. The number of homogenization cycles and the applied pressure play vital roles in the cellulose fibrillation. Thus, the higher the pressure, the more the micro fibrillation of cellulose fibers.

Herrick *et al.* and Turbak *et al.* (1985) first reported the procedure for microfibillation of cellulose[7]. They achieved the microfibrillation of cellulose *via* passing the dilute cellulosic wood pulp-water suspensions through a mechanical homogenizer, in which a large pressure drop facilitated microfibrillation.[7] Another study was done by Meng Li *et al.* (2014). They employed the combination of chemical treatments and high-pressure homogenization to extract the cellulose nanofibers from de-pectinated sugar beet pulp. The diameter of the obtained cellulose nanofibers ranged from several nm to 70 nm. They showed that the crystallinity[18] of nanofibers increased significantly after treatments.[18] A similar work was reported by Chen *et al.* (2014). They fibrillated raw dried cotton fibers into individual cellulose nanofibers by chemical purification and pretreatment in a high-speed blender combined with high-pressure homogenization (nanofibrillation). They found that resultant nanofibers were in the range of 10–30 nm in diameter and had high aspect ratio.[19] Cellulose nanofibers were isolated from banana peels using a combination of chemical treatments by Pelissari *et al.* (2014) using alkaline treatment, bleaching, and acid hydrolysis. This was done by passing the suspensions of chemically treated fibers through a high-pressure homogenizer.[24]

4. Isolation of Cellulose Nanocrystals (CNCs)

CNC can be obtained from various starting materials, including algae cellulose, cotton linters, wood pulp, and microcrystalline cellulose *via* acid hydrolysis of the macrocellulose. It has been shown that the nanocrystal could be extracted from purified cellulose fibers after complete dissolution of the non-crystalline fractions by chemical hydrolysis. The amorphous region in the cellulose microfibrils could be hydrolyzed by the strong acid treatment, leaving behind a highly crystalline residue. The resulting rod-like crystalline cellulose is known as cellulose nanocrystals, and usually their diameter ranges from 5 to 20 nm depending on the nature of the cellulosic plant source, purification pre-treatment, and controlled conditions of acid hydrolysis.[25,26]

4.1 *Mechanical Size Reduction*

The first step prior to extracting the CNCs from plant source material is size reduction and washing of the raw lignocellulosic biomass.[8] Normally, the cleaned raw fibers are broken down by milling or grinding. The resultant milled powder has uniform size of millimeter level, which is more effective material for chemical treatment.

4.2 *Alkali Treatment*

This is the process of treating the raw lignocellulosic mass with a concentrated strong base solution (aqueous KOH or NaOH solution) in order to remove the alkali-soluble materials like hemicelluloses and other impurities covering the external surface of the fiber cell wall.[27,28] The main mechanism behind the alkaline treatment is the interruption of OH bonding in the fiber network structure, leading to the separation of the interfibrillar regions from the cellulose fibers.[29] The mechanism is as follows.

$$[\text{Fiber-OH} + \text{NaOH} \rightarrow \text{Fiber-O-Na}^{+} \text{ (alkoxide} + H_2O]$$

Generally, alkali treatments can be divided into two steps: (1) alkali solution heating and (2) alkali cooling by autoclave. The former is carried out using a combination of high temperature (70–90°C) and mechanical stirring. The latter is a cooling process of the fibers done using a combination of high pressure homogenization and high temperature (160°C).[30,31]

4.3 *Bleaching Treatment*

This process is called delignification and is an additional important step for the further removal of residual cementing material, mainly lignin in the alkali-treated fibers. Various bleaching treatments have been reported from various cellulosic sources.[32] Almost all alkali-treated fibers are bleached by boiling the fibers with sodium chlorite ($NaClO_2$) solution under acidic condition. The acidic condition is created by acetate buffer solution which is made up of NaOH and glacial acetic acid, diluted with distilled water. When put in an acidic buffer solution, sodium chlorite would break down into chlorine dioxide (ClO_2) under the presence of buffer salts which help to liberate the chlorine dioxide from sodium chlorite during the bleaching process. The common function of these acids is their ability to release hydronium ions (H^+) for the hydrolytic cleavage of glycosidic bonds in cellular molecular chains within amorphous regions along with the cellulose fibrils, therefore breaking down the hierarchical structure of the nanofibril bundles into crystalline nanocrystals.[33]

4.4 *Acid Hydrolysis*

Among the acid hydrolysis treatments, sulfuric acid (H_2SO_4) treatment of cellulosic materials has been widely used to extract the CNCs. Acid hydrolysis of cellulose microfibrils is a well-known effective process used to obtain the well-defined cellulose nanocrystals. In fact, most of the crystalline regions are insoluble in acids.[34] A lot of acids have been used for the acid hydrolysis of cellulose microfibrils, including H_2S_{04}, HCl, oxalic acid, hydrobromic acid (HBr), and HNO_3.[35–37] Danial *et al.* (2015) extracted the CNCs from waste paper by acid hydrolysis using 60% (V/V) H_2SO_4 solution at 45°C with constant stirring. They obtained the CNCs with length ranging from 100–300 nm and they observed that newspaper can serve as a precursor for the effective production of CNCs.[38]

Similar results were reported by Hanieh *et al.* (2012) using H_2SO_4 acid hydrolysis of kenaf fibers. They extracted the CNCs from kenaf bast fibers by mechanical stirring using 65 wt% of H_2SO_4 at various reaction times. They concluded that H_2SO_4 (65 wt%), 45°C, and 40 minutes represent the optimal preparation condition for the effective hydrolysis of kenaf bast fibers.[39] Cellulose nanocrystals have been successfully extracted by Jiang *et al.* (2015) from tomato peels. They used acidified sodium chlorite or chlorine-free alkaline peroxide routes. They isolated CNCs at a 15.7% yield by H_2SO_4 acid hydrolysis (64% H_2SO_4, 8.75 ml/g, 45°C, and 30 minutes). They obtained spindle-shaped CNCs with an average thickness of 3.3 nm and width of 7.2 nm. They demonstrated that pure cellulose was effectively isolated from under-utilized food processing residues.[40] Lamaming *et al.* (2015) also reported the extraction of CNCs from oil palm trunk by sulfuric acid hydrolysis. They obtained CNCs with an average diameter of 7.67 nm and length of 397.03 nm.[41] Bettaieb *et al.* (2015) extracted CNCs from the *Posidonia oceanica* leaves by sulfuric acid hydrolysis. They obtained CNCs with an average diameter of 6.8 nm and

520 nm length, and a similar result was observed by Fortunati *et al.* (2015).[42,43] Mohammed *et al.* (2015) have extracted CNCs from recycled newspaper by sulfuric acid hydrolysis and investigated the thermal stability of the extracted CNCs. They observed that the thermal stability of the extracted CNCs was affected due to the dehydration reaction by the sulphate groups from the sulfuric acid hydrolysis process.[44]

5. Nanocellulose Aerogel for Thermal Insulation Application

The world temperature has been rising day by day due to global warming. Introduction of thermal insulation in buildings is one of the solutions to maintain comfort. The greenhouse gas effect has been warming up the earth and threatening human life. It was found that CO_2 emission from buildings contributed more than 30% of greenhouse gas emission in developed countries in 2005. Buildings also accounted for 39% of energy consumption the same year. Improving thermal insulation of buildings is one of the most effective solutions for this issue. Therefore, there have been many efforts in developing new insulation materials. In this context, a lot of investigations and studies have been reported on the fabrication and applicability of nanocellulose aerogels for thermal insulation.

Wicklein *et al.* (2015) reported on the thermal insulating and fire-retardant lightweight anisotropic foams based on nanocellulose and graphene oxide. Freeze-casting was employed for the fabrication of foams based on nanocellulose and graphene oxide *via* controlled freezing of an aqueous suspension of CNF, GO, and SEP in a mold placed on top of a cold finger immersed in a liquid-nitrogen bath. They found that the fabricated foams/aerogels showed an excellent combustion resistance and exhibit a thermal conductivity of 15 mW m^{-1} K^{-1}, which was about half that of expanded polystyrene. Finally, they concluded that the freeze casting suspensions of cellulose nanofibers, graphene oxide, and sepiolite nanorods produces super-insulating, fire-retardant, and strong anisotropic foams that could perform better than traditional polymer-based insulating materials.[45]

A cost-effective and scalable recipe for fabricating biodegradable cellulose aerogels from paper waste has been realized by Nguyen *et al.* (2014). They studied advanced thermal insulation and absorption

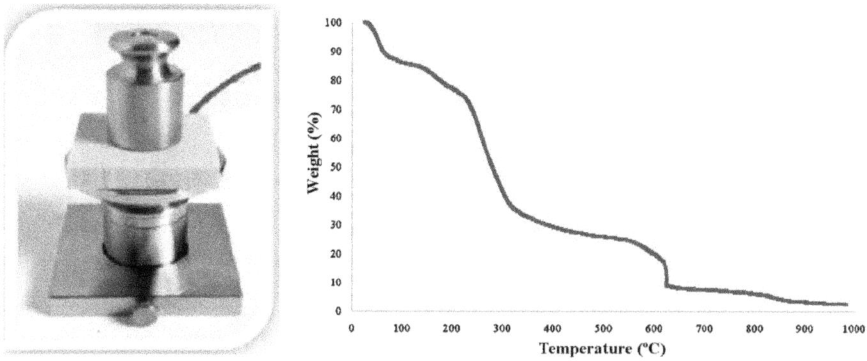

(a) (b)

Fig. 3. (a) A setup of thermal conductivity measurement. (b) TGA result of recycled cellulose aerogel.[46]

Source: Reproduced with permission from Elsevier Ltd.

properties of recycled cellulose aerogels. They demonstrated that the fabricated cellulose aerogel had extreme low thermal conductivity of 0.029–0.032 $Wm^{-1}K^{-1}$. They investigated the thermal insulation ability of the fabricated recycled cellulose aerogel using the C-Therm TCi Thermal Conductivity Analyzer System, which is shown in Fig. 3. The measured thermal conductivity of the material was 0.032 $Wm^{-1}K^{-1}$, which was comparable to that of good insulation materials such as silica aerogel (0.026 $Wm^{-1}K^{-1}$), wool (0.03–0.04 $Wm^{-1}K^{-1}$), and Aspen Aerogel products (0.021 $Wm^{-1}K^{-1}$). They concluded that the low thermal conductivity value and the low cost of paper waste make the recycled cellulose aerogel promising for thermal insulation applications.[46]

Another interesting work was reported by Hayase *et al.* They prepared organic–inorganic hybrid aerogels composed of PMSQ and a small number of cellulose nanofibers (CNFs). They found that the obtained PMSQCNF composite aerogels showed thermal conductivity as low as pure silica aerogels. They have successfully obtained composite aerogels with low density (0.020 g cm^{-3} at lowest) and low thermal conductivity (15 mW m^{-1} K^{-1}). Thermal conductivity of PMSQCNF aerogel panels was comparable to or even lower than that of other composite aerogels reported elsewhere and conventional thermal insulators such as mineral wools and polymer foams, as shown in Fig. 4. They concluded that the

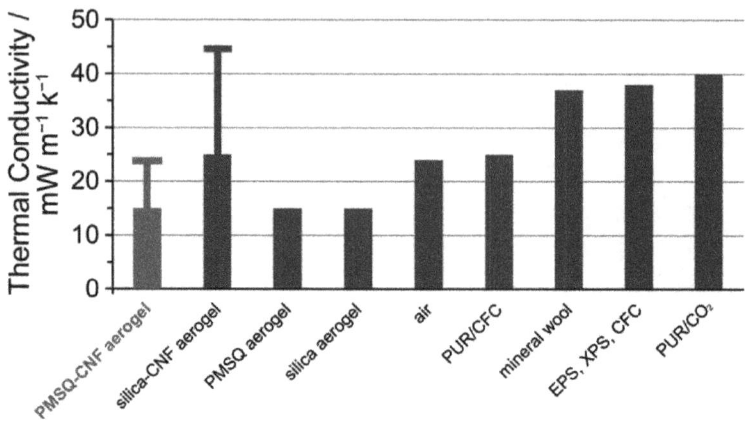

Fig. 4. Comparison of the thermal insulation properties of aerogels and some commercially available insulating materials. The thin bars on the top of PMSQ-CNF aerogel and silica-CNF aerogel show the ranges of the values. PUR: polyurethane foam. CFC: chlorofluorohydrocarbons. EPS, XPS: expanded and extruded polystyrene.[47]

Source: Reproduced with permission from American Chemical Society.

PMSQCNF composite aerogels would be a candidate for practical high-performance thermal insulating materials.[47]

6. Nanocellulose Aerogel as Efficient Flame-Retardant Material

Cellulose aerogels are ultralight three-dimensional porous materials that combine the excellent properties of highly porous aerogels, such as high specific surface area, low density, low thermal conductivity, and high insulation, with the fantastic properties of sustainable biopolymers such as biodegradability and biocompatibility. Cellulose aerogel is considered one of the most promising heat insulating materials due to its abundant internal pores and good thermal insulation performance. However, nanocellulose aerogels are easy to ignite, which is the main limitation to its wide applicability in many commercial applications. In this context, researches have been more focused to develop the flame-retardant nanocellulose aerogels while maintaining their unique properties for various technological applications.

He *et al.* (2018) reported a facile strategy to construct flame-retardant, sound-adsorptive, and mechanically enhanced cellulose-based composite

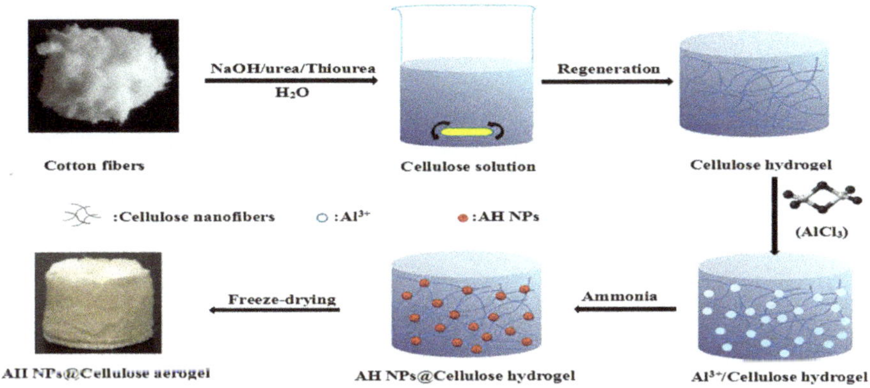

Fig. 5. Schematic illustration of the *in situ* synthesis of AH NPs@cellulose composite aerogels.[48]

Source: Reproduced with permission from American Chemical Society.

aerogels by the incorporation of aluminum hydroxide nanoparticles (AH NPs) into cellulose gels *via* an *in situ* sol-gel process followed by freeze-drying to coat AH NPs on cellulose composite aerogels (AH NPs@ cellulose composite aerogels) as shown in Fig. 5.

The prepared composite cellulose aerogels showed excellent flame retardancy, the peak of heat release rate (PHRR) of the composite aerogels decreased significantly from 280 W/g of the control sample to 22 W/g, and total heat release (THR) of the composite aerogels decreased remarkably from 13.2 kJ/g to 1.6 kJ/g as shown in Fig. 6.

Moreover, the incorporation of AH NPs composite aerogels exhibited remarkable mechanical properties; the compressive strength of the composite aerogels increased significantly from 0.08 MPa to 1.5 MPa. In addition, AH NPs composite cellulose aerogels have excellent sound absorption at high frequencies, with a maximum sound absorption coefficient of 1. In summary, they concluded that the demonstrated work was a facile strategy to rationally construct flame-retardant, mechanically robust, high-efficiency sound-adsorption and superamphiphobic cellulose-based composite aerogels, which have promising applications in the future as green engineering materials.[48]

Yuan *et al.* (2018) prepared cellulose-silica composite aerogels by *in situ* formation of silica nanoparticles *via* a two-step sol-gel process in cellulose gel, which was prepared by dissolving cotton pulp in 1-allyl-methylimidazole chloride ionic liquid (AmimCl) and then regenerating it from AmimCl/water bath. The results showed that the composite aerogels

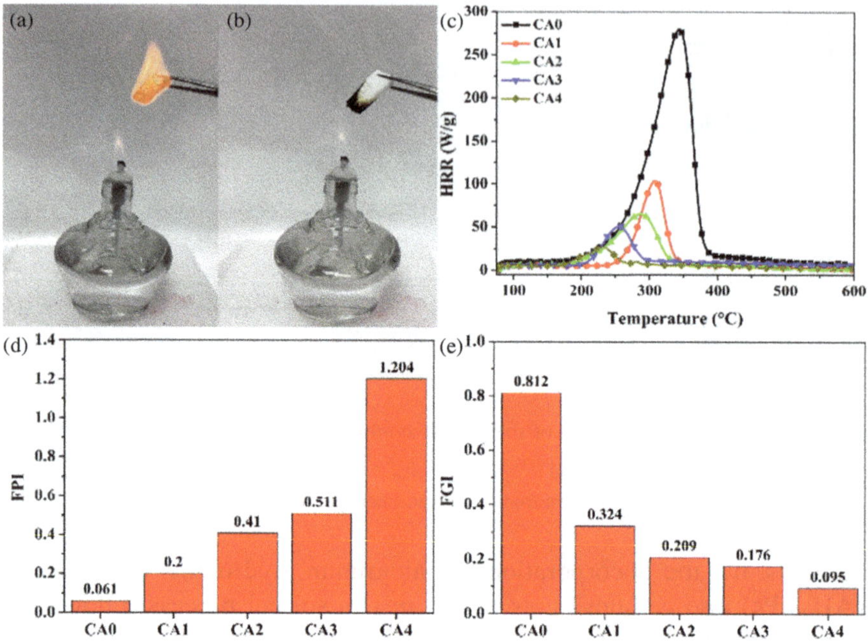

Fig. 6. Optical images of cellulose aerogels (a) and AH NPs@cellulose composite aerogels (b) captured after ignition by an alcohol burner. HRR curves (c) of cellulose aerogel and AH NPs@cellulose composite aerogels measured by MCC. MCC fire performance index (d) and fire growth index (e) for all samples.[48]

Source: Reproduced with permission from American Chemical Society.

displayed an enhanced transparency, compressive properties, and thermal and thermal-oxidative stability with increasing silica content. The found that the incorporation of silica nanoparticles also improved the mesoporous characteristics of aerogels, including specific surface area and mesopore volume, and significantly delayed the decomposition of cellulose, suppressing the heat release during combustion. The composite aerogels with high silica content (33.6% or more) exhibited good transparency with light transmittance as high as 78.4% at 800 nm, even higher than the neat cellulose aerogels, and presented excellent flame-retardant performance, achieving self-extinguishment after ignition. They concluded that the transparent cellulose-silica composite aerogels with enhanced mechanical performance and improved flame retardancy might show great potential in a wide variety of applications.[49]

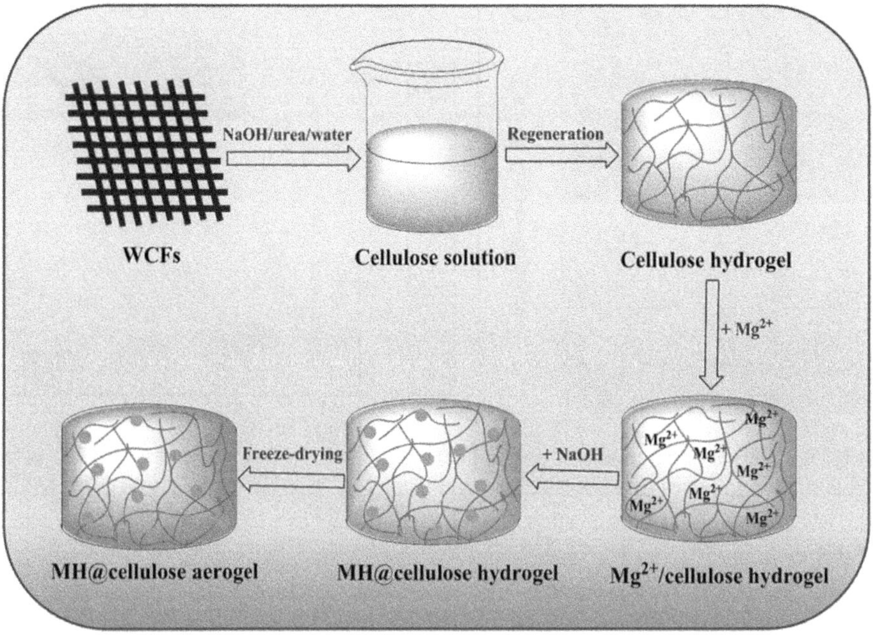

Fig. 7. Schematic illustration of the *in situ* synthesis of MH NPs @ cellulose composite aerogel.[50]

Source: Reproduced with permission from American Chemical Society.

Another interesting work was reported by Han *et al.* (2015) who fabricated flame-retardant cellulose aerogels from waste cotton fabrics by *in situ* synthesis of magnesium hydroxide nanoparticles (MH NPs) in cellulose gel nanostructures, followed by freeze-drying, as shown in Fig. 7. They also found that the three dimensionally nanoporous cellulose gel prepared from the NaOH/urea solution could serve as scaffold/template for the nonagglomerated growth of MH NPs.

They showed that the prepared hybridized cellulose aerogels have excellent flame retardancy, which could be extinguished within 40 s as shown in Fig. 8. Moreover, the thermal conductivity of the composite aerogel was increased moderately from 0.056 to 0.081 W m^{-1} k^{-1} as the specific surface area decreased slightly from 38.8 to 37.6 cm^2 g^{-1}, which indicated that the excellent heat insulating performance of cellulose aerogel was maintained. They concluded that the demonstrated work was

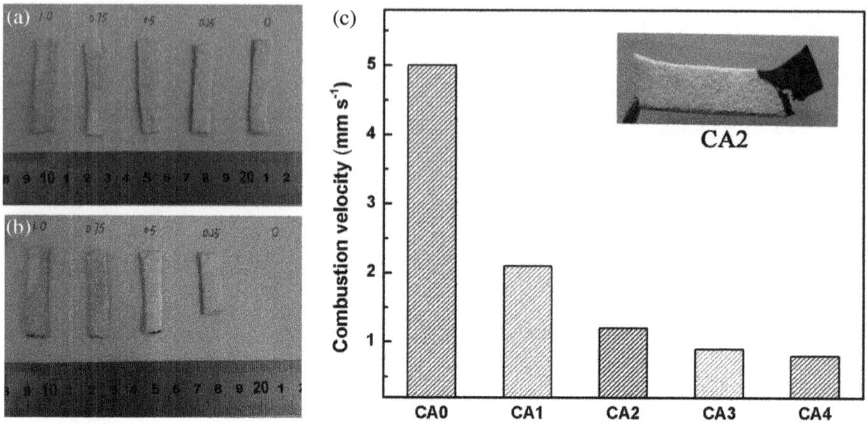

Fig. 8. Digital photographs of original aerogels (CA1, CA2, CA3, and CA4 were aerogels with different MH NPS loadings and CA0 was aerogel without MH NPS before (a) and after (b) 10 s of combustion. Combustion velocity as a function of MH NPs content (c).[50]

Source: Reproduced with permission from American Chemical Society..

suitable for industrial-scale production and has great potential in the future as green building materials.[50]

Bendahou *et al.* (2015) prepared Aerogel monoliths consisting of cellulose microfibers, cellulose nanofibers, and nanozeolites. They have showed that the fabricated aerogel monoliths had tunable thermal conductivity and mechanical properties. They employed the hot filament technique to measure the thermal conductivity as shown in Fig. 9. They reported the thermal conductivity of 18 $mWm^{-1}\ K^{-1}$ for the fabricated aerogel monoliths and confirmed that the fabricated aerogel had super insulation properties. They concluded that this study certainly opens a new investigation field to optimize the thermal conductivity properties of hybrid nanocellulose-based aerogels.[51]

7. Nanocellulose Aerogel as Substrates for EMI Shielding Application

It is very important to develop highly efficient electromagnetic interference (EMI) shielding materials to prevent the interference or malfunction

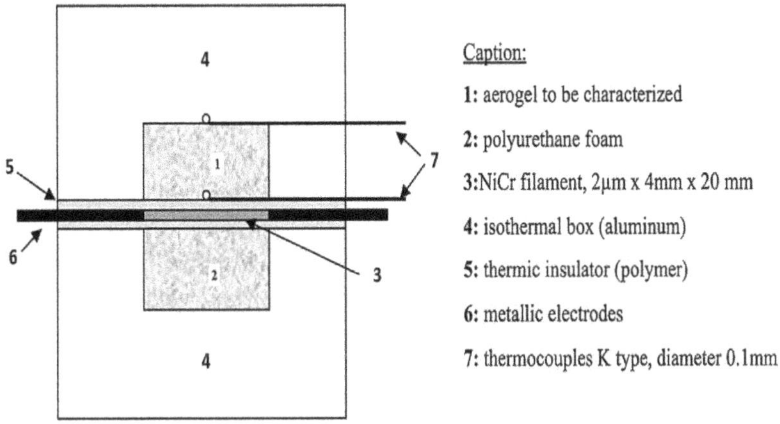

Fig. 9. Scheme of the thermal conductivity measurement system.[51]
Source: Reproduced with permission from Elsevier Ltd.

of highly sensitive precision electronic instruments and prevent the health threats to human beings caused by the emerging undesirable but unavoidable EMI due to the use of modern electronics packed with highly integrated circuits and the widespread use of transient power sources.[52]

Huang *et al.* reported an ultra-light and highly conductive cellulose composite aerogel fabricated *via* simple, efficient, and environmentally benign strategy. The fabricated scaffold structure was well designed from nanofibrillar networks to nanosheet networks by controlling the concentration of cellulose in the sodium hydroxide/urea solution. The obtained conductive aerogel was first reported as an electromagnetic interference shielding material, and it exhibited an electromagnetic interference (EMI) shielding effectiveness of −20.8 dB and a corresponding specific EMI shielding effectiveness as high as −219 dB cm^3 g^{-1} with microwave absorption as the dominant EMI shielding mechanism in the microwave frequency range of 8.2–12.4 GHz at a density of as low as 0.095 g cm^{-3}. They concluded that the demonstrated green conductive aerogel has the potential to be used as lightweight shielding material against electromagnetic radiation, especially for aircraft and spacecraft applications.[53]

Ultralight cellulose fiber/thermally reduced graphene oxide (CF/RGO) hybrid aerogel with excellent electromagnetic interference (EMI) shielding capability was reported by Wan *et al.* (2017). They fabricated

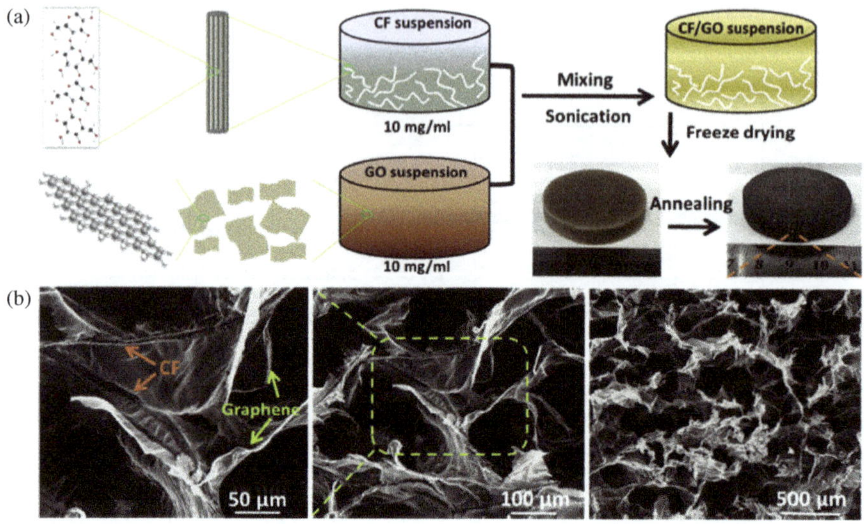

Fig. 10. (a) Schematic illustration of the fabrication process of hybrid aerogel and (b) SEM images of CF2G1-1000H aerogel with different magnifications.[54]

Source: Reproduced with permission from Elsevier Ltd.

the hybrid aerogel *via* lyophilization and carbonization process as shown in Fig. 10. They found that the CF/RGO aerogel with 5 mm thickness exhibited high EMI shielding effectiveness (SE) of ~47.8 dB after annealing at 1000°C with 5% hydrogen–argon mixture. The superior SE was mainly ascribed to the cellular structure and good electrical conductivity of the fabriacted hybrid aerogel. They also investigated the effects of RGO and annealing conditions on EMI shielding and mechanical properties of the fabricated composite aerogel. Finally, they concludes that the demonstrated work will be an effective platform to fabricate low-cost and volume-preserving porous carbon material with superior and tunable EMI shielding performance for potential applications in aerospace applications and in wearable electronic devices.[54]

Another work was reported by Wan *et al.* (2016) wherein they fabricated graphene oxide/cellulose aerogels nanocomposite for electromagnetic interference shielding. They fabricated the hybrid aerogels consisting of graphene oxide (GO) and cellulose *via* a solution mixing–regeneration–freeze drying process as shown in Fig. 11. There were strong hydrogen bond interactions between GO sheets and cellulose molecules as shown in Fig. 11.

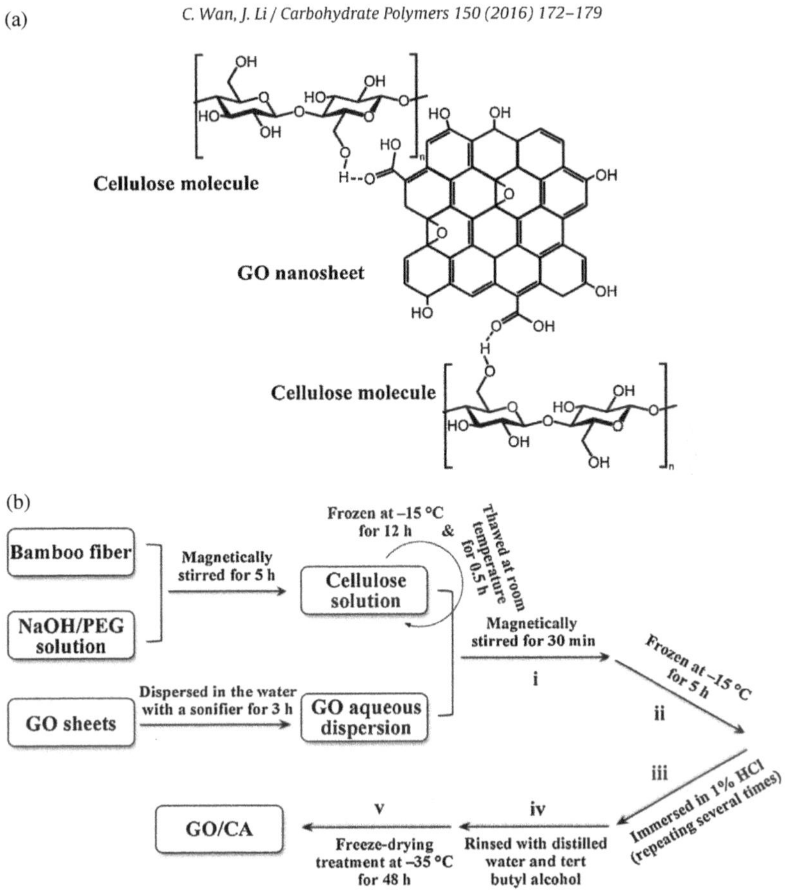

Fig. 11. Schematic illustration of hydrogen bond interactions between GO sheets and cellulose molecules (a) and fabrication of GO/CA nanocomposite (b).

Source: Reproduced with permission from Elsevier Ltd.

They found that, after the reduction of GO in the hybrid aerogels by l-ascorbic acid and the subsequent pyrolysis of the aerogels, the resultant product displayed some interesting characteristics, including good electro-magnetic interference (EMI) shielding capacity (SEtotal = 58.4 dB) (as shown in Fig. 12) high electrical conductivity (19.1 Sm^{-1}), hydrophobicity, and fire resistance. They concluded that the demonstrated work helps to facilitate the development of both cellulose- and GO-based materials for some advanced applications such as EMI protection, electrochemical devices, water-proofing agents, and fire retardants.[55]

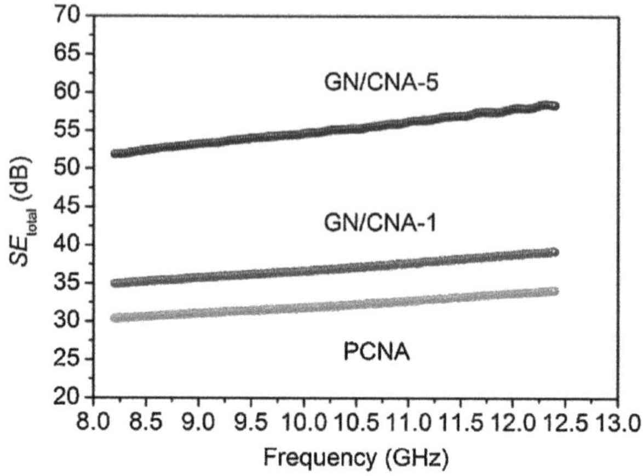

Fig. 12. EMI shielding effectiveness of GN/CNA-1, GN/CNA-5, and PCNA: (a) SEtotal, (b) SER, (c) SEA.[55]

Source: Reproduced with permission from Elsevier Ltd.

8. Nanocellulose Aerogel as Efficient Media for Oil Absorption

Due to the increasing incidents of oil leakage and industrial wastewater discharge, water pollution has attracted considerable attention because of its disastrous effect on the environment and ecosystems. Various techniques such as burning, hydrocarbon degrading, and physical absorption have been developed to rectify this global issue.[56–58] Among these approaches, physical absorption *via* absorbents has gained the most interest due to its low cost and the fact that it does not generate by-products. Many absorbents have been made from mineral materials,[59,60] carbon fibers,[61] polydivinylbenzene[62] etc. However, the abovementioned materials usually suffer from low absorption capacities, complicated fabrication processes, environmental incompatibility, and insufficient buoyancy. In this context, nanocellulose aerogels have gained considerable interest due to their enhanced adsorption capability, eco-friendly nature, and sustainability.[63–65] Moreover, the nanocellulose aerogels also possess high compression strength and high flexibility, which could affect the reusability of the aerogels.

Zhou *et al.* (2016) fabricated mechanically robust aerogels with super hydrophobicity, low density ($\leq$5.08 mg/cm^3), and high porosity

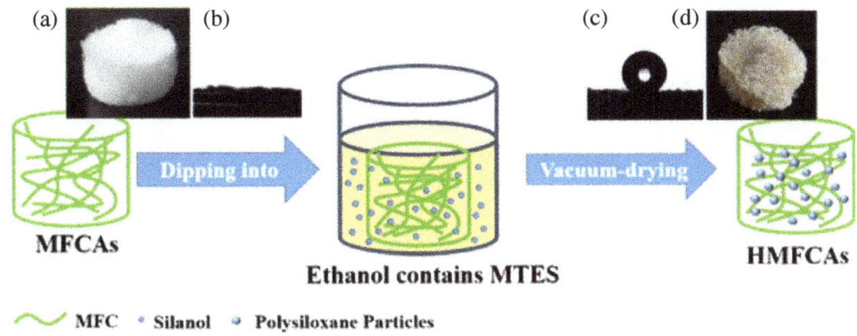

Fig. 13. HMFCAs preparation: (a, d) MFCAs and HMFCAs, (b, c) water contact angle of MFCAs and HMFCAs.[66]

Source: Reproduced with permission from American Chemical Society.

(≥99.68%) from microfibrillated cellulose *via* a facile freeze-drying process followed by a silanization reaction as shown in Fig. 13.

They found that, after silanization process, the polysiloxane particles were successfully coated over the surface of the aerogel, which was resulted in the super hydrophobicity (water contact angle up to 151.8°) and excellent absorption capacities for various oils and organic solvents (220 g/g for dichloromethane) as shown in Fig. 14. They conclude that the demonstrated low-cost, facile, and environmentally friendly preparation method could open up new opportunities for the fabrication of advanced absorbents with oils and organic pollutant cleaning properties.[66]

Korhonen *et al.* (2011) demonstrated such a material, based on functionalized cellulose nanofibrils. They prepared highly porous nanocellulose aerogels shown in Fig. 15 *via* vacuum freeze drying from microfibrillated cellulose hydrogels.

They showed that by functionalizing the native cellulose nanofibrils of the aerogel with a hydrophobic titanium dioxide oleophilic coating, they could selectively absorb oils from water. Because of the low density and ability to absorb non-polar liquids and oils (up to nearly all of its initial volume), the surface-modified aerogels allow the collection of organic contaminants from the water surface, as shown in Fig. 16. They showed that the absorption was close to the overall volume of the aerogel (80–90% vol/vol) and that the mass-based adsorption capacity varied from 20 to 40 (wt/wt) depending on the density of the liquid. They concluded that the nanocellulose-based aerogels will be a good promising candidate as oil absorbents for future sustainable applications.[67]

Fig. 14. Removal of gasoline and 1,2-dichloroethane (dyed with Sudan III) from the surface and bottom of water with HMFCA.[66]

Source: Reproduced with permission from American Chemical Society.

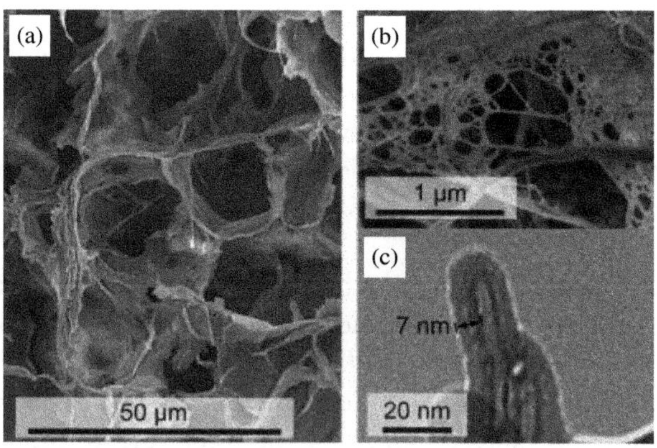

Fig. 15. Microscopic structure of native nanocellulose aerogels. SEM micrographs of (a) freeze-dried nanocellulose aerogels with fibrils packed into sheets, which are connected to form an open, porous aerogel structure, and (b) a magnification of a sheet, which is composed of fibrils making the structure also nanoporous. (c) TEM micrograph of a nanocellulose fibril with a uniform 7 nm TiO_2 coating.[67]

Source: Reproduced with permission from American Chemical Society.

(a) (b)

Fig. 16. (a) TiO_2-coated aerogels are hydrophobic and oleophilic: the water and glycerol stay as droplets (colored with reactive blue dye for clarity), whereas paraffin oil and mineral oil are readily absorbed. (b) Coated aerogel floating on water.[67]

Source: Reproduced with permission from American Chemical Society.

Another study by Jin *et al.* (2011) demonstrated that super hydrophobic and super oleophobic nanocellulose aerogels, consisting of fibrillar networks and aggregates with structures at different length scales, support considerable load on a water surface and also on oils. They showed that the load carrying capacity of the aerogels might be a few orders of magnitude larger than the weight of the aerogel itself. They suggested that this would open up a new platform for carriers or for coatings that have large load bearing capability on various organic and aqueous liquids, including oil-polluted water.[68] Another interesting work was done by Meng *et al.* (2015), and they reported a sponge-like nanocellulose-based carbon aerogels for oil absorption. They prepared carbon aerogel material *via* heat treatment of cellulose microfibril aerogel. They showed that this aerogel was useful as a highly porous oil absorbent (99% of porosity) with ultra-light density (0.01 g/cm^3), hydrophobic properties (144° static contact angle), fast absorption rate, and reusability. Their investigations revealed that carbon aerogel, which was heat treated at 700°C, had higher oil absorption capacity for various types of oils. This was due to the 3D-network structure formed by the entanglement of carbonized cellulose fibrils and the large surface energy of carbonized fibers. They concluded that the unique properties of carbon aerogel formed from nanocellulose, along with the advantages of using it being natural, renewable, low cost, and sustainable, recommend it for a wide range of oil-spill clean-up applications.[69] Feng *et al.* (2015) reported the advanced fabrication and oil absorption properties of super-hydrophilic recycled cellulose aerogels. They coated the recycled cellulose aerogels with methyl/trimethoxy silane (MTMS) *via* chemical vapor deposition yielded and obtained a product with very stable super hydrophobicity for over five months and excellent oil absorption capacities up to 95 gg^{-1} when using the 0.25 wt% cellulose aerogel. It was shown that the initial cellulose fiber concentration significantly affects the oil absorption capability of the developed cellulose aerogels. They concluded that the demonstrated super-hydrophobic recycled cellulose aerogels could be one of the very promising sorbents for oil spill cleaning.[70] Zhang *et al.* (2016) fabricated a highly efficient cellulose-based aerogel with high mechanical strength *via* a novel method. The prepared aerogel possessed a perfect 3D skeleton and interconnected pores like a honey comb, exhibiting excellent oil/water selectivity. In addition, the demonstrated aerogels exhibited stable super hydrophobic and super oleophilic properties even after strong mechanical abrasion. Moreover, they showed that demonstrated cellulose aerogels can be reused for up to 30 cycles.[71] Wang *et al.* (2015) reported an ultra-lightweight, elastic, cost-effective, highly recyclable super absorbent from microfibrillated cellulose fibers for oil spillage clean-up. They demonstrated a simple method to produce a low-cost highly recyclable super absorbent material from renewable cellulose fibers *via* simple and environmentally friendly microfibrillation treatment and freeze-drying. The demonstrated hydrophobic superabsorbent could selectively absorb oil from an oil–water mixture and showed an ultra-high absorption capacity of 88–228 gg^{-1}, which was comparable to those of other novel carbon-based super absorbents. More importantly, the superabsorbent showed excellent flexibility and could be repeatedly squeezed without structure failure (more than 30 times).[72]

9. Conclusion

Cellulose aerogels are ultralight, three-dimensional porous materials that combine the excellent properties of highly porous aerogels, such as high specific surface area, low density, low thermal conductivity, and high insulation, with the fantastic properties of sustainable biopolymers, such as biodegradability and biocompatibility. Due to the above-mentioned unique properties, nanocellulose aerogel are attractive in various applications.

Nanocellulose possesses a combination of properties suitable to produce ultralight, strong, and flexible foams and aerogels for various technological applications. Various processing techniques have been developed that facilitate the production of nanocellulose-based foams and aerogels with a tunable porosity and pore architecture. Nanocellulose aerogel studies on the properties and structure of the wet foams are sparse, and a fundamental understanding of the stability of the foams is still lacking. Studies on how the stability and dynamics of the air–liquid interface is influenced by nanocellulose could provide important insights into how the stability of wet foams can be enhanced. We firmly believe that the research on nanocellulose-based foams and aerogels will continue to grow and look forward to the development of novel process routes and lightweight materials with outstanding properties for various target applications.

References

(1) Chen, A. J. W., Boudreau, M., Watson, R. T. Information systems and ecological sustainability. *J. Syst. Inf. Technol.* **2008**.

(2) Munhoz, D., Gregolin, J. A. R., Faria, L. I. L. De, Andrade, T. F. Automotive materials: Current status, technology trends and challenges. *SAE Conf.* **2007**.

(3) Mohanty, A. K., Misra, M., Drzal, L. T. Sustainable bio-composites from renewable resources: Opportunities and challenges in the green materials world. *J. Polym. Environ.* **2002**, *1–2*, 19–26.

(4) Gopakumar, D. A., Thomas, S., Grohens, Y. Nanocelluloses as innovative polymers for membrane applications. In *Multifunctional Polymeric Nanocomposites Based on Cellulosic Reinforcements*, **2016**, 253–275.

(5) Gopakumar, D. A., Manna, S., Pasquini, D., Thomas, S., Grohens, Y. Nanocellulose: Extraction and application as a sustainable material for wastewater purification. In *New Polymer Nanocomposites for Environmental Remediation*, **2018**, 469–486.

(6) Lavoine, N., Bergström, L. Nanocellulose-based foams and aerogels: Processing, properties, and applications. *J. Mater. Chem. A.* **2017**, *31*, 16105–16417.

(7) Turbak, A. F. A., Snyder, F., Fred, W., Sandberg, K., Albin, F. T., Snyder, K., Sandberg, R. Suspensions containing microfibrillated cellulose. *J. Appl. Polym. Sci. Appl. Polym. Symp.* **1983**.

(8) Ng, H. M., Sin, L. T., Tee, T. T., Bee, S. T., Hui, D., Low, C. Y., Rahmat, A. R. Extraction of cellulose nanocrystals from plant sources for application as reinforcing agent in polymers. *Compos. Part B Eng.* **2015**, *75*, 176–200.

(9) Wang, B., Sain, M. Isolation of nanofibers from soybean source and their reinforcing capability on synthetic polymers. *Compos. Sci. Technol.* **2007**, *67* (11–12), 2521–2527.
(10) Alemdar, A., Sain, M. Isolation and characterization of nanofibers from agricultural residues — Wheat straw and soy hulls. *Bioresour. Technol.* **2008**, *99* (6), 1664–1671.
(11) Chakraborty, A., Sain, M., Kortschot, M. Cellulose microfibrils: A novel method of preparation using high shear refining and cryocrushing. *Holzforschung* **2005**, *1*, 102–107.
(12) Abe, K., Iwamoto, S., Yano, H. Obtaining cellulose nanofibers with a uniform width of 15 Nm from wood. *Biomacromol.* **2007**, *8* (10), 3276–3278.
(13) Iwamoto, S., Nakagaito, A. N., Yano, H., Nogi, M. Optically transparent composites reinforced with plant fiber-based nanofibers. *Appl. Phys. A Mater. Sci. Process.* **2005**, *81* (6), 1109–1112.
(14) Ma, Z., Kotaki, M., Ramakrishna, S. Electrospun cellulose nanofiber as affinity membrane. *J. Memb. Sci.* **2005**, *265* (1–2), 115–123.
(15) Henriksson, M., Henriksson, G., Berglund, L. A., Lindström, T. An environmentally friendly method for enzyme-assisted preparation of microfibrillated cellulose (MFC) nanofibers. *Eur. Polym. J.* **2007**, *43* (8), 3434–3441.
(16) Pääkko, M., Ankerfors, M., Kosonen, H., Nykänen, A., Ahola, S., Österberg, M., Ruokolainen, J., Laine, J., Larsson, P. T., Ikkala, O., *et al.* Enzymatic hydrolysis combined with mechanical shearing and high-pressure homogenization for nanoscale cellulose fibrils and strong gels. *Biomacromol.* **2007**, *6*, 1934–1941.
(17) Tibolla, H., Pelissari, F. M., Menegalli, F. C. Cellulose nanofibers produced from banana peel by chemical and enzymatic treatment. *LWT — Food Sci. Technol.* **2014**, *2*, 1311–1318.
(18) Li, M., Wang, L., Li, D., Cheng, Y.-L., Adhikari, B. Preparation and characterization of cellulose nanofibers from de-pectinated sugar beet pulp. *Carbohydr. Polym.* **2014**, *102*, 136–143.
(19) Chen, W., Abe, K., Uetani, K., Yu, H., Liu, Y., Yano, H. Individual cotton cellulose nanofibers: Pretreatment and fibrillation technique. *Cellulose* **2014**, *3*, 1517–1528.
(20) Deepa, B., Abraham, E., Cherian, B. M., Bismarck, A., Blaker, J. J., Pothan, L. A., Leao, A. L., de Souza, S. F., Kottaisamy, M. Structure, morphology and thermal characteristics of banana nano fibers obtained by steam explosion. *Bioresour. Technol.* **2011**, *102* (2), 1988–1997.
(21) Chirayil, C. J., Joy, J., Mathew, L., Mozetic, M., Koetz, J., Thomas, S. Isolation and characterization of cellulose nanofibrils from helicteres isora plant. *Ind. Crops Prod.* **2014**, *59*, 27–34.
(22) Cherian, B. M., Leão, A. L., de Souza, S. F., Thomas, S., Pothan, L. A., Kottaisamy, M. Isolation of nanocellulose from pineapple leaf fibres by steam explosion. *Carbohydr. Polym.* **2010**, *3*, 720–725.

(23) Fortunati, E., Luzi, F., Jiménez, A., Gopakumar, D. A., Puglia, D., Thomas, S., Kenny, J. M., Chiralt, A., Torre, L. Revalorization of sunflower stalks as novel sources of cellulose nanofibrils and nanocrystals and their effect on wheat gluten bionanocomposite properties. *Carbohydr. Polym.* **2016**, *149*, 357–368.

(24) Pelissari, F. M., Sobral, P. J. D. A., Menegalli, F. C. Isolation and characterization of cellulose nanofibers from banana peels. *Cellulose* **2014**, *1*, 417–432.

(25) Floros, M., Hojabri, L., Abraham, E., Jose, J., Thomas, S., Pothan, L., Leao, A. L., Narine, S. Enhancement of thermal stability, strength and extensibility of lipid-based polyurethanes with cellulose-based nanofibers. In *Polymer Degradation and Stability*, **2012**, *10*, 1970–1978.

(26) Siqueira, G., Bras, J., Dufresne, A. Cellulosic bionanocomposites: A review of preparation, properties and applications. *Polymers*. **2010**, *4*, 728–765.

(27) Savadekar, N. R., Mhaske, S. T. Synthesis of nano cellulose fibers and effect on thermoplastics starch based films. *Carbohydr. Polym.* **2012**, *1*, 146–151.

(28) Zainuddin, S. Y. Z., Ahmad, I., Kargarzadeh, H., Abdullah, I., Dufresne, A. Potential of using multiscale kenaf fibers as reinforcing filler in Cassava starch-kenaf biocomposites. *Carbohydr. Polym.* **2013**, *2*, 2299–2305.

(29) Petersen, K., Væggemose Nielsen, P., Bertelsen, G., Lawther, M., Olsen, M. B., Nilsson, N. H., Mortensen, G. Potential of biobased materials for food packaging. *Trends in Food Sci. Technol.* **1999**, *2*, 52–68.

(30) Karimi, S., Tahir, P. M., Karimi, A., Dufresne, A., Abdulkhani, A. Kenaf bast cellulosic fibers hierarchy: A comprehensive approach from micro to nano. *Carbohydr. Polym.* **2014**, *101*, 878–885.

(31) Shin, H. K., Pyo Jeun, J., Bin Kim, H., Hyun Kang, P. Isolation of Cellulose fibers from kenaf using electron beam. *Radiat. Phys. Chem.* **2012**, *8*, 936–940.

(32) Panaitescu, D. M., Frone, A. N., Nicolae, C. Micro- and nano-mechanical characterization of polyamide 11 and its composites containing cellulose nanofibers. *Eur. Polym. J.* **2013**, *12*, 3857–3866.

(33) Tee, T. T., Sin, L. T., Gobinath, R., Bee, S. T., Hui, D., Rahmat, A. R., Kong, I., Fang, Q. Investigation of nano-size montmorillonite on enhancing polyvinyl alcohol-starch blends prepared *via* solution cast approach. *Compos. Part B Eng.* **2013**, *47*, 238–247.

(34) Alemdar, A., Sain, M. Biocomposites from wheat straw nanofibers: Morphology, thermal and mechanical properties. *Compos. Sci. Technol.* **2008**, *2*, 557–565.

(35) Maiti, S., Jayaramudu, J., Das, K., Reddy, S. M., Sadiku, R., Ray, S. S., Liu, D. Preparation and characterization of nano-cellulose with new shape from different precursor. *Carbohydr. Polym.* **2013**, *98* (1), 562–567.

(36) Sonia, A., Priya Dasan, K., Alex, R. Celluloses Microfibres (CMF) Reinforced poly (Ethylene-Co-Vinyl Acetate) (EVA) composites: Dynamic

mechanical, gamma and thermal ageing studies. *Chem. Eng. J.* **2013**, *228*, 1214–1222.

(37) Lee, S. Y., Mohan, D. J., Kang, I. A., Doh, G. H., Lee, S., Han, S. O. Nanocellulose reinforced PVA composite films: Effects of acid treatment and filler loading. *Fibers Polym.* **2009**, *10* (1), 77–82.

(38) Danial, W. H., Abdul Majid, Z., Mohd Muhid, M. N., Triwahyono, S., Bakar, M. B., Ramli, Z. The reuse of wastepaper for the extraction of cellulose nanocrystals. *Carbohydr. Polym.* **2015**, *118*, 165–169.

(39) Kargarzadeh, H., Ahmad, I., Abdullah, I., Dufresne, A., Zainudin, S. Y., Sheltami, R. M. Effects of hydrolysis conditions on the morphology, crystallinity, and thermal stability of cellulose nanocrystals extracted from kenaf bast fibers. *Cellulose* **2012**, *3*, 855–866.

(40) Jiang, F., Hsieh, Y. Lo. Cellulose nanocrystal isolation from tomato peels and assembled nanofibers. *Carbohydr. Polym.* **2015**, *122*, 60–68.

(41) Lamaming, J., Hashim, R., Sulaiman, O., Leh, C. P., Sugimoto, T., Nordin, N. A. Cellulose nanocrystals isolated from oil palm trunk. *Carbohydr. Polym.* **2015**, *127*, 202–208.

(42) Bettaieb, F., Khiari, R., Hassan, M. L., Belgacem, M. N., Bras, J., Dufresne, A., Mhenni, M. F. Preparation and characterization of new cellulose nanocrystals from marine biomass posidonia oceanica. *Ind. Crops Prod.* **2015**, *72*, 175–182.

(43) Fortunati, E., Luzi, F., Puglia, D., Petrucci, R., Kenny, J. M., Torre, L. Processing of PLA nanocomposites with cellulose nanocrystals extracted from Posidonia Oceanica waste: Innovative reuse of coastal plant. *Ind. Crops Prod.* **2015**, *67*, 439–447.

(44) Mohamed, M. A., Salleh, W. N. W., Jaafar, J., Asri, S. E. A. M., Ismail, A. F. Physicochemical Properties of "Green" Nanocrystalline cellulose isolated from recycled newspaper. *RSC Adv.* **2015**, *5* (38), 29842–29849.

(45) Wicklein, B., Kocjan, A., Salazar-Alvarez, G., Carosio, F., Camino, G., Antonietti, M., Bergström, L. Thermally insulating and fire-retardant lightweight anisotropic foams based on nanocellulose and graphene oxide. *Nat. Nanotechnol.* **2015**, *3*, 277–283.

(46) Nguyen, S. T., Feng, J., Ng, S. K., Wong, J. P. W., Tan, V. B. C., Duong, H. M. Advanced thermal insulation and absorption properties of recycled cellulose aerogels. *Colloids Surfaces A Physicochem. Eng. Asp.* **2014**, 128–134.

(47) Hayase, G., Kanamori, K., Abe, K., Yano, H., Maeno, A., Kaji, H., Nakanishi, K. Polymethylsilsesquioxane-cellulose nanofiber biocomposite aerogels with high thermal insulation, bendability, and superhydrophobicity. *ACS Appl. Mater. Interfaces* **2014**, *6* (12), 9466–9471.

(48) He, C., Huang, J., Li, S., Meng, K., Zhang, L., Chen, Z., Lai, Y. Mechanically resistant and sustainable cellulose-based composite aerogels

with excellent flame retardant, sound-absorption, and superantiwetting ability for advanced engineering materials. *ACS Sustain. Chem. Eng.* **2018**, *1*, 927–936.
(49) Yuan, B., Zhang, J., Mi, Q., Yu, J., Song, R., Zhang, J. Transparent cellulose-silica composite aerogels with excellent flame retardancy *via* an *in situ* sol-gel process. *ACS Sustain. Chem. Eng.* **2017**, *11*, 11117–11123.
(50) Han, Y., Zhang, X., Wu, X., Lu, C. Flame retardant, heat insulating cellulose aerogels from waste cotton fabrics by *in situ* formation of magnesium hydroxide nanoparticles in cellulose gel nanostructures. *ACS Sustain. Chem. Eng.* **2015**, *8*, 1853–1859.
(51) Bendahou, D., Bendahou, A., Seantier, B., Grohens, Y., Kaddami, H. Nano-fibrillated cellulose-zeolites based new hybrid composites aerogels with super thermal insulating properties. *Ind. Crops Prod.* **2015**, *65*, 374–382.
(52) Chung, D. D. Electromagnetic interference shielding effectiveness of carbon materials. *Carbon N.Y.* **2001**, *2*, 279–285.
(53) Huang, H. D., Liu, C. Y., Zhou, D., Jiang, X., Zhong, G. J., Yan, D. X., Li, Z. M. Cellulose composite aerogel for highly efficient electromagnetic interference shielding. *J. Mater. Chem. A* **2015**, *9*, 4983–4991.
(54) Wan, Y. J., Zhu, P. L., Yu, S. H., Sun, R., Wong, C. P., Liao, W. H. Ultralight, super-elastic and volume-preserving cellulose fiber/graphene aerogel for high-performance electromagnetic interference shielding. *Carbon N.Y.* **2017**, 629–639.
(55) Wan, C., Li, J. Graphene oxide/cellulose aerogels nanocomposite: Preparation, pyrolysis, and application for electromagnetic interference shielding. *Carbohydr. Polym.* **2016**, 172–179.
(56) Jiang, F., Hsieh, Y. Lo. Amphiphilic superabsorbent cellulose nanofibril aerogels. *J. Mater. Chem. A* **2014**, *18*, 6337–6342.
(57) Ventikos, N. P., Vergetis, E., Psaraftis, H. N., Triantafyllou, G. A high-level synthesis of oil spill response equipment and countermeasures. *J. Hazard. Mater.* **2004**, *1*, 51–58.
(58) Yuan, J., Liu, X., Akbulut, O., Hu, J., Suib, S. L., Kong, J., Stellacci, F. Superwetting nanowire membranes for selective absorption. *Nat. Nanotechnol.* **2008**, *6*, 332–336.
(59) Carmody, O., Frost, R., Xi, Y., Kokot, S. Adsorption of hydrocarbons on organo-clays-implications for oil spill remediation. *J. Colloid Interface Sci.* **2007**, *1*, 17–24.
(60) Chen, D., Zhu, G., Lin, J., Liu, J., Huang, S. Renewable Reduced Graphene Oxide-Based Oil-Absorbent Aerosols: Preparation and essential oils absorption ability. *ACS Sustain. Chem. Eng.* **2015**, *7*, 1428–1433.
(61) Inagaki, M., Kawahara, A., Nishi, Y., Iwashita, N. Heavy oil sorption and recovery by using carbon fiber felts. *Carbon N.Y.* **2002**, *9*, 1487–1492.

(62) Zhang, Y., Wei, S., Liu, F., Du, Y., Liu, S., Ji, Y., Yokoi, T., Tatsumi, T., Xiao, F. S. Superhydrophobic nanoporous polymers as efficient adsorbents for organic compounds. *Nano Today* **2009**, *2*, 135–142.
(63) Zhang, Z., Sèbe, G., Rentsch, D., Zimmermann, T., Tingaut, P. Ultralightweight and flexible silylated nanocellulose sponges for the selective removal of oil from water. *Chem. Mater.* **2014**, *26* (8), 2659–2668.
(64) Chen, W., Li, Q., Wang, Y., Yi, X., Zeng, J., Yu, H., Liu, Y., Li, J. Comparative study of aerogels obtained from differently prepared nanocellulose fibers. *Chem. Sus. Chem.* **2014**, *7* (1), 154–161.
(65) Pääkkö, M., Vapaavuori, J., Silvennoinen, R., Kosonen, H., Ankerfors, M., Lindström, T., Berglund, L. A., Ikkala, O. Long and entangled native cellulose i nanofibers allow flexible aerogels and hierarchically porous templates for functionalities. *Soft Matter.* **2008**, *12*, 2492–2499.
(66) Zhou, S., Liu, P., Wang, M., Zhao, H., Yang, J., Xu, F. Sustainable, reusable, and superhydrophobic aerogels from microfibrillated cellulose for highly effective oil/water separation. *ACS Sustain. Chem. Eng.* **2016**, *4* (12) 6409–6416.
(67) Korhonen, J. T., Kettunen, M., Ras, R. H. A., Ikkala, O. Hydrophobic nanocellulose aerogels as floating, sustainable, reusable, and recyclable oil absorbents. *ACS Appl. Mater. Interfaces.* **2011**, *3* (6) 1813–1816.
(68) Jin, H., Kettunen, M., Laiho, A., Pynnönen, H., Paltakari, J., Marmur, A., Ikkala, O., Ras, R. H. A. Superhydrophobic and superoleophobic nanocellulose aerogel membranes as bioinspired cargo carriers on water and oil. *Langmuir.* **2011**, *27* (5), 1930–1934.
(69) Meng, Y., Young, T. M., Liu, P., Contescu, C. I., Huang, B., Wang, S. Ultralight carbon aerogel from nanocellulose as a highly selective oil absorption material. *Cellulose* **2015**, *22* (1), 435–447.
(70) Feng, J., Nguyen, S. T., Fan, Z., Duong, H. M. Advanced fabrication and oil absorption properties of super-hydrophobic recycled cellulose aerogels. *Chem. Eng. J.* **2015**, 168–175.
(71) Zhang, H., Li, Y., Xu, Y., Lu, Z., Chen, L., Huang, L., Fan, M. Versatile fabrication of a superhydrophobic and ultralight cellulose-based aerogel for oil spillage clean-up. *Phys. Chem. Chem. Phys.* **2016**, *40*, 28297–28306.
(72) Wang, S., Peng, X., Zhong, L., Tan, J., Jing, S., Cao, X., Chen, W., Liu, C., Sun, R. An ultralight, elastic, cost-effective, and highly recyclable superabsorbent from microfibrillated cellulose fibers for oil spillage cleanup. *J. Mater. Chem. A* **2015**, *3* (16), 8772–8781.

https://doi.org/10.1142/9781786349477_0016

Chapter 16

Applications of Bacterial Cellulose

Fazli Wahid[*,†] **and Cheng Zhong**[*,†,‡]

[*]*State Key Laboratory of Food Nutrition and Safety, Tianjin University of Science and Technology, Tianjin, PR China*

[†]*Key Laboratory of Industrial Fermentation Microbiology, (Ministry of Education), Tianjin University of Science and Technology, Tianjin, PR China*

[‡]*czhong@tust.edu.cn*

Abstract

Over the past couple of decades, the microbial cells-produced cellulose, termed as bacterial cellulose (BC) or bacterial nanocellulose (BNC), has received immense consideration for its applications in different fields. The unique morphological features, surface chemistry, mechanical strength, and inertness make BC an ideal biomaterial. BC, both alone and in the form of composites with other materials ranging from natural and synthetic polymers, nanomaterials, and clay materials, have found applications in different fields. This chapter provides an overview of applications of BC in the biomedical field, including as wound dressings, antimicrobials, drug delivery system, blood vessels, optics, and bone and cartilage implants. It further discusses its applications in food industry as packaging material as well as a dietary food, food additive, and stabilizer. The non-toxic, hydrophilic, high liquid absorbing capacity, and biocompatible nature enable BC to find applications in cosmetics industry.

Keywords: Bacterial cellulose; properties; biomedical applications; food industry; cosmetics

Based on the unique properties of bacterial cellulose (BC) such as high degree of crystallinity, high purity, high density, high water-binding capacity, good shape retention properties, and high surface area, it has many applications in a variety of contexts such as food, pharmaceuticals, wastewater treatment, broadcasting, and so forth [1–3]. These distinctive features have facilitated the use of BC in the production of different products. Although BC is mainly prepared on the laboratory scale for research purposes, there are also some marketable outlets of BC. In addition to commercial nata de coco, a German-based company (Fzmb GmbH) is one of the largest manufacturers of BC for biomedical and cosmetic applications. Furthermore, Prima Cel™, a BC-based wound dressing is produced by Xylos Co., USA. Biofill® and Gengiplex® are other brands of BC-based materials used for tissue regenerations. There are many food industries in Asian countries that are producing BC for food and packing applications. Sony Corporation (Japan) in association with Ajinomoto (Japan) and other companies are developing BC-based diaphragms for audio speakers. Ajinomoto (Japan) also provides wet BC. The objective of this chapter is to highlight the applications of BC in biomedical fields, food, and cosmetics, as summarized in Fig. 1 and discussed in the following sections [4, 5].

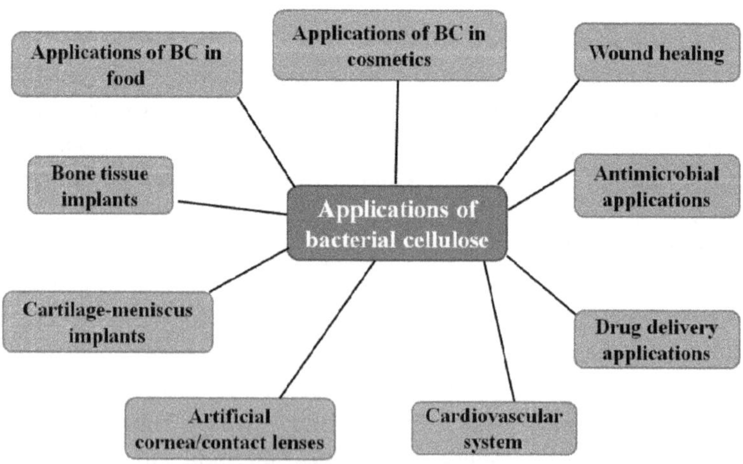

Fig. 1. Applications of BC in different fields.

1. Wound Dressing

Wound dressings have been used to treat severe skin injuries and burns for centuries. In early ages, the major role of the wound dressings was considered as a wound protection in the healing process. The main function of conventional dressings (natural and synthetic bandages) was to keep the wound dry. The absorption and evaporation of exudates, as well as the inhibition of bacterial invasion, were considered to play an important role in effective wound healing. However, this opinion of wound supervision has been changed considerably in the past few decades. Nowadays, dressings are considered as a dynamic part of the healing process, which are developed to control infection and provide a promising healing environment. A moist and warm environment encourages fast and effective healing, particularly essential in chronic wounds [6, 7].

Currently, the international market offers various kinds of wound dressing materials, developed from synthetic or natural polymers, and their combinations. These are applied in several systems (films, hydrogels, hydrocolloids, and foams); these materials may contain drugs, peptides, growth factors, and other active agents that accelerate wound healing [7]. However, an "ideal dressing" demands a moist environment, effective oxygen circulation, and thermal insulation. Additionally, it must assist in the inhibition of bacterial invasion, absorb exudates, and ensure liquid drainage. Moreover, it must easy to be applied and simple to be removed, and it must be biocompatible [8]. This variety of required characteristics for wound dressing materials is summarized in Fig. 2. Unfortunately, none of the present supplies is capable of realizing all requests of an "ideal dressing" [9]. However, a variety of materials have been used as wound dressings. Among them, hydrogels have been evaluated for treating chronic and burn wounds. These three-dimensional polymer networks are filled with a considerable amount of liquids and offer a supportive environment for tissue engineering. These materials cover the wound surface, protecting them from bacterial invasion, and allow the transportation of water and oxygen. BC is a naturally derived hydrogel that is extensively used for wound dressing. The use of BC as wound dressing materials is governed by its unique characteristics: it has a high water-holding capacity, flexibility, and tensile strength, great compatibility with living tissues, and a characteristic permeability for gases and liquids [10]. BC in its pristine form or modified to acquire the essential characteristics is used for wound dressing (Fig. 2). Its high surface area

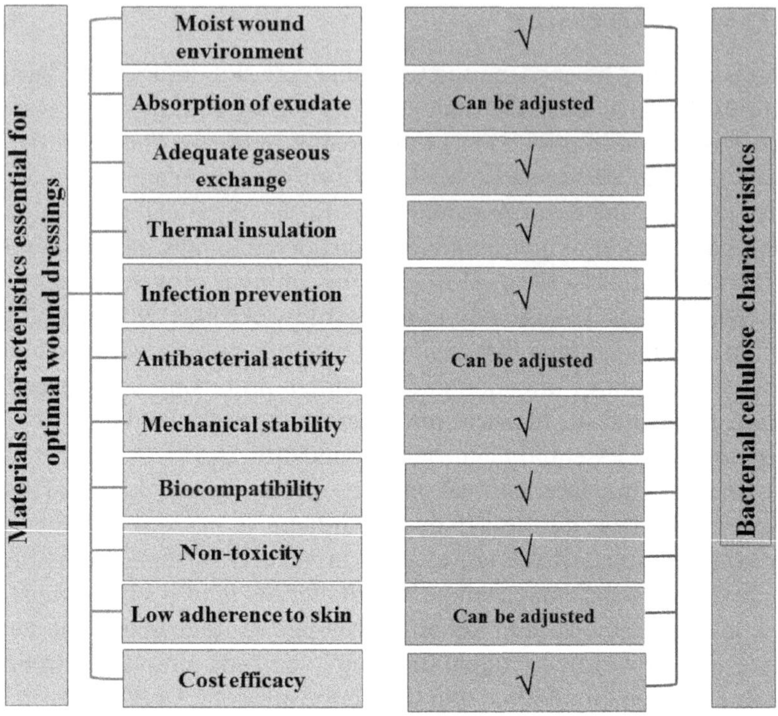

Fig. 2. An overview of the general requirements of a wound dressing material with respect to characteristics of BC.

Source: Reproduced from Ref. [23].

and porosity show the possibility to introduce and release medicine and other functional materials [11]. Chemically reactive sites within the structure of BC provide an additional possibility to introduce specific non-native functionalities [12]. The incorporation of other agents that can induce wound healing and the designing of BC-based composites can all contribute to the development of new dressing materials with enhanced features for wound treatment [13–15]. One of the first proposed and direct applications of the BC-based membrane as a wound dressing is shown in Fig. 3.

Some wound dressings based on BC are available commercially, such as Bioprocess®, BioFill®, Geniflex®, and XCell®. However, in terms of an "ideal dressing", the biomembrane BioFill® is among the first marketable products that realizes the major requirements, including permeability, low cost, elasticity, durability, acting as a physical barrier for bacteria, and

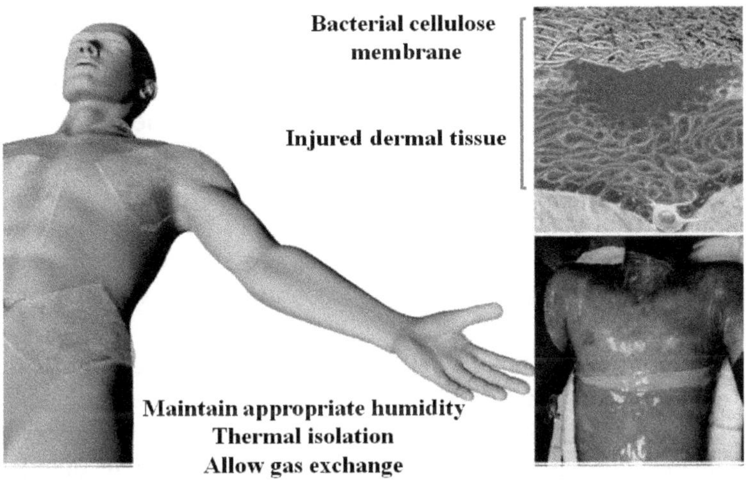

Fig. 3. A schematic representation of the BC membrane as a wound dressing.
Source: Reproduced from Ref. [17].

hemostatic activity. Moreover, the efficacy of BioFill® to enhance the healing process and provide pain relief has been confirmed in more than 300 cases [16]. Nevertheless, the pain-relieving mechanism of these wound dressings is not completely recognized. Some researchers have suggested that the healing mechanism is due to the capture of ions by hydrogen bonds or the three-dimensional network of BC mimicking the skin surface, and making a favorable environment for healing and regeneration [17]. It is worth mentioning that the wet BC membranes are representing a new class of wound dressings in the treatment of burn wounds. These kinds of membranes provide a favorable moist microenvironment for rapid wound cleansing, and subsequently faster-wound healing.

BC can be modified to enable the design of new nanocomposites with better properties and improved functionalities. In this context, different BC-based nanocomposites have been reported for wound dressings. For example, BC/gelatin [18] and BC/collagen [19] nanocomposites exhibited increased mechanical properties that have the potential for wound dressing applications. Saibuatong and Phslaphong [20] developed BC/aloe vera nanocomposite films. The films were obtained by adding 30% (v/v) of aloe gel in the culture medium, resulting in reinforced BC fibers with enhanced mechanical strength, crystallinity, and water absorption as compared to pure BC films. One of the most important characteristics of designing a wound dressing material is the improvement of the water-holding capacity of the final product. Accordingly, a BC/chitosan composite was designed that was applied to treat wounds that were hard to heal, bedsores, burns, and skin ulcers that need frequent dress changes [21]. Lin and coworkers [22] reported wound healing efficacy *in vivo* of BC/chitosan composites using rat models. They observed that the composite did not show any toxic effect on animal cells. Furthermore, a faster epithelialization was found in wounds treated with the BC/chitosan composite as compared to pure BC or commercially available dressing materials.

Another important characteristic of designing wound dressing materials is antibacterial activity. However, BC does not possess any antibacterial properties. Therefore, to introduce the antibacterial activity to BC films, various approaches have been used to modify it. This detail will be discussed in the next section.

In summary, BC itself has great potential in biomedical and biotechnological fields due to its distinctive nanoporous, three-dimensional fibrous structure. This structure as well as its transparency, mechanical stability, and biocompatibility make BC a suitable material for wound dressing. BC-based dressings allow proper gaseous exchange, prevent bacterial invasion, provide a moist atmosphere, and also absorb excess wound exudates. The unavailability of another biomaterial that addresses the challenges required for a modern wound dressing makes BC intensively used as a primary material for wound care.

2. Antimicrobial Applications

The emergence of pathogen resistance to antibiotics is a major challenge in biomedical fields [24, 25]. However, new antimicrobial materials should not be entirely consisting of antibiotics, but other antimicrobial materials can also be designed. One of such technique can be the use of BC-based materials. However, BC itself does not show antimicrobial properties [26]. To impart antimicrobial functionality to BC, several people have modified BC with various antimicrobial agents [27]. Normally, the introduction of other materials into BC is not a complicated process. The large surface area and the presence of hydroxyl groups in BC have great adsorption capability of large and small molecules such as polysaccharides, proteins, quaternary ammonium compounds, and nanoparticles [28, 29].

Antibacterial BC/chitosan nanocomposites were developed by supplementing deacetylated chitosan nanocrystals to the culture medium. The nanocomposites were also prepared by an *ex situ* method *via* mixing an aqueous suspension of chitosan nanocrystals with BC. The results suggested using BC/chitosan nanocomposite as an effective antimicrobial material [30]. In another similar study, BC/chitosan composites were prepared by *in situ* method [31], and the resulting composite material had *N*-acetyl glucosamine and glucosamine units incorporated into the BC film. These materials showed a variety of important features such as good wet-state mechanical properties, moisture-retaining properties, bacteriostatic activity, and the release of oligosaccharides under lysozyme action. These characteristics make BC/chitosan composite a unique antimicrobial material for treating burns, ulcers, and wounds.

A study reported that the addition of poly(vinylalcohol) (PVA) at concentrations as low as 0.5% can enhance the permeability of BC for biomedical applications. The pore size as well as the mechanical properties of the composite were improved compared to the pure BC. In their study, a BC-based protecting material possessing an effective pore size of 50–70 nm with antiviral function was developed by combining with chitosan. As the majority of the viruses possess an average size greater than 50 nm, these composites (with pore size 50–70 nm) have the ability to filter most of the viruses, thus inhibiting viral infection. Moreover, BC/PVA films incorporated with sorbic acid have also been prepared for antimicrobial applications [32]. In the developed composite films, crushed BC was used as reinforcing fibers, PVA was used as a polymeric matrix, and sorbic acid was used as an antibacterial agent. The authors suggested that the BC/PVA/sorbic acid composites can have potential applications as antimicrobial films. In another study, silver nanoparticles (Ag-NPs) were introduced to BC/PVA membrane for antimicrobial applications [33]. The developed BC/PVA/Ag-NPs showed antibacterial activity against *Escherichia coli* (*E. coli*). Several other studies presented the incorporation of Ag-NPs containing BC-based materials for antimicrobial applications. For example, Ag-NPs were incorporated into the BC membrane by reducing silver salt by using a reducing agent sodium borohydride ($NaBH_4$). Stable Ag-NPs were formed due to the interaction between hydroxyl groups in cellulose and Ag-NPs. The nanocomposite membrane showed antibacterial activity greater than 99.99% against *E. coli* and *Staphylococcus aureus* (*S. aureus*) bacteria [34]. Similarly, the antibacterial activity of BC/Ag-NPs was tried against *Klebsiella pneumoniae*

(*K. pneumoniae*), *S. aureus,* and *Bacillus subtilis (B. subtilis*). The nanocomposite membranes were effective against the tested bacteria due to the presence of Ag-NPs even at concentrations as low as 5.0×10^{-4} % (w/w). These characteristics of the BC/Ag-NPs membrane make it an effective agent for wound dressing with antimicrobial activity.

An environmentally friendly hydrothermal method was suggested to produce a BC/Ag-NPs by using BC as a reducing agent without the addition of any other reducing agent [35]. A narrow size distribution (17.1 ± 5.9 nm) was obtained on the BC matrix having 1.78% (w/w) content of Ag-NPs under optimum condition. The resultant nanocomposite showed a sustainable release of Ag-NPs and extended antibacterial activity against *S. aureus* [36]. In another study, BC/Ag-NPs antimicrobial nanocomposite was prepared by *in situ* synthesis of Ag-NPs in the BC matrix, by reducing silver salt by triethanolamine (TEA). The resultant nanocomposite membrane revealed strong antimicrobial activity against tested bacteria [37]. Similarly, BC/Ag-NPs membranes were developed by using TEMPO (2,2,6,6-tetramethylpiperidine-1-oxyl)-oxidized BC membrane. Ag-NPs were incorporated into BC by *in situ* method using different reducing agents such as $NaBH_4$, and sodium citrate. The prepared nanocomposite exhibited effective antibacterial activity against the tested bacteria [38].

ZnO nanoparticles (ZnO-NPs) have also been incorporated into BC by various research groups for antimicrobial activity. For example, a BC/ZnO nanocomposite membrane was reported which showed antibacterial properties against *E. coli* and *S. aureus*. The composite was fabricated by ultrasonic-assisted *in situ* precipitation of ZnO-NPs on the surface of BC [39]. In another study, ZnO-NPs were incorporated into regenerated BC, the nanocomposite exhibited ZnO-NPs content-dependent antibacterial activity [40]. Xiong *et al.* [41] reported that the morphology of *in situ* preparation of ZnO-NPs could be controlled by the hydroxyl groups of BC due to the electrostatic interactions with zinc ions. They observed flower-like aggregates of ZnO-NPs formed on the BC, where spindle-shaped ZnO-NPs were formed without BC. The incorporation of CuO [42, 43] and TiO_2 [44] nanoparticles have also been reported to incorporate into BC membranes for antimicrobial applications.

3. Drug Delivery Applications

An effective drug delivery system can be affected by many factors. One of these is to identify a suitable material for research and designing of novel drug delivery systems. BC is a biopolymer that can justify the standards to be considered as a drug delivery material. Recently, various nanocellulose-based drug delivery systems have been proposed for pharmaceutical applications. Several BC-based nanocomposites systems for drug delivery have been designed by combining different materials such as inorganic nanoparticles, polymers, and small molecules with the BC membrane [16]. Among them, the BC/polyacrylic acid (PAA) composite has been widely investigated for drug delivery applications. For instance, BC/PAA composites were developed by polymerization initiated by irradiating with electron beams [45]. The composites exhibited sensitivity to pH and presented maximum swelling at pH 7, which was investigated for controlled delivery of bovine serum albumin (BSA). The release of the drug was also investigated in simulated gastric fluid (SGF) for two hours and simulated intestinal fluid (SIF) without enzymes until the maximum drug was released. About 15% of drug release was observed in SGF after 2 hours, while the drug release rate was considerably higher in SIF. The drug release was also dependent on the radiation doses. For the lowermost radiation doses, complete release was attained after 8 hours, whereas the complete release took about 13–14 hours for the highest radiation doses. The difference in the release rate of the drug in SGF and SIF was due to the difference in the swelling ratio of the materials. The pH-responsive drug release behavior of the prepared material was similar to that of the pure BC membrane [45]. A similar study was conducted by another group of researchers who used BC/PAA nanocomposite hydrogels for the oral delivery of proteins. A minor cumulative release (less than 10%) of BSA was found in SGF. The results revealed the possibility of BC/PAA nanocomposite hydrogel to protect BSA from the severe acidic conditions of the stomach. Moreover, the *ex vivo* penetration tests revealed an increased BSA permeation across the intestinal mucosa. Furthermore, the BC/PAA revealed excellent biocompatibility with no sign of toxicity [46]. A group of researchers investigated the BC-based drug delivery system for serum albumin. They found that the pristine BC exhibited more protein uptake as compared to freeze-dried BC and that the albumin exhibited biological stability throughout materials processing [47].

BC membranes have also investigated for transdermal diclofenac delivery. The drug-loaded membranes were homogenous, flexible, and showed significant swelling behavior. *In vitro* diffusion experiments showed that diclofenac loaded membranes had similar diffusion rates to commercially available patches but lower than the commercial gel [48].

Other model studies also investigated the possible effective delivery systems based on BC. For example, paracetamol tablets were coated with BC by a spray coating technique and the release of drug was investigated *in vitro*. It was found that BC presented a remarkable capability to form soft, foldable, and flexible films without the presence of any plasticizer and was capable of delivering paracetamol [49].

Researchers also used BC membranes for topical delivery of ibuprofen and lidocaine [50, 51]. A burst release profile was observed in phosphate buffer solution (pH 7.4) at 32°C, where over 90% drug was released within 20 minutes. The transdermal delivery of lidocaine of different kinds (BC membrane, aqueous solution and, gel) was investigated *in vitro* using human epidermis. The results revealed that the penetration rate of lidocaine associated with BC membranes was considerably lower than other conventional delivery systems (aqueous solution and gel). Almeida and coworkers [52] prepared the BC-glycerin membranes system for topical drug delivery. The *in vivo* study using human subjects revealed good skin tolerance after a single application of BC-glycerin membranes. Moreover, the composite membrane exhibited a moisturizing effect that could be used for the treatment of skin diseases, such as atopic dermatitis and psoriasis.

Conclusively, simple approaches of drug loading, comparable drug release profile, and ease of application showed the capabilities of the BC membrane as a drug delivery system for different drugs.

4. Cardiovascular System

Blood circulates in the body through blood vessels under normal physiological conditions. Nevertheless, the blood vessels can be injured or blocked [53], which sometimes need to be substituted by another blood vessel collected from the patient's own body or the donor or by artificial blood vessels. There are several problems with using the patient's own vessels, because the double surgery may lead to other health problems such as blood loss, multiple surgeries, and surgical complications. The donor approach has also some drawbacks in the form of rejection of tissues and the development of arteriosclerosis [54]. In order to solve these problems, researchers are working to develop artificial blood vessels. Similarly, the rejection among the natural tissue and synthetic grafts is a major issue for the failure of presently accessible substituents [55, 56]. Synthetic grafts, such as Dacron (polyester) and expanded polytetrafluoroethylene (ePTFE), induce thrombosis, and hence they are not appropriate substituents for small blood vessels (diameter less than 6 mm) [5].

BC and its composites because of their unique structural features such as fibrous nature, non-toxicity, foldability, and moldability are excellent candidates for artificial blood vessels [57]. The designing of BC-based artificial blood vessels, BASYC®, indicates a breakthrough in biomedical applications. BASYC® was used to replace the carotid artery in five mice and exhibited one-year patency. Similarly, BASYC® has demonstrated patency for 90 days in pig prostheses. The prosthesis was epithelialized and developed three distinct layers of the structure, similar to natural blood vessels [58]. Moreover, BASYC® alone or loaded with a neuroregenerative drug also revealed excellent results in the innervation of tissues in the animal model after neurosurgery [59]. In addition, Brown and coworkers [60] fabricated BC/fibrin composites cross-linked by glutaraldehyde, which showed similar mechanical properties to that of natural blood vessels. The viscoelastic properties of the BC/fibrin composites were comparable with natural blood vessels (coronary artery).

The substitution of a heart valve with a suitable artificial valve is called a cardiac prosthesis. This is an innovative medical therapy that minimizes the mortality rate in cardiac patients [61]. Various biomedical materials have been used in such medical practices. However, these materials have restrictions for long-term use because of the oxidative reactions with blood, and they also have some other serious side effects [62]. Thus, research is in progress to find novel materials for developing artificial heart valves. Recently, BC and its composites have been examined for developing artificial heart valves [63, 64]. Various groups of researchers have been comprehensively studying BC in combination with PVA as blood-contacting materials. In the first attempt, BC/PVA composites were synthesized with lower values of tensile strength and Young's modulus [65]. By introducing PVA, the tensile strength of the composite was decreased five times, while Young's modulus was reduced from 3.5 GPa to 1 GPa. The hemocompatibility of BC/PVA composites was also examined by measuring the activation Factor-XII and adhesion and activation of platelets [66]. The results demonstrated that the BC/PVA composites have a fair hemocompatibility due to the low activation of Factor-XII and platelets. BC and BC/PVA composites exhibited superior platelet adhesion and activation profiles compared to ePTFE. The prepared composites with similar properties to heart valve were hemocompatible and hence showed the possibility of being used in cardiac prosthesis [62]. Similarly, biofabricaiton of PVA-based nanocomposite reinforced with BC has also been reported, which was produced by the modification of culture medium with water-soluble PVA. The PVA/BC composites showed synergistic properties contributed by both polymers [67]. To minimize the loss during the purification process and to improve the functional properties of nanocomposites, PVA was cross-linked with glyoxal. The reinforcement of BC with PVA produced nanocomposites with excellent dimensional, mechanical, and thermal properties. The composites were used for the construction of heart valve leaflets. These nanocomposites showed mechanical properties similar to the porcine-based heart valves.

5. Artificial Cornea/Contact Lenses

Corneal disease is the main cause of vision loss across the globe because the transparency of the cornea is lost in the pathological conditions [68]. In order to deal with this problem, keratoplasty is a common practice to substitute the damaged cornea. However, corneal neovascularization, xerophthalmia, the shortage of donors, and graft rejection are related to adverse effects [69]. Recently, tissue engineering techniques have been used to design a corneal replacement. BC is a biocompatible polymer and holds unique material properties and thus has been explored as a novel scaffold for preparing artificial cornea and for other corneal tissue engineering uses [70]. It has been revealed that BC scaffolds can be used for maintaining, growth, and proliferation of corneal stromal cells. It is further indicated that BC scaffolds supported corneal stromal cells, which presented its potential to be used for cornea production.

Another major vision problem is the refractive error (RE); more than 2.3 billion people suffer from this problem worldwide [71]. RE may result in poor vision such as astigmatism, presbyopia, and myopia, with severe social and economic implications, if not treated properly. RE can be diagnosed and corrected very simply by using refractive surgical strategies or devices, such as contact lenses and spectacles. Contact lenses are extensively used to treat vision problems in humans. The contact lenses are commercially available in both soft and hard forms. Generally, the soft variety of contact lenses are made from soft polymers, while hard varieties are made from glass or stiff polymers, such as poly(hydroxyethyl methacrylate) (PHEMA), poly(methyl methacrylate) (PMMA), and polycarbonates [72].

BC has been used to develop contact lenses because of its high degree of transparency. The BC-based contact lenses were produced from a solution of BC in 1-butyl-3-methylimidazolium chloride. The prepared viscous solution was added into a mold and a clear contact lens was detached from the mold. The residual solvent was diffused from the contact lens into the water, and the formed contact lens was transparent and exhibited stability for about eight weeks [72]. Another transparent nanocomposite hydrogel was developed by combining water-soluble polymer PHEMA and water-soluble nanofibers of BC. The developed material possessed high water content, mechanical integrity, and good strength [73]. The resulting BC/PHEMA nanocomposite hydrogels have the capability to be used in contact lenses.

In view of the aforementioned literature, BC alone or combined with other materials could be used as the artificial cornea and contact lenses.

6. Cartilage-Meniscus Implants

The repairing of the cartilage defects is a major clinical requirement because it has a very limited regeneration capacity. The materials required for artificial cartilage not only need to be strong but also tough to withstand biodegradation, whereby a material implanted in a living body deteriorates by a biological mechanism with the passage of time. The requirements of bio-mimic scaffolds have led to the consideration of BC as a scaffold material. The *in vitro* macrophage screening showed that BC did not stimulate the pro-inflammatory cytokines. Chondrocytes seeded on the BC membrane indicated proliferation and the production of collagen type II, revealing the suitability of the BC membrane as a bio-mimicking scaffold [74]. BC/PVA nanocomposites were prepared by the addition of BC less than 1% into PVA followed by exposure of the blend to thermal treatment [75]. The resultant BC/PVA nanocomposites exhibited tunable mechanical properties, and the elastic modulus of the nanocomposite was similar to natural articular cartilage. Therefore, BC/PVA nanocomposites could be a potential substitute for articular cartilage and other orthopedic defects such as injured vertebral discs. Similarly, due to the good mechanical properties, foldability, biocompatibility, and low price, an artificial endocardium (AE) was developed. In this research, the wet BC membranes were crushed, followed by the addition of sodium alginate, PVA, and carboxymethyl cellulose (CMC) to obtain a moldable liquid. The blended liquid was added to molds and freeze-dried to obtain

AE. The resultant AE showed a strength of 4–5 N mm^{-2} at the breakpoint, hook strength of 5–8 N mm^{-2}, and the elongation at break point was ranging from 45 to 98%, while the scaffold showed zero water permeability. All these characteristics were similar to the rabbit endocranium. In addition, the scaffold did not show any tearing or off-clip while stitching with sutures.

One of the most common problems is meniscal lesions caused by trauma or degenerative diseases. Partial or total meniscectomy can cause changes in the lesion area, which may lead to osteoarthritis. For the symptomatic treatment of middle-aged patients, the meniscus allograft was preformed after they have gone through the total or partial meniscectomy [76]. The mechanical properties of BC were compared with real pig meniscus and traditional collagen meniscus implants [77]. The Young's modulus of BC was similar to pig meniscus and was five fold higher than that of the collagen implant. Due to the biocompatibility of BC, which also showed cell migration, and tunable meniscus shape, it could be a promising material for meniscus implants.

7. Bone Tissue Implants

Bone is a composite material comprising of an inorganic phase (calcium hydroxyapatite) and an organic phase (collagen and nanocollagenous proteins). Bone damage and bone loss may occur as a result of infectious diseases, genetic conditions, tumors, and trauma. Synthetic biomaterials such as ceramics, polymers, and metals have advantages for treating bone defects [78]. Nevertheless, the practical applications of natural materials due to their biocompatibility, biodegradability, and chemical functionality as a scaffold for tissue engineering is a major concern.

A biomaterial is a systematically and pharmacologically inert substance which is developed for implantation within a living system [79]; it works as a scaffold to maintain and develop the cells or help in tissue regeneration [80]. Guided bone regeneration (GBR) is a technique that induces the growth of bone tissue by osteopromoting fillers and membranes. The BC fibers were applied as a base for hydroxyapatite (HAp) growth to formulate filler materials for GBR. The phosphorylation of BC fibers was carried out by treating with phosphoric acid, followed by preconditioning with calcium ions for HAp nucleation. The growth of HAp was enhanced by using simulated body fluid (SBF). After 14 days, the BC-HAp retained a Ca-to-P ratio as high as 1.45 ± 0.92, reaching the standard of HAp. A greater Ca-to-P ratio on the surface of the pellicle was observed, indicating the deposition of HAp crystals. The results showed the potential to form 3-dimensional samples and laid the foundation for the preparation of the BC-HAp scaffold for GBR. A pathway of biomimetic mineralization of BC is shown in Fig. 4 [81].

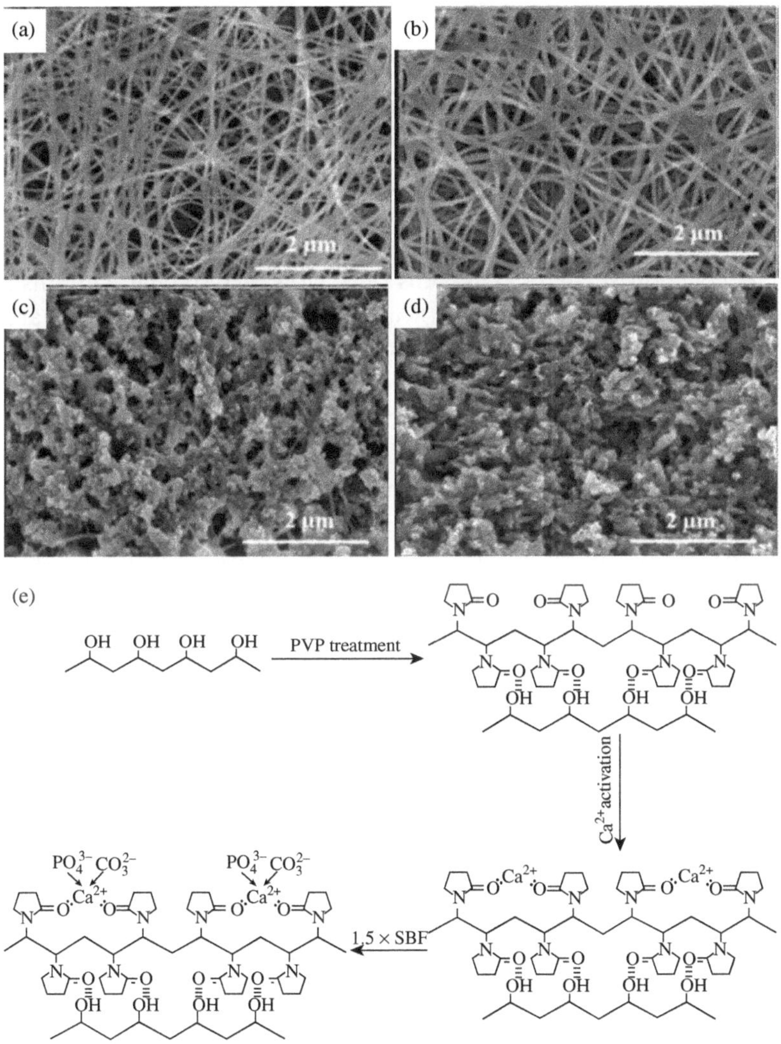

Fig. 4. FE-SEM micrographs of (a) BC, (b) BC-PVP, (c) 5-PVP-HAp-BC, and (d) 7-PVP-HAp-BC, and (e) schematic representation of the preparation of PVP-BC-Hap. *Source*: Reproduced with permission from Ref. [81].

A 3D network based on BC/nanocomposites was fabricated with HAp *via* a biomimetic route. In this process, the phosphorylated BC was soaked in SBF, leading to the development of uniform HAp crystals, having structural resemblances comparable to biological apatite [82].

8. Applications of BC in Cosmetics

Cosmetics are the materials used for improving the organoleptic characteristics of the human body. These products are applied to the human body for changing its appearance, cleaning or beautifying the body parts, and enhancing the attraction without any impact on the normal function and structure of the body [83]. Nowadays, people are using cosmetics to improve their beauty without considering their toxic effects, for example, toxicity concerns related to parabens [84]. However, to deal with this problem, natural products consisting of natural and herbal ingredients are recommended. In this regard, to enhance the stability of oil-in-water (O/W) emulsions, cellulose fibers are applied in cosmetics without the addition of any surfactant. BC has also been reported as a unique non-allergic biopolymer for cosmetic applications [83]. Due to non-toxicity and the capability to keep the skin moist, BC facial masks are attracting great attention as cosmetics materials for dry skin (Fig. 5). In a study, a volunteer group was asked to apply the translucent BC facial masks on their faces for

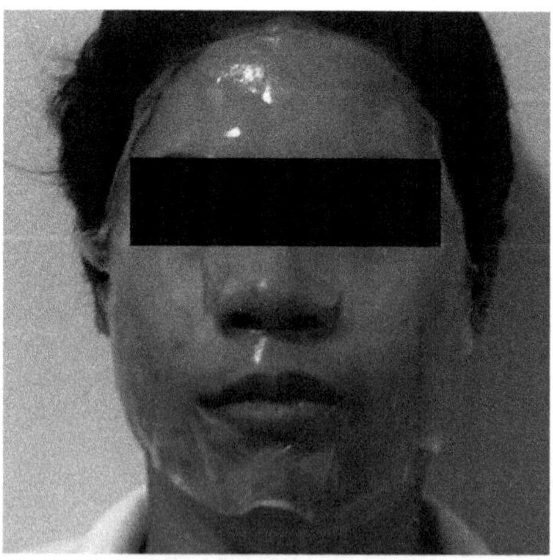

Fig. 5. The appearance of the BC mask when applied onto the skin.

Source: Figure reproduced from Ref. [85] under the Creative Commons Attribution License.

25 minutes, while another group was asked to put moist towels on their faces for the same period of time [85]. The following week, the groups used the alternate treatment. The skin elasticity, texture, moisture content, and desquamation levels were evaluated by routine skin testing procedures before and 5 minutes after removing the masks. The satisfaction of users regarding the BC mask was also evaluated. The BC masks significantly increased the moistness of the skin in comparison to moist towels with a single use. The rate of satisfaction of BC facial masks was 4 out of 5. Moreover, the other characteristics of the skin were not obviously affected [85]. Similarly, BC facial masks loaded with or without glycerin were also examined for human skin irritation [52]. There was no significant difference in the transepidermal water loss; however, mild skin irritation was found in a few subjects. Furthermore, the incorporation of glycerin improved the moisture content of the skin, suggesting its potential as a moisturizing facial mask.

A BC-based facial mask was developed containing openings for mouth, nose, and eyes. The authors revealed that these masks could be promising for prolonged or repeated use for skin nutrition, skin beauty, and cosmetic effects. Likewise, a BC-based facial mask with sodium bicarbonate (5 g), ascorbic acid (0.5 g), monohydrated citric acid (4 g), and salicylic acid (0.05 g) was developed for brightening purposes [86]. In this work, the authors also used BC masks with an anti-wrinkle constituent obtained from thermal plankton.

A personal cleaning formulation was developed, comprising of a liquid matrix, i.e., water, foaming surfactant, and a structuring agent, consisting of both BC and cationic polymers (cationic cellulose derivatives or cationic starch derivatives or a combination of both) [87]. These components were mixed in the liquid medium having a pH value less than ca. 4 or 7. The formulation delivered good foaming and simple wash-off properties without any slimy or filmy hand feel. The existence of particulate matter improved cleaning with conditioning advantages, and without any damage or irritation to the skin. The pH value lower than ca. 4 was particularly preferred for a formulation containing salicylic acid. This formulation can be used as a body wash for sensitive skin. Moreover, such formulations could be also used to clear and prevent skin pimples and blemishes.

9. Applications of BC in Food

BC fibers are classified as "generally recognized as safe" (GRAS) by the Food and Drug Administration (FDA) [88]. It has diverse potential in food industries due to its high purity, variability in shapes and texture, easy production process, and the ability to acquire *in situ* changes, i.e., color and flavor of culture medium. One of the first applications of BC in food is the production of "nata de coco" a Filipino traditional dessert. In the production of "nata de coco", BC is obtained by the fermentation of coconut water. Then it is chopped into small pieces and immersed in the sugar syrup [89]. The "nata de coco" is a well-liked food and is spreading across the globe in the form of candies [90].

Compared to plant cellulose (PC), BC holds higher water content and cation exchange capabilities with cholesterol-lowering effects. Therefore, low cholesterol, fat-free, and low-calorie food products can be made from BC [91]. BC gel is a tough material for biting, but by treating with calcium chloride and alginate it may become edible. The incorporation of these materials makes BC edible by holding the water in its gelatinous form, which makes the BC gel easy for biting. These evidences make BC a novel material for processed food [92].

BC has also been evaluated as a promising gelling, suspending, thickening, and stabilizing material in the food industry [88]. It works as a filler to reinforce the fragile food hydrogels, enhancing the value of paste food by reducing their stickiness. In addition, BC (0.2–0.3%) enhances the strength of tofu (food made from soy milk by coagulation and pressing), providing a better texture and firmness. BC has also improved stiffness and brittleness of Kamaboko (processed Japanese seafood) [93]. Similarly, BC-containing food items can sustain humidity for at least one month of storage period. The contour of an ice-cream containing BC was sustained for a minimum period of 60 minutes after removing from the freezer, while it would have melted in the same duration without the use of BC [88]. These evidences suggesting the use of BC in processed foods to improve their stability over a range of temperature, pH, and freeze–thaw environments.

BC can also work as food packaging material to increase the shelf-life of the food products. Antimicrobial agents, oxygen and ethylene scavengers, and moisture removal are used in the BC-based active packaging system [94]. In addition, modified BC (BCM) with tunable properties has been developed by heterogeneous esterification with hexonoyl chloride. The BCM exhibited an increased hydrophobicity, while the structure of pure BC was maintained. The resultant MBC showed a 50% decrease in the permeability of both gas and water [94]. Moreover, Jipa and coworkers [95] designed biodegradable BC and sorbic acid (BC/SA)-based films by the addition of antimicrobial SA. The results revealed the antimicrobial properties of the developed films. From these results, it can be concluded that BCM is a promising composite material with potential applications in the packaging industry, with durability and antimicrobial properties.

References

[1] N. Lin and A. Dufresne, Nanocellulose in biomedicine: Current status and future prospect. *European Polymer Journal,* 59 (2014) 302–325.

[2] S. Khattak, F. Wahid, L.-P. Liu, S.-R. Jia, L.-Q. Chu, Y.-Y. Xie, Z.-X. Li, and C. Zhong, Applications of cellulose and chitin/chitosan derivatives and composites as antibacterial materials: Current state and perspectives. *Applied Microbiology and Biotechnology*, 103 (2019) 1989–2006.

[3] R.R. McCarthy, M.W. Ullah, P. Booth, E. Pei, and G. Yang, The use of bacterial polysaccharides in bioprinting. *Biotechnology Advances*, 37 (2019) 107448.

[4] C.-C. Ng and Y.-T. Shyu, Development and production of cholesterol-lowering Monascus-nata complex. *World Journal of Microbiology and Biotechnology*, 20 (2004) 875–879.

[5] H. Ullah, F. Wahid, H.A. Santos, and T. Khan, Advances in biomedical and pharmaceutical applications of functional bacterial cellulose-based nano-composites. *Carbohydrate Polymers*, 150 (2016) 330–352.

[6] I. Sulaeva, U. Henniges, T. Rosenau, and A. Potthast, Bacterial cellulose as a material for wound treatment: Properties and modifications. A review. *Biotechnology Advances,* 33 (2015) 1547–1571.

[7] N. Bergstrom, S.D. Horn, R.J. Smout, S.A. Bender, M.L. Ferguson, G. Taler, A.C. Sauer, S.S. Sharkey, and A.C. Voss, The national pressure ulcer long-term care study: Outcomes of pressure ulcer treatments in long-term care. *Journal of the American Geriatrics Society*, 53 (2005) 1721–1729.

[8] M.A. Fonder, G.S. Lazarus, D.A. Cowan, B. Aronson-Cook, A.R. Kohli, and A.J. Mamelak, Treating the chronic wound: A practical approach to the care of nonhealing wounds and wound care dressings. *Journal of the American Academy of Dermatology*, 58 (2008) 185–206.

[9] G. Lagana and E.H. Anderson, Moisture dressings: The new standard in wound care. *The Journal for Nurse Practitioners*, 6 (2010) 366–370.

[10] W. Czaja, A. Krystynowicz, S. Bielecki, and R.M. Brown, Microbial cellulose — the natural power to heal wounds. *Biomaterials*, 27 (2006) 145–151.

[11] N. Shah, M. Ul-Islam, W.A. Khattak, and J.K. Park, Overview of bacterial cellulose composites: A multipurpose advanced material. *Carbohydrate Polymers*, 98 (2013) 1585–1598.

[12] I. Siró and D. Plackett, Microfibrillated cellulose and new nanocomposite materials: A review. *Cellulose*, 17 (2010) 459–494.
[13] W. Hu, S. Chen, J. Yang, Z. Li, and H. Wang, Functionalized bacterial cellulose derivatives and nanocomposites. *Carbohydrate Polymers*, 101 (2014) 1043–1060.
[14] D. Klemm, F. Kramer, S. Moritz, T. Lindström, M. Ankerfors, D. Gray, and A. Dorris, Nanocelluloses: A new family of nature-based materials. *Angewandte Chemie International Edition*, 50 (2011) 5438–5466.
[15] Y. Li, H. Jiang, W.F. Zheng, N.Y. Gong, L.L. Chen, X.Y. Jiang, and G. Yang, Bacterial cellulose-hyaluronan nanocomposite biomaterials as wound dressings for severe skin injury repair (vol. 3, p. 3498, 2015). *Journal of Materials Chemistry B*, 7 (2019) 1962–1962.
[16] H.G. de Oliveira Barud, R.R. da Silva, H. da Silva Barud, A. Tercjak, J. Gutierrez, W.R. Lustri, O.B. de Oliveira, and S.J.L. Ribeiro, A multipurpose natural and renewable polymer in medical applications: Bacterial cellulose. *Carbohydrate Polymers*, 153 (2016) 406–420.
[17] W.K. Czaja, D.J. Young, M. Kawecki, and R.M. Brown, The future prospects of microbial cellulose in biomedical applications. *Biomacromolecules*, 8 (2007) 1–12.
[18] J. Wang, Y.Z. Wan, H.L. Luo, C. Gao, and Y. Huang, Immobilization of gelatin on bacterial cellulose nanofibers surface *via* crosslinking technique. *Materials Science and Engineering: C.*, 32 (2012) 536–541.
[19] M.G. Albu, Z. Vuluga, D.M. Panaitescu, D.M. Vuluga, A. Căşărică, and M. Ghiurea, Morphology and thermal stability of bacterial cellulose/collagen composites. *Central European Journal of Chemistry*, 12 (2014) 968–975.
[20] O.-A. Saibuatong and M. Phisalaphong, Novo aloe vera–bacterial cellulose composite film from biosynthesis. *Carbohydrate Polymers*, 79 (2010) 455–460.
[21] M. Ul-Islam, N. Shah, J.H. Ha, and J.K. Park, Effect of chitosan penetration on physico-chemical and mechanical properties of bacterial cellulose. *Korean Journal of Chemical Engineering*, 28 (2011) 1736.
[22] W.-C. Lin, C.-C. Lien, H.-J. Yeh, C.-M. Yu, and S.-H. Hsu, Bacterial cellulose and bacterial cellulose–chitosan membranes for wound dressing applications. *Carbohydrate Polymers*, 94 (2013) 603–611.
[23] I. Sulaeva, U. Henniges, T. Rosenau, and A. Potthast, Bacterial cellulose as a material for wound treatment: Properties and modifications. A review. *Biotechnology Advances*, 33 (2015) 1547–71.
[24] V.W. Ng, J.M. Chan, H. Sardon, R.J. Ono, J.M. Garcia, Y.Y. Yang, and J.L. Hedrick, Antimicrobial hydrogels: A new weapon in the arsenal against multidrug-resistant infections. *Advanced Drug Delivery Reviews*, 78 (2014) 46–62.

[25] F. Wahid, H.-S. Wang, C. Zhong, and L.-Q. Chu, Facile fabrication of moldable antibacterial carboxymethyl chitosan supramolecular hydrogels cross-linked by metal ions complexation. *Carbohydrate Polymers*, 165 (2017) 455–461.
[26] G. Yang, J. Xie, F. Hong, Z. Cao, and X. Yang, Antimicrobial activity of silver nanoparticle impregnated bacterial cellulose membrane: Effect of fermentation carbon sources of bacterial cellulose. *Carbohydrate Polymers,* 87 (2012) 839–845.
[27] F. Wahid, H. Bai, F.-P. Wang, Y.-Y. Xie, Y.-W. Zhang, L.-Q. Chu, S.-R. Jia, and C. Zhong, Facile synthesis of bacterial cellulose and polyethyleneimine based hybrid hydrogels for antibacterial applications. *Cellulose*, 27 (2019) 369 383.
[28] F. Wahid, J.-J. Yin, D.-D. Xue, H. Xue, Y.-S. Lu, C. Zhong, and L.-Q. Chu, Synthesis and characterization of antibacterial carboxymethyl Chitosan/ZnO nanocomposite hydrogels. *International Journal of Biological Macromolecules*, 88 (2016) 273–279.
[29] F. Wahid, X.-H. Hu, L.-Q. Chu, S.-R. Jia, Y.-Y. Xie, and C. Zhong, Development of bacterial cellulose/chitosan based semi-interpenetrating hydrogels with improved mechanical and antibacterial properties. *International Journal of Biological Macromolecules*, 122 (2019) 380–387.
[30] N. Butchosa, C. Brown, P.T. Larsson, L.A. Berglund, V. Bulone, and Q. Zhou, Nanocomposites of bacterial cellulose nanofibers and chitin nanocrystals: Fabrication, characterization and bactericidal activity. *Green Chemistry*, 15 (2013) 3404–3413.
[31] D. Ciechańska, Multifunctional bacterial cellulose/chitosan composite materials for medical applications. *Fibres & Textiles in Eastern Europe, Nr,* 4(48), (2004) 69–72.
[32] L.-M. Dobre, A. Stoica-Guzun, M. Stroescu, I.M. Jipa, T. Dobre, M. Ferdeş, and Ş. Ciumpiliac, Modelling of sorbic acid diffusion through bacterial cellulose-based antimicrobial films. *Chemical Papers*, 66 (2012) 144–151.
[33] M.L. Dobre and A. Stoica-Guzun, Antimicrobial Ag-Polyvinyl Alcohol-Bacterial Cellulose Composite Films. *Journal of Biobased Materials and Bioenergy*, 7 (2013) 157–162.
[34] R. Jung, Y. Kim, H.S. Kim, and H.J. Jin, Antimicrobial properties of hydrated cellulose membranes with silver nanoparticles. *Journal of Biomaterials Science-Polymer Edition*, 20 (2009) 311–324
[35] H.S. Barud, T. Regiani, R.F.C. Marques, W.R. Lustri, Y. Messaddeq, and S.J.L. Ribeiro, Antimicrobial bacterial cellulose-silver nanoparticles composite membranes. *Journal of Nanomaterials*, (2011).
[36] G. Yang, J. Xie, Y. Deng, Y. Bian, and F. Hong, Hydrothermal synthesis of bacterial cellulose/AgNPs composite: A "green" route for antibacterial application. *Carbohydrate Polymers*, 87 (2012) 2482–2487.

[37] H.S. Barud, T. Regiani, R.F.C. Marques, W.R. Lustri, Y. Messaddeq, and S.J.L. Ribeiro, Antimicrobial bacterial cellulose-silver nanoparticles composite Membranes. *Journal of Nanomaterials*, 2011 (2011) 1–8.

[38] J. Feng, Q. Shi, W. Li, X. Shu, A. Chen, X. Xie, and X. Huang, Antimicrobial activity of silver nanoparticles *in situ* growth on TEMPO-mediated oxidized bacterial cellulose. *Cellulose*, 21 (2014) 4557–4567.

[39] C. Katepetch, R. Rujiravanit, and H. Tamura, Formation of nanocrystalline ZnO particles into bacterial cellulose pellicle by ultrasonic-assisted *in situ* synthesis. *Cellulose*, 20 (2013) 1275–1292.

[40] M. Ul-Islam, W.A. Khattak, M.W. Ullah, S. Khan, and J.K. Park, Synthesis of regenerated bacterial cellulose-zinc oxide nanocomposite films for biomedical applications. *Cellulose*, 21 (2014) 433–447.

[41] G. Xiong, H. Luo, J. Zhang, J. Jin, and Y. Wan, Synthesis of ZnO by chemical bath deposition in the presence of bacterial cellulose. *Acta Metallurgica Sinica (English Letters)*, 27 (2014) 656–662.

[42] I.M.S. Araújo, R.R. Silva, G. Pacheco, W.R. Lustri, A. Tercjak, J. Gutierrez, J.R.S. Júnior, F.H.C. Azevedo, G.S. Figuêredo, M.L. Vega, S.J.L. Ribeiro, and H.S. Barud, Hydrothermal synthesis of bacterial cellulose–copper oxide nanocomposites and evaluation of their antimicrobial activity. *Carbohydrate Polymers*, 179 (2018) 341–349.

[43] Y.-Y. Xie, X.-H. Hu, Y.-W. Zhang, F. Wahid, L.-Q. Chu, S.-R. Jia, and C. Zhong, Development and antibacterial activities of bacterial cellulose/graphene oxide-CuO nanocomposite films. *Carbohydrate Polymers*, 229 (2020).

[44] S. Khan, M. Ul-Islam, W.A. Khattak, M.W. Ullah, and J.K. Park, Bacterial cellulose-titanium dioxide nanocomposites: Nanostructural characteristics, antibacterial mechanism, and biocompatibility. *Cellulose*, 22 (2015) 565–579.

[45] M.C.I. Mohd Amin, N. Ahmad, N. Halib, and I. Ahmad, Synthesis and characterization of thermo- and pH-responsive bacterial cellulose/acrylic acid hydrogels for drug delivery. *Carbohydrate Polymers*, 88 (2012) 465–473.

[46] N. Ahmad, M.C.I.M. Amin, S.M. Mahali, I. Ismail, and V.T.G. Chuang, Biocompatible and mucoadhesive bacterial cellulose-g-poly(acrylic acid) hydrogels for oral protein delivery. *Molecular Pharmaceutics*, 11 (2014) 4130–4142.

[47] A. Mueller, Z. Ni, N. Hessler, F. Wesarg, F.A. Mueller, D. Kralisch, and D. Fischer, The biopolymer bacterial nanocellulose as drug delivery system: Investigation of drug loading and release using the model protein albumin. *Journal of Pharmaceutical Sciences*, 102 (2013) 579–592.

[48] N.H.C.S. Silva, A.F. Rodrigues, I.F. Almeida, P.C. Costa, C. Rosado, C.P. Neto, A.J.D. Silvestre, and C.S.R. Freire, Bacterial cellulose membranes as transdermal delivery systems for diclofenac: *In vitro* dissolution and permeation studies. *Carbohydrate Polymers*, 106 (2014) 264–269.

[49] M. Amin, A.G. Abadi, N. Ahmad, H. Katas, and J.A. Jamal, Bacterial Cellulose film coating as drug delivery system: Physicochemical, thermal and drug release properties. *Sains Malaysiana*, 41 (2012) 561–568.
[50] E. Trovatti, N.H.C.S. Silva, I.F. Duarte, C.F. Rosado, I.F. Almeida, P. Costa, C.S.R. Freire, A.J.D. Silvestre, and C.P. Neto, Biocellulose membranes as supports for dermal release of lidocaine. *Biomacromolecules*, 12 (2011) 4162–4168.
[51] F.G. Torres, S. Commeaux, and O.P. Troncoso, Biocompatibility of bacterial cellulose based biomaterials. *Journal of Functional Biomaterials*, 3 (2012) 864–878.
[52] I.F. Almeida, T. Pereira, N.H.C.S. Silva, F.P. Gomes, A.J.D. Silvestre, C.S.R. Freire, J.M. Sousa Lobo, and P.C. Costa, Bacterial cellulose membranes as drug delivery systems: An *in vivo* skin compatibility study. *European Journal of Pharmaceutics and Biopharmaceutics*, 86 (2014) 332–336.
[53] B. Adams, Q. Xiao, Q. Xu, Stem Cell Therapy for Vascular Disease. *Trends in Cardiovascular Medicine*, 17 (2007) 246–251.
[54] S. Parveen, K. Krishnakumar, and S. Sahoo, New era in health care: Tissue engineering. *Journal of Stem Cells & Regenerative Medicine*, 1 (2006) 8–24.
[55] J.H. Campbell, J.L. Efendy, and G.R. Campbell, Novel vascular graft grown within recipient's own peritoneal cavity. *Circulation Research*, 85 (1999) 1173–1178.
[56] E.R. Edelman, Vascular tissue engineering — Designer arteries. *Circulation Research*, 85 (1999) 1115–1117.
[57] S. Zang, R. Zhang, H. Chen, Y. Lu, J. Zhou, X. Chang, G. Qiu, Z. Wu, and G. Yang, Investigation on artificial blood vessels prepared from bacterial cellulose. *Materials Science and Engineering: C*, 46 (2015) 111–117.
[58] D.A. Schumann, J. Wippermann, D.O. Klemm, F. Kramer, D. Koth, H. Kosmehl, T. Wahlers,and S. Salehi-Gelani, Artificial vascular implants from bacterial cellulose: Preliminary results of small arterial substitutes. *Cellulose*, 16 (2009) 877–885.
[59] D. Klemm, D. Schumann, U. Udhardt, and S. Marsch, Bacterial synthesized cellulose — artificial blood vessels for microsurgery. *Progress in Polymer Science*, 26 (2001) 1561–1603.
[60] E.E. Brown, M.-P.G. Laborie, and J. Zhang, Glutaraldehyde treatment of bacterial cellulose/fibrin composites: Impact on morphology, tensile and viscoelastic properties. *Cellulose*, 19 (2012) 127–137, 2012.
[61] S. Takatani, Cardiac prosthesis as an advanced surgical therapy for end-stage cardiac patients: Current status and future perspectives. *Journal of Medical and Dental Sciences*, 47 (2000) 151–165.
[62] H. Mohammadi, Nanocomposite biomaterial mimicking aortic heart valve leaflet mechanical behaviour. *Proceedings of the Institution of Mechanical Engineers, H: J. Eng. in Medicine*, 225 (2011) 718–722.

[63] F.K. Andrade, J.P. Silva, M. Carvalho, E.M.S. Castanheira, R. Soares, and M. Gama, Studies on the hemocompatibility of bacterial cellulose. *Journal of Biomedical Materials Research Part A.*, 98A (2011) 554–566.
[64] H. Fink, L. Faxälv, G.F. Molnár, K. Drotz, B. Risberg, T.L. Lindahl, and A. Sellborn, Real-time measurements of coagulation on bacterial cellulose and conventional vascular graft materials. *Acta Biomaterialia*, 6 (2010) 1125–1130.
[65] A. Leitão, J. Silva, F. Dourado, and M. Gama, Production and characterization of a new bacterial Cellulose/Poly(Vinyl Alcohol) nanocomposite. *Materials*, 6 (2013) 1956.
[66] A.F. Leitão, S. Gupta, J.P. Silva, I. Reviakine, and M. Gama, Hemocompatibility study of a bacterial cellulose/polyvinyl alcohol nanocomposite. *Colloids and Surfaces B: Biointerfaces*, 111 (2013) 493–502.
[67] C. Castro, A. Vesterinen, R. Zuluaga, G. Caro, I. Filpponen, O.J. Rojas, G. Kortaberria, and P. Gañán, *In situ* production of nanocomposites of poly(vinyl alcohol) and cellulose nanofibrils from Gluconacetobacter bacteria: Effect of chemical crosslinking. *Cellulose*, 21 (2014) 1745–1756.
[68] J.P. Whitcher, M. Srinivasan, M.P. Upadhyay, Corneal blindness: A global perspective. *Bulletin of the World Health Organization*, 79 (2001) 214–221.
[69] H.-Y. Wang, R.-H. Wei, and S.-Z. Zhao, Evaluation of corneal cell growth on tissue engineering materials as artificial cornea scaffolds. *International Journal of Ophthalmology*, (2013) 873–878.
[70] H. Jia, Y. Jia, J. Wang, Y. Hu, Y. Zhang, and S. Jia, Potentiality of bacterial cellulose as the scaffold of tissue engineering of cornea, *2009 2nd International Conference on Biomedical Engineering and Informatics*, (2009) 1–5.
[71] K.S. Naidoo and J. Jaggernath, Uncorrected refractive errors. *Indian Journal of Ophthalmology*, 60 (2012) 432–437.
[72] D.J. Levinson, Glonek, and Thomas, Microbial cellulose contact lens, US Patent US7832857 B2, (2010).
[73] X. Li, Wan, Wankei, Panchal, Chandrakant J., Transparent bacterial cellulose nanocomposite hydrogels, Patent, US 8940337 B2, (2015).
[74] A. Svensson, E. Nicklasson, T. Harrah, B. Panilaitis, D.L. Kaplan, M. Brittberg, and P. Gatenholm, Bacterial cellulose as a potential scaffold for tissue engineering of cartilage. *Biomaterials*, 26 (2005) 419–431.
[75] L.E. Millon, C.J. Oates, W. Wan, Compression properties of polyvinyl alcohol — bacterial cellulose nanocomposite. *Journal of Biomedical Materials Research Part B: Applied Biomaterials*, 90B (2009) 922–929.
[76] T.G. Tienen, N. Verdonschot, R.G.J.C. Heijkants, P. Buma, J.G.F. Scholten, A.V. Kampen, and R.P.H. Veth, Prosthetic replacement of the medial meniscus in cadaveric knees: Does the prosthesis mimic the functional behavior

of the native meniscus? *The American Journal of Sports Medicine*, 32 (2004) 1182–1188.

[77] A. Bodin, S. Concaro, M. Brittberg, and P. Gatenholm, Bacterial cellulose as a potential meniscus implant. *Journal of Tissue Engineering and Regenerative Medicine*, 1 (2007) 406–408.

[78] X. Zheng Shu, Y. Liu, F.S. Palumbo, Y. Luo, and G.D. Prestwich, *In situ* crosslinkable hyaluronan hydrogels for tissue engineering. *Biomaterials*, 25 (2004) 1339–1348.

[79] M.R. Rogel, H. Qiu, and G.A. Ameer, The role of nanocomposites in bone regeneration. *Journal of Materials Chemistry*, 18 (2008) 4233–4241.

[80] M.S. Cortizo and M.S. Belluzo, Biodegradable Polymers for Bone Tissue Engineering, in S.N. Goyanes, N.B. D'Accorso (eds.), *Industrial Applications of Renewable Biomass Products: Past, Present and Future*, Springer International Publishing. *Cham*, (2017) 47–74.

[81] N. Yin, S.-y. Chen, Y. Ouyang, L. Tang, J.-X. Yang, and H.-P. Wang, Biomimetic mineralization synthesis of hydroxyapatite bacterial cellulose nanocomposites. *Progress in Natural Science: Materials International*, 21 (2011) 472–477.

[82] Y.Z. Wan, Y. Huang, C.D. Yuan, S. Raman, Y. Zhu, H.J. Jiang, F. He, and C. Gao, Biomimetic synthesis of hydroxyapatite/bacterial cellulose nanocomposites for biomedical applications. *Materials Science and Engineering: C*, 27 (2007) 855–864.

[83] D.R.A.B. Norhasliza Hasan, Suryani Kamarudin, Application of bacterial cellulose (BC) in natural facial scrub, *International Journal on Advanced Science Engineering Information Technology*, 4 (2012).

[84] P.D. Darbre and P.W. Harvey, Paraben esters: Review of recent studies of endocrine toxicity, absorption, esterase and human exposure, and discussion of potential human health risks. *Journal of Applied Toxicology*, 28 (2008) 561–578.

[85] T. Amnuaikit, T. Chusuit, P. Raknam, and P. Boonme, Effects of a cellulose mask synthesized by a bacterium on facial skin characteristics and user satisfaction, *Med Devices (Auckl)*, 4 (2011) 77–81.

[86] J.-y. Legendre, Assembly comprising a substrate comprising biocellulose, and a powdered cosmetic composition to be brought into contact with the substrate, US Patent, US 20090041815 A1, (2009).

[87] B.P. Heath, Coffindaffer, Timothy Woodrow, Kyte III, Kenneth, Smith III, Edward Dewey, Mcconaughy, Shawn D., Personal cleansing compositions comprising a bacterial cellulose network and cationic polymer, US Patent, US 8097574 B2, (2012).

[88] Z. Shi, Y. Zhang, G.O. Phillips, and G. Yang, Utilization of bacterial cellulose in food. *Food Hydrocolloids*, 35, 539–545.

[89] D. Klemm, B. Heublein, H.-P. Fink, and A. Bohn, Cellulose: Fascinating biopolymer and sustainable raw material. *Angewandte Chemie International Edition*, 44 (2005) 3358–3393.
[90] H. Ullah, H.A. Santos, and T. Khan, Applications of bacterial cellulose in food, cosmetics and drug delivery. *Cellulose*, 23 (2016) 2291–2314.
[91] C.-F. Chau, P. Yang, C.-M. Yu, and G.-C. Yen, Investigation on the Lipid- and Cholesterol-Lowering Abilities of Biocellulose. *Journal of Agricultural and Food Chemistry*, 56 (2008) 2291–2295.
[92] A. Okiyama, M. Motoki, and S. Yamanaka, Bacterial cellulose II. Processing of the gelatinous cellulose for food materials. *Food Hydrocolloids*, 6 (1992) 479–487.
[93] A. Okiyama, M. Motoki, and S. Yamanaka, Bacterial cellulose IV. Application to processed foods. *Food Hydrocolloids*, 6 (1993) 503–511.
[94] L.C. Tomé, L. Brandão, A.M. Mendes, A.J.D. Silvestre, C.P. Neto, A. Gandini, C.S.R. Freire, and I.M. Marrucho, Preparation and characterization of bacterial cellulose membranes with tailored surface and barrier properties. *Cellulose*, 17 (2010) 1203–1211.
[95] I.M. Jipa, A. Stoica-Guzun, and M. Stroescu, Controlled release of sorbic acid from bacterial cellulose based mono and multilayer antimicrobial films. *LWT — Food Science and Technology*, 47 (2012) 400–406.

https://doi.org/10.1142/9781786349477_bmatter

Index

F

G

H

M

N

O

P

R

Lightning Source UK Ltd.
Milton Keynes UK
UKHW051526160421
381970UK00005B/41